Communications in Computer and Information Science   539

Commenced Publication in 2007
Founding and Former Series Editors:
Alfredo Cuzzocrea, Dominik Ślęzak, and Xiaokang Yang

More information about this series at http://www.springer.com/series/7899

Tadeusz Morzy · Patrick Valduriez
Ladjel Bellatreche (Eds.)

New Trends in Databases and Information Systems

ADBIS 2015 Short Papers and Workshops
BigDap, DCSA, GID, MEBIS, OAIS, SW4CH, WISARD
Poitiers, France, September 8–11, 2015
Proceedings

 Springer

Editors
Tadeusz Morzy
Poznan University of Technology
Poznan
Poland

Patrick Valduriez
INRIA
Montpellier
France

Ladjel Bellatreche
National Engineering School for Mechanics
  and Aerotechnics
Poitiers
France

ISSN 1865-0929   ISSN 1865-0937   (electronic)
Communications in Computer and Information Science
ISBN 978-3-319-23200-3   ISBN 978-3-319-23201-0   (eBook)
DOI 10.1007/978-3-319-23201-0

Library of Congress Control Number: 2015947123

Springer Cham Heidelberg New York Dordrecht London
© Springer International Publishing Switzerland 2015

Printed on acid-free paper

Springer International Publishing AG Switzerland is part of Springer Science+Business Media
(www.springer.com)

Preface

This volume contains a selection of the papers presented at the 19th East-European Conference on Advances in Databases and Information Systems (ADBIS 2015), held during September 8–11, 2015, at Futuroscope, Poitiers, France.

The ADBIS series of conferences aims at providing a forum for the presentation and dissemination of research on database theory, development of advanced DBMS technologies, and their advanced applications. ADBIS 2015 in Poitiers continued after St. Petersburg (1997), Poznan (1998), Maribor (1999), Prague (2000), Vilnius (2001), Bratislava (2002), Dresden (2003), Budapest (2004), Tallinn (2005), Thessaloniki (2006), Varna (2007), Pori (2008), Riga (2009), Novi Sad (2010), Vienna (2011), Poznan (2012), Genoa (2013), and Ohrid (2014). This edition was special, as it was the first time that ADBIS took place in France. The conferences are initiated and supervised by an international Steering Committee consisting of representatives from Armenia, Austria, Bulgaria, Czech Republic, Estonia, Finland, Germany, Greece, Hungary, Israel, Italy, Latvia, Lithuania, Poland, Russia, Serbia, Slovakia, Slovenia, and the Ukraine.

This volume contains 18 papers selected as short contributions to be presented at ADBIS conference as well as papers contributed by all associated satellite events. An introductory chapter summarizes the main issues and contributions of all the events whose papers are included in this volume. Each of the satellite events complementing the main ADBIS conference had its own international Program Committee, whose members served as the reviewers of papers included in this volume. The volume is divided into six parts, one devoted to ADBIS short contributions and each other part to a single satellite event.

The selected short papers span a wide spectrum of topics related to the development of advanced database applications covering different types of data (tweet data, time series, data streams, services, workflow instance data, semantic data, etc.), different management issues (ETL, querying, optimizing, data analysis, recommending, etc.), different types of deployment platforms (cloud and distributed), and different storage layouts (column store and row store).

The Second International Workshop on Big Data Applications and Principles (BigDap 2015) aimed at promoting and sharing research and innovation on big data. The workshop allowed researchers and practitioners to share their experiences on designing and developing big data applications, to discuss the big data open issues, and to share new ideas and techniques for big data management and analysis. Nine papers were selected for presentation at BigDap 2015 and are included in this volume.

The First Workshop on Data-Centered Smart Applications (DCSA 2015) sought to bring together computer science and information systems experts, and more precisely data scientists, involved in smart applications engineering. The goal is to exchange experiences, build a state of the art of realizations and challenges, and reuse and adapt solutions that have been proposed in other domains. There are no conferences focusing on data-centered smart applications, except the conferences and workshops on big data and smart cities in their own right. Four papers were selected for presentation at DCSA 2015 and are included in this volume.

The 4th International Workshop on GPUs in Databases (GID 2015) aimed at providing a forum for discussion for those interested in the contribution of graphics of processing units (GPU) in database environments. The idea of using GPU in databases allows us to achieve high speed-ups in many applications domains of databases, especially in the big data era. Two papers were selected for presentation at GID and are included in this volume.

The Second International Workshop on Managing Evolving Business Intelligence Systems (MEBIS 2015) gathered business intelligence architects, engineers, decision makers, and researchers to share their vision and solutions for handling dynamics in business intelligence systems and their impact on the whole life cycle of decision system design. Four papers were selected for presentation at MEBIS and are included in this volume.

The Fourth International Workshop on Ontologies Meet Advanced Information Systems (OAIS 2015) aimed at presenting new and challenging issues in the contribution of ontologies for designing high-quality information systems and advanced database applications and presenting new research and technological developments that use ontologies throughout the life cycle of information systems. Five papers were selected for presentation at OAIS and are included in this volume.

The First International Workshop on Semantic Web for Cultural Heritage (SW4CH 2015) aimed to bring together computer scientists, and more precisely data scientists, involved in Semantic Web solutions for cultural heritage. The goal is to exchange experiences, build a state of the art of realizations and challenges, and reuse and adapt solutions that have been proposed in other domains. Nine papers were selected for presentation at SW4CH and are included in this volume.

The Workshop on Information Systems for Alarm Diffusion (WISARD 2015) gathered researchers and practitioners working in the area of information systems, database, document management, metadata, workflow management, conceptual modeling, design, software, etc., to exchange and to share information, to network, to swop points of view, and to collaborate in project proposals. WISARD discusses questions about liveness, dynamicity, reactivity of alarm systems. Six papers were selected for presentation at WISARD and are included in this volume.

We would like to express our gratitude to every individual who contributed to the success of ADBIS 2015. Firstly, we thank all authors for submitting their research paper to the conference. However, we are also indebted to the members of the community who offered their precious time and expertise in performing various roles ranging from organizational to reviewing roles – their efforts, energy, and degree of professionalism deserve the highest commendations. Special thanks to the Program Committee members and the external reviewers for their support in evaluating the papers submitted to ADBIS 2015, ensuring the quality of the scientific program. Thanks also to all the colleagues, secretaries, and engineers involved in the conference organization, as well as the workshop organizers. We would like to thank Dr. Mickaël Baron, from LIAS/ISAE-ENSMA for his endless help and support. A special thank you to the members of the Steering Committee, an in particular, its chair, Leonid Kalinichenko and his vice chair Yannis Manolopoulos, for all their help and guidance.

Finally, we thank Springer for publishing the proceedings containing invited and research papers in the LNCS series. The Program Committee work relied on Easy-Chair, and we thank its development team for creating and maintaining it; it offered a great support throughout the different phases of the reviewing process. The conference would not have been possible without our supporters and sponsors:

- Région Poitou Charentes
- ISAE-ENSMA
- Poitiers University
- INFORSID Association
- CRITT Informatique, Futuroscope
- LIAS laboratory

Last, but not least, we thank the participants of ADBIS 2015 for sharing their works and presenting their achievement, thus providing a lively, fruitful, and constructive forum, and giving us the pleasure of knowing that our work was purposeful.

September 2015

Ladjel Bellatreche
Tadeusz Morzy
Sofian Maabout
Boris Novikov
Athena Vakali
Patrick Valduriez
Bernhard Thalheim

Organization

General Chair

Ladjel Bellatreche   LIAS/ISAE-ENSMA, Poitiers, France

Program Committee Co-chairs

Patrick Valduriez   Inria of Montpellier, France
Tadeusz Morzy   Poznan University, Poland

Workshop Co-chairs

Athena Vakali   Aristotle University of Thessaloniki, Greece
Bernhard Thalheim   Kiel University, Germany

Doctoral Consortium Co-chairs

Sofian Maabout   Labri/Bordeaux, France
Boris Novikov   St. Petersburg University, Russia

Publicity Chair

Selma Khouri   LIAS/ISAE-ENSMA, France

WebSite Chair

Mickaël Baron   LIAS/ISAE-ENSMA, Poitiers, France

Proceedings Technical Editor

Stéphane Jean   LIAS/ISAE-ENSMA, Poitiers, France

Local Organizing Committee Chair

Patrick Girard   LIAS/ISAE-ENSMA, France

Local Organizing Committee

Mickaël Baron   LIAS/ISAE-ENSMA, Poitiers, France
Frédéric Carreau   LIAS/ISAE-ENSMA, Poitiers, France
Brice Chardin   LIAS/ISAE-ENSMA, Poitiers, France

Zoé Faget	LIAS/ISAE-ENSMA, Poitiers, France
Patrick Girard	LIAS/ISAE-ENSMA, Poitiers, France
Laurent Guittet	LIAS/ISAE-ENSMA, Poitiers, France
Stéphane Jean	LIAS/ISAE-ENSMA, Poitiers, France
Yassine Ouhammou	LIAS/ISAE-ENSMA, Poitiers, France
Claudine Rault	LIAS/ISAE-ENSMA, Poitiers, France
Okba Barkat	LIAS/ISAE-ENSMA, Poitiers, France
Selma Bouarar	LIAS/ISAE-ENSMA, Poitiers, France
Ahcène Boukorca	LIAS/ISAE-ENSMA, Poitiers, France
Lahcène Brahimi	LIAS/ISAE-ENSMA, Poitiers,France
Zouhir Djilani	LIAS/ISAE-ENSMA, Poitiers, France
Géraud Fokou	LIAS/ISAE-ENSMA, Poitiers, France
Nadir Guetmi	LIAS/ISAE-ENSMA, Poitiers, France
Yves Mouafo	LIAS/ISAE-ENSMA, Poitiers, France
Guillaume Phavorin	LIAS/ISAE-ENSMA, Poitiers, France

Supporters

Région Poitou Charentes
ISAE-ENSMA
Poitiers University
INFORSID Association
CRITT Informatique, Futuroscope
LIAS laboratory

Steering Committee

Paolo Atzeni	Italy
Andras Benczur	Hungary
Albertas Caplinskas	Lithuania
Barbara Catania	Italy
Johann Eder	Austria
Theo Haerder	Germany
Marite Kirikova	Latvia
Hele-Mai Haav	Estonia
Mirjana Ivanovic	Serbia
Hannu Jaakkola	Finland
Mikhail Kogalovsky	Russia
Yannis Manolopoulos	Greece
Rainer Manthey	Germany
Manuk Manukyan	Armenia
Joris Mihaeli	Israel
Tadeusz Morzy	Poland
Pavol Navrat	Slovakia
Boris Novikov	Russia

Mykola Nikitchenko   Ukraine
Jaroslav Pokornyv   Czech Republic
Boris Rachev   Bulgaria
Bernhard Thalheim   Germany
Gottfried Vossen   Germany
Tatjana Welzer   Slovenia
Viacheslav Wolfengagen   Russia
Robert Wrembel   Poland
Ester Zumpano   Italy

Program Committee

Reza Akbarinia   Inria, France
Paolo Atzeni   Università Roma Tre, Italy
Andreas Behrend   University of Bonn, Germany
Ladjel Bellatreche   ISAE-ENSMA, France
Omar Boucelma   Aix-Marseille University, France
Mahdi Bohlouli   University of Siegen, Germany
Albertas Caplinskas   Institute of Mathematics and Informatics, Italy
Barbara Catania   DISI-University of Genoa, Italy
Wojciech Cellary   Poznan School of Economy, Poland
Ricardo Rodrigues Ciferri Federal University of São Carlos, Brazil
Alfredo Cuzzocrea   University of Trieste, Italy
Todd Eavis   Concordia University, Canada
Johann Eder   Alpen Adria Universität Klagenfurt, Germany
Markus Endres   University of Augsburg, Germany
Pedro Furtado   University of Coimbra/CISUC, Portugal
Johann Gamper   Free University of Bozen-Bolzano, Italy
Jérôme Gensel   Grenoble University, France
Shahram Ghandeharizadeh   University of Southern California, USA
Matteo Golfarelli   DISI - University of Bologna, Italy
Goetz Graefe   Hewlett-Packard Laboratories, USA
Dawid Gross-amblard   IRISA, Rennes University, France
Jarek Gryz   York University, Canada
Mohand-Said Hacid   University of Claude Bernard Lyon 1 - UCBL, France
Theo Härder   TU Kaiserslautern, Germany
Mirjana Ivanovic   University of Novi Sad, Faculty of Sciences, Serbia
Hannu Jaakkola   Tampere University of Technology, Finland
Leonid Kalinichenko   Russian Academy of Science, Russia
Ahto Kalja   Küberneetika Instituut, Estonia
Kalinka Kaloyanova   University of Sofia - FMI, Bulgaria
Mehmed Kantardzic   University of Louisville, USA
Marite Kirikova   Riga Technical University, Latvia
Mikhail Kogalovsky   Market Economy Institute of the Russian Academy
                          of Sciences, Russia
Christian Koncilia   Alpen-Adria University Klagenfurt, Austria

Robert Wrembel   Poznan Unviersity of Technology, Institute of
                            Computing Science, Poland
Vladimir Zadorozhny   University of Pittsburgh, USA

Additional Reviewers

Fabian Benduhn   Magdeburg University, Germany
Jevgeni Marenkov   Tallinn University of Technology, Estonia
Sonja Ristic   University of Novi Sad, Serbia
Giorgos Giannopoulos   National Technical University of Athens, Greece
Karoly Bosa   Johannes Kepler University Linz, Austria
Grégory Smits   IRISA, France
Fatma Slaimi   LSIS, Marseille, France
Olga Gkountouna   National Technical University of Athens (NTUA),
                            Athens, Greece
John Liagouris   University of Hong Kong, SAR China
Panagiotis Symeonidis   Aristotle University, Thessaloniki, Greece
Konstantinos Theocharidis   IMIS, Research Center Athena, Greece
Mustafa Al-Hajjaji   University of Magdeburg, Germany
Sebastian Dorok   University of Magdeburg, Germany
Loredana Tec   AIT Austrian Institute of Technology GmbH, Vienna,
                            Austria
Anton Dignos   University of Zürich, Switzerland
Felix Kossak   Software Competence Center Hagenberg GmbH,
                            Hagenberg, Austria
Amel Mammar   Telecom/Telecom SudParis, France
Sahar Vahdati   University of Bonn, Germany
Nabil Hameurlain   University of Pau, France
Tarmo Robal   Tallinn University of Technology, Estonia
Hala Skaf-Molli   LINA, Nantes University, France
Zoltan Miklos   Inria, Rennes, France
Farida Semmak   Université Paris-Est, France
Christophe Gnaho   Université Paris Est, France
Lorena Paoletti   Universidad de Santiago de Chile, Chile
Gilles Nachouki   LINA, Nantes University, France
Irina Astrova   Tallinn University of Technology, Estonia
Shuaiqiang Wang   University of Jyvaskyla, Finland
Zoé Faget   LIAS/ISAE-ENSMA, France
Vladimir Ivančević   University of Novi Sad, Serbia
Saulius Gudas   Vilnius University, Lithuania
Dirk Habich   Technische Universität Dresden, Germany
Slavica Kordić   University of Novi Sad, Serbia
Eike Schallehn   Otto-von-Guericke-Universität Magdeburg, Germany
Vladimir Dimitrieski   University of Novi Sad, Serbia
Christian Koncilia   Alpen-Adria-Universität Klagenfurt, Australia
Ioannis N. Athanasiadis   Hellenic Open University, Kozani, Greece

BigDap 2015 - Second International Workshop on Big Data Applications and Principles

Co-chairs

Elena Baralis   Politecnico di Torino, Italy
Tania Cerquitelli   Politecnico di Torino, Italy
Pietro Michiardi   EURECOM, France

Program Committee

Fernando Arias   EMC, Spain
Juliette Dromard   LAAS-CNRS, France
Paolo Garza   Politecnico di Torino, Italy
Mirko Kampf   Cloudera Inc., USA
Miguel Angel López Peña   Satec, Spain
Alberto Mozo   Universidad Politécnica de Madrid, Spain
Jose Maria Ocon Quintana   SATEC, Spain
Philippe Owezarski   LAAS-CNRS, France
Elisa Quintarelli   Politecnico di Milano, Italy
Xiaolan Sha   EURECOM, France
Bo Zhu   Universidad Politecnica de Madrid, Spain

DCSA 2015 - First International Workshop on Data-Centered Smart Applications

Co-chairs

Ajantha Dahanayake	Prince Sultan University, Saudi Arabia
Bernhard Thalheim	Christian Albrechts University, Germany

Program Committee

Henk Sol	Groningen University, The Netherlands
Elio Masciari	Università della Calabria Via Pietro Bucci, Italy
Barbara Catania	University of Genoa, Italy
Areej Alwabil	Prince Sultan University, Suadi Arabia
Mario Pichler	Software Competence Center, Austria
Samira Si-Said	Conservatoire National des Arts et Métiers, France
Jari Porras	Lappeenranta University, Finland
Isabella Watteau	Conservatoire National des Arts et Métiers, and ESSC, France
Qing Wang	Australian National University, Australia
Margita Kon-Popovska	Ss Cyril and Methodious University, Macedonia
Antje Dusterhoft	Applied University of Wismar, Germany

GID 2015 - 4th International Workshop on GPUs in Databases

Co-chairs

Witold Andrzejewski	Poznan University of Technology, Poland
Krzysztof Kaczmarski	Warsaw University of Technology, Poland
Tobias Lauer	Offenburg University of Applied Sciences, Germany

Program Committee

Artur Gramacki	University of Zielona Góra, Poland
Bingsheng He	Nanyang Technological University, Singapore
Ming Ouyang	University of Massachusets, Boston, USA
Gunter Saake	University of Magdeburg, Germany
Peter Sestoft	IT University of Copenhagen, Denmark
Krzysztof Stencel	University of Warsaw, Poland
Jens Teubner	Technische Universität Dortmund, Germany
Paweł Wojciechowski	Poznan University of Technology, Poland

MEBIS 2015 - Workshop on Managing Evolving Business Intelligence Systems

Co-chairs

Selma Khouri — National Engineering School for Mechanics and Aerotechnics (ISAE-ENSMA), France and National High School of Computer Science (ESI), Algeria

Robert Wrembel — Poznan University of Technology, Poland

Program Committee

Alberto Abelló	Universitat Politècnica de Catalunya, Spain
Faten Atigui	Conservatoire National des Arts et Métiers, France
Jorge Bernardino	ISEC-Polytechnic Institute of Coimbra, Portugal
Bartosz Bębel	Poznan University of Technology, Poland
Jérôme Darmont	Université Lyon 2, France
Todd Eavis	Concordia University, Canada
Cécile Favre	Université Lyon 2, France
Pedro Furtado	University of Coimbra, Portugal
Matteo Golfarelli	University of Bologna, Italy
Marcin Gorawski	Silesian University of Technology, Poland
Christian Koncilia	University of Klagenfurt, Austria
Patrick Marcel	François-Rabelais de Tours, France
Elsa Negre	Université Paris-Dauphine, France
Torben Bach Pedersen	Aalborg University, Denmark
Stefano Rizzi	University of Bologna, Italy
Oscar Romero	Universitat Politècnica de Catalunya, Spain
Olivier Teste	IRIT - Université Toulouse 3 Paul Sabatier, France
Alejandro Vaisman	Universidad de Buenos Aires, Argentina
Panos Vassiliadis	University of Ioannina, Greece
Hannes Voigt	Technische Universität Dresden, Germany
Esteban Zimányi	Université Libre de Bruxelles, Belgium

OAIS 2015 - 4th International Workshop on Ontologies Meet Advanced Information Systems

Co-chairs

Ladjel Bellatreche   LIAS/ISAE-ENSMA, France
Yamine Ait Ameur   IRIT-ENSEIHT, France

Program Committee

Brice Chardin   ISAE-ENSMA, Poitiers, France
Oscar Romero   Barcelona, Spain
Selma Khouri   ESI, Algeria and ISAE-ENSMA, France
Sofian Maabout   LABRI, Bordeaux, France
Robert Wrembel   Poznan University of Technology, Poland
Idir Ait-Sadoune   Supélec, Paris, France
Stéphane Jean   University of Poitiers, France
Sadok Benyahia   Tunis, Tunisia
Yamine Ait-Ameur   ENSEIHT, Toulouse, France
Mahmoud Barhamgi   University of Lyon 1, Lyon, France
Dickson Chiu   University of Hong Kong, Hong Kong, SAR China
Boris Vrdoljak   Zagreb, Croatia
Brahim Medjahed   University of Michigan, USA
Djamal Benslimane   University of Lyon 1, Lyon, France
Germain Forestier   University of Haute-Alsace, Mulhouse, France
Amine Abdelmalek   University of Saida, Saida, Algeria
Rim Moussa   Inria, France

SW4CH 2015 - First International Workshop on Semantic Web for Cultural Heritage

Co-chairs

Béatrice Bouchou Markhoff   LI, University François Rabelais de Tours, France
Stéphane Jean   LIAS/ISAE-ENSMA and University of Poitiers, France

Program Committee

Mirian Halfeld Ferrari Alves   LIFO, University of Orléans, France
Katja Hose   Aalborg University, Denmark
Cvetana Krstev   University of Belgrade, Serbia
Patrick Le Boeuf   Bibliothèque nationale de France
Denis Maurel   LI, University François Rabelais Tours, France
Nizar Messai   LI, University François Rabelais Tours, France
Isabelle Mirbel   WIMMMICS, University Nice Sophia Antipolis, France
Alessandro Mosca   SIRIS Academic, S.L., Barcelona, Spain
Cheikh Niang   Laboratoire de Recherche des Monuments Historiques, France
Martin Rezk   Free University of Bozen-Bolzano, Italy
Xavier Rodier   Laboratoire Archéologie et Territoires, University François Rabelais Tours, France
Simon Scerri   Fraunhofer Institute for Intelligent Analysis and Information Systems, Germany
Dusko Vitas   University of Belgrade, Serbia

WISARD 2015 - Workshop on Information Systems for Alarm Diffusion

Co-chairs

Rémi Delmas   ONERA, Toulouse, France
Thomas Polacsek   ONERA, Toulouse, France
Florence Sèdes   IRIT, Toulouse, France

Program Committee

Fabien Autrel   Telecom Bretagne, France
Rocio Abascal-Mena   UAM, Mexico
Ikram Amous-Ben Amor   MIRACL, Sfax, Tunisia
Sadok Ben Yahia   University of Tunis, Tunisia
Raffaele Campagnuolo   ESA, Noordwijk, The Netherlands
Lilia Gzara, G-SCOP   Grenoble, France
Sergio Ilarri   University of Zaragoza, Spain
Asanobu Kitamoto   NII, Tokyo, Japan
Rose Lema   UAM, Mexico
Michael Mrissa   LIRIS, Lyon, France
TV Prabhakar   IIT Kanpur, India
Arnaud Quirin   Centro de Telecomunicacions de Galicia, Vigo, Spain

Introduction

New Trends in Databases and Information Systems: Contributions from ADBIS 2015

Yamine Aït Ameur[1], Witold Andrzejewski[2], Elena Baralis[3],
Ladjel Bellatreche[4], Béatrice Bouchou Markhoff[5], Tania Cerquitelli[3],
Ajantha Dahanayake[6], Rémi Delmas[7], Stéphane Jean[4],
Krzysztof Kaczmarski[8], Selma Khouri[4,9], Tobias Lauer[10],
Sofian Maabout[11], Pietro Michiardi[12], Tadeusz Morzy[13],
Boris Novikov[14], Thomas Polacsek[7], Florence Sèdes[15],
Bernhard Thalheim[16], Athena Vakali[17], Patrick Valduriez[18],
and Robert Wrembel[13]

[1] IRIT-ENSEIHT, France
[2] Poznan University of Technology, Poland
[3] Politecnico di Torino, Italy
[4] LIAS/ISAE-ENSMA, Poitiers University, France
[5] LI, University François Rabelais de Tours, France
[6] Prince Sultan University, Saudi Arabia
[7] ONERA, Toulouse, France
[8] Warsaw University of Technology, Poland
[9] National High School of Computer Science (ESI), Algeria
[10] Offenburg University of Applied Sciences, Germany
[11] Labri, Bordeaux, France
[12] EURECOM, France
[13] Poznan University of Technology, Poland
[14] St Petersburg University, Russia
[15] IRIT, Toulouse, France
[16] Kiel University, Germany
[17] Aristotle University of Thessaloniki, Greece
[18] INRIA of Montpellier, France

Abstract. Research on database and information technologies has been rapidly evolving over the last few years. Advances concern several dimensions: data types, new management issues, new deployment platforms and new storage layouts. The 19 East-European Conference on Advances in Databases and Information Systems (ADBIS 2015), held on September 8-11, 2015, at Futuroscope, Poitiers, France, and associated satellite events aimed at covering some emerging issues concerning new trends in database and information system research. The aim of this paper is to present such events, their motivations and topics of interest, as well as briefly outline the papers selected for presentations. The selected papers will then be included in the remainder of this volume.

1 Introduction

The East-European Conference on Advances in Databases and Information Systems (ADBIS) aims at providing a forum where researchers and practitioners in the fields of databases and information system can interact, exchange ideas and disseminate their accomplishments and visions. Inaugurated 19 years ago, ADBIS originally included communities from Central and Eastern Europe, however, throughout its lifetime it has spread and grown to include participants from many other countries throughout the world. The ADBIS conferences provide an international platform for the presentation of research on database theory, development of advanced DBMS technologies, and their advanced applications. The ADBIS series of conferences aims at providing a forum for the presentation and dissemination of research on database theory, development of advanced DBMS technologies, and their advanced applications. ADBIS 2015 in Poitiers continues after St. Petersburg (1997), Poznan (1998), Maribor (1999), Prague (2000), Vilnius (2001), Bratislava (2002), Dresden (2003), Budapest (2004), Tallinn (2005), Thessaloniki (2006), Varna (2007), Pori (2008), Riga (2009), Novi Sad (2010), Vienna (2011), Poznan (2012), Genoa (2013) and Ohrid (2014). This edition is special, as it is the first time that ADBIS takes place in France. The program of ADBIS 2015 included keynotes, research papers, two tutorials, and thematic workshops. While papers accepted at the ADBIS main conference span a wide spectrum of topics in the field of databases and information systems, ranging from Database Theory & Access Methods, User Requirements and Database Evolution, Multidimensional Modeling and OLAP, ETL, Advanced Design Modeling, Ontologies, Time Series Processing, Performance & Tuning, to Advanced Query Processing, Approximation & Skyline, Preferences & Recommender Systems, Confidentiality & Trust, and data quality, the general idea behind each satellite event was to collect contributions from various subdomains of the broad research areas of databases and information systems, representing new trends in these two important areas. More precisely, the following satellite events have been organized:

- 2nd International Workshop on Big Data Applications and Principles (BigDap 2015), organized by Elena Baralis (Politecnico di Torino, Italy), Tania Cerquitelli (Politecnico di Torino, Italy) and Pietro Michiardi (EURECOM, France).
- Workshop on Data Centered Smart Applications (DCSA 2015), organized by Ajantha Dahanayake (Prince Sultan University, Saudi Arabia) and Bernhard Thalheim (Christian Albrechts University, Germany).
- Fourth International Workshop on GPUs in Databases (GID 2015), organized by Witold Andrzejewski (Poznan University of Technology, Poland), Krzysztof Kaczmarski (Warsaw University of Technology, Poland) and Tobias Lauer (Offenburg University of Applied Sciences, Germany).
- Workshop on Managing Evolving Business Intelligence Systems (MEBIS 2015), organized by Selma Khouri (National Engineering School for Mechanics and Aerotechnics (ISAE-ENSMA), France and National High School of Computer Science (ESI), Algeria) and Robert Wrembel (Poznan University of Technology, Poland).

- Fourth International Workshop on Ontologies Meet Advanced Information Systems (OAIS 2015), organized by Ladjel Bellatreche (LIAS/ISAE-ENSMA, France) and Yamine Aït Ameur (IRIT-ENSEIHT, France).
- First InternationalWorkshop on SemanticWeb for Cultural Heritage (SW4CH 2015), organized by Béatrice Bouchou Markhoff (LI, University François Rabelais de Tours, France) and Stéphane Jean (LIAS/ISAE-ENSMA and University of Poitiers, France).
- Workshop on Information Systems for AlaRm Diffusion (WISARD 2015), organized by Rémi Delmas (ONERA, Toulouse, France), Thomas Polacsek (ONERA, Toulouse, France), Florence Sédes (IRIT, Toulouse, France).

The main ADBIS conference as well as each of the satellite events had its own international program committee, whose members served as the reviewers of papers included in this volume. This volume contains papers selected as short contributions to be presented at the ADBIS 2015 main conference as well as papers contributed by all satellite events listed above. In the following, for each event, we present its main motivations and topics of interest and we briefly outline the papers selected for presentations. The selected papers will then be included in the remainder of this volume. Some acknowledgements from the organizers are finally provided.

2   ADBIS Selected Short Contributions

Description. The ADBIS main conference was chaired by Patrick Valduriez (INRIA, France) and Tadeusz Morzy (Poznan University, Poland). The main conference attracted 135 paper submissions from 39 different countries representing all continents. All papers were evaluated by at least three reviewers. After rigorous reviewing process by the members of the International Program Committee consisting of 77 reviewers from 22 countries, the 31 papers included in this LNCS proceedings volume were accepted as full contributions, making an acceptance rate of 23% and 18 papers were selected as short contributions and included in this volume.

Selected Papers. The selected 18 short papers span a wide spectrum of topics in the database field and related technologies.

The selected short papers span a wide spectrum of topics related to the development of advanced database applications covering different types of data (tweet data, time series, data streams, services, work-flow instance data, semantic data, etc.), different management issues (ETL, querying, optimizing, data analysis, recommending, etc.), different types of deployment platforms (Cloud and Distributed) and different storage layouts (column store and row store). Specifically, the paper entitled: *Revisiting the Definition of the Relational Tuple Calculus*, by Bader Albdaiwi and Bernhard Thalheim revisits logical relational query languages, mainly the tuple relational calculus (TRC) to deal with the finiteness problem. ETL (Extract, Transform, Load) and data quality are addressed as well. Specifically, the paper titled *Using a Domain-Specific Language to Enrich ETL Schemas* (Orlando Belo, Claudia Gomes, Bruno Oliveira, Ricardo Marques and Vasco Santos) presents a way for enriching data migration conceptual schemas in

BPMN using a domain-specific language (DSL). It demonstrates how to convert such enriched schemas to a first correspondent physical representation (a skeleton), having the possibility to be executed and validated in a conventional ETL implementation tool like Kettle. The paper titled *Avoiding Ontology Confusion in ETL Processes* (Selma Khouri, Sabrina Abdellaoui and Fahima Nader) proposes a semantic ETL approach which considers both canonical and non-canonical layers of ontologies. Under an ontological scope, the main contribution is centered on a clear identification of their ontological dependencies. Capturing these dependencies ensures a high quality of the target warehouse. The paper *AutoScale: Automatic ETL scale process* (Pedro Furtado) presents an architecture for efficiently increasing the scalability of ETL processes. The primary objective is to more or less automate the mechanism by which additional computational processes can be spawned in response to the incoming workload. User supported configuration is kept to a minimum. The paper *Relational-Based Sensor Data Cleansing* (Nadeem Iftikhar, Xiufeng Liu and Finn Ebertsen Nordbjerg) deals with an important problem which is data cleaning. It presents a simple approach to clean sensor data, relying on relational technology.

Selected papers also address advanced design modeling. The paper *OLAP4-Tweets: Multidimensional Modeling of tweets* (Maha Ben Kraiem, Jamel Feki, Kas Khrouf, Franck Ravat and Olivier Teste) presents the implementation of a data warehouse for tweets analyzing. A typical fact constellation schem with denormalized dimensions is proposed. The usage of the obtained warehouse with some typical analtyical application is well described. The paper *Data Warehouse Design Methods Review: Trends, Challenges and Future Directions for the Healthcare Domain* (Christina Khnaisser, Luc Lavoie, Hassan Diab and Jean-François Ethier) presents an interesting review of methods for designing a data warehouse for a medical domain. It come up with a classification of a number of design warehouse approaches in the light of some requirements that the authors have derived from an analysis of the healthcare domain. The paper *Towards A Generic Approach for the Management and the Assessment of Cooperative Work* (Amina Cherouana, Amina Aouine, Abdelaziz Khadraoui and Latifa Mahdaoui) presents a generic approach for managing cooperative work processes. A model based on activity theory and design patterns and using the Model Driven Engineering principles is given. A well detailed example illustrating the use of this model in the domain of legal ontology development in Algeria is presented. Advanced query processing is addressed as well in advanced database applications and Big Data context, where C-store is mainly used. Specifically, the paper *Incrementally Maintaining Materialized Temporal Views in Column-oriented NoSQL Databases with Partial Deltas* addresses an interesting topic which is maintaining materialized temporal views (MTV) in the context of column-oriented NoSQL DB systems for data warehouses. It appears that generating complete change data history (complete delta), as it is usually done, to update MTV is not suitable in the context of temporal column-oriented DB systems due to performance issues and missing values. An algorithm to build the complete delta from a partial one is provided. The paper *Towards self-management in a distributed column-store system* (Chernishev George) presents a general approach to self-management and tuning for column-store systems. A large portion of the paper is devoted to describe the general issue of tuning and self-tuning, with respect to traditional database systems. The paper *A Requirements Specification*

Framework for Big Data Collection and Capture (Noufa Alnajran and Ajantha Dahanayake) contributes to big data collection through the development of a conceptual model. An overview of data collection techniques is presented. The paper *Continuous Query Processing over Data, Streams and Services: Application to Robotics* (Vasile-Marian Scuturici, Yann Gripay, Jean-Marc Petit, Yutaka Deguchi and Einoshin Suzuki). This paper describes an approach how to integrate database management technologies in robotics. More specifically, it shows how to use database functionalities in order to facilitate the development of robot applications. The basic idea of this work to integrate database functionalities, in particular query processing, sounds appealing and might have some potential to use synergies between two completely different research areas.

Personalisation and recommender systems are also addressed. The paper *Database Querying in the Presence of Suspect Values* (Olivier Pivert and Henri Prade) presents a database model relying on the notion of possibilistic certainty to deal with suspect values. Algebra to query instances of such model is also provided. Algebraic operators are clearly described and examples are given to explain the introduced concepts. The paper *Query Skylines for Optimization and Approximate Evaluation* (Anna Yarygina and Boris Novikov) addresses the problem of efficient approximate query evaluation, by modeling it as a multiobjective optimization problem, with two parameters: the usage of computational resources and result quality. Experimental results are also provided. The paper *Viewer Behavior Prediction in Social-TV Recommender Systems: Survey and Challenges* (Meriam Bambia and Rim Faiz) provides an outline of the state-of-the-art on TV recommender systems and open research issues in the area of context-awareness in order to improve recommendation in social TV (netflix, videos on youtube, etc.). The paper *Generalized Bichromatic Homogeneous Vicinity Query Algorithm in Road Network Distance* (Yutaka Ohsawa, Htoo Htoo, Naw Jacklin Nyunt and Myint Myint Sein) proposes an enhancement method by a single source multi-targets A* algorithm (SSMTA*), applied in an integrated and adaptive framework for several vicinity type of queries like: set k-nearest neighbour query (kNN), ordered k-nearest neighbor query (ordered-kNN), bichromatic reverse k-nearest neighbor query (BRkNN), and distance range query. The paper also provides an extensive set of experiments with a real network interest objects sets and computer generated rival object sets data. The experiments show that the proposed algorithm is faster than the usual nearest neighbour search.

Finally, new trends in Web Content and Cloud Computing are also addressed. The paper *Towards Multilayer Multimedia Exploration* (Juraj Mosko, Jakub Lokoc, Tomas Grosup, Premysl Cech, Tomas Skopal and Jan Lansky) presents a multilayer structure that enables exploration of multimedia collections in different levels of details. Each layer can be indexed by a separate similarity index structure that improves the scalability of the whole structure. Defined operations zoom-in/out and pan enable horizontal and vertical browsing. Implemented user study compares 3-layer principles of proposed structure and related 2-layer structure that corresponds to standard k-NN browsing. The paper *SLAOntology-Based Elasticity in Cloud Computing* (Taher Labidi, Achraf Mtibaa and Faiez Gargouri) presents a concept of the platform for Ontology based SLA that considers elasticity strategies by semantic meaning.

3   Workshop on Managing Evolving Business Intelligence Systems (MEBIS)

Descritpion. Workshop on Managing Evolving Business Intelligence Systems (ME-BIS 2015), organized by Selma Khouri (National Engineering School for Mechanics and Aerotechnics (ISAE-ENSMA), France and National High School of Computer Science (ESI), Algeria) and Robert Wrembel (Poznan University of Technology, Poland). Nowadays, a business intelligence (BI) technology is a worldwide accepted and obligatory component of information systems deployed in companies and institutions. From a technological point of view, BI includes a few layers. The first one is an ETL layer, whose goal is to integrate data coming from heterogeneous, distributed, and autonomous data sources, which include operational databases and other storage systems. The integrated data are stored in a central database, called a data warehouse (DW), located in the second layer. Based on the central DW, smaller thematically oriented data warehouses can be built. They are called data marts (DM). Data stored in a DW or DM are analysed by multiple applications, located in the third layer. An inherent feature of the data sources that fed the BI architecture with data is that in practice they evolve in time independently of the BI architecture. The evolution of the data sources can be characterized by content changes, i.e., insert/update/delete data, and structural (schema) changes, i.e., add/modify/drop a data structure or its property. Handling content changes at a DW can be implemented by means of: (1) temporal extensions, (2) materialized views, and (3) data versioning mechanisms. The propagation of structural changes into the BI architecture is much more challenging, as it requires modifications in all the layers in this architecture, forcing the layers to evolve. Three basic approaches to handling the evolution of data warehouses were proposed by research communities. These approaches are classified as: (1) schema evolution, (2) schema versioning, and (3) data warehouse versioning. Even though a decent contribution was made in this research area, there still exist multiple open research and technological problems, like querying heterogeneous DW versions, efficient indexing multi-version data, integrity constraints for multi-version DW schemas, modeling multiversion DWs. To the best of our knowledge, no solutions were proposed so far to support the evolution of the ETL layer and the analytical applications layer. We observe that the research on evolving BI systems is still very active. Recently R. Kimball proposed the extension to his concept of slowly changing dimensions, extending it with SCD types 4 to 7. The international research conferences and journals publish papers on evolving BI systems, e.g., DaWaK 2014, EDA 2014, EDA 2015, TIME 2014, Information Systems 2015. Handling multiple and evolving states of some entities is a more general problem. Intensive research is conducted also in the areas of versioning XML documents and versioning ontologies. Last but not least, some NoSQL storage systems support versioning of data, e.g., HBase, Cassandra. The aim of this workshop is twofold. First, to gather those researchers and possibly industry developers who focus on handling dynamics in business intelligence systems, in order to discuss their achievements and open issues. Second, to communicate to the BI community still existing open issues and to inspire them to conduct research in this area. The MEBIS workshop would be the second in the series of workshops devoted to research on evolving BI systems. The first one was organized in conjunction with ADBIS 2009, also by Robert Wrembel. The program committee was composed of 20 members.

Selected papers. This edition of MEBIS includes the following four papers. *E-ETL Framework: ETL Process Reparation Algorithms using Case-based Reasoning* (A. Wojciechowski) contributes a framework, called E-ETL, for semiautomatic reparation of ETL workflows in response to structural changes in data sources. The proposed framework suggests modifications to an ETL workflow using the case-based reasoning method. E-ETL is implemented as a software that is external to ETL design environments and that communicates with them via API. *Handling Evolving Data Warehouse Requirements* (D. Solodovnikova, L. Niedrite, and N. Kozmina) presents a method for propagating changing business requirements into a data warehouse. The method is supported by meta-models that store data about the evolution of a logical and physical level of a data warehouse design. A data warehouse evolves by means of a schema versioning mechanism. *Querying Multiversion Data Warehouses* (W. Ahmed and E. Zimányi) addresses the problem of querying heterogeneous data warehouse versions, i.e., versions that have different schemas. The authors analyze an impact of various schema changes on OLAP queries and propose a method for processing queries that refer to multiple heterogeneous versions of a data warehouse. *CUDA-Powered CTBE Algorithm for Zero-Latency Data Warehouse* (M. Gorawski, D. Lis, and A. Gorawska) proposes an algorithm for refreshing a data warehouse in real- or almost-real time. Their previously developed algorithm, called Choose Transaction by Election (CTBE) was implemented in a CUDA architecture. Its performance was compared to a standard (CPU-based) architecture, showing much better performance.

4 Fourth International Workshop on GPUs in Databases (GID 2015)

Descritpion. Fourth International Workshop on GPUs in Databases (GID 2015), organized by Witold Andrzejewski (Poznań University of Technology, Poland), Krzysztof Kaczmarski (Warsaw University of Technology, Poland) and Tobias Lauer (Offenburg University of Applied Sciences, Germany). The ADBIS Workshop on GPUs in Databases GID is devoted to all subjects related to utilization of Graphics Processing Units in database environments. This is already a 4th edition of this workshop. Initially, at the time of the 1st edition, the concept of using GPUs in databases was relatively young. Since that time this idea has slowly gained recognition and gradually, received a lot of attention from the scientific community. The main reason is that a lot can be gained from using GPUs in database related applications. GPUs can be utilized as co-processors for computing aggregations in data warehouses or as components for high performance in-memory databases and, last but not least, they can provide means, which allow achieving high performance in data mining computations. These results require the scientific community to redesign almost every aspect of database management systems. New query execution algorithms have to be developed to make efficient use of the specific computing capabilities of GPUs. This in turn leads to the necessity of developing new query optimization algorithms, which are able to generate query execution plans taking into account the trade-off between speed-up that can be gained and costs, such as host< — >device data transfer bottleneck. New data storage structures and supporting indexes have to be developed as well. Finally, novel transaction scheduling techniques are needed to provide correct transaction processing. While a lot of research has already been done on this subject, the work is still far from being finished. Each new generation of graphics cards introduces new features and new capabilities that provide possibilities for designing better and faster algorithms. Modern IT solutions make heavy use of virtualization techniques. Providing viable means of GPU virtualization would strongly accelerate the deployment of GPU based algorithms in commercially available database applications. Finally, state-of-the-art algorithms are always a subject for improvement. The intention of the GID workshop is to provide a discussion forum for the scientists working on providing GPU based algorithms for database applications as well as industry. Presentation of practical and theoretical research creates chances for fruitful cooperation between these two communities. The program committee was composed of 8 members. **Selected papers.** This edition of GID workshop includes two papers. *Big Data Conditional Business Rule Calculations in Multidimensional In-GPU-Memory OLAP Databases* (Alexander Haberstroh and Peter Strohm) deals with the problem of generating virtual cubes in limited memory conditions by accelerating computation of conditional business rules on GPUs. *Optimizing Sorting and Top-k Selection Steps in Permutation Based Indexing on GPUs* (Martin Krulis, Hasmik Osipyan and Stéphane Marchand-Maillet) presents a very interesting algorithm for finding top-k nearest pivot points on GPUs which is an important step for permutation index construction.

5   Workshop on Information Systems for AlaRm Diusion (WISARD)

Descritpion. The first International Workshop on Information Systems for AlaRm Diffusion (WISARD), organized by Rémi Delmas (ONERA, Toulouse, France), Thomas Polacsek (ONERA, Toulouse, France), Florence Sèdes (IRIT, Toulouse, France). The aim of this workshop is to gather researchers and practitioners working in the area of Information Systems, DataBase, Document management, metadata, workflow management, conceptual modeling, design, software, etc., to exchange and to share information, to network, to give points of view and to collaborate in project proposals. This workshop will discuss questions about liveness, dynamicity, reactivity of alarm systems: how to ensure the warning information reach the right destination at the right moment and in the right location, still being relevant for the recipient, in spite of the various and successive filters of confidentiality, privacy, firewall policies, etc.? Also relevant in this context are technical contingency issues: material failure, defect of connection, break of channels, independence of information routes and sources. Alarms with crowd media, (mis-)information vs. rumours: how to make the distinction? The program committee was composed of 12 members.

Selected papers. Six papers are selected that span the wide spectrum of exchange policies, Alarms, Diffusion Systems and user profiles. The paper *Abduction for Analysing Data Exchange Policies* (Laurence Cholvy) addresses the question of checking the quality of data exchange policies which exist in organizations in order to regulate data exchanges between members. More particularly, we address the question of generating the situations compliant with the policy but in which a given property is unsatisfied. We show that it comes to a problem of abduction and we propose to use an algorithm based on the SOL-resolution. Our contributions are illustrated on a case study.

The paper *ADMAN: an Alarm-based mobile Diabetes MANagement system for mobile geriatric teams* (Dana Al Kukhun, Bouchra Soukkarieh and Florence Sédes) introduces ADMAN an alarm-based diabetes management system for the disposal of Mobile Geriatric Teams MGT. The system aims at providing a form of remote monitoring in order to control the diabetes rate for elder patients. The system is multidimensional in a way that it resides at the patient mobile machine from a side, the doctors mobile machine from another side and can be connected to any other entity related to the MGT thats handling his case (e.g. dietitian). The paper *An Architectural Roadmap Towards Building an Alarm Diffusion System (ADS)* (Sumit Kalra, Saurabh Srivastava and Prabhakar) aims at providing a software architecture perspective towards ADS. It looks at both functional and quality requirements for an ADS and also attempts to identify certain quality attributes specific to an ADS and tries to provide a set of architectural tactics to realise them. It also proposes a Reference Architecture for designing such systems. Ample examples to support our inferences and take a deeper look at a case study of the Traffic Collision Avoidance System (TCAS) in aircrafts are also given. The paper *Information exchange policies at an organisational level: formal expression and analysis* (Claire Saurel) starts from the formal logical framework PEPS intended to define and analyse information exchange policies for critical information systems. It introduces in PEPS a layer PEPSORG to express organisational information exchange policies at an abstract level, together with some exchange rights transmission axioms between roles or organisations. Some properties are defined within this organisational layer, in particular about information permeability through organisations. More efficient expression and analysis of organisational exchange policies are expected with this contribution.

The paper *Critical Information Diffusion Systems* (Rémi Delmas and Thomas Polacsek) points that fact that today, individuals, companies, organizations and national agencies are increasingly interconnected, forming complex and decentralized information systems. In some of these systems, the very fact of exchanging information can constitute a safety critical concern in itself. Take for instance the prevention of natural disasters. We have a set of actors who share their observation data and information in order to better manage crises by warning the more competent authorities. The aim of this article is to find a definition to such kind of systems we name Critical Information Diffusion Systems. In addition, we see why Critical Information Diffusion Systems need information exchange policies.

The paper *A case study on the influence of the user profile enrichment on Buzz propagation in social media: Experiments on Delicious* (Manel Mezghani, Sirinya On-At, André Peninou, Marie Françoise Canut, Corinne Zayani, Ikram Amous-Ben

Amor and Florence Sédes) points the fact that the user is the main contributor for creating information in social media. In these media, users are influenced by the information shared through the network. In a social context, there are so-called Buzz, which is a technique to make noise around an event. This technique offers that several users will be interested in this event at a time t. A Buzz is then popular information in a specific time. A Buzz may be a fact (true information) or a rumour (fake, false information). We are interested in studying Buzz propagation through time in the social network Delicious. Also, The authors study the influence of enriched user profiles proposed in the literature to propagate the Buzz in the same social network. The authors state a case study on some information of the social network Delicious. This latter contains social annotations (tags) provided by the user. These tags contribute to influence the other users to follow this information or to use it. This study relies on three main axes: 1) they focus on tags considered as Buzz and analyse their propagation through time. 2) They consider a user profile as the set of tags provided by him. We will use the result of our previous work on temporal user profile enrichment, in order to analyse the influence of this enrichment in the Buzz propagation. 3) They analyse each enriched user profile in order to show if the enrichment approach anticipate the Buzz propagation. So, this approach gives adapted results (e.g. recommendation) to the user.

6   2nd International Workshop on Big Data Applications and Principles (BigDap)

Introduction. 2nd International Workshop on Big Data Applications and Principles (BigDap 2015), organized by Elena Baralis (Politecnico di Torino, Italy), Tania Cerquitelli (Politecnico di Torino, Italy), Pietro Michiardi (EURECOM, France). Large volumes of data (Big Data) are being produced by various modern applications (e.g., smart cities, computer network traffic, e-commerce, social networks) at an ever increasing rate. To deal with such huge data volumes, innovative solutions for data models, algorithms, and architectures have to be designed providing the necessary scalability and flexibility for novel big data analytics applications. These challenges have been attracting great attention from both academia and industry. The BigDAP 2015 workshop aims at promoting and sharing research and innovative applications on Big Data. The workshop allows researchers and practitioners to describe their experiences in designing and developing big data applications, to discuss the big data open issues and to share new ideas and techniques for big data management and analysis. Topics of interests for this workshop include all facets of a Big Data application and range from big data models, algorithms, and architectures, cloud computing techniques, programming models and environments, data integrity and privacy, to search and mining. BigDAP 2015 has been sponsored by FP7 ONTIC project (Online Network Traffic Characterization, http://ict-ontic.eu), funded by the European Commission. Therefore, the application of scalable Big Data analytics to network traffic characterization is a key topic in the workshop.

The workshop is organized as a keynote presentation given by Daniele Quercia on Computational Urban "Science" and a selection of nine papers. These papers address different interesting research issues, application domains, and experience reports on big data management and analysis. In the era of Big data, Information and Communication Technologies have supported cities in becoming smart. A smart city is a urban environment in which social and environmental resources are combined to make cities more livable by increasing urban prosperity and competitiveness. Thus, the smart city applications are an interesting domain where Big Data methodologies can effectively contribute to enhance user life.

The BigDap workshop includes a keynote presentation held by Daniele Quercia. He is a computer scientist and has been named one of Fortune magazine's 2014 Data All-Stars. He spoke about "happy maps" at TED. He is interested in the relationship between on-line and off-line worlds, and his work has been focusing on the area of urban informatics. He was Research Scientist at Yahoo Labs, a Horizon senior researcher at The Computer Laboratory of the University of Cambridge, and Post-doctoral Associate at the Massachusetts Institute of Technology. He received his Ph.D. from UC London. His thesis was sponsored by Microsoft Research Cambridge and was nominated for BCS Best British Ph.D. dissertation in Computer Science. **Keynote Presentation. Computational Urban "Science"!, by Daniele Quercia.** Some of you have used large quantities of online data to study social dynamics in new ways. That tremendous effort resulted in the emergence of a new research area called "computational social science". Consider the specific case of online-networked individuals (e.g., users of Twitter, Instagram, Flickr). Can their social dynamics be used to build better tools for future cities? To answer that question, a few years ago, our research started to focus on understanding how people psychologically experience the city. We used computer science tools to replicate 1970s social science experiments at scale, at web scale. The result of that research has been the creation of new maps, maps where one does not only find the shortest path but also the most enjoyable path[1]. What if we had a mapping tool that would return the most enjoyable routes based not only on aesthetics but also based on smell and sound? This talk will address that question by showing how a creative use of social media can tackle hitherto unanswered research questions (e.g., how to capture smellscapes and soundscapes of entire cities[2]).

Selected papers. The workshop contains six research papers (Afrati et al., Zhu et al., Ordozgoiti et al., Dromard et al., Gomez Canaval et al., Bidoit et al.), two application papers (Attanasio et. al. and Bascuñana et al.), and a review paper (Apiletti et al.). Four research papers proposed parallel and distributed algorithms on either Spark or Map Reduce frameworks (Afrati et al., Zhu et al., Ordozgoiti et al., Dromard et al.). Afrati et al. proposed a n-way join algorithm to extract meaningful information from multiple data sources that provide information seemingly about the same attributes. The proposed algorithm has been implemented on top of the MapReduce framework. Zhu et al. proposed a parallel implementation of a subspace clustering algorithm, based on SUBCLU, on the Spark platform. Ordozgoiti et al., instead, proposed a parallel and scalable implementation of the state-of-the-art unsupervised feature selection CSSP (Column Subset Selection Problem) algorithm. The latter two contributions exploit

---

[1] http://www.ted.com/talks/daniele quercia happy maps?language=en

[2] http://researchswinger.org/smellymaps/index.html

Spark in-memory primitives to achieve a good scalability. Dromard et al. presented a scalable implementation of an existing unsupervised network anomaly detector (UNADA) to efficiently analyze large collections of network traffic in real-time. Two research papers presented distributed engines to effectively analyze large volumes of data (Gomez Canaval et al., Bidoit et al.). The NPEPE engine has been presented by Gomez Canaval et al. to solve NP-complete problems in an efficient way. NPEPE exploits a platform to deploy and run massive networks of NPEPs (Networks of Polarized Evolutionary Processors). The Andromeda system, presented by Bidoit et al., is an engine to efficiently process both queries and updates on large XML documents. Statical and dynamical partitioning strategies of the input data have been proposed to balance the computing load among machines in a cluster. The workshop also includes two application papers that exploit Big Data strategies to analyze a large volume of network data, i.e., social networks (Attanasio et al.), mobile communications (Bascuñana et al.). Attanasio et al. proposed a scalable platform to extract real-time information about an ongoing crisis from social networks. Specifically, to estimate people reactions, messages posted on social networks are analyzed to drive the decision making during a crisis. Bascuñana et al. introduced the concept of Adaptive Quality of Experience (AQoE) in the mobile communication networks. It relies on innovative real-time data analysis techniques able to measure and anticipate the users' Quality of Experience (QoE). The QoE model, the metrics to evaluate and predict QoE and which data analytics algorithms might be exploited have been discussed. Finally, a review of scalable state-of-the-art frequent itemset algorithms has been discussed by Apiletti et al.

7   Fourth International Workshop on Ontologies in Advanced Information Systems (OAIS)

Description. Fourth International Workshop on Ontologies in Advanced Information Systems (OAIS 2015), organized by Ladjel Bellatreche (LIAS/ISAE-ENSMA, France), Yamine Aït Ameur (IRIT-ENSEIHT, France). Information Systems are record sensitive and crucial data to support day-to-day company applications and decision-making processes. Therefore, these systems often contain most of company product and process knowledge. Unfortunately, this knowledge is implicitly encoded within the semantics of the modelling languages used by the companies. The explicit semantics is usually not recorded in such models of information systems. References to ontologies could be considered as an added value for handling the explicit semantics carried by the concepts, data and instances of models Thus, developing new user Interfaces or reconciling data and/or models with external ones often require some kind of reverse engineering processes for making data semantic explicit. Nowadays, ontologies are used for making explicit the meaning of information in some research and application domains. Ontologies are now used in a large spectrum of fields such as: Semantic Web, information integration, database design, e-Business, data warehousing, data mining, system interoperability, formal verification. They are also used to provide information system user knowledge-level interfaces. Over the last five year, a number of interactions between ontologies and information systems have emerged. New methods have been proposed to embed within database both ontologies and data, defining new ontology-based database. New languages were developed in order to facilitate exchange both ontology and data. Other languages dedicated for querying data at the ontological level were proposed (e.g., RQL, SOQA-QL, or OntoQL). Various approaches have been designed to deal with semantic integration of heterogeneous information sources and system interoperability using ontologies either in data sources or in mediators. In some domains, like products modelling, ontologies were published as standards. These ontologies are actually used to define worldwide exchange consortium for sharing information in various application domains. Due to these recent developments, most commercial database systems offer solutions for managing data and ontologies. All these motivations let to the organisation of OAIS 2015 that continues to attract scientists, engineers, educators, industry people, decision makers, and others to share their insight, vision and understanding of the ontologies challenges in Advanced Information Systems. The program committee was composed of 17 members.

Selected papers. Five papers were selected covering hot topics: trajectory data, Cloud, Linked Data, and Fuzzy inference. The paper *An ontology-based approach for handling explicit and implicit knowledge Application to trajectory data using domain, temporal and spatial semantics* (Rouaa Wannous, Cécile Vincent, Jamal Malki and Alain Bouju) presents an ontology-based approach for handling explicit and implicit knowledge. The main idea of this paper is to transform data and semantics models into ontologies to analyse seal trajectories. The paper *Interpretation of DD-LOTOS specification by C-DATA** (Maarouk Toufik Messaoud, Djamel Eddine Saidouni, Mahdaoui Rafik and Houassi Hichem) presents a formal specification and compilation of concurrent systems DD-LOTOS). Then, the specification is translated to C-DATA*, a semantic model than can be inputed into verification tools by the means of theory of timed automata. The paper *Mobile Co-Authoring of Linked Data in the Cloud* (Moulay Driss Mechaoui, Nadir Guetmi and Abdessamad Imine) proposes a cloud SaaS application used for real-time collaborative editing of linked data represents as a RDF graph in a mobile environment. Two layers model is proposed: (i) cloning engine that enables users to clone their mobile devices in the cloud to delegate the overload of collaborative tasks and provides peer-to-peer networks where users can create ad-hoc groups and (ii) collaborative engine that allows updating freely and concurrently a shared RDF graph in peer- to-peer fashion without requiring a central server. The paper *Ontology based Linkage between Enterprise Architecture, Processes, and Time* (Marite Kirikova, Ludmila Penicina and Andrejs Gaidukovs) proposes to model time (using W3C Time Ontology) as one of the sub-models of enterprise architectures, and to relate time to other models of the enterprise architecture, especially to the business process model. The papers *Fuzzy Inference-based Ontology Matching Using Upper Ontology* (Alsayed Algergawy and Samira Babalou) presents an ontology matching approach under using a fuzzy approach. It describes a global large size bio-ontologies. Several small ontologies are defined in the same domain and are matched with the global large ontology. The matching approach is a two step approach. First a fuzzy inference approach, and second the composition of the obtained results in order to fix the proposed matching. An experimentation using both Dolce and OpenCyc and its result are provided as well.

8   First International Workshop on Data Centered Smart Applications (DCSA)

Description. First Internation Workshop on Data Centered Smart Applications (DCSA 2015) is organized by Ajantha Dahanayake (Prince Sultan University, Saudi Arabia), Bernhard Thalheim (Christian Albrechts University, Germany). Massive data collections and the usage of that data for smart applications has become the key to the process for improving the efficiency, reliability, and security of a traditional application. There are many issues that should be taken into consideration at the first step: (a) Modeling, analytics and design of data and smart applications, (b) Metadata, ontologies, vocabularies perspectives, (c) Semantic web for smart applications, (d) Applications of existing technologies, etc. The classical approach where people had to learn how equipment can be used is not appropriate any-more. Applications are only accepted if they are natural for the users. In the past, research tried participatory design based on some knowledge of the main users, their way of operating systems and their main desires and demands. This research and technology development must however be extended by tools that can be used in an intuitive way, within many cultures, within a variety of deployment scenarios, within a group of people deploying the techniques, within different collaboration scenarios, and within different levels of attention. Cities, regions, application will not become directly intelligent and useful. Smart application can be based on the *HOME* approach: high quality content, often updated, minimal effort, e.g. download and processing time and space, and ease of use. Therefore, we need applications that do not require additional training, are simple to operate, are obvious to operate, are simple for everybody, straightforward, within expectation, provide services and features context-sensitive help, selection of wording, simple dialogues and tasks, and are simple to remember, error robust, reliable, of high quality. Applications are becoming *SMART* if they are simple in any step of usage, are motivational for any user, are *Attainable* for the goals the user has in mind, tend to be *Rewarding* since they seem to be worthwhile, right time, match efforts and needs, and *Time-efficient* within the limits and expectations of all members of the community of practice. This set of criteria can be considered as a lacuna of challenges that novel systems must meet.

The concept of smart applications includes various aspects such as environmental sustainability, social sustainability, regional competitiveness, natural resources management, cyber security, and quality of life improvement. With the massive deployment of networked smart devices and sensors, unprecedented large amounts of sensory data can be collected and processed by advanced computing paradigms which are today the enabling techniques for smart applications. For example, given historical environmental, population and economic information, salient modelling and analytics are needed to simulate the impact of potential application development strategies, which will be critical for intelligent decision-making. Analytics are indispensable for discovering the underlying structure from retrieved data in order to design the optimal policies for real time automatic control in the cyberphysical smart system. The uncertainties and security concerns in the data collected from heterogeneous resources aggravate the problem, making smart planning, operation, monitoring and control highly challenging. To address some of those concerns if not all the 1st International Workshop on Data Centered Smart Applications (DCSA 2015) is organized with the aim to provide a platform for exchanging empirical and theoretical research results about DCSA problems and solutions.

We received eight submissions for the workshop and could select four of them for the workshop. We are thankful to the reviewers and to the authors who made a good revision of their papers. *A Mutual Resource Exchanging Model and its Applications to Data Analysis in Mobile Environment* (Naofumi Yoshida). In this paper, a mutual resource exchanging model in mobile computing environment and its application to data analysis are introduced. Resource exchanging is a key issue in data analysis because the only way for efficient use of limited resources in mobile environment is exchanging them each other. Also this paper presents the applicability of this model by showing applications of (1) a universal battery, (2) bandwidth sharing, (3) a 3D movie production by collective intelligence, and, (4) mobile data analysis for motion sensors. By showing results of data analysis, the applicability of this model will be clarified.

Gamification in Saudi Society: A Framework to Develop Human Values for Early Generations (Alia AlBalawi, Bariah AlSaawi, Ghada AlTassan, and Zaynab Fakeerah). In a technology era, where the evolution is faster than we can imagine, it is without no doubts that our daily life is mostly driven by the technology, in learning, arranging to do lists, in our cars, mobile phones and even in games. In fact if we want to talk about gaming nowadays, it is also used not just for entertainment and there is a new revolution on a young field called gamification, where it combines learning with entertainment. It is used in many fields such as social media, schools, sports, marketing and even in NASA, so you can imagine how gamification is spreading like fire in the past few years that even oxford has added the word gamification to its dictionary. We are aiming to use gamification in teaching values to the young generation. *Detection of trends and opinions in geo-tagged social text streams*, (Jevgenij Jakunschin, Andreas Heuer, and Antje Raab-Dusterhoft). This paper, describes an application of social media, database and data-mining techniques for the analysis of convicting trends and opinions in a spatial area. This setup was used to demonstrate the distribution of interests during a global event and can be used for several social media data mining tasks, such as trend prediction, sentiment analysis and social-psychological feedback tracing. To this end the application clusters trends in social text media streams, such as Twitter and detects the different opinion differences within a single trend based on the temporal, spatial and semantic-pragmatic dimensions. The data is stored in a multidimensional space to detect correlations and combine similar trends into clusters, as it is expanded over time. The results of this work provide a system, with the focus to trace several clusters of conflicts within the same trend, as opposed to the common approach of tag-based filtering and sorting by occurrence count.

Software Architecture for Collaborative Crowd-storming Applications (Nouf Jaafar and Ajantha Dahanayake). Diversity in crowdsourcing systems in terms of processes, participants, workflow, and technologies is quite problematic; as there is no standard guidance to inform the designing process of such crowdsourcing ideation system. To build a well-engineered crowdsourcing system with different ideation and collaboration components, a software architecture model is needed to guide the design of the collaborative ideation process. Within this general context, this paper is focused to create a vision for the architectural design for crowdsourcing collaborative idea generation, with an attempt to provide the required features for coordinating and aggregating ideation of individual participants into collective solutions.

9   First International Workshop on Semantic Web for Cultural Heritage (SW4CH)

The aim of the Semantic Web for Cultural Heritage workshop (SW4CH) is to bring together interdisciplinary research teams involved in Semantic Web solutions for Cultural Heritage. The interdisciplinarity is inherent here, as the Semantic Web is a matter of Computer Scientists, and importantly Data Scientists, while the Cultural Heritage field brings together actors from various Humanities and Social Sciences disciplines. In this novel collaboration space, it is crucial to exchange experiences, build states of the art of realizations and challenges, etc.

The Université Francois Rabelais de Tours and the Universite de Poitiers are more and more involved in such interdisciplinary projects. The CESR (Centre of Advanced Studies of the Renaissance) and the CESCM (Centre of Advanced Studies of Medieval Civilisation) in Poitiers are two internationally renowned laboratories that develop knowledge and stimulating reflection on Western civilisation between the end of the Roman Empire and the Galilean revolution. For instance, the CESR (Universite de Tours/CNRS), a DARIAH laboratory and one of the principal European laboratories in the field of the Renaissance heritage, leads for instance the Intelligence des Patrimoines (IPat) programme, in which several projects are underway, involving the collection of large quantities of heterogeneous data. The CITERES laboratory (Centre for Cities, Territories, Environment and Societies, Universite de Tours/CNRS), whose general aim is analysing the spatial and territorial dynamics of societies, also develops numerous projects where data scientists are more and more important. The Computer Laboratory LI of Tours naturally cooperates in those projects that offer great applications for researchers in data science.

Nowadays, Cultural Heritage is gaining a lot of attention from academic and industry perspectives. Scientific researchers, organizations, associations, schools are looking for relevant technologies for accessing, integrating, sharing, annotating, visualizing, analyzing the mine of cultural collections by considering profiles and preferences of end users. Most cultural information systems today process data based on the syntactic or structural level, without leveraging the rich semantic structures underlying the content. Moreover, they use multiple thesauri, or databases, without a formal connection between them. Our invited talk, given by Martin Rezk from the Free University of Bozen-Bolzano, Italy, is about *Integrating cultural data using an ontology-based framework*. In this talk he will introduce an ontology-based data access framework that allows to virtually integrate different databases by means of a conceptual layer (an ontology), that provides a convenient query vocabulary to the user, and a unified view of the underlying data. The ontology is connected to the data sources through a declarative specification given in terms of mappings. He will illustrate how to integrate cultural data by relying on this OBDA framework.

The web has widen the views of the human and social sciences by opening new distributed horizons, and revealing little known data, as well as new forms of accessing and sharing it. During the last decades, Semantic Web solutions have been proposed to explicit the semantic of data sources and make their content machine understandable and interoperable. By analyzing the most important conferences and workshops related to the Semantic Web, four main categories of topics have been identified: (i) the development of Ontologies and vocabularies dedicated to the studied domain, (ii) explicitation of collection semantics, (iii) usage of Semantic Web Cultural Heritage and (iv) applications related to Cultural Heritage. The 9 selected articles reflect these important topics, some of them being parts of important European projects:

Knowledge Representation in EPNet is about the building of an innovative Virtual Research Environment for scholars interested in the history of the Roman Empire. Based on their EPNet CRM and Ontology modeling, the authors explain how they have been able to integrate together the Heidelberg, Pleiades, and the EPNet dataset, using the -ontop- software.

Designing for Inconsistency ? The Dependency-based PERICLES Approach, presents a key outcomes of the PERICLES FP7 project, the Linked Resource Model, for modeling contextual and environmental dependencies as a set of evolving linked resources. This contributes to long-term digital preservation. The proposed model is evaluated in the domain of digital video art.

In A Semantic exploration method based on an ontology of 17th century texts on theatre: la Haine du theatre, the authors propose a method to semantically explore a collection of texts with an ontology, which has been built from the same corpus.

Combining semantic and collaborative recommendations to generate personalized museum tours deals with a recommender system for mobile devices, that combines a semantic representation of museum knowledge using ontologies and thesauruses with a semantically-enhanced collaborative filtering method.

A Novel Vision for Navigation and Enrichment in Cultural Heritage Collections, introduces a framework for semantic enrichment of CH collections, that relies on the creation of thematic knowledge bases, that aggregate information by exploiting structured resources and by extracting new relationships from streams and textual documents. The presented application is to transform library records into semantic web data, based on the FRBR model.

Application of CIDOC-CRM for the Russian Heritage Cloud platform describes the motivation for choosing CIDOC-CRM ontology for the online representation of cultural heritage data, and its usage based on class templates, within the Russian Heritage Cloud project.

The real life use of the CIDOC Conceptual Reference Model and extensions to this model such as the FRBRoo is challenging due to their complexity as well as their extensive and detailed nature. The authors of *A Pattern-based Framework for Best Practice Implementation of CRM/FRBRoo* present a framework for sharing best practice knowledge related to the use of these models, which is inspired by the use of design patterns in software engineering.

All the preceding applications, semantic enrichments, explorations, navigations, recommendations, etc. are enhanced with NLP supports, in particular when the recognized named entities can be linked with some knowledge about the referred entities. In *Improving Retrieval of Historical Content with Entity Linking*, the relevance of Entity Linking for cultural heritage institutions is evaluated through a case-study involving the semantic enrichment of historical periodicals, in order to improve the search experience of end-users with the mapping of entities to the Linked Open Data (LOD) cloud.

Disambiguation of Named Entities in cultural heritage texts using Linked Data sets, proposes a graph-based algorithm for the disambiguation of authors' names in 19th century essays. It leverages knowledge from different Linked Data sources in order to collect candidates for each mentions, then performs fusion of DBpedia and BnF individuals into a single graph, and finally decides the best referent using centrality.

We are very grateful to the international program committee members and the external reviewers, for their precise and objective work in reviewing papers. We received 12 papers, with authors from 11 countries, and special thanks are due to all them, for submitting papers that perfectly matched the workshop topics. Hoping that SW4CH 2015 will bring new collaborations and will open new horizons.

10   Conclusion

ADBIS 2015 organizers and ADBIS 2015 satellite events organizers would like to express their thanks to everyone who contributed to the volume content. We thank the authors, who submitted papers to the various events organized in the context of ADBIS 2015. Special thanks go to the Program Committee members as well as to the external reviewers of the ADBIS 2013 main conference and of each satellite event, for their support in evaluating the submitted papers, providing comprehensive, critical, and constructive comments and ensuring the quality of the scientific program and of this volume. We all hope you will find the volume content an useful overview of new trends in the areas of databases and information systems that may further stimulate new ideas for further research and developments by both the scientific and industrial communities. Enjoy the reading!

Invited Talk

Integrating Cultural Data Using an Ontology-Based Framework

Martin Rezk

Faculty of Computer Science,
Free University of Bozen-Bolzano, Piazza Domenicani 3, Bolzano, Italy
mrezk@inf.unibz.it

Abstract. In this talk we will introduce an ontology-based data access framework that allows to virtually integrate different databases by means of a conceptual layer (an ontology). The ontology provides a convenient query vocabulary to the user, and a unified view of the underlying data. The ontology is connected to the data sources through a declarative specification given in terms of mappings. I will illustrate how to integrate cultural data by relying on a OBDA framework. In particular, I will concentrate on the following crucial questions:

- How this paradigm can contribute to ease the access of scholars to cultural heritage data: integrating temporal and spatial data, cross-linking datasets.
- What is the theory behind it.
- How to map available data sources to an ontology.
- How to query the underlying data sources using the terms in the ontology.
- How to check consistency of the data sources w.r.t. the ontology.

CV. Martin Rezk received his PhD in Computer Science from the Free University of Bozen-Bolzano, Italy, in 2012. Currently he is a postdoc in the Faculty of Computer Science, Free University of Bozen-Bolzano, working in KRDB Research Centre for Knowledge and Data. He leads the development of the OBDA framework Ontop. His research interests include data integration, ontology-based data access, query optimisation, and knowledge representation and reasoning. His articles are published in respected journals and conferences, such as ISWC, Journal of Web Semantics, EDBT, etc. He has been involved in interdisciplinary research for some time now, exploring the practical applications of OBDA.

Contents

ADBIS Short Papers

Revisiting the Definition of the Relational Tuple Calculus 3
  Bader AlBdaiwi and Bernhard Thalheim

A Requirements Specification Framework for Big Data Collection
and Capture 12
  Noufa Al-Najran and Ajantha Dahanayake

AutoScale: Automatic ETL Scale Process. 20
  Pedro Martins, Maryam Abbasi, and Pedro Furtado

Using a Domain-Specific Language to Enrich ETL Schemas 28
  Orlando Belo, Claudia Gomes, Bruno Oliveira, Ricardo Marques,
  and Vasco Santos

Continuous Query Processing Over Data, Streams and Services:
Application to Robotics 36
  Vasile-Marian Scuturici, Yann Gripay, Jean-Marc Petit,
  Yutaka Deguchi, and Einoshin Suzuki

Database Querying in the Presence of Suspect Values 44
  Olivier Pivert and Henri Prade

Context-Awareness and Viewer Behavior Prediction in Social-TV
Recommender Systems: Survey and Challenges 52
  Mariem Bambia, Rim Faiz, and Mohand Boughanem

Generalized Bichromatic Homogeneous Vicinity Query Algorithm
in Road Network Distance 60
  Yutaka Ohsawa, Htoo Htoo, Naw Jacklin Nyunt, and Myint Myint Sein

OLAP4Tweets: Multidimensional Modeling of Tweets 68
  Maha Ben Kraiem, Jamel Feki, Kaîs Khrouf, Franck Ravat,
  and Olivier Teste

Data Warehouse Design Methods Review: Trends, Challenges and Future
Directions for the Healthcare Domain 76
  Christina Khnaisser, Luc Lavoie, Hassan Diab,
  and Jean-Francois Ethier

Incrementally Maintaining Materialized Temporal Views
in Column-Oriented NoSQL Databases with Partial Deltas   88
  Yong Hu and Stefan Dessloch

Towards Self-management in a Distributed Column-Store System   97
  George Chernishev

Relational-Based Sensor Data Cleansing   108
  Nadeem Iftikhar, Xiufeng Liu, and Finn Ebertsen Nordbjerg

Avoiding Ontology Confusion in ETL Processes   119
  Selma Khouri, Sabrina Abdellaoui, and Fahima Nader

Towards a Generic Approach for the Management and the Assessment
of Cooperative Work   127
  Amina Cherouana, Amina Aouine, Abdelaziz Khadraoui,
  and Latifa Mahdaoui

MLES: Multilayer Exploration Structure for Multimedia Exploration   135
  Juraj Moško, Jakub Lokoč, Tomáš Grošup, Přemysl Čech,
  Tomáš Skopal, and Jan Lánský

SLA Ontology-Based Elasticity in Cloud Computing   145
  Taher Labidi, Achraf Mtibaa, and Faiez Gargouri

Bi-objective Optimization for Approximate Query Evaluation   153
  Anna Yarygina and Boris Novikov

**Second International Workshop on Big Data Applications and Principles
(BigDap 2015)**

Cross-Checking Data Sources in MapReduce   165
  Foto Afrati, Zaid Momani, and Nikos Stasinopoulos

CLUS: Parallel Subspace Clustering Algorithm on Spark   175
  Bo Zhu, Alexandru Mara, and Alberto Mozo

Massively Parallel Unsupervised Feature Selection on Spark   186
  Bruno Ordozgoiti, Sandra Gómez Canaval, and Alberto Mozo

Unsupervised Network Anomaly Detection in Real-Time on Big Data   197
  Juliette Dromard, Gilles Roudière, and Philippe Owezarski

NPEPE: Massive Natural Computing Engine for Optimally Solving
NP-complete Problems in Big Data Scenarios   207
  Sandra Gómez Canaval, Bruno Ordozgoiti Rubio, and Alberto Mozo

Andromeda: A System for Processing Queries and Updates
on Big XML Documents 218
   Nicole Bidoit, Dario Colazzo, Carlo Sartiani, Alessandro Solimando,
   and Federico Ulliana

Fast and Effective Decision Support for Crisis Management by the
Analysis of People's Reactions Collected from Twitter 229
   Antonio Attanasio, Louis Jallet, Antonio Lotito, Michele Osella,
   and Francesco Ruà

Adaptive Quality of Experience: A Novel Approach to Real-Time Big Data
Analysis in Core Networks. 235
   Alejandro Bascuñana, Manuel Lorenzo, Miguel-Ángel Monjas,
   and Patricia Sánchez

A Review of Scalable Approaches for Frequent Itemset Mining 243
   Daniele Apiletti, Paolo Garza, and Fabio Pulvirenti

**First International Workshop on Data Centered Smart Applications
(DCSA 2015)**

A Mutual Resource Exchanging Model and Its Applications to Data
Analysis in Mobile Environment. 251
   Naofumi Yoshida

Detection of Trends and Opinions in Geo-Tagged Social Text Streams 259
   Jevgenij Jakunschin, Andreas Heuer, and Antje Raab-Düsterhöft

Software Architecture for Collaborative Crowd-Storming Applications. 268
   Nouf Jaafar and Ajantha Dahanayake

Gamification in Saudi Society: A Framework to Develop Human Values
for Early Generations. 279
   Alia AlBalawi, Bariah AlSaawi, Ghada AlTassan, and Zaynab Fakeerah

Fourth International Workshop on GPUs in Databases (GID 2015)

Big Data Conditional Business Rule Calculations in Multidimensional
In-GPU-Memory OLAP Databases 291
   Alexander Haberstroh and Peter Strohm

Optimizing Sorting and Top-k Selection Steps in Permutation
Based Indexing on GPUs. 305
   Martin Kruliš, Hasmik Osipyan, and Stéphane Marchand-Maillet

First International Workshop on Managing Evolving Business Intelligence Systems (MEBIS 2015)

E-ETL Framework: ETL Process Reparation Algorithms
Using Case-Based Reasoning   321
    Artur Wojciechowski

Handling Evolving Data Warehouse Requirements   334
    Darja Solodovnikova, Laila Niedrite, and Natalija Kozmina

Querying Multiversion Data Warehouses   346
    Waqas Ahmed and Esteban Zimányi

CUDA-Powered CTBE Algorithm for Zero-Latency Data Warehouse   358
    Marcin Gorawski, Damian Lis, and Anna Gorawska

Fourth International Workshop on Ontologies Meet Advanced Information Systems (OAIS 2015)

Mobile Co-Authoring of Linked Data in the Cloud   371
    Moulay Driss Mechaoui, Nadir Guetmi, and Abdessamad Imine

Ontology Based Linkage Between Enterprise Architecture, Processes,
and Time   382
    Marite Kirikova, Ludmila Penicina, and Andrejs Gaidukovs

Fuzzy Inference-Based Ontology Matching Using Upper Ontology   392
    S. Hashem Davarpanah, Alsayed Algergawy, and Samira Babalou

An Ontology-Based Approach for Handling Explicit and Implicit
Knowledge over Trajectories   403
    Rouaa Wannous, Cécile Vincent, Jamal Malki, and Alain Bouju

Interpretation of DD-LOTOS Specification by C-DATA*   414
    *Maarouk Toufik Messaoud, Saidouni Djamel Eddine, Mahdaoui Rafik,
and Houassi Hichem*

First International Workshop on Semantic Web for Cultural Heritage (SW4CH 2015)

Knowledge Representation in EPNet   427
    Alessandro Mosca, José Remesal, Martin Rezk, and Guillem Rull

A Pattern-Based Framework for Best Practice Implementation
of CRM/FRBRoo   438
    Trond Aalberg, Audun Vennesland, and Maliheh Farrokhnia

Application of CIDOC-CRM for the Russian Heritage Cloud Platform   448
Eugene Cherny, Peter Haase, Dmitry Mouromtsev, Alexey Andreev, and Dmitry Pavlov

Designing for Inconsistency – The Dependency-Based PERICLES
Approach   458
Jean-Yves Vion-Dury, Nikolaos Lagos, Efstratios Kontopoulos, Marina Riga, Panagiotis Mitzias, Georgios Meditskos, Simon Waddington, Pip Laurenson, and Ioannis Kompatsiaris

A Semantic Exploration Method Based on an Ontology of 17th Century
Texts on Theatre: la *Haine du Théâtre*.   468
Chiara Mainardi, Zied Sellami, and Vincent Jolivet

Combining Semantic and Collaborative Recommendations to Generate
Personalized Museum Tours   477
Idir Benouaret and Dominique Lenne

A Novel Vision for Navigation and Enrichment in Cultural Heritage
Collections   488
Joffrey Decourselle, Audun Vennesland, Trond Aalberg, Fabien Duchateau, and Nicolas Lumineau

Improving Retrieval of Historical Content with Entity Linking   498
Max De Wilde

Disambiguation of Named Entities in Cultural Heritage Texts
Using Linked Data Sets   505
Carmen Brando, Francesca Frontini, and Jean-Gabriel Ganascia

**First International Workshop on Information Systems
for AlaRm Diffusion (WISARD 2015)**

Abduction for Analysing Data Exchange Policies   517
Laurence Cholvy

ADMAN: An Alarm-Based Mobile Diabetes Management System
for Mobile Geriatric Teams   527
Dana Al Kukhun, Bouchra Soukkarieh, and Florence Sèdes

An Architectural Roadmap Towards Building an Alarm Diffusion System . . .   536
Sumit Kalra, T.V. Prabhakar, and Saurabh Srivastava

Information Exchange Policies at an Organisational Level:
Formal Expression and Analysis   547
Claire Saurel

Critical Information Diffusion Systems 557
    Rémi Delmas and Thomas Polacsek

A Case Study on the Influence of the User Profile Enrichment on Buzz
Propagation in Social Media: Experiments on *Delicious*............... 567
    *Manel Mezghani, Sirinya On-At, André Péninou,
    Marie-Françoise Canut, Corinne Amel Zayani, Ikram Amous,
    and Florence Sedes*

Author Index 579

ADBIS Short Papers

Revisiting the Definition of the Relational Tuple Calculus

Bader AlBdaiwi[1]([⊠]) and Bernhard Thalheim[2]

[1] Computer Science Department, Kuwait University, Kuwait City, Kuwait
bdaiwi@cs.ku.edu.kw
http://www.cs.ku.edu.kw/people/faculty/albdaiwi-b
[2] Department of Computer Science,
Christian-Albrechts-University Kiel, 24098 Kiel, Germany
thalheim@is.informatik.uni-kiel.de
http://www.is.informatik.uni-kiel.de/~thalheim

Abstract. The tuple relational calculus has been based on the classical predicate logics. Databases are however not exactly representable in this calculus. They are finite. This finiteness results in a different semantics. The result of a query must be finite as well and must be based on the values in the database and in the query. In this case, negation and disjunction of query expression must be defined in a different way. The classical theory has developed restrictions to the tuple relational calculus such as safe formulas.

This paper takes a different turn. We introduce a different definition of the tuple relational calculus and show that this calculus is equivalent to the relational algebra and thus equivalent to the safe tuple relational calculus.

1 Introduction

Database technology is considered to be one of the best-settled and best defined areas of Computer Science. The theoretical underpinning of relational databases resulted in superficial effective systems and brought the victory over old network and hierarchical technologies. Nowadays, database theory is considered to be the showcase of a best-developed and best-integrated coexistence of theory and technology.

Beside the overwhelming success of database theory many problems remain to be open. There are lacunas of uninvestigated areas [17]. There are however also problems that have not yet found an appropriate and satisfying solution. There are also statements that database technology must be redefined for more complex applications such as big data and NoSQL computations. These statements and the problems hint on some pitfalls of current technology and compel a revision of current theory.

The tuple relational calculus is considered to be the basis for set-based querying and for SQL query formulation. It has been defined based on direct application of first-order predicate calculus. Starting with the papers by E.F. Codd

© Springer International Publishing Switzerland 2015
T. Morzy et al. (Eds): ADBIS 2015, CCIS 539, pp. 3–11, 2015.
DOI: 10.1007/978-3-319-23201-0_1

[4–6] this approach has been accepted by almost all researchers[1]. The finiteness property of databases[2] combined with the potential infiniteness of the data domains results in clumsy definitions of tuple relational queries. The choice of first-order predicate logic based on the canonical definition and interpretation might not be the best option.

The Approach Used in This Paper

Logicians in the past knew that the right logic in the best-fitted structure, in the best complexity and micro-structure must be selected for a solution of a problem. We define another predicate calculus that preserves the finiteness of a database, that is equivalent to the relational algebra, and that is as convenient as the classical tuple relational calculus. This predicate calculus does not allow to define arbitrary formulas. It allows to define only those formulas that are computing a finite set for a given finite database.

2   Known Notions of the Relational Tuple Calculus

It is a common belief that predicate logic in its canonical interpretation and treatment is the best for database theory, starting with Codd's papers [4–6] who first exploited the equivalence of set theory and first order predicate logic for application to database management.

A fundamental result of database theory is the equivalence of the relational algebra, the safe domain relational calculus, the safe tuple relational calculus and Datalog [1,11,13,14,18]. The relational calculi have been defined through the direct application of the canonical definition scheme of first-order predicate logics. It has been quickly discovered that the relational calculi do not preserve finiteness of a database.

A *relational data scheme* $\mathcal{DD} = (U, \underline{D}, dom)$ uses *domains* $D_i \in \underline{D}$ with a potentially infinite structure of values, with operations such as $\circ, +, -, \times$, mandatory predicates $=, \neq$ and optional predicates such $<, =, >, \leq, \geq$, a finite set U of attribute names and an assignment function dom that assigns attributes to domains.

A relational database schema consists of *relation types* R_i which are triples $(attr(R_i), id(R_i), \Sigma(R_i))$, where R_i is a relation name, $attr(R_i)$ is a set of attributes from U, $id(R_i)$ is a subset of $attr(R_i)$ called primary key, and $\Sigma(R_i)$ is a set of constraints defined on R.

Consider, for instance, a small database schema with $attr(\text{Movie}) = \{$ title, year $\}$ and $attr(\text{Performance}) = \{$ cinema, movie, date, time $\}$ for the assignments $dom(\text{cinema}) = STRING$, $dom(\text{time}) = TIME$ and $dom(\text{date}) = DATE$. Let us use the variables m for movie titles, d for dates, t for time and

---

[1] The only exception is the book [2] where the authors defined a typed variant of the tuple relational calculus.

[2] A database is a finite collection of objects. A query returns finite many objects.

y for year (more precisely, m, d, t and y represent values in dom(movietitles), dom(date), dom(time) and dom(dom(year), respectively.

The question *'Which movies are performed in UCI?'* can be expressed by the query

$$\{(m)|\exists d\,\exists t\,\exists y\,\text{Performance}(UCI, m, d, t) \;\wedge\; Movie(m, y)\}.$$

The question *'In which cinemas has not been played the movie Casablanca?'* might be expressed by the query

$$\{(c)|\forall d\,\forall t\,\neg(\text{Performance}(c, Casablanca, d, t)\}\,.$$

Its result is however an infinite set which is neither desired nor in accordance with definitions of a query.

The query

$$\{(c, d)|\exists t\,\text{Performance}(c, Albatros, June\,17,\,2015, t)$$
$$\vee\,\text{Performance}(UCI, Casablanca, d, t)\}$$

also produces undesired answers.

For instance, if our database contains a performance of *Casablanca* on *July 1, 2015* (any time), then the query will contain all pairs $(c, July\,1, 2015)$ with arbitrary values for the cinema c.

Similarly, if our database contains a performance of *Albatros* in the *Cinemax* cinema (any time), then the answer relation will contain all pairs $(Albatros, d)$ with arbitrary values for the date d.

The problem in the last two cases is that the answer relation depends on the domains D_i. More precisely, it depends on those values in D_i that do neither occur in the database nor in the given query q. Note that it is not possible to produce infinite query results in the relational algebra, because the evaluation of algebra expressions relies only on the values that occur in the database or the query.

A first brute-force solution that has classically been considered is to restrict the set of formulas to *domain-independent formulas*. It has however a flaw that completely depreciates its value [1, 13, 14, 18]:

Theorem 1. *The problem in determining whether a tuple relational expression is domain independent is undecidable.*

The next step was to introduce *safe formulas*. The positive result is now [1, 13, 14, 18]:

Theorem 2. *Every safe formula is domain independent.*

This explicit solution shows a path that might be a better solution to the problem of well-defined tuple relational expressions despite the following discouraging result [1, 13, 14, 18].

Theorem 3. *Whether a given domain independent formula is equivalent to a safe formula is undecidable.*

A solution with similar flaws is based on *relationally restricted formulas*.

A third solution considers a transformation of a formula to formulas over *active domains*.

The main drawback of the three solutions is the brute-force utilisation of the predicate calculus. The solutions resulted in rather problematic and difficult to check definitions.

[2] uses another definition frame

$$\{\mathcal{T} \mid \mathcal{L} \mid \alpha\}$$

for a *tuple relational calculus with domain restrictions*. It turns however out that [2]:

Theorem 4. *The tuple relational calculus with domain restrictions has not the same expressive power as the relational algebra.*

This insufficiency is caused by problems of properly expressing union, intersection, and difference in this calculus.

The clever idea of this proposal is however the definition frame that we shall use in the sequel:

$$\{target\ structure \mid context \mid conditions\} \quad .$$

We might now ask ourselves why such clumsy definitions are given. Or we ask whether there is a more clever definition for relational calculi. The main drawback is the brute-force utilisation of the predicate calculus that resulted in problematic definitions.

3   The Algebraic Tuple Calculus over a Schema

Let us first remember a very general form of definition: structural recursion [16]. This definition form allows also to define more advanced formulas.

Given a relational data scheme $\mathcal{DD} = (U, \underline{D}, dom)$. Given additional domains $D_1^*, ...D_p^*$ and names $C_1^*, ...C_p^*$ with $dom(C*_i) = D_i^*$. Let $U^* = \{C_1, ...C_p\}$.

The *extended domain* $dom(X)$ for the ordered set $X = \{A_1, ..., A_m\}$ with $X \subseteq U \cup U^*$ is defined as the Cartesian product $dom(A_1) \times ... \times dom(A_m)$. The extended domain is a set. We could also consider other collections such as multisets (bags), lists and powersets. Since the relational database model uses set semantics we restrict consideration to sets.

Extended domains can be associated by mappings $h : X_1 \to X_2$ that map a singleton value from $dom(X_1)$ to a value $dom(X_2)$.

For each extended domain $dom(X)$, union $\cup^X$, intersection $\cap^X$ and the difference $\backslash^X$ are defined in the classical set-theoretic manner. Furthermore, we denote the empty set by $\emptyset^X$. The powerset $\mathfrak{P}(dom(X))$ forms an algebra with $\cup^X, \cap^X, \backslash^X$, and $\emptyset^X$. Let us denote the type of this powerset by 2^X .

Given two sets $X_1, X_2 \subseteq U \cup U^*$ and

- a value $e \in dom(X_2)$,
- a function $g : X_1 \to X_2$ that maps values from $dom(X_1)$ to values from $dom(X_2)$, and

– a function $h : X_2 \times 2^{X_2} \to 2^{X_2}$ that maps a singleton value and a set of values on X_2 to a set of values on X_2.

The triple (e, g, h) is called *recursion frame* on X_1, X_2.

The structural recursion $\texttt{src}[e, g, h]$ is inductively defined on the domains $dom(X_1)$ and $dom(X_2)$ as follows:

$\texttt{src}[e, g, h](\emptyset^{X_1}) = e$,

$\texttt{src}[e, g, h](\{x\}) = g(x)$ for $x \in dom(X_1)$,

$\texttt{src}[e, g, h](\{d\} \cup^{X_2} M) = h(g(d), \texttt{src}[e, g, h](M))$

     for $d \notin M$ and $M \subseteq dom(X_2)$.

We introduce now a typed and algebraic tuple calculus that does not have these drawbacks:

– identical copy expression:

$$\{t \mid T \mid T^C\} \qquad \text{for each type T in schema } \mathcal{S} ,$$

– selection expression:

$$\frac{\{t \mid T \mid \beta\}}{\{t \mid T \mid \beta \wedge \alpha\}}$$

for a predicate α defined on T,

– join expression:

$$\frac{\{t_1 \mid T_1 \mid \alpha_1\} , \{t_2 \mid T_2 \mid \alpha_2\}}{\{t_1 \bowtie^{T_1 \cup T_2} t_2 \mid T_1 \cup T_2 \mid \alpha_1 \wedge \alpha_2\}}$$

for a predicates α_1, α_2 defined T_1, T_2, respectively, and the operation $\bowtie^{T_1 \cup T_2} = \texttt{srec}[\emptyset^{T_1 \cup T_2}, \bowtie^{T_1 \cup T_2}, \cup^{T_1 \cup T_2}]$ via

$$t_1 \bowtie^{T_1 \cup T_2} t_2 \quad := \quad t \in dom(T_1 \cup T_2) \mid t|_{T_1} = t_1 \wedge t|_{T_2} = t_2 ,$$

– union expression:

$$\frac{\{t \mid T \mid \alpha_1\} , \{t \mid T \mid \alpha_2\}}{\{t \mid T \mid \alpha_1 \vee \alpha_2\}}$$

for a predicates α_1, α_2 defined on the same type T ,

– difference expression:

$$\frac{\{t \mid T \mid \alpha_1\} , \{t \mid T \mid \alpha_2\}}{\{t \mid T \mid \alpha_1 \wedge \neg \alpha_2\}}$$

for a predicates α_1, α_2 defined on the same type T ,

– renaming expression:

$$\frac{\{t \mid T \mid \alpha\}}{\{\rho(t) \mid \rho(T) \mid \rho(\alpha)\}}$$

for a function ρ that renames T into $\rho(T)$ and $\rho^T = \texttt{srec}[\emptyset^{\rho(T)}, \rho^{\rho(T)}, \cup^{\rho(T)}]$ via

$$\rho^{\rho(T)}(t) \quad := \quad t' \in dom(\rho(T)) \mid \rho(t) = t'$$

- projection expression:

$$\frac{\{t \mid T \mid \alpha\}}{\{\pi_X(t) \mid \mathcal{C} \mid \alpha\}}$$

$\pi_X = \mathtt{srec}[\emptyset^X, \pi_X, \cup^X]$ and $X \subseteq T$ via a restriction of t to $\pi_X(t)$.

The expressions use the triple definition that allows to maintain the type structure together with definition of the expressions via the frame

$$\{target\ structure \mid context \mid conditions\} \quad .$$

The main trick of this construction is the invariance of the type correspondence. Any other expression is not defined. Therefore, only those expressions can be built that are type-safe.

The algebraic tuple calculus is now inductively defined for a database schema $\mathcal{S}$, the relational data scheme $\mathcal{DD} = (U, \underline{D}, dom)$, and $U^* = \{C_1, ...C_p\}$:

- The identical copy expression is an expression in the algebraic tuple calculus over $\mathcal{S}$.
- Given expressions E_1, E_2 in the algebraic tuple calculus over $\mathcal{S}$. Any selection, union, difference, renaming and projection expression over over $\mathcal{S}$ is an expression in the algebraic tuple calculus over $\mathcal{S}$.
- Nothing else is an expression in the algebraic tuple calculus over $\mathcal{S}$.

Proposition 1. *The algebraic tuple calculus is correct, i.e. it does not result in undefinable expressions. Any expression is evaluated in a finite relational database to a finite set of objects within the context of the expression.*

The proof of this proposition is based on inductive application of the definitions and the closure condition that no other expression is allowed unless it has been defined within the algebraic tuple calculus. Since this proof is obvious we omit it.   □

We observe now based on the construction of expressions:

Lemma 1. *The algebraic operations selection, join, union, difference, renaming, and projections are directly expressible within the algebraic tuple calculus.*

Lemma 2. *Expressions in the algebraic tuple calculus can directly expressed through algebraic operations selection, join, union, difference, renaming, and projections.*

Theorem 5. *The algebraic tuple calculus has the same expressive power as the relational algebra.*

The proof of this theorem is obvious. The two lemmata stated a direct equivalence of expressions in the relational algebra and in the algebraic tuple calculus. The selection, join, union, difference, renaming, and projections operations form a complete set of operations in the relational algebra. Any algebraic expression from the relational algebra can be directly expressed by an expression.   □

Based on [13] we can conclude:

Corollary 1. *The following three query languages for the relational database model are equivalent:*

- *The relational algebra.*
- *The algebraic tuple calculus.*
- *The safe formula calculus.*

4 The General Tuple Calculus

Structural recursion is more powerful. We can also define an extended and general tuple calculus with aggregation functions. For instance, the aggregation function count is defined by $src[0, g, h]$ for $\mathbb{N}$, the value 0, the function g that maps values to the value 1 and the function h that sums the values. Another example is the aggregation function sum. If X_1 is an arithmetic type, Id denotes the identical function for X_1, 0 is the zero in $dom(X_1)$, and $+^{X_1}$ denotes the sum function on X_1 then $src[0, Id, +^{X_1}]$ is the generalised aggregation function sum^{X_1}.

We define now the general tuple calculus as follows for a database schema $\mathcal{S}$, the relational data scheme $\mathcal{DD} = (U, \underline{D}, dom)$, and $U^* = \{C_1, ...C_p\}$:

- Given ordered sets $X \subset U$, $Y \subset U \cup U^*$, *recursion frame* (e, g, h) on X, Y, a relation type R from $\mathcal{S}$, and a relation R^C.
  The expression
  $$\{t \mid Y \mid src[e, g, h](R^C)\} \quad .$$
  is called (atomic) general tuple expression on Y over $\mathcal{S}$.
- Given an ordered set $Y \subset U \cup U^*$, a *recursion frame* (e, g, h) on X, Y, and a general tuple expression over $\mathcal{S}$
  $$X^C = \{t \mid X \mid \alpha\} \quad .$$
  Then the expression
  $$\{t \mid Y \mid src[e, g, h](X^C)\}$$
  is a general tuple expression.
- Nothing else is a general tuple expression on Y over $\mathcal{S}$.

We abbreviate the construction in the second step by

$$\frac{\{t \mid X \mid \alpha\}}{\{t \mid Y \mid \beta\}}$$

for (e, g, h) and $X, Y \subset U \cup U^*$.

Theorem 6. *Given a relational database schema $\mathcal{S}$ and a finite database $\mathfrak{D}_{\mathcal{S}}^C$ over a schema $\mathcal{S}$. Any general tuple expression over $\mathcal{S}$ yields a finite set for $\mathfrak{D}_{\mathcal{S}}^C$.*

The proof is based on the definition frame for structural recursion and on the inductive definition of the general tuple expressions.

We notice that the calculus of general tuple expressions is far more powerful than the classical relational calculus. We may define nesting and unnesting, aggregation and also views within this calculus. The SQL query frame can easily be expressed within this calculus.

5   Conclusion

This article introduces a tuple relational calculus that has the same expressive power as the relational algebra. We concentrate our consideration to this equivalence. We could have however also defined more expressive definitions for entity-relationship schemata or for nested relational databases. In this case, we extend the definitions by an explicit definition of new operations. Also, we may use the same approach to define a calculus for object-oriented databases [15]. A similar extension can be defined for aggregation operations. In this case, we use the frame that has been introduced in [12].

The main revision of database classics and the contribution of this paper are a more flexible treatment of predicate calculus and the definition frame

$$\{target\ structure \mid context \mid conditions\}$$

for formulas. This definition frame can also be used for introduction of a calculus that expresses XML queries. The definition frame has already implicitly used in papers that tried to develop a theory of functional dependencies and keys in XML databases [7–10]. This paper makes this frame explicit.

One minor but not less important achievement of the presented approach is that it can directly be extended to bag semantics or to integrated bag and set semantics based on the approach used in [3]. Therefore, this approach would allow to define a calculus for SQL queries.

References

1. Abiteboul, S., Hull, R., Vianu, V.: Foundations of databases. Addison-Wesley, Reading (1995)
2. Atzeni, P., Ceri, S., Paraboschi, S., Torlone, R.: Database systems: concepts, languages & architectures. McGraw-Hill, London (1999)
3. Beeri, C., Thalheim, B.: Identification as a primitive of database models. In: Proc. FoMLaDO 1998, pp. 19–36. Kluwer, London (1999)
4. Codd, E.F.: A relational model for large shared data banks. CACM **13**(6), 197–204 (1970)
5. Codd, E.F.: Relational completeness of data base sublanguages. In: Rustin, R. (ed.) Data base systems, pp. 65–98. Prentice-Hall, Englewood Cliffs (1971)
6. Codd, E.F.: Further normalization of the database model. In: Rustin, R. (ed.) Data Base Systems, pp. 33–64. Prentice-Hall, Englewood Cliffs (1972)
7. Hartmann, S., Hoffmann, A., Link, S., Schewe, K.-D.: Axiomatizing functional dependencies in the higher-order entity-relationship model. Inf. Process. Lett. **87**(3), 133–137 (2003)
8. Hartmann, S., Köhler, H., Link, S., Thalheim, B.: Armstrong databases and reasoning for functional dependencies and cardinality constraints over partial bags. In: Lukasiewicz, T., Sali, A. (eds.) FoIKS 2012. LNCS, vol. 7153, pp. 164–183. Springer, Heidelberg (2012)
9. Hartmann, S., Köhler, H., Link, S., Trinh, T., Wang, J.: On the notion of an XML key. In: Schewe, K.-D., Thalheim, B. (eds.) SDKB 2008. LNCS, vol. 4925, pp. 103–112. Springer, Heidelberg (2008)

10. Hartmann, S., Link, S., Schewe, K.-D.: Functional dependencies over XML documents with dtds. Acta Cybern. **17**(1), 153–171 (2005)
11. Kandzia, P., Klein, H.-J.: Theoretische Grundlagen relationaler Datanbanksysteme. BI Wissenschaftsverlag, Mannheim (1993)
12. Lenz, H.-J., Thalheim, B.: A formal framework of aggregation for the olap-oltp model. J. UCS **15**(1), 273–303 (2009)
13. Levene, M., Loizou, G.: A guided tour of relational databases and beyond. Springer, Berlin (1999)
14. Maier, D.: The theory of relational databases. Computer Science Press, Rockville (1983)
15. Schewe, K.-D., Thalheim, B.: Readings in object-oriented databases. Reprint, BTU-Cottbus,   accessible   through   http://www.is.informatik.uni-kiel.de/ ~thalheim. Collection of papers by Beeri, C., Schewe, K.-D., Schmidt, J.-W., Stemple, D., Thalheim, B., Wetzel, I. (1998)
16. Thalheim, B.: Entity-relationship modeling - Foundations of database technology. Springer, Berlin (2000)
17. Thalheim, B.: Open problems of information systems research and technology. In: Kobyliński, A., Sobczak, A. (eds.) BIR 2013. LNBIP, vol. 158, pp. 10–18. Springer, Heidelberg (2013)
18. Yang, C.-C.: Relational Databases. Prentice-Hall, Englewood Cliffs (1986)

A Requirements Specification Framework for Big Data Collection and Capture

Noufa Al-Najran and Ajantha Dahanayake(✉)

Prince Sultan University – College for Women, King Abdullah Road,
Riyadh 11586 Saudi Arabia
it.nouf@hotmail.com, adahanayake@psu.edu.sa

Abstract. The ad hoc processes of data gathering used by most organizations nowadays are proving to be inadequate in a world that is expanding with infinite information. As a consequence, users are often unable to obtain relevant information from large-scale data collections. The current practice tends to collect bulks of data that most often: (1) containing large portions of useless data; (2) leading to longer analysis time frames and thus, longer time to insights. The premise of this paper is; that big data analytics can only be successful when they are able to digest captured data and deliver valuable information. Therefore, this paper introduces 'big data scenarios' to the domain of data collection. It contributes to a paradigm shift of big data collection through the development of a conceptual model. In time of mass content creation, this model aids in a structured approach to gathering scenario-relevant information from various domain contexts.

Keywords: Big data scenarios · Information filtering · Big data collection · Big data analytics · Scenario-based data collection · Software requirements engineering

1   Introduction

Today, for business and many other purposes, everyone is dealing with data in one way or another. People communicate through social networks and generate content such as blog posts, photos and videos. Wireless sensors and RFID readers create signals, and servers continuously create log messages. Scientists make scientific experiments and create detailed measurements and marketers' record information about sales, suppliers, customers, and etc. This rapid growth of data is the reason behind the evolution of big data [1]. However, these huge volumes of data are evolving in a great pace, making the process of retrieving relevant and valuable information for decision making very difficult [2].

In the past, excessive data volume was a storage issue, but with decreasing storage costs, organizations tend to acquire and store all the available data through data streaming, whether it matches their organizational needs or not [3]. This leads to the size of datasets getting so huge that efficiency becomes a big challenge for current data

© Springer International Publishing Switzerland 2015
T. Morzy et al. (Eds): ADBIS 2015, CCIS 539, pp. 12–19, 2015.
DOI: 10.1007/978-3-319-23201-0_2

analytical technologies [4]. This is unfortunate because analytics consume a lot of time trying to figure out matching patterns in the data and may not come up with answers to important questions in a timely manner. Organizations are stuck with the ever-growing volume of data, and miss out opportunities to take actions for critical business decisions [2]. Therefore, it is imperative that businesses, organizations, and associations find better approaches for information filtering, which would effectively decrease the information overload and improve the precision of results [5]. They need to separate the meaningful information from the chatter and focus on what counts. Thus, the real issue behind big data value does not only include the acquisition and storage of the massive volumes of data; rather it lies in the process of acquiring only what is suspected of being relevant for further analysis [6]. When the amount of data to be analysed is reduced, the managing of their storage, merging, analysing, and governing different varieties of the data is expected to be simpler and more controllable [5].

Big Data analytics can only be effective when the underlying data collection processes are able to leverage the relevant information to a particular scenario [7]. Thus by improving the usefulness of the analysis results. Therefore, this research looks into the question: *"How can we improve the ad hoc process of data collection that hinders the efficiency of extracting value from large datasets in a timely manner?"* This research endeavours to answer this question through the introduction of a Requirements Specification Framework, for collecting the requirements of a data collection process in order to assist users in understanding what they require to know before attempting to collect the data.

2   Related Works

In terms of 'big data collection', much research is conducted in this field but there is no clear and sufficient information on how to determine relevancy within structured, semi structured and unstructured data in all the available universes of information [1].

In [8], the authors emphasized that data analysis based on spatial and temporal relationships yields new knowledge discovery in multi-database environments. They have developed a novel approach to data analysis by turning topsy-turvy the analysis task. This approach provides the data collector concept, that the analysis task drives the features of data collectors. These collectors are small databases which collect the data of interest.

Nakanishi emphasizes in [9] that most current data analytics and data mining methods are insufficient for big data environments. Therefore, they have designed and proposed a model that creates axes for correlation measurement on big data analytics. This model maps the Bayesian network to measure correlation mutually in the coordination axes. It contributes to a shift in the domain of big data analytics.

The authors in [10], studied different big data types and problems. They developed a conceptual framework that classifies big data problems according to the format of the data that must be processed. It maps the big data types with the appropriate combinations of data processing components. These components are the processing and analytic tools in order to generate useful patterns from this type of data.

IBM in [11] provides a means of classifying big data business problems according to a specified criteria. They have provided a pattern-based approach to facilitate the task of defining an overall big data architecture. Their idea of classifying data is to map each problem with its suitable solution pattern and provides an understanding of how a structured classification approach can lead to an analysis of the need and a clear vision of what needs to be captured.

There are several traditional approaches and technologies that may possibly lead to have a control on limiting or reducing unwanted data such as:

- *Visualization and manual Data Collection* [13]. However, several challenges emerged as a result of this process. These include the possibility for correct misses/false alarms and errors in categorizing the data and can be very time consuming.
- *Machine Learning and Data Mining techniques* [14]. However, data mining can only be applied to structured data that can be stored in a relational database.
- *Collaborative Filtering (CF)* is a common web technique for providing personalized recommendations, such as the ones generated by Amazon (based on the user profile and transaction history). In spite of the technique's effectiveness, it rises privacy issues as some customers don't prefer to have their preferences or habits widely known, along with other associated challenges such as data sparsity, scalability, and synonymy [15].
- *Contextual Approach* uses semantic technologies such as an NLP, annotation, and classification to handle information integration (depending on the context of the web page at that moment in time) and querying of distributed data. For query representation, SPARQL language is specifically designed for the semantic technology and enables constructing sophisticated queries to search for different types of data [16]. This approach is efficient in terms of its high precision in controlling unwanted data, as it takes into account the important factors such as keywords, synonyms and antonyms. However, it requires a different infrastructure and highly skilled experts to deal with the complicated technology.

Backward Analysis
According to [17], "Backward analysis is the process of defining the properties of the input, given or based on the context and properties of the output". This concept is utilized in optimizing the process of data collection. Analysing the properties of the scenario at hand and determining the relevant elements that, when collected, will probably reveal hidden value, should be done prior to the data collection process. Comprehensive backward analysis will eliminate the chance of being overwhelmed by bulks of irrelevant data. This will help users and businesses to generate fast management decisions and answer mission critical questions. Therefore, collecting data upon prior analysis needs to a particular business scenario eliminates the presence of unrelated data. Therefore, the effectiveness of the final insights derived from the analytics depends on the quality more than the quantity of the data that will form the foundation for the analytic techniques [1].

Much research is conducted around big data scenarios and around data collection [12]. However, there is no clear and sufficient information that links the two fields together. Therefore, the innovativeness of this research lies in the development of a scenario-based big data collection framework that performs as the Requirements Engineering phase for big data capturing. The framework links the two or more aspects together to provide a well-defined approach for identifying the properties of the scenario context in which the data collection process will take place. This research studies the requirements specification of the big data collection process and makes it more tailored to the business needs, in order to decreases the analysis time and increases the value of the results by making faster management decisions.

3   The Big Data

According to the Gartner group [19]: "Big Data are high-*Volume*, high-*Velocity*, and/or high-*Variety* information assets that require new forms of processing to enable enhanced decision making, insight discovery and process optimization". Yet there are two more equally important characteristics to consider, which are *Veracity* and *Value* [20].

Characteristics of Big Data spans across five dimensions:

- *Volume*: How big the data is growing.
- *Velocity*: How fast the data is being generated.
- *Variety*: Variations of data types and structures.
- *Veracity*: Trustworthiness, validity and quality of the data.
- *Value*: The success of big data drives businesses in terms of better and faster management decisions and financial performance.

Following are some companies which are taking advantage of big data to leverage their performance:

- **Amazon** uses big data to build and power their recommender system that suggests products to people through their purchase history and clickstream data.
- **Samsung Inc.** uses big data on its new smart TVs to enhance their content recommendation engine, and thus, provide the customer with more accurate and user specific recommendations.
- **Progressive Insurance Inc.** relies on big data to decide on competitive pricing and capture customer driving behaviour.
- **LexisNexis Risk Solutions Inc.** uses big data to help financial organizations and other clients detect and reduce fraud through identifying individuals, including family relationships.

4   Data Collection

According to [21], the process of data collection is defined as: "The process of gathering and measuring information on variables of interest, in an established systematic fashion that enables one to answer stated research questions, test hypotheses, and evaluate outcomes". Clearly, those huge volumes of continuously generated data are more than what conventional technologies can sustain. Hence, the lack of effective processes for information collection and management in organizations adapting to big data solutions, can result in a negative impact.

Acquiring the data that holds useful information from tremendous amounts of available data with the rapid increase of online information is a non-trivial task. Collecting scenario-based relevant data from all the available information sources, poses several challenges:

- Integrating multi-disciplinary methods aiming to locate useful data in the large volume, messy, often schema-less, and complex world of big data.
- Understanding big data analyzing techniques as well as big data capturing techniques to be able to select the right one for the scenario being processed. And consolidating the possible factors that can have a control over reducing the unwanted data.
- The ability to develop a simple yet comprehensible and powerful approach to guide and streamline the data collection process, based on the properties of the given scenario.
- Collecting big data requires experts with technical knowledge who can map the right data to the right analytical technique, and execute complex data queries.

5   Data Collection Requirements Modeling

The main inspiration of this research comes from the W*H Conceptual Model for Services [18]. The authors in their research have studied the concept of 'services' as a design artefact. They have aimed to merge the gap between main service design initiatives and their abstraction level interpretations. In order to address their research goal, the authors have developed an inquiry-based conceptual model for service systems designing. This model formulates the right questions that specify service systems innovation, design and development.

By using W*H model, the identification of the main factors that govern the data collection phase of a big data analytics solution is discussed in relation to W*H model. The foundation of the W*H model is based on a set of primary, secondary, and additional questions that support completeness, simplicity, and correctness into service systems innovation, design, and development [18]. In the W*H model, the service is primarily declared by answering the following questions:

- **Wherefore?** (The ends)
  It defines the benefit a potential user may obtain when using the service. This factor is based on the answers for the following questions: *why, whereto, for when, for which reason.*
- **Whereof?** (The sources)
  It defines a general description of the environment for the service.
- **Wherewith?** (The supporting means)
  It identifies the aspects that must be known to potential users in case of utilizing the service.
- **Worthiness?** (The surplus value)
  It defines the value of service utilization for the potential user.

The focus of this paper is identifying the scenario specific to the Data Collection phase for a Big Data Analytics solution. It aims to accelerate the analysis time through data reduction by focusing on retrieving data from the source that meets the scenario. Due to the sheer volume, velocity, and variety of big data, it is challenging to minimize the amount of data to be collected. The framework developed shall be applied during the data collection phase; as the initial process in a big data analytics solution. Figure1 provides a diagram that presents the mapping of the primary phase, which must exist on top of the data collection process.

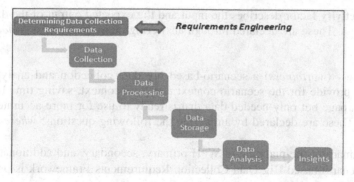

Fig. 1. Requirements Engineering Phase in a Big Data Analytics Life Cycle

The Characteristics of Big Data Scenarios

Scenarios are to be characterized according to their specific properties, the domain that it belongs to, the temporal and spatial factors, the search patterns such as keywords, phrases or named entities, the analysing technique and the capturing technique [11]. They may be composed of other scenarios. In this case, some properties of the scenarios may contradict, while others may be the same. Nested scenarios are beyond the scope of this study. A full description of the scenario-based data collection process is governed by the following factors:

The scenario or purpose - (wherefore) of the data collection and thus the insights a potential domain expert may obtain from analysing the collected data. The scenario description governs the data collection process. It allows to characterize the data collection. This characterization is based on the answers for the following questions: *why, whereto, for when, for which reason.* They define the potential and the capability of the scenario-based data collection process.

The sources - (*whereof*) factor determines the data source that is most likely to generate relevant information about the given scenario among the available data sources. It describes the provider of the data to be collected that is relevant to the given scenario, the consumer of the processed data, the classification of content format and how much data is expected. This classification can help to understand how the data is acquired, and how it will be analysed. These are declared through answering questions such as: *where from, to whom, in what format* and *how much.*

The search patterns - (*by-what*) captures the Supporting needs and Activity factors. It determines which part-of-speech (POS), phrases and keywords correspond to the scenario at hand and must be contained within the data that we want to pull out.

The **Supporting needs** factor describes the analytical technique that can analyse the collected data, the capturing technique that can be utilized to capture the right data, the frequency of the arriving data, the environment at which the data collection process is implemented, whether the data is processed in real-time or batched for later processing. These are declared through answering questions such as: *what, how, whence, where and whether.*

The **Activity** factor describes the input and the expected output of the data collection process. These are declared through answering questions such as: *what-in, what-out.*

The value - (*worthiness*) a scenario-based big data collection and analytics is expected to provide for the scenario context and time context, saving time by not collecting garbage but only needed data that is ready to use for more accurate real-time analysis. These are declared by answering the following questions: *where about and when.*

A summary of the mapping of W*H primary, secondary, and additional questions to the Scenario-based Big Data Collection Requirements Framework is presented in table 3.

Table 1. Mapping W*H Model Qs to Scenario-based Data Collection Qs

W*H Model Questions		Scenario-based Data Collection	
Ends or Purpose	*Why?*	Scenario	*Why?*
	Where to?		*Where to?*
	For when?		*For when?*
	For which reason?		*For which reason?*
Supporting means	*Wherein?*	Search Patterns	*What?*
	Wherefrom?		*How?*
	For what?		*Whence?*
	Where?		*Whether?*
	Whence?		*What in?*
	What?		*What out?*
	How?		
Sources	*By whom?*	Sources	*Where from?*
	To whom?		*To whom?*
	Whichever?		*In what format?*
	What in?		*Where?*
	What out?		*How much?*
Surplus value	*Where at?*	Value	*Where about?*
	Where about?		
	Wither?		*When?*
	When?		

6   Conclusion and Further Research Directions

The main contribution of this research focused on improving the current process of big data collection. Based on a study of the scientific research materials and the literature exploration, it has been observed that the concept of "Backward Analysis", which is performing Reverse Engineering, can add a positive impact to the process of data collection. A data collection that considers the properties of the scenario and the required output before attempting to collect any data (input). The research has studied the scenario-based factors that govern the data collection process and organized them in the form of primary, secondary and additional questions. These questions form the kernel of the Requirements Specification Framework developed as a structured, well-defined approach for scenario-based big data collection process.

References

1. Santovena, Z.A.: Big data: evolution, components, challenges and opportunities. Massachusetts Institute of Technology (2013)
2. Economist Intelligence Unit: The Deciding Factor: Big Data & Decision Making. Capgemini (2012)
3. META: 3D Data Management: Controlling Data Volume, Velocity, and Variety. META Group (2001)
4. Martin, G.: Profit from Big Data. White paper, HP Corp. (2013)
5. Hermansen, S.W.: Reducing big data to manageable portions. In: SESUG, USA
6. Akerkar, R.: Big Data computing, 1st edn. Chapman and Hall/CRC (2013)
7. EY: Big Data, Changing the way business compete and operate. Insights on Governance, Risk and Compliance (2014)
8. Thalheim, B., Kiyoki, Y.: Analysis-driven data collection, integration and preparation for visualisation. In: EJC 2010, pp. 142–160 (2012)
9. Nakanishi, T.: A data-driven axes creation model for correlation measurement on big data analytics. In: Proceedings of 24th International Conference on Information Modelling and Knowledge Bases (EJC 2014) (2014)
10. Al-Najran, N., Al-Swilmi, M., Dahanayake, A.: Conceptual framework for big data analytics solutions. In: Proceedings of 24th International Conference on Information Modelling and Knowledge Bases (EJC 2014) (2014)
11. Mysore, D., Khupat, S., Jain, S.: Big Data architecture and patterns, Part1: Introduction to Big Data classification and architecture. IBM Corp (2013)
12. Claire, B.B.: Managing semantic big data for intelligence. In: CEUR Workshop Proceedings of the STIDS, vol. 1097, pp. 41–47 (2013)
13. Angela, C.: Challenges of Capturing Relevant Data. Umati Project (2013)
14. Neck, F., Andersen, D.G.: Challenges and Opportunities in Internet Data Mining. Carnegie Mellon University, Pittsburgh, pp. 15213–3890 (2006)
15. Su, X., Khoshgoftaar, T.M.: A Survey of Collaborative Filtering Techniques. Advances in Artificial Intelligence 6(4) (2009)
16. Dimitre, D., Roopa, P., Abir, Q., Jeff, H.: ISENS: A System for Information Integration, Exploration, and Querying of Multi-Ontology Data Sources. IEEE Computer Society, ICSC, pp. 330–335 (2009)
17. Backward analysis: The Free On-line Dictionary of Computing (n.d.)
18. Dahanayake, A., Thalheim, B.: W*H: the conceptual model for services. In: ESF 2014 workshop on Correct Software for Web Application. Sringer-Verlage (2014)
19. Regina, C., Beyer, M., Adrian, M., Friedman, T., Logan, D., Buytendijk, F., Pezzini, M., Edjlali, R., White, A., Laney, D.: Top 10 Technology Trends Impacting Information Infrastructure. Gartner publication (2013)
20. Hitzler, P., Janowicz, K.: Linked Data, Big Data, and the 4th Paradigm. Semantic Web Journal 4(3), 233–235 (2013)
21. Punch, K.F.: Introduction to Social Research: Qualitative and Quantitative Approaches. Sage, Britain (2005)

AutoScale: Automatic ETL Scale Process

Pedro Martins$^{(\boxtimes)}$, Maryam Abbasi, and Pedro Furtado

Department of Computer Sciences, University of Coimbra, Coimbra, Portugal
{pmom,maryam,pnf}@dei.uc.pt

Abstract. In this paper we investigate the problem of providing automatic scalability and data freshness to data warehouses, when at the same time dealing with high-rate data efficiently. In general, data freshness is not guaranteed in those contexts, since data loading, transformation and integration are heavy tasks that are performed only periodically, instead of row by row.

Many current data warehouse deployments are designed to be deployed and work in a single server. However, for many applications problems related with data volume processing times, data rates and requirements for fresh and fast responses, require new solutions to be found.

The solution is to use/build parallel architectures and mechanisms to speed-up data integration and to handle fresh data efficiently.

Desirably, users developing data warehouses need to concentrate solely on the conceptual and logic design (e.g. business driven requirements, logical warehouse schemas, workload and ETL process), while physical details, including mechanisms for scalability, freshness and integration of high-rate data, should be left to automated tools.

We propose a universal data warehouse parallelization solution, that is, an approach that enables the automatic scalability and freshness of any data warehouse and ETL process. Our results show that the proposed system can handle scalablity to provide the desired processing speed.

Keywords: Algorithms · Architecture · Scalability · ETL · Freshness · High-rate · Performance · Scale · Parallel processing

1 Introduction

Extract - Transform - Load (ETL) tools are special purpose software artifacts used to populate a data warehouse with up-to-date, clean records from one or more sources. The majority of current ETL tools organize such operations as a workflow. At the logical level, the E (extraction) can be considered as a capture of data flow from the sources, normally more than one with high-rate throughput. Then we have T representing (transformations and cleansing) were data must be distributed and replicated across many processes. The end part, L (load) into the data warehouse is where the data is stored to be queried and analyzed. When implementing these type of systems beside the necessity to know all of

© Springer International Publishing Switzerland 2015
T. Morzy et al. (Eds): ADBIS 2015, CCIS 539, pp. 20–27, 2015.
DOI: 10.1007/978-3-319-23201-0_3

these steps, the user is required to be aware of the scalability issues which the ETL+Q (queries) might be represent.

When defining the ETL+Q the user must have in mind the existence of data sources, where and how the data is extracted to be transformed (e.g. completed, cleaned, validated), the loading into the data warehouse and finally the data warehouse schema, each of these steps requires different processing capacity, resources and data treatment (e.g. distribution of data across nodes, replication). However, the ETL is never so linear and it is more complex than it seems. Most of the times because the data volume is too large and one single extraction node is not sufficient. Thus more nodes must be added to extract the data and extraction policies from the sources must be created (e.g. round-robin OR on-demand). Consequently the other phases, transformation and load must be scaled.

In this paper we studied how to provide for the user automatic ETL+Q scalability with ingress high-data-rate in big data warehouses. We propose a software tool, AutoScale, to parallelize and scale the entire ETL+Q process. The user can only focus on the conceptual design of the ETL+Q processes to be executed. The paralelization of the system and scalability to provide performance and data freshness is all assured by the proposed framework, where the user besides the location of sources, transformation process scripts and data warehouse schema must also provide some minor parametric execution configurations and information.

The presented results prove that the proposed framework is able to scale-out when more resources are necessary.

2 Related Work

Works in the area of ETL scheduling includes efforts towards the optimization of the entire ETL workflows [8] and of individual operators in terms of algebraic optimization; e.g., joins or data sort operations). The work [4] deals with the problem of scheduling ETL workflows at the data level and in particular scheduling protocols and software architecture for an ETL engine in order to minimize the execution time and the allocated memory needed for a given ETL workflow. A second aspect in ETL execution that the authors address is how to schedule flow execution at the operations level (blocking, non-parallelizable operations may exist in the flow) and how we can improve this with pipeline parallelization [3].

The Aurora system [1]can execute more than one continuous query for the same input stream(s). Every stream is modeled as a graph with operators (a.k.a. boxes). Scheduling each operator separately is not very efficient, so sequences of boxes are scheduled and executed as an atomic group. The Aurora stream manager has three techniques for scheduling operators in streams [7], each with the goal of minimizing one of the following criteria: execution time, latency time, and memory. This work focuses on the aspect of real-time resource allocation.

The work [5] focuses on finding approaches for the automatic code generation of ETL processes which is aligning the modeling of ETL processes in data warehouse with MDA (Model Driven Architecture) by formally defining a set of QVT (Query, View, Transformation) transformations.

Related problems studied in the past include the scheduling of concurrent updates and queries in real-time warehousing and the scheduling of operators in data streams management systems. However, we argue that a fresher look is needed in the context of ETL technology. The issue is no longer the scalability cost/price, but rather the complexity it adds to the system. The point that previews presented recent works in the field do not address is the automatic scalability to make ETL scalability easy and automatic. The authors focus on mechanisms to improve scheduling algorithms and optimizing work-flows and memory usage. In our work we assume that scalability in number of machines and quantity of memory is not the issue, we focus in offering automatic ETL scalability, without the nightmare of operators relocation and complex execution plans. Thus in our work we focus on automatic scalability, given a single server ETL process, our proposed system (AutoScale) will scale it automatically. AutoScale based on generic ETL process thought for a single server (no concern with performance or scalability) will scale it to provide the users desired performance with minimum complexity and implementations. In addition, we also support queries execution.

3   Architecture

In this section we describe the main components of the propose architecture, AutoScale, for automatic ETL+Q scalability.

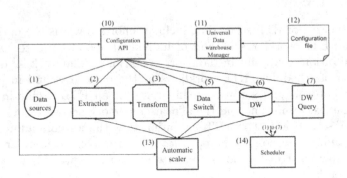

Fig. 1. Automatic ETL+Q scalability

Figure 1 shows the main components to achieve automatic scalability.

- All components from (1) to (7) are part of the Extract, Transform, Load and query (ETL+Q) process.
- The "Automatic Scaler" (13), is the node responsible for performance monitoring and scaling the system when is necessary.

- The "Configuration file" (12) represents the location where all user configurations are defined by the user.
- The "Universal Data Warehouse Manager" (11), based on the configurations provided by the user and using the available "Configurations API" (10), sets the system to perform according with the desired parameters and algorithms. The "Universal Data Warehouse Manager" (11), also sets the configuration parameters for automatic scalability at (13) and the policies to be applied by the "Scheduler" (14).
- The "Configuration API" (10), is an access interface which allows to configure each part of the proposed Universal Data Warehouse architecture, automatically or manually by the user.
- Finally the "Scheduler" (14), is responsible for applying the data transfer policies between components (e.g. control the on-demand data transfers).

All these components when set to interact together are able to provide automatic scalability to the ETL and to the data warehouses processes without the need for the user to concern about its scalability or management.

Paralelization Approach. Figure 2 depicts the main processes needed to support total ETL+Q scalability.

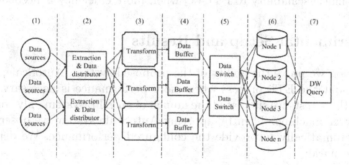

Fig. 2. Total automatic ETL+Q scalability

(1) Represents the data sources from where data is extracted into the system.
(2) The data distributor(s) is responsible for forwarding or replicating the raw data to the transformer nodes. The distribution algorithm to be used is configured and enforced in this stage. The data distributors (2) should also be parallelizable if needed, for scalability reasons.
(3) In the transformation nodes the data is cleaned and transformed to be loaded into the data warehouse. This might involve data look-ups to in-memory or disk tables and further computation tasks. In Figure 2 the transformation (3) is parallelized for scalability reasons.
(4) The data buffer can be in memory, disk file (batch files) or both. In periodically configured time frames/periods, data is distributed across the data warehouse nodes.

(5) The data switches are responsible to distribute (pop/extract) data from the "Data Buffers" and set it for load into the data warehouse, which can be a single-node or a parallel data warehouse depending on configured parameters (e.g. load time, query performance).

(6) The data warehouse can be in a single node, or parallelized by many nodes. If parallelized, the "Data Switch" nodes will manage data placement according to configurations (e.g. replication and distribution). Each node of the data warehouse loads the data independently from batch files.

(7) Queries (7) are rewritten and submitted to the data warehouse nodes for computation. The results are then merged, computed and returned.

The main concept we propose is the automatic ETL+Q scalability where user designs a logical view of the ETL+Q and data warehouse (single server), without worrying about scalability. Additionally, he specifies some parameters needed by the automatic scalability mechanisms, such as: extraction frequency; maximum extraction time; maximum transformation buffer queues load; maximum supported data-rates; data warehouse load frequency; maximum data warehouse load time.

Based on user configuration parameters, the system scales automatically, figure 1, (10) to (13). All the components in Figure 2 interact together for providing automatic scalability to ETL+Q when more efficiency is needed.

4   Experimental Setup and Results

In this section we prove the ability of the proposed system, AutoScale, to automatic scale-out the ETL process when more performance is necessary. For the purpose of these tests we simulated the launch of an ETL system only concerning a single server machine. Then by applying AutoScale system we observed how it scales automatically to provide the configured performance. We defined the following scenario:

- Our sources are based on the TPC-H benchmark [2] generator, which generate data at the highest possible rate (on each node) and through a web service make it available for extraction.
- The transformation process consisted on transforming the TPC-H relational model into a star-schema, which resulted in the recreation of the SSB benchmark [6]. This process involves heavy computational memory and temporary storage systems, look-ups and data transformation to assure consistency.
- The data warehouse tables schema consists on the same schema form SSB benchmark. Replication and partitioning is assured by the proposed system, whereas, dimension tables are replicated and fact tables are partitioned across the data warehouse nodes.
- The E (extraction), T (transformation) and L (load) were set to perform every 2 seconds, and can not last more than 1 second. Thus the ETL process will last at the worst case 3 seconds total.
- The load process was made in batches of 10MB maximum.

The experimental tests were performed using 12 computers, denominated as nodes, with the following characteristics: Processor Intel Core i5-5300U Processor (3M Cache, up to 3.40 GHz); Memory 16GB DDR3; Disk: western digital 1TB 7500rpm; Ethernet connection 1Gbit/sec; Connection switch SMC SMCOST16, 16 Ethernet ports, 1Gbit/sec;

Software installed/used - The 12 nodes were formatted before the experimental evaluation and installed with: Windows 7 enterprise edition 64 bits; Java JDK 8; Netbeans 8.0.2; Oracle Database 11g Release 1 for Microsoft Windows (X64) - used in each data warehouse node; PostgreSQL 9.4 - used for look ups during the transformation process; TPC-H benchmark - representing the operational log data used at the extraction nodes. This is possible since TPC-H data is still normalized; SSB benchmark - representing the data warehouse. The SSB is the star-schema representation of TPC-H data;

We tested the full system running, without queries, testing all parts involved in the automatic Extraction, Transformation and Load process. The data-rate was gradually increased at the sources (adding more data sources when necessary). The next base configuration were used:

- Each time we increment the data rate, we let the system run for 60 minutes, to allow it to stabilize and scale.
- The data-rate was increased from 1.000 rows per second, up to, 500.000 rows per second (50 tests, each test 60 minutes).
- Each part of the ETL process was not allowed to exceed more than 1 second. Thus the entire ETL process should take 3 seconds.
- Processing nodes were added, by the system request, gradually when the queues reached a limit size, or the processing time was exceeded.
- For the load method we considered a data warehouse without indexes or views and the data load was made in batch files with 10MB size;

Figure 3 and 4 show AScale, scaling-out and scaling-in automatically, respectively, to deliver the configured near-real-time ETL time bounds, while the data rate increases/decreases. The charts show: the X axis represents the data-rate, from 10.000 to 500.000 rows per second; the Y axis is the ETL time expressed in seconds; the system objective was set to deliver the ETL process in 3 seconds; the charts show the scale-out and scale-in of each part of the AScale, obtained by adding and removing nodes when necessary; A total of 7 data sources were used/removed gradually, each one delivering a maximum average of 70.000 rows/sec; AScale used a total of 12 nodes to deliver the configured time bounds.

Near-real-time scale-out results in Figure 3 show that, as the data-rate increases and parts of the ETL pipeline become overloaded, by using all proposed monitoring mechanisms in each part of the AScale framework, each individual module scales to offer more performance where and when necessary.

Near-real-time scale-in results in Figure 4 show the instants when the current number of nodes is no longer necessary to ensure the desired performance, leading to some nodes removal (i.e. set as ready nodes in stand-by, to be used in other parts).

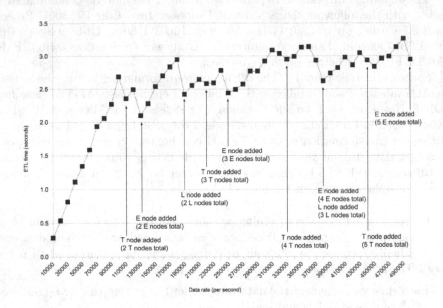

Fig. 3. Near-real-time, full ETL system scale-out

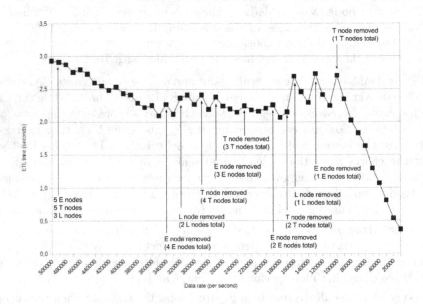

Fig. 4. Near-real-time, full ETL system scale-in

5   Conclusions and Future Work

In this work we propose mechanisms and algorithms to achieve automatic scalability for complex ETL+Q, offering the possibility to the users to think solely in the conceptual ETL+Q models and implementations for a single server.

The tests demonstrate that AutoScale is able to scale-out. Other algorithms were developed and testes obtaining positive results. Also extended tests were performed to the framework modules, but due to space limitations we present only the most relevant chart, including all modules working together.

Future work will involve scale-in approaches and deal with real-time ETL for querying and at the same time providing data freshness. Other direction of future work would be to compare AutoScale framework using Map-Reduce modules implementation, both in terms of approach, implementation curve, usability and performance with similar resources.

Acknowledgement. This project is part of a larger software prototype, partially financed by CISUC research group from the University of Coimbra and by the Foundation for Science and Technology.

References

1. Babcock, B., Babu, S., Datar, M., Motwani, R., Thomas, D.: Operator scheduling in data stream systems. The VLDB Journal-The International Journal on Very Large Data Bases **13**(4), 333–353 (2004)
2. T.P.P. Council: Tpc-h benchmark specification 2008. http://www.tcp.org/hspec.html
3. Halasipuram, R., Deshpande, P.M., Padmanabhan, S.: Determining essential statistics for cost based optimization of an etl workflow. In: EDBT, pp. 307–318 (2014)
4. Karagiannis, A., Vassiliadis, P., Simitsis, A.: Scheduling strategies for efficient etl execution. Information Systems **38**(6), 927–945 (2013)
5. Muñoz, L., Mazón, J.-N., Trujillo, J.: Automatic generation of etl processes from conceptual models. In: Proceedings of the ACM Twelfth International Workshop on Data Warehousing and OLAP, pp. 33–40. ACM (2009)
6. O'Neil, P.E., O'Neil, E.J., Chen, X.: The star schema benchmark (ssb). Pat (2007)
7. Simitsis, A., Vassiliadis, P., Sellis, T.: State-space optimization of etl workflows. IEEE Transactions on Knowledge and Data Engineering **17**(10), 1404–1419 (2005)
8. Simitsis, A., Wilkinson, K., Dayal, U., Castellanos, M.: Optimizing etl workflows for fault-tolerance. In: 2010 IEEE 26th International Conference on Data Engineering (ICDE), pp. 385–396. IEEE (2010)

Using a Domain-Specific Language to Enrich ETL Schemas

Orlando Belo[1(✉)], Claudia Gomes[1], Bruno Oliveira[1],
Ricardo Marques[2], and Vasco Santos[1]

[1] ALGORITMI R&D Centre, Department of Informatics, School of Engineering,
University of Minho, Campus de Gualtar 4710-057, Braga, Portugal
obelo@di.uminho.pt
[2] WeDo Technologies, Centro Empresarial de Braga, 4705-319, Ferreiros, Braga, Portugal

Abstract. Today it is easy to find a lot of tools to define data migration schemas among different types of information systems. Data migration processes use to be implemented on a very diverse range of applications, ranging from conventional operational systems to data warehousing platforms. The implementation of a data migration process often involves a serious planning, considering the development of conceptual migration schemas at early stages. Such schemas help architects and engineers to plan and discuss the most adequate way to migrate data between two different systems. In this paper we present and discuss a way for enriching data migration conceptual schemas in BPMN using a domain-specific language, demonstrating how to convert such enriched schemas to a first correspondent physical representation (a skeleton) in a conventional ETL implementation tool like Kettle.

Keywords: Data warehousing systems · Data migration schemas · ETL conceptual modeling · BPMN specification models · Domain-Specific language · ETL skeletons · Kettle

1   Introduction

Nowadays, companies implemented very specialized data warehouses [6], trying to cover their information needs on decision-making processes. The process of populating a data warehouse - ETL (Extract-Transform-Load) [5] – is a sophisticated data migration process, involving a large range of tasks and operators articulated often in a complicated coordination workflow. ETL projects are quite complex and difficult to accomplish. Planning and designing assume very relevant roles on these projects in order to ensure the success of building a data warehouse. ETL tools provide today mature technologies. Most of them provide support in all stages of the ETL life cycle, from conceptual to logical, and from logical to physical ETL. However, most of them still use proprietary notations or follow other specific methodologies that cannot be generalized or easily used. Recent works have evolved around the idea of conceptually modeling ETL using BPMN patterns [7], as a part of a high-level abstraction layer of a set of commonly used tasks in the implementation of an ETL system. Despite the interesting results of these works [11] [1] [2], they do not yet solved the problem of

© Springer International Publishing Switzerland 2015
T. Morzy et al. (Eds): ADBIS 2015, CCIS 539, pp. 28–35, 2015.
DOI: 10.1007/978-3-319-23201-0_4

using all the potential of a conceptual model. This could be easily achieved if one used a custom language – a *Domain Specific Languages* (DSL) - with the ability to describe the semantics of ETL patterns when integrated in a BPMN conceptual model. According to [12] a DSL is «a high-level software implementation language that supports concepts and abstractions that are related to a particular (application) domain». Thus, a DSL has a very expressive power for describing the specificities of a particular domain such as the case of the description of ETL patterns in BPMN [3][4]. DSL are built by defining a grammar for the language – the logic between the components and rules – and for the components. Then, the language will be used to build or describe ETL patterns.

In this paper we present and discuss a way for enriching data migration conceptual schemas in BPMN using a DSL, especially designed and implemented to add semantics to BPMN tasks and processes. Additionally, we also demonstrate how to convert such enriched schemas to a first correspondent physical representation (a skeleton), having the possibility to be executed and validated in Kettle [8], a conventional ETL implementation tool from Pentaho. The paper is structured as follows. Section 2 describes and exemplifies how the DSL we designed and its grammar were created to describe complementary operational semantics on BPMN tasks and processes. After that, we demonstrate the use of the DSL on the specification of particular instantiations of processes in a simplified version of an ETL system (section 3). Finally, in section 4, we present some conclusions and final remarks.

2   Integrating DSL Specifications in BPMN Models

In BPMN [3] a task or an activity representing a pattern is just a visual component having a label and a small set of properties, but nothing to define its structure or its behavior. There are other approaches that propose specific oriented languages, like BPEL (Business Process Execution Language) [9] to solve this problem. This language can be seen as an extension to other programming languages, but it is focused especially on web services. A language like this can be used to validate and execute BPMN models, converting each BPMN component into some XML representation that BPEL recognizes. However, this conversion is not simple because: 1) not all elements in BPMN have a direct translation in BPEL, which provokes a certain loss of accuracy in the BPEL model; 2) there is no automated way to translate BPMN to BPEL, meaning that you have to know the elements in BPEL and make the writing of the BPEL model yourself; and 3) the tools that provide support for BPEL force the user to think at a very low level detail regarding the BPEL programming, requiring specialized users to make the conversion of a BPMN model to BPEL. Thus, we can see why BPEL is not a very suitable solution for the problem we want to approach, which stresses the need of a specific language that could add more operational semantic to an ETL conceptual model in BPMN. Generically speaking, ETL tasks can be organized into three ETL pattern categories: extractors, transformers and loaders. Each category entails the execution of various tasks and their properties, relating to different states of the development cycle of an ETL system. Therefore, each category has different properties that characterize the different states of the development of an ETL system. The first category is responsible for identifying and extracting the data that were modified in information sources. To perform this task it is necessary to indicate the data sources' paths and the data structures that will receive data. In this category we can find patterns such as sequential log files readers or *Change Data Captures* (CDC). Data sources usually are highly heterogeneous, which impose frequently the application of data format recognition mechanisms to detect the kind of information sources – e.g. a relational database, a text file, a CSV file, or a XML file, among others. The second category includes all patterns with the ability to perform transformation tasks, such as data filtering, data quality assurance, data merging, splitting or replacement operations, or the simple conversion of an operational key. Here, some of the most common patterns are Data Quality Enhancement (DQE), Surrogate Key Pipelining (SKP), or *Error Handling Hub* (HHS). Finally, the third category: the loaders. This category includes the patterns and the operations that are responsible for loading data into a specific data repository like a data warehouse. *Slowly Changing Dimension* (SCD) or *Intensive Data Loading* (IDL) patterns are two of the most relevant elements of this category.

```
-- An excerpt of the DSL's grammar.
ETL:
   'use' (elements+=Pattern)+'on sources{' (source+=Data)+ '}'
   ('and target{' target=Data '}')? ('with mapping source{' (map+=Data)+'}')?
   ('options{'opt = Option '}')?;
-- Pattern categories.
Pattern:
   Gather | Transform | Load;
-- Gathers configuration
Gather:
   'Gather'('sort by'  sort = STRING ',')? //(gatherFields+=Fields)*;
-- Information sources configuration
Data:
   'BEGIN' (Path | Source)','type=' type=STRING'END';
Option:
   'parallel operation=' p=STRING ','number of threads=' t=STRING (','
   'number of rows in batch=' r=STRING)?;
Path:
   'data=' src=STRING;
Source:
   'name=' n=STRING ',' 'server=' s=STRING ',' 'database=' db=STRING ','
   ('table=' tb+=STRING ('{'(fields+=Fields)+'}')? ',')*
   'technology=' tech=STRING','access=' a=STRING','user=' user=STRING ','
   'password=' pwd=STRING(', port=' p=STRING)?;
Fields:
   ('source' | 'target')?'fields{' (fd+=Field)+'}';
(...)
-- Loaders configurations.
Load:
   'Load from'('each record')?;
```

Fig. 1. An excerpt of the DSL's grammar

Considering this, we designed and built a specific DSL taking into account all the properties and characteristics of each pattern category. Looking at the DSL's grammar (Fig. 1), we can see how to configure a data source, or even the destiny of the data collected, using an extensive list of attributes (source's attributes). For example, in case of the source or the destiny be a relational source, we need to specify the connection name (*name*), the server identification (*server*), the name of the database (*database*), and others. Otherwise, the source or the destiny of the data can be set through the path of a file (*data*) containing the data required and the source type. In short, with this kind of approach we will be able to build ETL conceptual models, enrich them with task behavior descriptions using the DSL, converting the enriched models to an intermediate standard format, and produce a physical skeleton to be imported by a conventional ETL implementation tool. Fig. 2 shows an application of the DSL in the configuration of a gather pattern.

```
-- Configuration of a gather pattern.
use Gather
on sources{BEGIN data=CDR_Calls.csv,type=CSV input END}
and target{BEGIN server=localhost, database=CALLSR,table= MSISDN_ACTIVITY
   {fields{id,type,caller_id,called_id,caller_number,called_number,time_start,
   time_end,}}, technology=MySQL,access=Native,user=root,
   password=Saphira23,type=relational END}
options{loop_criteria=FOR_EACH_ROW}
```

Fig. 2. An excerpt of the behavior of an ETL task using the DSL

3   Using a DSL on Process Instantiation

Using a DSL language on the specification of data-driven workflows allows for the formalization of a set of abstract and self-contained constructors, which simplify process representation across several detail levels. To achieve the necessary adequacy, both in terms of business level and physical level, any physical model should be developed following the sequence of operations showed in Fig. 3. This way, business and physical requirements associated with the execution of a given process can be easily represented, and posteriorly used by a commercial tool to enable its real execution.

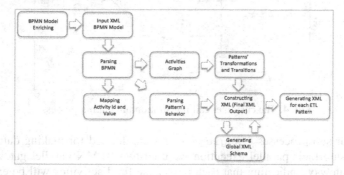

Fig. 3. Steps for producing a physical model based on a BPMN conceptual description

To start, we need to design and produce a conceptual representation in BPMN. Both semantic and physical metadata should be addressed by the DSL, providing a way to distinguish and support all the business rules involved in the process. The BPMN configuration is described using the DSL, indicating the data required by each task. The CSV Gather case (Fig. 2) shows the logical metadata involved with, the field mappings between the source CSV and the target relational schema, and the execution support metadata describing all the connection parameters of the physical schemas as well as all the specific execution requirements. To provide a clearer picture of how BPMN can be used on early ETL development stages and how their artifacts can be mapped to execution primitives, we selected a very common data migration application case on the telecommunications domain. To support it we used a *Call Detail Records* (CDR) file from a mobile phone company having data about customers' calls. This file will act as a source data for a specific database table that identifies callers, number of calls and total call duration. The migration process (Fig. 4) aims to handle the referred data (stored in a CSV file), in which every row has a record of a call phone done by a specific customer. The record's attributes are: DATE, the date of the call; TIME, a timestamp; DURATION, the time spent with the call; MSISDN of the source customer that depends if the call direction (incoming or outgoing); OTHER_MSISDN of the target customer that depends of the call direction (incoming or outgoing); and DIRECTION of the call. The MSISDN belongs to the called customer and the OTHER_MSISDN to the caller customer, while the value 'O' represents an outgoing call, i.e. the MSISDN belongs to the caller and OTHER_MSISDN to the called customer.

The process initiates extracting data from the CSV file using a BPMN activity. The name of this activity (as many others) includes the # character to identify that the activity represents a container of tasks configured through the DSL proposed. In fact, the Extract Source activity configuration was already presented before in Fig. 2. The output data from each source is stored in a separated repository. The logic of this process is handled internally by each pattern.

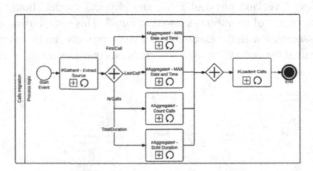

Fig. 4. The customer call registration populating process

Next, four sub-processes (#Aggregate#) were defined for making data aggregation. To ensure their parallel execution we used two BPMN parallel gateways: 1) a divergent gateway, indicating that each one of the fired activities will be executed in parallel; and 2) a convergent gateway to force the completion of all parallel flows. Finally, the process ends with the execution of a #Loader# container configured to load data into the target data structure. Additionally, a loop marker was used for each activity, meaning that each one should be executed iteratively based on some criteria. For example, the gather specification presented in Fig. 2 describes (using the options block) the loop_criteria statement with a FOR_EACH_ROW value, which means that the internal activity tasks should be executed for each row in the process.

For compatibility reasons, the DSL descriptions for specifying ETL patterns behavior were integrated in the BPMN model using text annotations.

The BPMN process serialization enables the representation of all BPMN elements including DSL into a *XML Process Definition Language* (XPDL) standard format [10]. The XPDL is used by some software products to interchange business process definitions between different tools, providing an XML-based format with the ability to support every aspect of the BPMN process definition notation, including the graphical descriptions of the diagram, as well as its executable properties. Thus, there is no need to extend the XPDL meta-model to ensure compatibility with other existing tools. The XPDL file is used as input of the developed parser to interpret not only the DSL code but also the logic of the workflow, providing the necessary means to preserve the details of the ETL model configuration about the process control, data transformation requirements and configuration parameters – e.g. the access to a database or a file, or the identification of a source or a target data repository (Fig. 5). Having this, it is possible to generate a physical model and provide automatically a data migration package configured accordingly.

To filter metadata enabling its transformation to a recognized specific data format by an ETL implementation tool we designed and implemented a specific parser. This parser interpreted DSL specifications and translated them to an intermediary specification format recognized by a data migration tool. To do this, we first need to convert the model, exporting it to a known serialization format. The parser generates two distinct structures: 1) a map, whose key is the identification of the activity, and a value, which is a class embodying all the details about the XML node activity; and 2) a graph, containing the information about all the existing connections in the BPMN model. BPMN tools use standard formats for process interchange, but data migration tools use their own file formats. This compromises system flexibility. To avoid this we used Acceleo (http://www.eclipse.org/acceleo/), a model-to-code transformation module. Under the MDA (*Model-Driven Architecture*) provided by Eclipse, we also selected and applied some specific Ecore models to describe each pattern skeleton and to identify process instantiation metadata. Next, a standard transformation template was built to encapsulate the conversion process and to transform the internal structure of the pattern into a XML serialization format supported by the implementation tool - Kettle. This process is executed in two steps: 1) the verification of all patterns included in the BPMN conceptual model and their conversion to a single XML representation; and 2) the generation of the general XML Kettle schema. To establish the mapping rules between a BPMN conceptual model specification and Kettle package representation it is necessary to follow a set of generic rules that are explicitly defined in what we call a transformation template. On this template: 1) the three main patterns classes - gather, transformer, and loader - are mapped directly to specific Kettle jobs or transformations; 2) the internal tasks are created inside each top-level package as well as the details related to the execution of the process. In Fig. 6 we can see four distinct examples of packages that were generated based on a BPMN model specification. The top package level (Fig. 6a) represents the three main clusters that could be explicitly represented at a conceptual level using BPMN. This layer represents a transformation that corresponds to the data extraction activity represented in the BPMN process presented before in Fig. 4. The associated model (Fig. 6b) includes a specific *CSV input* task that is responsible to load data to a *copy rows to result* task.

Rows are stored in memory since all jobs and transformations were configured to execute one row each time and the case presented does not process large volumes of data at once. The *Transformers* job (Fig. 6c) is associated with the transformation that performs, for each application scenario, the aggregation tasks. The parallel configuration described in the BPMN model was translated enabling the parallel execution of each transformation that encapsulates the logic of each aggregation task. Finally, the Loaders transformation (Fig. 6d) from top-level package represents the data load to the target relational table.

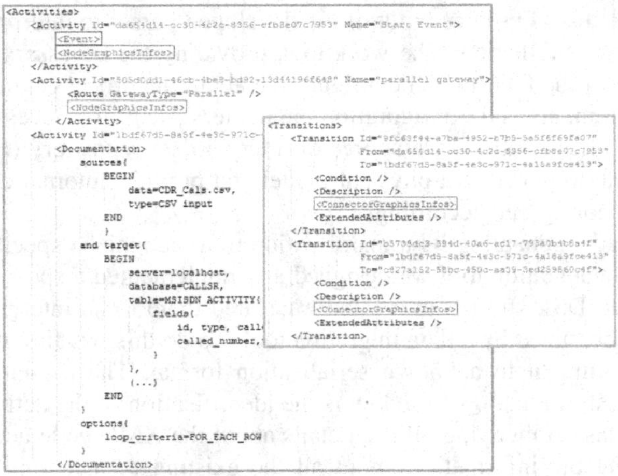

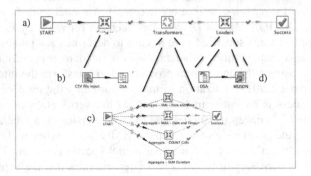

Fig. 5. A XPDL excerpt used for the serialization of a BPMN process (Fig. 4)

Fig. 6. The Kettle package generated from the BPMN model (Fig. 4)

4   Conclusions

Conceptual models are frequently discarded in favor of a more detailed logical model. In our opinion, ETL conceptual modeling is a very useful activity in the life cycle of a data warehousing system project. Based on BPMN and on a set of ETL patterns (characteristics and behavior description), we designed and implemented a specific ETL development process that enhances the importance of building ETL conceptual models as a way to establish a first executable version of the system – an ETL skeleton. To enrich the expressiveness of BPMN we designed and created a DSL especially oriented to describe the behavior of an ETL pattern when integrated in an ETL conceptual model. Results are quite satisfactory, and prove the viability of the approach. It was possible to demonstrate with a real application case that it is possible to use a lot of the material applied in the conceptual specification as a catalyst for a possible physical implementation. At this stage, the ETL skeletons that can be produced still are very primary physical models. In a near future we need to improve them, in order to allow for the generation of more effective physical.

Acknowledgment. This work was developed under the project RAID B2K - RAID Enterprise Platform / NUP: FCOMP-01-0202-FEDER-038584, a project financed by the Incentive System for Research and Technological Development from the Thematic Operational Program Competitiveness Factors.

References

1. El Akkaoui, Z., Zimanyi, E.: Defining ETL workflows using BPMN and BPEL. In: DOLAP 2009 Proceedings of the ACM Twelfth International Workshop on Data Warehousing and OLAP, pp. 41–48 (2009)
2. El Akkaoui, Z., Zimanyi, E., Mazón, J.-N., Trujillo, J.: A BPMN-Based Design and Maintenance Framework for ETL Processes. Int. J. Data Warehous. Min. **9** (2013)
3. BPMN, OMG, in Documents Associated With Business Process Model And Notation (BPMN), Version 2.0 (2011)
4. Deursen, A., Klint, P., Visser, J.: Domain-specific languages: an annotated bibliography. ACM SIGPLAN Notices **35**(6), 26–36 (2000)
5. Kimball, R., Caserta, J.: The Data Warehouse ETL Toolkit: Practical Techniques for Extracting, Cleaning, Conforming, and Delivering Data. Willey (2004)
6. Kimball, R., Ross, M., Thornthwaite, W., Mundy, J., Becker, B.: The Data Warehouse Lifecycle Toolkit, 2nd edn. Wiley, January 2008
7. Oliveira, B., Belo, O.: BPMN patterns for ETL conceptual modelling and validation. In: Chen, L., Felfernig, A., Liu, J., Raś, Z.W. (eds.) ISMIS 2012. LNCS, vol. 7661, pp. 445–454. Springer, Heidelberg (2012)
8. Pentaho: Pentaho Data Integration. http://www.pentaho.com/product/data-integration (accessed March 16, 2015)
9. Recker, J., Mendling, J.: On the translation between BPMN and BPEL: conceptual mismatch between process modeling languages. In: Latour, T., Petit, M. (eds.) Proceedings of the 18th International Conference on Advanced Information Systems Engineering, Luxembourg, pp. 521–32 (2006)
10. T.W.M.C. Specification: Workflow Management Coalition Workflow Standard Process Definition Interface – XML Process Definition Language (2012)
11. Trujillo, J., Luján-Mora, S.: A UML based approach for modeling ETL processes in data warehouses. In: Song, I.-Y., Liddle, S.W., Ling, T.-W., Scheuermann, P. (eds.) ER 2003. LNCS, vol. 2813, pp. 307–320. Springer, Heidelberg (2003)
12. Visser, E.: WebDSL: a case study in domain-specific language engineering. In: Lämmel, R., Visser, J., Saraiva, J. (eds.) Generative and Transformational Techniques in Software Engineering II. LNCS, vol. 5235, pp. 291–373. Springer, Heidelberg (2008)

Continuous Query Processing Over Data, Streams and Services: Application to Robotics

Vasile-Marian Scuturici[1]([✉]), Yann Gripay[1], Jean-Marc Petit[1], Yutaka Deguchi[2], and Einoshin Suzuki[2]

[1] Université de Lyon, CNRS, INSA-Lyon, LIRIS UMR5205,
69621 Villeurbanne, France
marian.scuturici@insa-lyon.fr
[2] Department of Informatics, ISEE, Kyushu University, Fukuoka, Japan

Abstract. Developing applications involving mobile robots is a difficult task which requires technical skills spanning different areas of expertise, mainly computer science, robotics and electronics. In this paper, we propose a SQL-like declarative approach based on data management techniques. The basic idea is to see a multi-robot environment as a set of data, streams and services which can be described at a high level of abstraction. A continuous query processing engine is used in order to optimize data acquisition and data consumption. We propose different scenarios to classify the difficulty of such an integration and a principled approach to deal with the development of multi-robot applications. We provide our first results using a SQL-like language showing that such applications can be devised easily with a few continuous queries.

1 Introduction

Mobile robots contain sensors and actuators working and communicating under real-world constraints. Developing applications on top of mobile robots is a difficult task which requires technical skills spanning different areas of expertise, mainly computer science, robotics and electronics. Actually, a typically mobile robot application is roughly built on top of some middleware, the semantic glue being implemented with an imperative language like C++ or Python or functional languages like LISP or Haskel. Moreover, robot environments generate more and more heterogeneous data (image, GPS, sensor values, . . .) that need to be integrated and analyzed to obtain useful patterns. In such data-centric applications, performance depends on many factors, ranging from raw data accesses to network latency, integration of heterogeneous data, data mining/machine learning techniques or dynamic (appearing/disappearing) distributed services in the environment of robots.

In this paper, we propose a declarative approach for robotics based on data management techniques. The basic idea is to see a multi-robot environment as

A part of this research was supported by a Bilateral Joint Research Project between Japan and France funded from JSPS and CNRS (CNRS/JSPS PRC 0672), and JSPS KAKENHI 24650070 and 25280085.

© Springer International Publishing Switzerland 2015
T. Morzy et al. (Eds): ADBIS 2015, CCIS 539, pp. 36–43, 2015.
DOI: 10.1007/978-3-319-23201-0_5

a set of data, streams and services which can be described at a high level of abstraction. We propose a principled approach to devise applications involving robots and we provide high level abstractions at two levels: (1) to represent a multi-robot environment using declarative data and service description languages and (2) to develop concise and elegant applications using declarative continuous queries on top of those descriptions to achieve the scenarios. We provide our first results using a SQL-like language showing that such applications can be devised with a few *continuous queries,* much easier to design than ad-hoc programming.

2 Related Works

In [3] the authors classify the robot programming into automatic programming (the robot learns from demonstration/experience), manual programming (the programmer writes the application code) and software architectures (the programmer reuses existing modules from existing frameworks, eventually enriched with some new code). Our interest is for applications built by programmers, which exclude automatic programming.

In this context, there are a lot of frameworks for helping the development of mobile robot applications, permitting to achieve rapid prototyping and to reuse as much as possible existing technologies [9]. Robotic frameworks such as ROS [11] or Player/Stage [5] bring software components which (partially) hide the complexity to integrate heterogeneous devices or robots, distribution mechanisms, scalability, diverse communication protocols, to mention a few. These components are integrated into a new application via a programming language. We note that there is no special attention to data management in such frameworks. We propose to go one step beyond by using a SQL-like declarative language to interact with the environment of robots, instead of procedural or functional languages.

Another line of research conducted in Robotics concerns the integration of Domain-Specific Languages (DSL) with Haskell in order to specify the behavior of mobile robots to build applications. For instance, we quote *Frob* [10], *GRL* [8] and *roshask* ([4]). Such languages permit to describe *what* an application does, not *how* it does it. An application is specified as a sequence of actions/tasks, and the implementation details concerning the communication with the sensors and the actuators are hidden to the programmer. Clearly, our objective is the same, i.e. simplifying the development of robot applications, but we use quite different technologies and concepts. We borrow declarative techniques from data management domain, known to be both efficient, scalable and robust.

3 Simple Scenario

For the sake of readability, we develop a simple scenario to explain our contribution and to convince the reader of the interest of our approach. More comprehensive scenarios have been developed but are not described here due to space constraints. Our implemented usecase is about the monitoring of a smart building. The building is equipped with some temperature sensors, cameras and mobile robots. We assume that every robot or device is connected to a wireless network. As mobile robots, we are using Eddie robots [1], embedding: five distance sensors, a Kinect camera [2], two odometers indicating the distance travelled by the robot wheels, a laptop with WiFi connection.

The Kinect sensor embedded algorithms produces skeletons corresponding to the persons situated in the field of view of this sensor. The skeletons are represented as vectors of 20 points, each point having four attributes *(X, Y, Z, State)*: the 3D coordinates of the sensor position and an attribute *State* specifying the confidence level: well detected (*Tracked*), predicted using the history (*Predicted*) or missing (*Missing*).

In this setting we are interested in writing a dashboard application which gives live information about the different offices occupancy in the building. Some events of interest have also to be triggered, for instance: signalling unusual postures of the persons inside the building. The application relies on the following queries:

- Q_1: display the temperature in a given office, if its temperature goes below 18°C (using one temperature sensor);
- Q_2: display all offices whose temperature has been less than 18°C in the last hour (using as many temperature sensors as offices);
- Q_3: trigger an event each time an unusual posture is detected (context sensitive, using multiple Kinects);
- Q_4: display the percentage of time that a specified office has been occupied in the last hour (position free, using one moving Eddie robot);

4   Declarative Data-Centric Development Approach

In this paper, we rely on enabling technologies based on UbiWare and SoCQ to get effective solutions. Both systems have been successfully applied in various Ambient Intelligence applications [6,13]. UbiWare is a middleware for ambient intelligence that exposes an API that greatly simplifies the coupling with other frameworks ([12]). SoCQ is a Pervasive Management System allowing to integrate data, streams and services in a unified data model ([7]).

The easiness of developing non trivial robot applications depends on many factors: (a) a middleware to handle diverse and heterogeneous devices and robots, (b) a network protocol used by all components, (c) a declarative system allowing to describe the robots environment, integrate data mining/machine learning modules and interact with the environment through continuous queries.

For a given environment, we propose a four-step procedure:

- Step 1: Setting up the robots/devices into UbiWare.
- Step 2: Description of the environment using a data and service description language (SoCQ).
- Step 3: Deployment of data processing modules on devices/robots, in charge of discovering relevant patterns from the robots' environment.

– Step 4: Application development using continuous queries in order to interact dynamically with the environment.

Figure 1 gives an overview of the main layers underlying our approach, devised to be extensible and scalable with respect to both the number of robots/devices and the heterogeneity/diversity of these components. Data mining techniques can be plugged as *data mining services* and become available to query processing. The integration of a query processing engine is the main advantage of our approach. Besides data processing optimizations, this module is responsible with ensuring the coherence of orders sent to actuators. Starting from the logical plan of continuous queries (where the orders are sent to generic actuators), the query engine produces physical execution plans containing associations between orders and concrete actuators.

In a nutshell, when a new hardware or a new software release has to be integrated, the first thing to do is to make them "UbiWare compatible". Then, abstractions needed by the developers to code their applications are described in a declarative manner: every type of data source and relevant service has to be specified at a high level of description. Finally, a set of continuous queries over SoCQ description are launched from a controller allowing to get the desired results over the running environment.

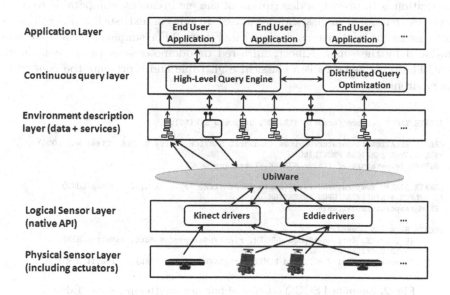

Fig. 1. Main layers of the architecture

Step 1: UbiWare Integration: UbiWare is driven by the idea that the environment contains an ever growing number of miscellaneous entities that expose interesting, but heterogeneous data. To tackle this problem, we abstract every entity as a *data service*, i.e., a producer as well as a supplier of data. We wrap each data service in a middleware nutshell that allows the "external world" (e.g., query engines) to query it via an API.

A data service is identified by a URI, it understands HTTP requests and delivers responses that respect the HTTP protocol. By using a HTTP protocol in a RESTful manner, UbiWare enables the integration of data services independently of the operating system and the used programming language [6].

A UbiWare data service exposes an API that permits to apply selection and projection (two basic relational operators) on the produced data. Using these functionalities, the upper layer (query engine) may reduce the data transfer between distributed entities.

For instance, let us consider a particular Kinect, denoted as K_0. At the lower layer of Figure 1, K_0 is a physical hardware component. A Kinect comes with a SDK allowing to develop applications using developer's favourite programming language: it corresponds to the second layer. Then, in order to be detected by upper layers, some small pieces of code have to be written to make them visible by the UbiWare middleware. UbiWare data services are automatically discovered by other network devices (via multicast messages [6]) using attribute naming convention. This functionality facilitates the transparent integration of an unknown number of similar devices in an application.

Step 2: Environment Description: One of the basic requirements of our proposition is to provide a description of the environment, comparable to a set of tables (relations) in classical databases. In our context, such a description is given by set of relations, as depicted in Figure 2. For simplicity, other streams produced by the sonar, Kinect, infra-red or odometer sensors are not shown here. The table Sensors is a classical SQL table and represents the *context* of the environment.

```
CREATE TABLE Sensors(SensorID SERVICE, OfficeID STRING);

CREATE RELATION temperatureDevices (SensorID SERVICE, Value NUMBER, Timestamp LONG)
AS DISCOVER SERVICES PROVIDING
 STREAM Temperature(): (Value NUMBER, Timestamp LONG);

CREATE STREAM temperatureStream (SensorID SERVICE, Value REAL, Timestamp LONG)
AS SELECT * STREAMING UPON INSERTION
 FROM temperatureDevices;

CREATE RELATION moveEddie (
 robotID SERVICE, leftWheelSpeed NUMBER, rightWheelSpeed NUMBER, result STRING)
AS DISCOVER SERVICES PROVIDING
 METHOD moveEddie(leftWheelSpeed NUMBER, rightWheelSpeed REAL) : (result STRING);
```

Fig. 2. Simplified SOCQ schema of our usecase (temperature, Eddie).

The relation temperatureDevices represents all temperature data sources available in the environment, regardless the number of existing devices. The scalability in the number of offices to be monitored is ensured since every data produced by a sensor is seen as a tuple in the stream temperatureStream.

The relation `moveEddie` allow to represent the behaviour of Eddie robots. The actuators are modelled with services and are also seen as relations. The robot Eddie can be moved as a result of move orders sent to the two wheels via the corresponding UbiWare driver. At the reception of an order with two parameters (`leftWheelSpeed`, `rightWheelSpeed`), Eddie moves until another moving command is received. The robot stops at the reception of a move command with the speed corresponding to 0 for each wheel.

Step 3: Integration of Data Processing Services: Besides devices and robots, some additional data processing services can be useful in implementing robot applications. Complex data mining/machine learning algorithms can be integrated in this way. For our usecase, a service which classifies a skeleton posture as usual/unusual is given in Figure 3.

```
CREATE RELATION postureService(ServiceID SERVICE,
  Head_X NUMBER, Head_Y NUMBER, Head_Z NUMBER, ..., Posture STRING)
AS DISCOVER SERVICES PROVIDING
  METHOD getPosture(Head_X NUMBER, Head_Y REAL, Head_Z NUMBER, ...): (Posture STRING);
```

Fig. 3. Integration of a classification service.

This step is mandatory and point out one of the main interest of our approach: to be able to reuse any piece of code implementing complex data mining algorithms.

Step 4: Interaction with the Environment - Application Development: This is the ultimate goal of our approach: interacting with the different robots and devices in order to provide useful functionalities. We argue that declarative continuous queries allow to significantly simplify the development of new applications. Application developers can focus on what they want to do and not on how they have to do it, since it is precisely the job of the query processing module to decide the best way to get answer from a pool of queries. Moreover, robots, devices and services may appear or disappear dynamically without impacting the existing code. The query language proposed by SoCQ is similar to SQL.

Q_1, Q_2. The code for the query Q_1 is given in Figure 4a and is clearly straightforward to write. The result of Q_1 is a single tuple if the conditions hold. The Q_2 query is continuous (Figure 4b), the time window of 3600 seconds is specified in the FROM clause, and it performs a join with the table Sensors. The results of Q_2 are produced continuously whenever the conditions hold.

Q_3. This query requires the usage of the data mining service postureService described in Step 3. For this service, the binding of its variables with concrete values from the skeletons produced by a Kinect is done by the WITH clause with the := operator (Figure 4c).

```
SELECT Value FROM temperatureStream WHERE Value <= 18 AND SensorID="URI3";  (a)

SELECT DISTINCT SensorID, OfficeID FROM Sensors, temperatureStream[3600]    (b)
WHERE Value <= 18 AND Sensors.SensorID = temperatureStream.SensorID;

SELECT SkeletonID, Posture, Timestamp                                       (c)
  FROM skeletonStream skS, postureService poS
  WHERE Posture = "unusual"
  WITH
     poS.Head_X := skS.Head_X, poS.Head_Y := skS.Head_Y,
     poS.Head_Z := skS.Head_Z, poS.Head_State := skS.Head_State,
     ...

SELECT COUNT(*)/(3600*30) AS occupancyPercent                              (d)
  WHERE KinectID = "URL_Kinect_3"
  FROM skeletonStream[3600]
  STREAMING UPON HEARTBEAT;

CREATE STREAM robotSupervision (HeadDistance NUMBER) AS                    (e)
SELECT skeleton.Head_X STREAMING UPON INSERTION
FROM skeleton[1], moveRobot MR
WITH
  MR.leftWheelSpeed := CASE WHEN Head_X > 0.2 THEN -0.1 ELSE 0,
  MR.rightWheelSpeed := CASE WHEN Head_X < -0.2 THEN 0.1 ELSE 0
USING MR.moveEddie;
```

Fig. 4. Q_1, Q_2, Q_3, q_4 and q'_4 queries.

Q_4. This query is more complex and is implemented with two continuous queries, q_4 and q'_4, running concurrently. The first query q_4 does not take into account the robot motion while the second query q'_4 does.

For q_4, the Kinect data service produces data skeletons at a rate of 30 frames/second and we consider that the room is occupied whenever a skeleton is produced. Then, in a time window of 1 hour, we want to compute the percentage of occupancy over 3600*30 possible frames. The results of q_4 are produced continuously whenever new tuples arrive into the system (cf **STREAMING UPON HEARTBEAT** clause).

The proposed solution for q_4 is not necessarily complete: A person can move in an office without being at every moment in front of the mobile robot. Therefore, q'_4 implements a "follow-me" functionality for mobile robots: A skeleton is represented by its 20 characteristic points. Each point has x, y, z coordinates in the $[-1, 1]$ interval. The value 0 corresponds to the center of the field of view. We want to keep a skeleton in a region situated in the front of the Kinect. We use the x coordinate of the point corresponding to the skeleton head and we define this region as $[-0.2; 0.2]$. The goal is to keep the head of the skeleton in this region, and in order to do it we turn continuously the mobile robot.

The relation moveEddie, already described in Figure 2, is connected to the software controlling the mobile robot, permitting to send commands to the left and the right wheels. To trigger the feedback from the current skeletons, q'_4 performs a join between the stream skeletonStream and moveEddie. For every new skeleton, the service is called in order to change the position of the robot. If the head is situated in the "good" region, the robot stops its motion.

5   Conclusions

In this paper, we introduce our first results to devise robot applications using a set of continuous queries defined in a declarative SQL-like language. We have pointed out how such techniques can be used in a principled manner on some simple scenarios. For any SQL-aware developers, the price to be paid to reach such a level of abstraction turns out to be very low: they do not need to worry anymore about low level programming details, specialized implementations using specific languages, subtle coupling of different systems, network protocols and so on. To the best of our knowledge, this paper is the first attempt to bridge the gap between robotics and data management principles.

The main perspective of this paper is to study in depth the challenges posed by multiple continuous query optimization techniques applied to robotics.

References

1. Eddie Robot Platform. http://www.parallax.com/product/28992
2. Kinect for Windows. http://www.microsoft.com/en-us/kinectforwindows/
3. Biggs, G., MacDonald, B.: A survey of robot programming systems. In: Proceedings of the Australasian conference on robotics and automation, pp. 1–3 (2003)
4. Cowley, A., Taylor, C.J.: Stream-oriented robotics programming: the design of roshask. In: IROS, pp. 1048–1054. IEEE (2011)
5. Gerkey, B., Vaughan, R.T., Howard, A.: The player/stage project: tools for multi-robot and distributed sensor systems. In: Proceedings of the 11th international conference on advanced robotics, vol. 1, pp. 317–323 (2003)
6. Gripay, Y., Laforest, F., Lesueur, F., Lumineau, N., Petit, J.-M., Scuturici, V.-M., Sebahi, S., Surdu, S.: Colistrack: testbed for a pervasive environment management system. In: EDBT, pp. 574–577 (2012)
7. Gripay, Y., Laforest, F., Petit, J.-M.: A simple (yet powerful) algebra for pervasive environments. In: EDBT, pp. 359–370 (2010)
8. Horswill, I.D.: Functional programming of behavior-based systems. Autonomous Robots 9(1), 83–93 (2000)
9. Kramer, J., Scheutz, M.: Development environments for autonomous mobile robots: A survey. Autonomous Robots 22(2), 101–132 (2007)
10. Peterson, J., Hager, G.D., Hudak, P.: A language for declarative robotic programming. In: Proceedings of the 1999 IEEE International Conference on Robotics and Automation, vol. 2, pp. 1144–1151. IEEE (1999)
11. Quigley, M., Conley, K., Gerkey, B., Faust, J., Foote, T., Leibs, J., Wheeler, R., Ng, A.Y.: Ros: an open-source robot operating system. In: ICRA workshop on open source software, vol. 3 (2009)
12. Scuturici, V.-M., Surdu, S., Gripay, Y., Petit, J.-M.: Ubiware: web-based dynamic data and service management platform for "AmI". In: 2012 - ACM/IFIP/USENIX 13th International Middleware Conference on Middleware, Montreal, QC, Canada, December 3–7, 2012
13. Surdu, S., Gripay, Y., Scuturici, V.-M., Petit, J.-M.: P-bench: benchmarking in data-centric pervasive application development. In: Hameurlain, A., Küng, J., Wagner, R., Amann, B., Lamarre, P. (eds.) TLDKS XI. LNCS, vol. 8290, pp. 51–75. Springer, Heidelberg (2013)

Database Querying in the Presence
of Suspect Values

Olivier Pivert[1]([✉]) and Henri Prade[2]

[1] University of Rennes 1 – Irisa, Lannion, France
`pivert@enssat.fr`
[2] IRIT – CNRS/University of Toulouse 3, Toulouse, France
`prade@irit.fr`

Abstract. In this paper, we consider the situation where a database may contain suspect values, i.e. precise values whose validity is not certain. We propose a database model based on the notion of possibilistic certainty to deal with such values. The operators of relational algebra are extended in this framework. A very interesting aspect is that queries have the same data complexity as in a classical database context.

1 Introduction

In many application contexts, databases appear to involve suspect values (i.e., values whose validity is dubious), for various reasons: i) some attribute values may have been produced by means of a prediction process, for instance using a technique aimed to estimate null values (in the sense of unknown but applicable), see e.g. [3,4], or ii) the database may result from the integration of multiple (more or less reliable, potentially conflicting) data sources [5], or iii) the database may have gone through an automated cleaning process [11] aimed to remove inconsistencies (and in general there are several ways of restoring consistency, even in simple cases, which is a source of potential errors).

It is of course important to deal with such suspect values with the required cautiousness, in particular when answering queries. A variety of uncertain database models have been proposed to represent and handle uncertain values. In these models, an ill-known attribute value is generally represented by a probability distribution (see, e.g. [7,12]) or a possibility distribution [1], i.e. a set of weighted candidate values. However, in many situations, it may be very problematic to quantify the level of uncertainty attached to the different candidate values. One may not even know the set of (probable/possible) alternative candidates. Then, using a probabilistic model in a rigorous manner appears quite difficult, not to say impossible. In this work, we assume that all one knows is that a given precise value is suspect, i.e. not totally certain, and we show that a database model based on the notion of possibilistic certainty is a suitable tool for representing and handling suspect data. The remainder of the paper is structured as follows. Section 2 briefly presents the three-valued fragment of possibility theory that will be used in our model. Section 3 presents the uncertain database model that we advocate for representing tuples that may involve

© Springer International Publishing Switzerland 2015
T. Morzy et al. (Eds): ADBIS 2015, CCIS 539, pp. 44–51, 2015.
DOI: 10.1007/978-3-319-23201-0_6

suspect attribute values. Section 4 gives the definitions of the algebraic operators in this framework. In Section 5, we discuss a way to make selection queries
more flexible, which makes it possible to discriminate the uncertain answers to a
query. Finally, Section 6 recalls the main contributions and outlines perspectives
for future work.

2   A Fragment of Possibility Theory with Three Certainty Levels

In possibility theory [6,13], each event E — defined as a subset of a universe
Ω — is associated with two measures, its possibility $\Pi(E)$ and its necessity
$N(E)$. Π and N are two dual measures, in the sense that $N(E) = 1 - \Pi(\overline{E})$
(where the overbar denotes complementation). This clearly departs from the
probabilistic situation where $Prob(E) = 1 - Prob(\overline{E})$. So in the probabilistic case,
as soon as you are not certain about E ($Prob(E)$ is small), you become rather
certain about $\overline{E}$ ($Prob(\overline{E})$ is large). This is not at all the situation in possibility
theory, where complete ignorance about E ($E \neq \emptyset$, $E \neq \Omega$) is allowed: This is
represented by $\Pi(E) = \Pi(\overline{E}) = 1$, and thus $N(E) = N(\overline{E}) = 0$. In possibility
theory, being somewhat certain about E ($N(E)$ has a high value) forces you to
have $\overline{E}$ rather impossible ($1 - \Pi$ is impossibility), but it is allowed to have no
certainty neither about E nor about $\overline{E}$. Generally speaking, possibility theory
is oriented towards the representation of epistemic states of information, while
probabilities are deeply linked to the ideas of randomness, and of betting in
case of subjective probability, which both lead to an additive model such that
$Prob(E) = 1 - Prob(\overline{E})$.

   In the following, we assume that the certainty degree associated with the
uncertain events considered (that concern the actual value of an attribute in a
tuple, for instance) is unknown. Thus, we use a fragment of possibility theory
where three values only are used to represent certainty : 1 (completely certain),
α (somewhat certain but not totally), 0 (not at all certain). The fact that one
uses α for every somewhat certain event does not imply that the certainty degree
associated with these events is the same; α is just a conventional symbol that
means "a certainty degree in the open interval (0, 1)". Notice that this corresponds to using three symbols for representing possibility degrees as well: 0, β
($= 1 - \alpha$), and 1 (but we are not interested in qualifying possibility).

3   The Database Model

In the database model introduced in [2] and detailed in [10], a certainty level
is attached to each ill-known attribute value (by default, an attribute value has
certainty 1). For instance, the tuple $\langle 037, John, (40, 0.7), (Engineer, 0.6) \rangle$ denotes
the existence of a person named $John$ for sure, whose age is 40 with certainty 0.7
(which means that the possibility that his age differs from 40 is upper bounded
by $1 - 0.7 = 0.3$ without further information on the respective possibility degrees

of other possible values), and whose job is *Engineer* with certainty 0.6. In the database model we introduce hereafter, the basic idea is also to represent the fact that an attribute value may not be totally certain, but we do not assume available any knowledge about the certainty level attached to a suspect value.

Let us consider a database containing suspect values. In the following, a suspect value will be denoted using a star, as in 17*. A value a^* means that it is somewhat certain (thus completely possible) that a is the actual value of the considered attribute for the considered tuple, but not totally certain (otherwise we would use the notation a instead of a^*).

In the model we propose, we restrict ourselves to the computation of the somewhat certain answers, since dealing with the answers that are only somewhat possible raises important difficulties.

The database model we propose relies on the fragment of possibility theory introduced in Section 2, where three values only are used to quantify certainty: 1 (completely certain), α (somewhat certain but not totally), 0 (not at all certain). The tuples or values that are not at all certain are discarded and do not appear in the database.

Uncertain tuples are denoted by α/t where α has the same meaning as above. α/t means that the existence of the tuple in the considered relation is only somewhat certain (thus, it is also possible to some extent that it does not exist). It is mandatory to have a way to represent such uncertain tuples since some operations of relational algebra (selection, in particular) may generate them. The tuples whose existence is completely certain are denoted by $1/t$. A relation of the model will thus involve an extra column denoted by N, representing the certainty attached to the tuples.

4   Algebraic Operators

In this section, we give the definition of the three main operators (projection, selection, join) of relational algebra in the certainty-based model defined above. We leave the set-oriented operators aside due to space limitation.

4.1   Selection

In the following, we denote by $c(t.A)$ the certainty degree associated with the value of attribute A in tuple t: $c(t.A)$ equals 1 if $t.A$ is a nonsuspect value, and it takes the (conventional) value α otherwise (with the convention $\alpha < 1$). It is the same thing for the certainty degree N associated with a tuple (the notation is then N/t).

Case of a condition of the form $A\,\theta\,q$ where A is an attribute, θ is a comparison operator, and q is a constant:

$$\sigma_{A\,\theta\,q}(r) = \{N'/t \mid N/t \in r \text{ and } t.A\,\theta\,q \text{ and } N' = \min(N,\,c(t.A))\} \qquad (1)$$

Table 1. Relation *Emp* (left), result of the selection query (right)

#id	name	city	job	N
37	John	Newton*	Engineer*	1
53	Mary	Quincy*	Clerk*	1
71	Bill	Boston	Engineer	1

#id	name	city	job	N
37	John	Newton*	Engineer*	α
71	Bill	Boston	Engineer	1

Example 1. Let us consider the relation *Emp* represented in Table 1 (left) and the selection query $\sigma_{job=\text{'Engineer'}}(Emp)$. Its result is represented in Table 1 (right).   ◇

Case of a condition of the form $A_1 \theta A_2$ *where* A_1 *and* A_2 *are two attributes and* θ *is a comparison operator:*

$$\sigma_{A_1 \theta A_2}(r) = \{N'/t \mid N/t \in r \text{ and } t.A_1 \theta t.A_2 \text{ and} \\ N' = \min(N, c(t.A_1), c(t.A_2))\}. \tag{2}$$

Case of a conjunctive condition $\psi = \psi_1 \wedge \ldots \wedge \psi_m$:

$$\sigma_{\psi_1 \wedge \ldots \wedge \psi_m}(r) = \{N'/t \mid N/t \in r \text{ and } \psi_1(t.A_1) \text{ and } \ldots \text{ and } \psi_m(t.A_m) \\ \text{and } N' = \min(N, c(t.A_i), \ldots, c(t.A_m))\}. \tag{3}$$

Case of a disjunctive condition $\psi = \psi_1 \vee \ldots \vee \psi_m$:

$$\sigma_{\psi_1 \vee \ldots \vee \psi_m}(r) = \{N'/t \mid N/t \in r \text{ and } (\psi_1(t.A_1) \text{ or } \ldots \text{ or } \psi_m(t.A_m)) \\ \text{and } N' = \min(N, \max_{i \text{ such that } \psi_i(t.A_i)} (c(t.A_i)))\}. \tag{4}$$

4.2   Projection

Let r be a relation of schema (X, Y). The projection operation is straightforwardly defined as follows:

$$\pi_X(r) = \{N/t.X \mid N/t \in r \text{ and} \\ \not\exists N'/t' \in r \text{ such that } sbs(N'/t'.X, N/t.X)\}.$$

The only difference w.r.t. the definition of the projection in a classical database context concerns duplicate elimination, which is here based on the concept of "possibilistic subsumption". Let $X = \{A_1, \ldots, A_n\}$. The predicate *sbs*, which expresses subsumption, is defined as follows:

$$sbs((N'/t'.X, N/t.X) \equiv \\ \forall i \in \{1, \ldots, n\}, t.A_i = t'.A_i \text{ and} \\ c(t.A_i) \le c(t'.A_i) \text{ and } N \le N' \text{and} \\ ((\exists i \in \{1, \ldots, n\}, c(t.A_i) < c(t'.A_i)) \text{ or } N < N'). \tag{5}$$

Example 2. Let us consider relation *Emp* represented in Table 2 (left) and the projection query $\pi_{\{city, job\}}(Emp)$. Its result is represented in Table 2 (right). ◇

Table 2. Relation *Emp* (left), result of the projection query (right)

#id	name	city	job	N
35	Phil	Newton	Engineer*	1
52	Lisa	Quincy*	Clerk*	α
71	Bill	Newton	Engineer	α
73	Bob	Newton*	Engineer*	α
84	Jack	Quincy*	Clerk	α

city	job	N
Newton	Engineer*	1
Newton	Engineer	α
Quincy*	Clerk	α

4.3 Join

The definition of the join in the context of the model considered is:

$$r_1 \bowtie_{A=B} r_2 = \{ \min(N_1, N_2, c(t_1.A), c(t_2.B))/t_1 \oplus t_2 \mid \\ \exists N_1/t_1 \in r_1, \ \exists N_2/t_2 \in r_2 \text{ such that } t_1.A = t_2.B \tag{6}$$

where $\oplus$ denotes concatenation.

Example 3. Consider the relations from Table 3 (top) and the query:

$$PersLab = Person \bowtie_{Pcity=Lcity} Lab$$

which looks for the pairs (p, l) such that p (somewhat certainly) lives in a city where a research center l is located. Its result appears in Table 3 (bottom).   ◇

Table 3. Relations *Person* (left), *Lab* (right), result of the join query (bottom)

#Pid	Pname	Pcity	N
11	John	Boston*	1
12	Mary	Boston	α
17	Phil	Weston*	α
19	Jane	Weston	1

#Lid	Lname	Lcity	N
21	BERC	Boston*	α
22	IFR	Weston	1
23	AZ	Boston	1

#Pid	Pname	Pcity	#Lid	Lname	Lcity	N
11	John	Boston*	21	BERC	Boston*	α
11	John	Boston*	23	AZ	Boston	α
12	Mary	Boston	21	BERC	Boston*	α
12	Mary	Boston	23	AZ	Boston	α
17	Phil	Weston*	22	IFR	Weston	α
19	Jane	Weston	22	IFR	Weston	1

In the case of a natural join (i.e., an equijoin on all of the attributes common to the two relations), one keeps only one copy of each join attribute in the resulting table. Here, this "merging" keeps the most uncertain value for each join attribute. This behavior is illustrated in Table 4.

Table 4. Result of the natural join query (assuming a common attribute *City*)

#Pid	Pname	City	#Lid	Lname	N
11	John	Boston*	21	BERC	α
11	John	Boston*	23	AZ	α
12	Mary	Boston*	21	BERC	α
12	Mary	Boston	23	AZ	α
17	Phil	Weston*	22	IFR	α
19	Jane	Weston	22	IFR	1

A crucial point is that the join operation does not induce intertuple dependencies in the result, due to the semantics of certainty. This is not the case when a probabilistic or a full possibilistic [1] model is used, and one then has to use a variant of c-tables [8] to handle these dependencies, which implies a non-polynomial complexity. On the other hand, since none of the operators of relational algebra induces intertuple dependencies in our model, the queries have the same data complexity as in a classical database context; see [10] for a more complete discussion.

5   Making Selection Queries More Flexible

If one assumes that the relation concerned by a selection is a base relation (i.e., where all the tuples have a degree $N = 1$), a tuple in the result is uncertain iff it involves at least one suspect value concerned by the selection condition. If such a tuple involves several such suspect values, it will be no more uncertain ($N = \alpha$) than if it involves only one. However, one may find it desirable to distinguish between these situations. For instance, considering the query

$$\sigma_{job='Engineer'\ and\ city='Boston'\ and\ age=30}(Emp)$$

the tuple ⟨John, Engineer*, Boston, 30⟩ could be considered more satisfactory (less risky) than, e.g., ⟨Bill, Engineer*, Boston*, 30⟩, itself more satisfactory than ⟨Paul, Engineer*, Boston*, 30*⟩.

For a selection condition $\psi = \psi_1 \wedge \ldots \psi_m$ and a tuple t, this amounts to saying that "every attribute value (certain and suspect) of t must satisfy the condition ψ_i that concerns it, and the less there are suspect values concerned by a ψ_i in t, the more t is preferred". In other words, the condition becomes:

$\psi_1 \wedge \ldots \wedge \psi_m$ and *as many* $(t.A_1, \ldots, t.A_m)$ *as possible* are totally certain.

In a user-oriented language based on the algebra described above, one may then introduce an operator IS CERTAIN (meaning "is totally certain"), in the same way as there exists an operator IS NULL in SQL.

The fuzzy quantifier [14] *as many as possible* (*amap* for short) corresponds to a function from $[0, 1]$ to $[0, 1]$. Its membership function μ_{amap} is such that:
i) $\mu_{amap}(0) = 0$, ii) $\mu_{amap}(1) = 1$, iii) $\forall x, y, x > y \Rightarrow \mu_{amap}(x) > \mu_{amap}(y)$.
Typically, we shall take $\mu_{amap}(x) = x$.

The selection condition as expressed above is made of two parts: a "value-based one" — that may generate uncertain answers —, and a "representation-based" one that generates gradual answers. A tuple of the result is assigned a satisfaction degree μ (seen as the complement to 1 of a suspicion degree), on top of its certainty degree N. For a conjunctive query made of m atomic conjuncts ψ_i, the degree μ associated with a tuple t is computed as follows:

$$\mu(t) = \mu_{amap}\left(\frac{\sum_{i=1}^{m} certain(t,\ i)}{m}\right) \tag{7}$$

where

$$certain(t,i) = \begin{cases} 1 \text{ if } \psi_i \text{ if of the form } A\ \theta\ q \text{ and } c(t.A) = 1, \\ 1 \text{ if } \psi_i \text{ if of the form } A_1\ \theta\ A_2 \text{ and } \min(c(t.A_1),\ c(t.A_2)) = 1, \\ 0 \text{ otherwise.} \end{cases}$$

In order to display the result of the query, one rank-orders the answers on N first, then on μ (in an increasing way in both cases).

Example 4. Let us consider the relation represented in Table 1 (top) and the selection query $\sigma_\psi(Emp)$ where ψ is the condition

job = 'Engineer' and $city$ = 'Boston' and age > 30 and
$\quad amap\ (job$ IS CERTAIN, $city$ IS CERTAIN, age IS CERTAIN)

Let us assume that the membership function associated with the fuzzy quantifier $amap$ is $\mu_{amap}(x) = x$. The result of the query appears in Table 5 (bottom). ◇

Table 5. Relation Emp (top), result of the selection query (bottom)

#id	name	city	job	age	N
38	John	Boston*	Engineer*	32	1
54	Mary	Quincy*	Engineer*	35	1
72	Bill	Boston	Engineer	40	1
81	Paul	Boston*	Engineer*	31*	1
93	Phil	Boston	Engineer	52*	1

#id	name	city	job	age	N	μ
72	Bill	Boston	Engineer	40	1	1
93	Phil	Boston	Engineer	52*	α	0.67
38	John	Boston*	Engineer*	32	α	0.33
81	Paul	Boston*	Engineer*	31*	α	0

This extended framework, where two degrees (N and μ) are associated with each tuple in the relations, can be easily made compositional. One just has to manage the degrees μ, in the definition of the algebraic operators, as in a gradual (fuzzy) relation context, see [9]. In base relations, it is assumed that $\mu(t) = 1\ \forall t$.

6   Conclusion

In this paper, we have proposed a database model and defined associated algebraic operators for dealing with the situation where some attribute values in a dataset are suspect, i.e., have an uncertain validity, in the absence of further information about the precise levels of uncertainty attached to such suspect values. The framework used is that of possibility theory restricted to a certainty scale made of three levels. It is likely that the idea of putting some kind of tags on suspect values/tuples/answers is as old as information systems. However, the benefit of handling such a symbolic tag in the framework of possibility theory is to provide a rigorous setting for this processing.

A very important point is that the data complexity of all of the algebraic operations is the same as in the classical database case, which makes the approach perfectly tractable. Moreover, the definitions of both the model and the operators are quite simple and do not raise any serious implementation issues.

References

1. Bosc, P., Pivert, O.: About projection-selection-join queries addressed to possibilistic relational databases. IEEE Trans. on Fuzzy Systems **13**(1), 124–139 (2005)
2. Bosc, P., Pivert, O., Prade, H.: A model based on possibilistic certainty levels for incomplete databases. In: Godo, L., Pugliese, A. (eds.) SUM 2009. LNCS, vol. 5785, pp. 80–94. Springer, Heidelberg (2009)
3. Chen, S.M., Chang, S.T.: Estimating null values in relational database systems having negative dependency relationships between attributes. Cybernetics and Systems **40**(2), 146–159 (2009)
4. Beltran, W.C., Jaudoin, H., Pivert, O.: Analogical prediction of null values: the numerical attribute case. In: Manolopoulos, Y., Trajcevski, G., Kon-Popovska, M. (eds.) ADBIS 2014. LNCS, vol. 8716, pp. 323–336. Springer, Heidelberg (2014)
5. Destercke, S., Buche, P., Charnomordic, B.: Evaluating data reliability: An evidential answer with application to a web-enabled data warehouse. IEEE Trans. Knowl. Data Eng. **25**(1), 92–105 (2013)
6. Dubois, D., Prade, H.: Possibility Theory. Plenum, New York (1988)
7. Haas, P.J., Suciu, D.: Special issue on uncertain and probabilistic databases. VLDB J. **18**(5), 987–988 (2009)
8. Imielinski, T., Lipski, W.: Incomplete information in relational databases. J. of the ACM **31**(4), 761–791 (1984)
9. Pivert, O., Bosc, P.: Fuzzy Preference Queries to Relational Databases. Imperial College Press, London (2012)
10. Pivert, O., Prade, H.: A certainty-based model for uncertain databases. IEEE Transactions on Fuzzy Systems (2015) (to appear)
11. Rahm, E., Do, H.H.: Data cleaning: Problems and current approaches. IEEE Data Eng. Bull. **23**(4), 3–13 (2000)
12. Suciu, D., Olteanu, D., Ré, C., Koch, C.: Probabilistic Databases. Synthesis Lectures on Data Management. Morgan & Claypool Publishers (2011)
13. Zadeh, L.: Fuzzy sets as a basis for a theory of possibility. Fuzzy Sets and Systems **1**(1), 3–28 (1978)
14. Zadeh, L.: A computational approach to fuzzy quantifiers in natural languages. Computing and Mathematics with Applications **9**, 149–183 (1983)

Context-Awareness and Viewer Behavior Prediction in Social-TV Recommender Systems: Survey and Challenges

Mariem Bambia[1,2(✉)], Rim Faiz[3], and Mohand Boughanem[2]

[1] LARODEC, ISG, University of Tunis, Le Bardo, Tunisia
[2] IRIT, University of Paul Sabatier, Toulouse, France
{meriam.bambia,bougha}@irit.fr
[3] LARODEC, IHEC, University of Carthage, Carthage Presidency, Tunis, Tunisia
Rim.Faiz@ihec.rnu.tn

Abstract. This paper surveys the landscape of actual personalized TV recommender systems, and introduces challenges on context-awareness and viewer behavior prediction applied to social TV-recommender systems. Real data related to the viewers behaviors and the social context have been picked up in real-time through a social TV platform. We highlighted the future benefits of analyzing viewer behavior and exploiting the social influence on viewers's preferences to improve recommendation in respect with TV contents' change.

Keywords: Social TV-recommender systems · Viewer behavior · Context-awareness · Social influence

1 Introduction

In the past few years, novel challenges are emerging as TV content consumption moves to the multiscreen and device ecosystems (e.g. Tablets and smartphones). Sites like YouTube, Twitter, Facebook, and Netflix have created a large number of multiscreen applications that complement TV viewing experience.

Moreover, viewers are no longer passive elements while watching TV. They can now rate TV programs, comment and suggest them to friends through social TV platforms and social networks. Obviously, by Nielsen's TV and Twitter relationship study [1] reported in 2013, more than 86% of viewers are accessing to Internet while watching TV shows.

However, the rapid growth of channels number, reported by NationMaster Statistics [2], have increased the alternative programs to watch and multiplied the choices of consumers for TV content consumption. Therefore, due to viewers' habits change according to the diversification of televisual contents, viewers are

---

[1] http://www.nielsensocial.com/
[2] http://www.nationmaster.com/country-info/stats/Media/
Television-broadcast-stations

© Springer International Publishing Switzerland 2015
T. Morzy et al. (Eds): ADBIS 2015, CCIS 539, pp. 52–59, 2015.
DOI: 10.1007/978-3-319-23201-0_7

having hard times to decide which program to watch among thousands of choices. Likewise, real-time prediction of viewer's preferences in certain circumstances becomes increasingly hard for TV producers.

Unfortunately, although the wealth of information gathered by social TV consumers, most existing personalized TV systems [e.g. [1], [2] and [3]] did not exploit real-time viewers behavior (e.g. social interactions on TV shows) and viewing context (e.g. social context) to predict their preferences. The proposed approaches used by these systems have also few consideration about solving recommender problems (e.g. no information about a new user known as "cold-start problem") which are the major cause of reducing recommender systems performance.

In this context, the open issue that inspired the research reported in this work is: How to learn viewer behavior change and to exploit the contextual information and the social influence in order to improve TV content recommendation.

The main contribution of this paper is to outline the state-of-the-art on TV recommender systems and to open research issues in the area of context-awareness in order to improve recommendation in social TV realm.

The rest of the paper is structured and organized as follows: Section 2 presents an overview on personalized TV- Recommender Systems. Section 3 discusses issues of the existing research. Section 4 opens relevant challenges on user behavior prediction models and Section 5 concludes the paper.

2   State-of-the-Art on Social TV-Recommender Systems

[4] defined an automatic recommendation scheme of TV program contents in sequence using sequential pattern mining based on user watching history. They proposed a weighted normalized and modified retrieval rank metric for similar user grouping taking into account the watching order and the weights of preferred TV program contents. Then, for each group of users, they defined a sequential pattern (a sequence of TV programs to recommend) which is constructed based on features such as the occurrence of frequently watched TV programs from the targeted users.

In [3], the user model is built by incorporating the time context and other features (i.e. the genre, the sub-genre and the viewing history). A smoothing function is introduced in order to aggregate the user preferences in each time slot with the preferences in the neighbor time slots.

A time-dependent profile technique was proposed by Oh et al. [2]. The construction of this profile is based on splitting each Watch Log into time slots and generating a time-dependent profile for each time slot. Henceforth, when a recommendation is issued, the system finds the corresponding profile based on the time stamp of the request. Unfortunately, this method poses a loss generality problem because some profiles will be totally generated by a specific program, because users are likely to watch the same programs. In consequence, the dependency for a specific program and time may incur the overspecialization issue.

Antonelli et al. [5] proposed a content-based recommender approach using the textual descriptors associated to TV contents extracted from newspaper articles.

They used matrix factorization technique to associate textual descriptors to TV contents.

Martinez et al. [1] introduced a personalized TV program recommendation system. To solve first-rater, cold-start, sparsity and overspecialization problems, they proposed a hybrid approach that combines content-filtering techniques with those based on collaborative filtering and provides advantages of any social networks such as comments, tags and ratings. They used vector space model to generate content-based recommendations. They used SVD (Singular Value Decomposition) to reduce the dimension of the active item's neighborhood, and to execute the item-based filtering with this low rank representation to generate its predictions.

Chang et al. [6] presented a TV program recommender framework integrating TV program content analysis module (e.g. TV program basic content information, watching statistics information, etc.), user profile analysis module (e.g. demographic information, watching histories, preferences) and user preference learning module (e.g. preferences of user implicit and explicit network).

Different architectures of personalized videos recommendation systems proposed in the literature were outlined by the survey presented by Asabere [7]. Likewise, several Social TV offerings and platforms were implemented in last few years (e.g. Netflix, GetGlue, GoogleTV, etc.), which allows the TV experience to move beyond the traditional confines of entertainment into a more holistic media. Obviously, according to "Netflix Challenges" [3], Netflix algorithms draw on the item-based collaborative filtering method and Matrix Factorization method to predict users' preferences.

3 Problem Description

We noticed that there are various problems regarding used recommender approaches that are summarized as following:

3.1 Collaborative Filtering Limitations

Almost of the approaches described above involved collaborative filtering methods in order to deal with content-based filtering issues. For instance, the method proposed by [2] poses a problem of overspecialization, in sense that such a viewer is likely to watch the same programs (i.e. with same categories) at a specific time which might degrade prediction accuracy.

As shown by Chen et al.[8], the usability of collaborative filtering approaches is confined by sparsity problem, since a user does not have enough co-rated TV programs with other users with similar preferences. New shows might also be problematic for collaborative filtering in TV recommender systems because if there are no ratings for a new TV content, thus, it may not be recommended.

Another issue of hybrid (combination of content-based and collaborative filtering recommenders) recommendation system is portfolio effect. In this case, a

---

[3] http://www.netflixprize.com/

viewer may get recommendations for programs being too similar to those he/she already knows [9]. Consequently, switching hybrid (content-based or collaborative filtering process) requires that the system be able to detect when one recommender filtering method should be preferred.

3.2 Viewers' Habits Change

Although the rich information related to viewer habits while watching his favorite TV show and that may be exploited to predict his/her preferences, user behavior prediction remains a difficult task for social TV systems due to the repeating change of his interactions in regard TV contents.

Indeed, it is most likely that the viewer behavior may change over time slots while watching the same TV program. For example, for the first episode of the TV program, the viewer tweeted and liked the show (Behavior 1) during respectively the 1rst and the 6th time slots. However, he/she recommend (Behavior 2) the 3rd episode of the same TV program to a friend.

Unfortunately, several features related to viewing experience, such as during of watching TV contents and users interactions in social networks, were not exploited by TV recommendation approaches.

3.3 Context-Awareness and Social Influence Realm

On the one hand, we are in the presence of different possible contexts that the viewer may experience. However, the contextual information exploited in social TV collaborative filtering systems are restricted to similar rating behavior with other viewers in a static way.

On the other hand, users preferences may change according to the presence of other group members. Obviously, a viewer might like horror movies while being with his friends, whereas, he might prefer comedy when being with his family. Moreover, [10] and [11] proved that a user's friends provide better recommendations than online recommender systems, since friends and family are likely to know his tastes better than any strangers and to have nearly similar interests.

Consequently, an effective Social TV recommender system must exploit the rich environment of viewers, capture all their social behavior, and analyze all the viewing experience and its change. Hence, this involves the need for a real-time recommendation system and the involvement of current viewing context experience.

Accordingly, we establish our key research question in this work: How to learn viewer behavior change and to exploit the contextual information and the social influence in order to improve TV content recommendation?

4 Challenges

In order to highlight the need for the involvement of the current viewing context and the modeling of the social influence on viewer' preferences, we picked up real-time data through Pinhole platform[4]. Pinhole is a Tunisian social TV platform that was created in 2012 and that allows users to watch TV programs, recommend them to their friends, and interact with on social networks (i.e. Facebook and Twitter). The viewings and the interactions of 7070 users on 781 TV programs are recorded in real-time. In practice, we aim to analyze these data which are likely to be valuable for improving the effectiveness of social TV-recommender systems.

However, it is also important to model the viewer behavior and his social interactions (e.g. tweet his preferred TV show or recommend it to friends) while watching TV in order to incorporate the current viewing context (e.g. the current location, the time slot) into the recommendation process and to emphasize the continuous variation of TV programs content.

4.1   Real-Time Viewing Context

In contrast to context-awareness, the concept of personalization usually involves explicitly stated data, mostly in form of a profile, and additional data of lower dynamic nature.

Contexts on all dynamic scales, from long term social relations to short term social situations, must be considered in Social TV area. Obviously, it is particularly important to merge the user short-term (real-time context) and long-term interests (past viewings) into a single user profile and evaluated how the integration of a temporal factor affects the quality of user models in the context of personalized search, as proposed by [12].

In order to predict viewers preference accurately, the entire real-time viewing context must be considered. For instance, on weekdays a user might prefer to watch world news (e.g., CNN or BBC) in the morning, the stock market report on weekends, and movies reviews on weekdays.

We argue that the most frequently used context is time and user location information. Nevertheless, accurate prediction of viewers' preferences undoubtedly depends upon other relevant contextual information provided by social TV platforms such as the viewer emotional behavior, his shared opinions, his interactions on social networks, the weather and the viewing occasion. These real-time contextual information must be exploited and learned in order to recommend programs under particular situations. For example, a TV viewer may like to watch a romantic movie around Valentine's Day.

As observed through the collected data, contextual information are obtained in different ways:

- Explicitly: context capturing relies on manual input from users. Registration modules are often used to capture information on preferred TV content of users (e.g. preferred categories).
- Implicitly: contextual information are captured automatically through the viewing experience features (e.g. the duration of watching a program).

---

[4] http://www.pinhole.tn/

– Inferred: contextual information are obtained by analyzing the user interactions, for instance, using opinion mining methods to estimate his current
mood based on his/her opinions shared while watching TV.

4.2 Social Context

We note that more than 82% of Pinhole users like such a TV show if it was
recommended by or watched with a friend. In consequence, the relevance of TV
shows depends also on the influence and the strong faith among viewers and
their explicit social networks (e.g. friends, family members or colleagues) on
their preferences. In this way, social filtering methods can be seen as a valuable
source of information about users that can benefit TV-recommender systems.
Nevertheless, many studies (such as [10], [13] and [14]) proved that social filtering
(e.g. group recommendation) is an efficient approach to cope with the sparseness
problem in collaborative filtering. This is considerably for taste related domain,
such as TV, cinema and music, which are strongly influenced by friends.

Generally, Social Filtering is associated with the integration of an underlying
social network into recommender system prediction models. Relations between
viewers or between viewers and items can be exploited together with context
approaches for recommending TV programs. The former study of [10] showed
that social filtering approaches work very well in taste related domains by focusing on the significance of the social context.

Otherwise, other studies showed also that in taste domains, users' preferences
are influenced by their social environment. This is mainly due to the fact that
users trust recommendations made by people they know such as their friends [[10]
and [14]]. For instance, while watching TV with a group of friends, some proposed
recommendation will be executed immediately. These social recommendations
may be considered as a significant source to enrich the viewing experience and
predict his preferences.

Social context considers also the crowd around the user. Sometimes a viewer
may invite other people to watch TV content together. In consequence, an effective recommender system must model the social influence on viewer preferences.

Therefore, social interactions around a viewer may influence how a program
is viewed and may reflect how he may interact with. The social influence may
encompass knowledge of social relations relevant for the current circumstances.

In consequence, the viewers behavior that makes up a social context is defined
as a natural resource that enriches the televisual experience. To address this
challenge, a predictive model may be proposed in order to extend the social
context by exploiting the social trust between the viewer and his explicit social
network and the presence of other persons while watching TV and benefit from
his interactions with received friends' recommendations.

4.3 Show's Profile Construction

TV programs content is changing daily. Though being interested with the whole
program, a viewer might not prefer the actual content.

In order to identify the target audience for such new content, profiles should be constructed for TV shows based on the provided real-time content descriptions, viewers' generated contents (e.g. tags and opinions) and the communities information on TV content. Then, it is significant to study the correlation between already seen shows and the intention to watch the same ones again. In this way, community detection (e.g. [15] and [16]) and conversation retrieval [17] techniques aim to find interactions among blogs and social networks which provide valuable and fresh information on new TV content for enriching shows' profiles.

5   Conclusion

This paper presents a survey of some of the researches conducted in the scientific and practical area of social TV-recommender system. This paper recommends the exploitation of rich information available on social TV platforms in order to improve recommendation of TV contents by:

- Introducing current viewing context features improved the recommendation of relevant TV contents in certain circumstances.
- Using social filtering methods, which measure social contextual trends, add better and more reliable results in the user behavior analysis.
- Enriching shows' profiles based on keywords related to viewers interactions enables defining relevant TV contents.

Acknowledgment. The authors would like to thank Mr. Nizar Menzli and Pinhole team for their kind collaboration and offering us the opportunity access to Pinhole Social TV Platform (www.pinhole.tn).

References

1. Barragns-Martnez, A.B., Arias, J.J.P., Vilas, A.F., Duque, J.G., Nores, M.L.: What's on tv tonight? an efficient and effective personalized recommender system of tv programs. IEEE Trans. Consumer Electronics **55**(1), 286–294 (2009)
2. Oh, J., Sung, Y., Kim, J., Humayoun, M., Park, Y.H., Yu, H.: Time-dependent user profiling for tv recommendation. In: Cloud and Green Computing, pp. 783–787 (2012)
3. Turrin, R., Pagano, R., Cremonesi, P., Condorelli, A.: Time-based TV programs prediction. In: 1st Workshop on Recommender Systems for Television and Online Video at ACM RecSys 2014 (2014)
4. Pyo, S., Kim, E., Kim, M.: Automatic and personalized recommendation of tv program contents using sequential pattern mining for smart tv user interaction. Multimedia Syst. **19**(6), 527–542 (2013)
5. Antonelli, F., Francini, G., Geymonat, M., Lepsøy, S.: Dynamictv: a culture-aware recommender. In: Proceedings of the Third ACM Conference on Recommender Systems, RecSys 1909, pp. 257–260. ACM, New York (2009)

6. Chang, N., Irvan, M., Terano, T.: A tv program recommender framework. In: KES 2013, pp. 561–570 (2013)
7. Asabere, N.: A survey of personalized television and video recommender systems and techniques. Information and Communication Technology Research **2**(7), 602–608 (2012)
8. Chen, Y., Wu, C., Xie, M., Guo, X.: Solving the sparsity problem in recommender systems using association retrieval. JCP **6**(9), 1896–1902 (2011)
9. Kuter, U., Golbeck, J.: Using probabilistic confidence models for trust inference in web-based social networks. ACM Trans. Internet Technol. **10**(2), 8:1–8:23 (2010)
10. Groh, G., Birnkammerer, S., Köllhofer, V.: Social recommender systems. In: Pazos Arias, J.J., Fernández Vilas, A., Díaz Redondo, R.P. (eds.) Recommender Systems for the Social Web. ISRL, vol. 32, pp. 3–42. Springer, Heidelberg (2012)
11. Jameson, A.: More than the sum of its members: challenges for group recommender systems. In: Proceedings of the Working Conference on Advanced Visual Interfaces, AVI 2004, pp. 48–54. ACM, New York (2004)
12. Kacem, A., Boughanem, M., Faiz, R.: Time-sensitive user profile for optimizing search personlization. In: Dimitrova, V., Kuflik, T., Chin, D., Ricci, F., Dolog, P., Houben, G.-J. (eds.) UMAP 2014. LNCS, vol. 8538, pp. 111–121. Springer, Heidelberg (2014)
13. Lathia, N., Hailes, S., Capra, L.: Trust-based collaborative filtering. In: Karabulut, Y., Mitchell, J., Herrmann, P., Jensen, C. (eds.) Trust Management II. IFIP, vol. 263, pp. 119–134. Springer, US (2008)
14. Groh, G., Ehmig, C.: Recommendations in taste related domains: collaborative filtering vs. social filtering. In: Proceedings of the 2007 International ACM Conference on Supporting Group Work, GROUP 2007, pp. 127–136. ACM, New York (2007)
15. Ying, J.C., Shi, B.N., Tseng, V.S., Tsai, H.W., Cheng, K.H., Lin, S.C.: Preference-aware community detection for item recommendation. In: 2013 Conference on Technologies and Applications of Artificial Intelligence, pp. 49–54 (2013)
16. Sahebi, S., Cohen, W.: Community-based recommendations: a solution to the cold start problem. In: Workshop on Recommender Systems and the Social Web (RSWEB), Held in Conjunction With ACM RecSys?11, October 2011
17. Belkaroui, R., Faiz, R., Elkhlifi, A.: Using social conversational context for detecting users interactions on microblogging sites. In: EGC 2015, Janvier 27–30, 2015, Luxembourg, pp. 389–394 (2015)

Generalized Bichromatic Homogeneous Vicinity Query Algorithm in Road Network Distance

Yutaka Ohsawa[1]([✉]), Htoo Htoo[1], Naw Jacklin Nyunt[1,2],
and Myint Myint Sein[2]

[1] Graduate School of Science and Engineering, Saitama University, Saitama, Japan
[2] University of Computer Studies, Yangon, Myanmar
ohsawa@mail.saitama-u.ac.jp

Abstract. This paper proposes a bichromatic homogeneous vicinity query method and its efficient algorithm. Several query algorithms have been proposed individually including set k nearest neighbor (NN) query, ordered NN query, bichromatic reverse k NN query, and distance range query. When these types of queries are performed in the road network distance, all take long processing time. The algorithm proposed in this paper gives a unified procedure that can be applied to these queries with a different query condition for each, and it reduces the processing time drastically. The basic idea of the algorithm is to expand the region on the road network gradually while verifying the query condition. It is the most time consuming process, and to improve the deficiency in verification step, an efficient road network distance search method is proposed. Through extensive experiments, the proposed algorithm significantly improves the performance in terms of processing time by nearly two orders of magnitude.

1 Introduction

In geographic information systems (GIS) and location based services (LBS), queries based on geographical proximity among data objects are important. In these types of queries, two types of data sets S and P are concerned, and thus they are called bichromatic queries. S is called a rival object set and P is called an interest object set. For a simple example, when an object q in S is selected, the query retrieves all objects in P whose nearest neighbor is q. This query is called bichromatic reverse nearest neighbor query (BRNN) [1]. The simple way for this query is to make Voronoi region [2] whose generator is q, and to find all data points in P that are included in the Voronoi region. In this paper, these types of queries are called the homogeneous vicinity queries (HVQ).

Fig. 1 shows an example of HVQ for BR3NN. In this figure, background lines show a road network, filled circles show objects belonging to S, white and filled triangles show objects belonging to P. When a query point q in S is specified, BRkNN is defined as follows.

$$BRkNN = \{p \in P | d_N(p, q) \le d_N(p, NN_k(p))\}$$

T. Morzy et al. (Eds): ADBIS 2015, CCIS 539, pp. 60–67, 2015.
DOI: 10.1007/978-3-319-23201-0_8

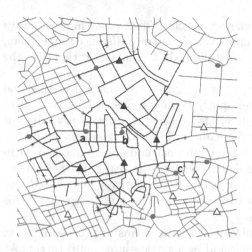

▲   Interest points included in the result

△   Interest points excluded in the result

●   Rival points

Fig. 1. An example of BR3NN HVQ

Here, $d_N(a, b)$ shows the road network distance between a and b, and $NN_k(p)$ shows the kth nearest neighbor of p considering only rival objects in S. In this figure, when q is set on b, b is included in 3NN of filled triangles, and they are included in BR3NN of b vice versa. Contrary, white triangles do not contain b in their 3NN, and hence, they are not members of BR3NN of b.

Several bichromatic queries similar to HVQ have been studied, and an efficient data structure, Voronoi diagram [2], targeting to the vicinity queries has been developed. For example, the region giving the same kNN query result set has been called order-k Voronoi region. In this Voronoi region, the kNN result is without sorted order in distances. On the other hand, the Voronoi region in which the result set in order has been called ordered order-k Voronoi region. To solve a trade area analysis or an optimal facility arrangement problem, the distance between facilities should be measured in road network distances. The result for several types of queries in the road network distance is apt to differ from the Euclidean distance. Therefore, several query algorithms in the road network distance have been proposed and applied to GIS and LBS.

Consequently, various vicinity queries in the road network distance have been emerged, including kNN query [3], RkNN query [4], range query [5], and high order network Voronoi diagram (NVD) [6] in the literature. However, the concept to expand search regions is basically implemented by Dijkstra's algorithm [7] in these queries, and it takes long processing time.

In this paper, a versatile method for several types of queries is introduced and applied to various bichromatic vicinity queries. Based on nearest neighbor queries, three types of vicinity queries are mainly focused. A query that finds all data points in P located in a specified order-k Voronoi region described above is called a set-kNN query in this paper. The other query based on ordered order-k Voronoi region is called an ordered kNN query. As in the example of HVQ for BR3NN, the bichromatic reverse k nearest neighbor (BRkNN) query result can be obtained by generating a region where a specified data point in S is always included in the kNN query result. Alternatively, based on a distance range query, a homogeneous distance range query which finds data points in P that give the same distance query result is studied.

In existing approaches, the search region is gradually enlarged in the road network distance starting from a given query point q by using Dijkstra's algorithm. In this node expansion step, it is also verified whether every adjacent network node to be expanded lies in the same region with the query point (q). In this case, if the verification result is true, adjacent nodes are expanded. Otherwise, adjacent nodes are not expanded to enlarge the region further. When the search region is large or the rival objects set S is sparsely distributed, the verifying process in the road network distance takes long processing time. This causes the performance deterioration.

To improve in the performance of each query in terms of processing time, this paper proposes an enhancement method by a single-source multi-targets A* algorithm (SSMTA*) [8], and introduces an integrated and adaptive framework for several vicinity queries. Moreover, to evaluate the efficiency of the proposed method, extensive experiments were conducted for set kNN query, ordered kNN query, BRkNN query and distance range query. Comparing to existing query approaches, the proposed method has a great efficiency in processing time especially when the rival objects are sparsely distributed.

The remainder of the paper is organized as follow. Section 2 discusses algorithms of the proposed method, and applies the method to various vicinity queries. The performance of the proposed method is evaluated in Section 3. Section 4 concludes this paper and describes future works.

2   Proposed Method

2.1   Generalized Homogeneous Vicinity Query

A homogeneous vicinity query (HVQ) is a unified query method on a road network for different types of queries with different query conditions. HVQ finds objects in a given interest objects set P whose vicinity condition is the same based on generated homogeneous vicinity region (HVR). This type of queries includes set kNN, ordered kNN, BRkNN and homogeneous distance range query. Especially, HVQ for set kNN is denoted by HVQ_s, for ordered kNN by HVQ_o, for BRkNN by HVQ_b, and for homogeneous distance range query by HVQ_r.

The relationship among homogeneous vicinity queries based on kNN queries satisfies the following relationship:

$$HVQ_o \subseteq HVQ_s \subseteq HVQ_b$$

For example, in Fig. 2 (a) and (b), $s1 \sim s4$ belong to a rival objects set S, and $a \sim d$ belong to an interest objects set P. Here, we deal with 2NN of each point. The vicinity query results for Fig. 2 are shown in Table 1.

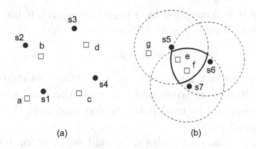

(a)   (b)

Fig. 2. Example of vicinity queries

Table 1. Vicinity Queries

Query type	Result set	Remark
Set 2NN query	$\{s1, s2\}$ for point a and b	result set in any order
Ordered 2NN query	$\{s1, s2\}$ for a, $\{s2, s1\}$ for b	result set in sorted order
BR2NN query	$\{a, b, c\}$	$s1$ as a query point
Distance range query	$\{s5, s6, s7\}$	within HVR for e and f

2.2 Homogeneous Vicinity Region (HVR)

In this paper, a large road network is considered and modeled as a directed graph $G(V, E, W)$, where V is a set of nodes (intersections), E is a set of edges (road segments), and W is a set of edge weights and $w(\in W) \geq 0$ stands. w is assumed the length of edge in the rest of the paper.

We define a homogeneous vicinity region (HVR) as a region that gives the same HVQ result. The basic principle how to generate HVR region for HVQ queries is followed by two steps below:

(1) Gradually expanding the search region from a query point q to adjacent nodes as the similar way in Dijkstra's algorithm.
(2) At every visited node in step (1), verifying whether expanded nodes meet the query condition or not.

In the above step (1), a best-first search is applied from a query point to a current node referring to priority queue (PQ). In the step (2) after verifying whether the current node meets the query condition, all adjacent nodes to the current node are inserted into PQ. Then, the further search proceeds from these nodes ahead. Contrary, if the query condition is not satisfied at the current node, node expansions from this node ahead are terminated. The query condition differs from each query type.

2.3 Basic Method

In this subsection, an algorithm how to generate vicinity region VR(q) is described. The process in this algorithm is to generate a region by expanding the search area gradually while the query condition is satisfied. The search starts at a query point q, and it is controlled by a best-first search using a priority queue (PQ). The record format in PQ is as follow:

$$< c, n, p, \ell(p, n) >$$

Here, n represents a current node, c is the cost (the distance from q to n), p is the previously visited node to n, and $\ell(p, n)$ is the road segment between p and n. The cost is assumed in road network distance, and the length of $\ell(p, n)$ is expressed as $d_N(p, n)$.

At first, records for two end nodes of a road segment on which q exists are inserted into PQ. Then, the record with the minimum cost is dequeued from PQ, and the nearest neighbor of the dequeued element is searched and checked whether the NN of it is q or not. If q is the NN of the dequeued element, it also lies in VR(q) where q is a generator of VR. Contrary, if the result is false, a border point on the current road segment is determined.

The purpose of bichromatic vicinity queries is to retrieve interest objects which are included in the VR(q). The interest objects on a line segment (ℓ) are added to the result set depending on the following two conditions. When the query condition is satisfied at the current node n, the whole ℓ belongs to VR(q), and all $p(\in P)$ on ℓ can be added into the result set. On the other hand, when the query condition is not satisfied at n, ℓ partially belongs to VR(q). Then, each interest object on ℓ is checked, and if it is in VR(q), it is added to the result set.

Algorithm 1 describes the above process in pseudocode.

Algorithm 1. Homogeneous Vicinity Query: HVQ

```
 1: function HVQ(q)
 2:     PQ.enqueue(< d_N(n_1, q), n_1, q, ℓ(n_1, q) >)
 3:     PQ.enqueue(< d_N(n_2, q), n_2, q, ℓ(n_2, q) >)
 4:     CS ← ∅
 5:     R ← ∅
 6:     T ← INITIALSET(q)
 7:     while PQ.size() > 0 do
 8:         r ← PQ.deleteMin()
 9:         if CS contains r then
10:             continue;
11:         end if
12:         CS ← CS ∪ r.ℓ
13:         if VERIFY(r.n,T) then
14:             ns ← AdjacentNode(r.n)
15:             for all n ∈ ns do
16:                 PQ ← PQ∪ < r.c + d_N(r.n, n), n, r.n, r.ℓ >
17:             end for
18:             R ← R ∪ ADDPONL(r.ℓ)
19:         else
20:             R ← R ∪ ADDPONLWITHCHECK(r.ℓ, q, T)
21:         end if
22:     end while
23:     return R
24: end function
```

2.4   Improvement in the Efficiency

The most time consuming step in Algorithm 1 is at the VERIFY procedure called for every node in region expansion. For a visiting node n, kNN candidates of

n are searched in Euclidean distance, and these candidates are verified in the road network distance. To verify in the road network distance, the A* algorithm can be applied. However, in the general HVQ, at least k number of objects are searched as targets at every node n. In such condition, even if A* algorithm is fast in practice, the processing time becomes long due to repeated searching in adjacent regions.

To improve in the efficiency in terms of processing time for the distance calculation in road network distances, the idea of the single- source multi- targets A* (SSMTA*) [8] algorithm has been applied to the proposed method. The original SSMTA* algorithm concurrently finds the shortest paths from a source node to multiple target nodes efficiently. Contrary, when it is applied to HVQ, target points are changed to search sequentially.

The comparison pair-wise A* algorithm (Basic) finds the shortest path from a start point s to a target point n repeatedly every time the target point is changed. Comparing to it, the processing time is considerably reduced in the improved method. Moreover, the update process for all records in the priority queue is taken place in the memory and it does not take long processing time.

2.5   Generalized HVQ

The algorithm for HVQ shown in Algorithm 1 is applicable to a variety of kNN queries including set kNN, ordered kNN and BRkNN. To adapt these queries, three functions, INITIALSET(q), VERIFY and ADDPONLWITHCHECK are needed to prepare for an individual query.

In set kNN query and ordered kNN query, INITIALSET(q) returns kNN rival objects to q as the result. In set kNN query, VERIFY(n,T) returns true if kNN results at n are exactly the same (order omitted) as the objects in the set T. In ordered kNN query, VERIFY returns ture when kNN results for n is exactly the same as the objects in T in order.

In BRkNN query, for INITIALSET(q), 1NN to q in rival objects is searched and returns the rival object (let it be s_q). This means that T holds only s_q. The VERIFY(n,T) returns true if s_q is included in kNN results of n.

The distance range query is also included as a variation of the homogeneous bichromatic vicinity query. In this query, INITIALSET(q) searches the rival objects located in the range whose distance from q is less than or equal to D (the radius of the range), and they are set to T. Distances from the current node n to each object in T are verified by VERIFY.

3   Experimental Results

Extensive experiments were conducted to evaluate the performance of the proposed method by using the real road map. The proposed method is independent of the size of a road map. In this experiment, a road map with 16,284 nodes and 24,914 links is used. We generated variety of rival object sets and interest object sets on the road network links by pseudorandom sequences. In these experiments, *Density* represents the density of objects and 0.01 for *Density* refers that an object exists once 100 road network links. The proposed method (Prop) and the comparison pair-wise A* method (Basic) to calculate road network distance were implemented in Java and evaluated on a PC with Intel Corei7-4770 CPU(3.4 GHz) and 32GB memory.

Fig. 3 (a) shows the processing time of HVQ_s varying density of the objects when k is fixed at 5. The processing time decreases according to the increase of the density of the data objects, because the region size for HVQ_s also decreases. Comparing two methods, the proposed method has 10 to 100 times faster result in the processing time. In Fig. 3 (b), when the density increases, the processing time decreases. This is because the region size monotonically decreases according to the increase of the density as same with HVQ_s.

Fig. 4 (a) compares the processing time varying the density of objects when k is set to 5 for BRkNN query. Fig. 4 (b) shows the processing time for distance range query varying the distance range D. When D increases, the region becomes

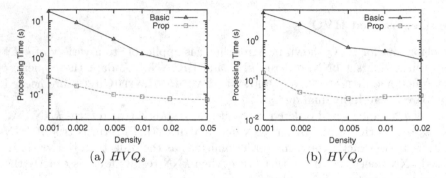

(a) HVQ_s　　　　　　　　　　(b) HVQ_o

Fig. 3. Processing time when k is 5

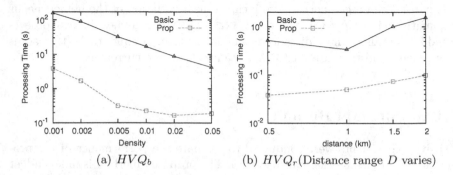

(a) HVQ_b　　　　(b) HVQ_r(Distance range D varies)

Fig. 4. Processing time for HVQ_b and HVQ_r

larger and the basic method takes long processing time. However, the proposed method reduces the processing time to less than 1/10 of the basic method.

4 Conclusion

In this paper, a generalized homogeneous vicinity query algorithm in road network distance was proposed. It is an integrated framework adaptable for several vicinity query methods including set kNN query, ordered kNN query, BRkNN query and distance range query. In the principle of the proposed algorithm, when a query point q is given, the search region is gradually expanded from q to its neighboring road network nodes by verifying these nodes meet the query condition. In verification step, the usual A* algorithm can be applied to calculate road network distances. However, the verification step takes long processing time. To improve the performance, a single source multi-targets A* algorithm was applied for the road network distance verification. With extensive experiments, the proposed method outperformed existing works in terms of processing time. Especially, when the density of distribution of rival objects is sparse, the proposed method is almost 100 times faster than the basic method. To apply the proposed method to various kinds of applications is future works.

Acknowledgments. The present study was partially supported by the Japanese Ministry of Education, Science, Sports and Culture (Grant-in-Aid Scientific Research (C) 15K00147).

References

1. ThaiTran, Q., Taniar, D., Safar, M.: Bichromatic reverse systems neighbor search in mobile systems. IEEE System Journal **4**(2), 230–242 (2010)
2. Okabe, A., Boots, B., Sugihara, K., Chiu, S.N.: Spatial Tessellations, Concepts and Applications of Voronoi Diagrams, 2nd edn. John Wiley & Sons (2000)
3. Kolahdouzan, M., Shahabi, C.: Voronoi-based K nearest neighbor search for spatial network databases. In: Proc. 30th VLDB, pp. 840–851 (2004)
4. Yiu, M.L., Papadias, D., Mamoulis, N., Tao, Y.: Reverse nearest neighbor in large graphs. IEEE Transaction on Knowledge and Data Engineering **18**(4), 1–14 (2006)
5. Cheema, M.A., Brankovic, L., Lin, X., Zhang, W., Wang, W.: Continuos monitoring of distance based range queries. IEEE Transactions on Knowledge and Data Engineering **23**, 1182–1199 (2011)
6. Furuta, T., Suzuki, A., Inakawa, K.: The kth nearest network voronoi diagram and its application to districting problem of ambulance systems. Discussion Paper, No.0501, Center for Management Studies, Nanzan Universitys (2005)
7. Dijkstra, E.W.: A note on two problems in connection with graphs. Numeriche Mathematik **1**, 269–271 (1959)
8. Htoo, H., Ohsawa, Y., Sonehara, N., Sakauchi, M.: Incremental single-source multi target a* algorithm for lbs based on road network distance. IEICE Transactions on Information and Systems **E96-D**(5), 1043–1052 (2013)

OLAP4Tweets: Multidimensional Modeling of Tweets

Maha Ben Kraiem[1,2(✉)], Jamel Feki[1], Kaîs Khrouf[1], Franck Ravat[2], and Olivier Teste[2]

[1] MIR@CL Laboratory, University of Sfax, Airport Road Km 4,
P.O. Box. 1088 3018, Sfax, Tunisia
Maha.benkraiem@yahoo.com, Jamel.Feki@fsegs.rnu.tn,
Khrouf.Kais@isecs.rnu.tn
[2] IRIT, University of Toulouse, 2, Rue Du Doyen Gabriel Marty 31042,
Toulouse Cedex 9, France
{Franck.Ravat,Olivier.Teste}@irit.fr

Abstract. Twitter, a popular microblogging platform, is at the epicenter of the social media explosion, with millions of users being able to create and publish short posts, referred to as tweets, in real time. The application of the OLAP (On-Line Analytical Processing) on large volumes of tweets is a challenge that would allow the extraction of information (especially knowledge) such as user behavior, new emerging issues, trends... In this paper, we pursue a goal of providing a generic multidimensional model dedicated to the OLAP of tweets. The proposed model reflects on some specifics such as recursive references between tweets and calculated attributes.

Keywords: Calculated attribute · Reflexive fact · OLAP model

1 Introduction

Since its introduction in 2006, Twitter has evolved into an extremely popular social network and it has revolutionized the ways of interacting and exchanging information on the Internet. By making its public stream available through a set of APIs, Twitter has triggered a wave of research initiatives aimed at analysing and knowledge discovering from the data about its users and their messaging activities. We notice that the majority of works provided in the literature of this domain (analysis of tweets) are intended to answer specific tasks or needs. However, very few studies were interested in the multidimensional analysis of data from tweets so far. If it incorporates all the data issued from a tweet, this modeling could be a judicious opportunity to explore the tweets through an OLAP process. Tweets can be represented in a multidimensional way by extracting this metadata and social aspect of the messages. For this reason, we focus on the data warehouse as a tool for the storage and analysis of multidimensional and historized data. It thus becomes possible to manipulate a set of measures according to different dimensions which may be provided with one or more hierarchies. OLAP tools provide means to query and to analyze the warehoused information and produce reports at different levels of detail. These reports are computed using aggregate functions. Moreover, data from tweets have particular specificities,

© Springer International Publishing Switzerland 2015
T. Morzy et al. (Eds): ADBIS 2015, CCIS 539, pp. 68–75, 2015.
DOI: 10.1007/978-3-319-23201-0_9

e.g. inter-tweet relationships, calculated attributes. Hence, we consider them during the phase of multidimensional modeling.

The rest of this paper has the following structure. A state of the art of related works to tweets will be presented in Section 2. In Section 3, we propose an extension of our model [1]. In section 4, results and analyses for testing this multidimensional model on data extracted from tweets are presented.

2   Related Work

Twitter has largely contributed to the appearance of new issues related to the modeling and manipulation of data. In this context, the analysis of textual content of tweets and their meta-data is a promised research topic that has attracted the attention of an increasing community of researchers and has given birth to novel analysis areas, such as Social Media Analysis. A spectacular novel area of data analysis is that of the detection of tweets topics. An approach for disambiguating and categorizing the entities in the tweets aimed at discovering topics is described in [2]. The results obtained are used for determining users' topic profiles, and the possibility of analyzing them using OLAP techniques is not considered. The real-time identification of emerging topics in tweets is studied in [3]. Bursty keywords are extracted first, and then grouped to identify trends; however, trends are analyzed using a front-end with limited flexibility.

To the best of our knowledge, few researches have focused on the multidimensional modeling of tweets. Among these works, the one of [4] defined a multidimensional star model for analyzing a large number of tweets. However the proposed model was dedicated to a particular trend. In order to do this, the authors proposed an adapted measure, called "TF-IDF $_{adaptive}$", which identifies the most significant words according to levels of hierarchies of the cube (the location dimension). Nevertheless, their case study deals with a specific area: the evolution of diseases, by adding to their multidimensional model a dimension called 'MotMesh' (MeshWords).

A work sharing some similarities with ours is the one in [5] that presents architecture to extract tweets from Twitter and load them to a data warehouse. Conceptual models for Twitter streams from both OLTP and OLAP points of view are also proposed. However, both models are focused on the inter-relationships between tweets and between users, and little attention is paid to the social aspect of tweets (Tweet/tweet response). In [6] the authors model non-onto and non-strict topic hierarchies as DAGs (Directed Acyclic Graph) of topics. The proposed solution has higher expressivity with respect to traditional hierarchies due to the presence of topic-oriented OLAP operators. A rather similar approach is proposed in the works of [7] where the authors propose Meta-stars to model topic hierarchies in ROLAP systems. Its basic idea is to use meta-modeling coupled with navigation tables and with traditional dimension tables: navigation tables support hierarchy instances with different lengths and with non-leaf facts, and allow different roll-up semantics to be explicitly annotated; meta-modeling enables hierarchy heterogeneity and dynamics to be accommodated; dimension tables are easily integrated with standard business hierarchies.

Further to this study, we may conclude that most of these works ensure a special treatment of tweets but do not offer tools for the decision-makers to manipulate the information contained in the combined meta-data associated with their tweets.

Hence, we aim at providing a generic multidimensional model supporting tweets i.e., independent of the special needs pre-defined a priori while considering the structural specificity and possibly semantic data.

3   Multidimensional Modeling of Tweets

In this section we introduce the multidimensional model dedicated to the OLAP of tweets. We then suggest some extensions to reflect the specificities of data extracted from tweets (Tweet/Tweet-responses and calculated attributes). The concepts of our multidimensional model are already presented in our previous work [1]. We only present the formalism of extended concepts.

We have extended the conventional concept of fact by adding a *reflexive relationship*, denoted $\mathcal{R}$, between *the fact instances* that allows connecting an instance of the fact to one or several instances of the same fact. This relationship will guarantee that every Tweet response added to the table corresponds to an existing Tweet and then analyses of linked tweets are possible. In addition to that, there exist data extracted directly from tweets (raw data) and calculated data. For this reason, we extend the concept of dimension.

- $\forall i \in [1...n]$, **a fact F_i** is defined by $(NameF_i ; M_i ; INS_i ; \mathcal{R}_i)$ where:
  - $NameF_i$ is the name identifying the fact F_i in the constellation,
  - $M_i = \{m_{i1},..., m_{ik}\}$ is a set of k measures of F_i,
  - $INS_i = \{ins_{i1},..., ins_{il}\}$ is the set of l instances of fact F_i,
  - $\mathcal{R}_i : INS_i \rightarrow INS_i$, as $\mathcal{R}_i(INS_i) \subseteq INS_i$.
- $\forall i \in [1...m]$, **a dimension D_i** is defined by $(NameD_i; A_i; A'_i; H_i)$ where:
  - $NameD_i$ is the name identifying the dimension in the constellation,

As a complement to the first proposal, including basic attributes, we extend the definition of the dimension by adding the *calculated attributes*.

  - $A_i = \{a_{i1},..., a_{iz}\}$ is the set of dimension attributes (parameters and weak attributes) extracted from *raw data*,
  - $A^r_i = \{a'_{i1},..., a'_{iq}\}$ is the set of dimension attributes extracted from *calculated data*,
  - $H_i = \{h_1,..., h_{ip}\}$ is the set of p hierarchies showing the arrangement of the attributes of D.

According to their calculation method, we distinguish two types of attributes: *ETL calculated* and *runtime calculated*. Values of ETL calculated attributes are directly computed by the ETL process, i.e., at loading; runtime *calculated* have their values derived at the time of analysis by applying various methods and functions.

ETL Calculated Data
- A new attribute *Tweet-Type[1]*. This helps classifying tweets into one of the following four types: "*Normal-Tweet*" (Every message comprising less than 140

---

[1] https://support.twitter.com/articles/119138-types-of-tweets-and-where-they-appear

characters posted on Twitter); *"Mention"* (Tweets containing the Twitter username of another user, prefixed by the "@" symbol); *"Responses"* (Tweet beginning with the username of another user, and is in response to one of his Tweets); *"Retweet"* (A tweet starting with the symbol RT).

- Use of additional data sources and APIs: Inclusion of external sources provides an opportunity to add new elements to the multidimensional model. Here are some examples of detection techniques relevant for enriching the Twitter data:

    - Language detection (*Tweet-Language*). Language detection APIs, such as the one offered by Google or JSON, provide such service. Once detected, the language information can be used for analysis.

    - Sentiment detection (*Tweet–Sentiment*) assesses the overall emotion of the content of a tweet (such as positive, negative or neutral). AlchemyAPI [9] is an example of platforms enabling this type of analysis.

    - Topic detection (*Tweet -Topic*) enriches the model with topic assignment. Twitter's own SearchAPI or AlchemyAPI can be used to retrieve daily trending topics.

The miscellaneous elements (*Tweet-Type, Tweet-Language, Tweet–Sentiment, Tweet–Topic*) are grouped into a single **Junk Dimension**, named *Tweet-Metadata*. [8] defines a "*Junk dimension as a convenient grouping of flags and attributes to get them out of a fact table into a useful dimensional framework*".

Runtime Calculated Data

- A new hierarchy: We introduce the additional hierarchy User-Category to the USER dimension based on the user's category in terms of the number of that user's followers and friends. There exist different methods dealing with the classification of the Twittos, but most of them agree on the prevailing role of the number of followers and friends. We define formula (1) in order to classify users into three categories.

    $$User\ Category = FollowerCount\ /\ FriendCount \quad (1)$$

    These categories are: i) **Information Seeker:** Person who posts rarely (*User Category < 0.8*), but follows other users' regularly, ii) **Information Sharing:** This user posts tweets frequently (*User Category > 1*) and has a large number of followers due to the valuable contents of his tweets, and iii) **Friendship relationship:** Equivalence between friends and subscribers (*0.8 ≤ User Category ≤ 1*).

- A new hierarchy *User-Activity:* It is added to the *User* dimension and represents the frequency of tweeting (*StatusCount*) relative to the period elapsed since the creation date of the user's account (*TimeDif*). We note that other studies in the literature have adopted the number of retweet to assess the activity of users [5]. However, with the number of retweet, we lose a part of the user activity (i.e., her/his own tweets). We define formula (2) to determine the user activity as

    $$User\ Activity = StatusCount\ /\ TimeDif \quad (2)$$

    This category should classify each user into one of the following four clusters: **"Old-Active"**, **"New-Active"**, **"Old-Passive"**, and **"New-Passive"** respectively for those users who registered long ago or recently and who tweet more or less frequently.

Figure 1 shows the multidimensional model of tweets enriched with these extensions. Graphically, calculated data are represented by dashed lines. The model for tweets presents some specifics. The cardinality 0 of a reflexive fact means that a tweet is not necessarily an answer to another tweet. The second specificity is relative to the possibility of having tweets without any associated locality (absence of the PLACE dimension). This aspect is taken into account by our model. Indeed, we defined a relationship of type 1:0 between the fact Activity-Tweet and the PLACE dimension.

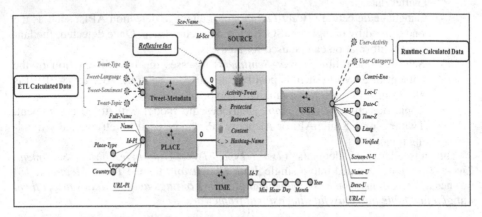

Fig. 1. Multidimensional constellation model dedicated for the OLAP of tweets.

4   Experimental Results

In order to evaluate our approach we have elaborated a software prototype called *OLAP4Tweets,* developed using JAVA and ORACLE 10g database, since it offers a stable database environment. Fig. 2 gives a functional view of the overall warehousing process of tweets, which is composed of four modules namely:

The ***data source*** is represented by the available Twitter APIs for data streaming. The dataset delivered by the Twitter Streaming API is semi-structured data file conform to JSON (JavaScript Object Notation) output format. Each tweet is streamed as an object containing 67 data fields [10].

The ***multidimensional schema design*** module is three steps based. It aims to propose a multidimensional model dedicated to conventional online analytical processing and, in addition, should allow more elaborate treatments of tweets.

The ***ETL*** module takes care of capturing the original data stream, bringing it into a format compliant with the target database and feeding automatically the various components of the multidimensional model (fact, dimensions, and parameters) issued from the tweets by using Hibernate software and Oracle 10g. In fact, we transform the proposed model into R-OLAP logical model according to the transformation rules for the denormalized R-OLAP model. For a reflexive fact, the primary key contains an additional attribute (Id-Activity-Twt) and a foreign key (Id-Activity-Twt-P) which can contain either a *null value,* or *only values from an existing tweet* (Id-Activity-Twt)

The **Querying module**. Once the multidimensional model is generated and loaded with data, the decision maker can perform OLAP analyses on tweets using the OLAP tool offered by the implementation platform (e.g., Oracle Discoverer).

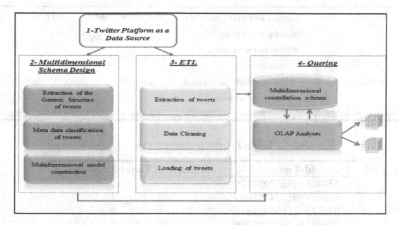

Fig. 2. Functional view of the overall warehousing process.

For our experiments, we consider the dataset obtained for the final show of *Arabs Got Talent* (Arab reality television talent show broadcast by MBC4 TV in the Arab world) which was held on March 7, 2015 at 18:00 GMT. We were able to crawl about half a million tweets encompassing (492.826 tweets) 3 hours starting from the beginning of the final show. These tweets are written in different languages. Notice that among these tweets, only 18570 were associated with a place, and 154,800 are response-tweets.

We start by studying the new type of *reflexive fact*. Figure 3 presents sample query using the reflexive fact Activity-Tweet. Table 1 shows the result of this query.

```
SELECT LEVEL, IDACTIV, CONTENTTWEET, SCREEN_N_U,
   IDACTIVREPONSE
FROM   ACTIVITY_TWEET A1, ACTIVITY_TWEET A2, DUSER D
WHERE  A1.Id-Activity-Twt = A2. Id-Activity-Twt-P
AND   A1.ID_U = D.ID_U
CONNECT BY PRIOR IDACTIV = '560673867825031297'
START WITH IDACTIVREPONSE IS NULL
ORDER BY IDACTIV
```

Fig. 3. Example of reflexive OLAP query

Conversational posts provide the building blocks of the social interaction between users which leads to the development of community, creation of interpersonal relationships, and the perception of reciprocity between Twitter users and their followers. These conversations are based on *Tweet-Response*. A conversation is intense if it is qualified by more than 5 replies. Table 2 shows all detected intense conversation.

Table 1. Result of the query

Level	IDACTIV	CONTENTTWEET	SCREEN-NAME-USER	IDACTIVREPONSE
1	560673867825031297	I would love to see more of this amazing performance! **#MarawaTheAmazing** #ArabsGotTalent	@MahmoudSNasser	NULL
2	560674721235677158	**@MahmoudSNasser** Always energetic performance	@jamietagg	560673867825031297
3	560874826145773641	**@jamietagg** Yes, Beautiful	@Whatweseee	560674721235677158

Table 2. Intense conversation

Id-Tweet	Number of Tweet Responses
530401489252937729	5
530428702124146689	7
530775621556002816	9
530391418645516288	6
530460963141844994	27

A Tweet-Content (*Content*) field must contain some content with a maximum length of 140 characters. This lengthy field is fundamental for the semantic analysis as they deliver valuable information about users and their opinions. We have used *AlchemyAPI* to semantically analyze the Tweet-Content. Table 3 shows the distribution of results for the Sentiment Analysis performed on the dataset of our experiment. Fig. 4 depicts sentiments across the top contestants with a variety of talents during the final show.

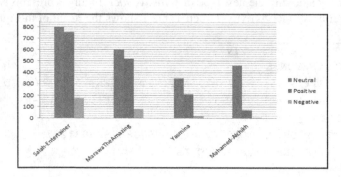

Fig. 4. Sentiment distribution for top contestants tweeted during the final show

Table 3. Sentiment Analysis statistics

Sentiment	Tweet count
Negative	27,858
Neutral	74,247
No Sentiment	324,725
Positive	65,996

5   Conclusion

In this work we have extracted multidimensional data cubes, for OLAP (On-Line Analytical Processing) analyses, from the Twitter social network. We have proposed a multidimensional model dedicated to the OLAP of the data exchanged through tweets and then we have ensured that this model is generic, that is, not limited to a set of pre-determined analytical requirements. Besides, we also have considered the specificities of data tweets: reflexive links between tweets and tweets-response. For that purpose, we have extended the classical concept of multidimensional fact by the proposal of a new type of fact named reflexive fact. We also handled the process of adding new elements such as dimension and hierarchies. Furthermore, our approach was tested on the dataset of the Twitter's public stream with a focus on getting more insight into the content. For this purpose, we presented further examples for sentiment Analysis. We currently continue to perform other OLAP experiments on a larger number of tweets.

Several perspectives for this work are possible. It would be interesting to define new OLAP operators that take into consideration the specificities of this new multi-dimensional model (Reflexive–Fact). We also expect to exploit the "Text Mining" techniques in order to extract knowledge from tweets and strengthen more semantics in the generic model here proposed.

References

1. Ben Kraiem, M., Feki, J., Khrouf, K., Ravat, F., Teste, O.: OLAP of the tweets: from modeling toward exploitation. In: 8th International Conference on Research Challenges in Information Science (IEEE RCIS 2014), Marrakesh, Morocco, May 28–30, 2014, pp. 45–55 (2014)
2. Michelson, M., Macskassy, S.: A. Discovering users' topics of interest on Twitter: a first look. In: Proc. AND, pp. 73–80 (2010)
3. Mathioudakis, M., Koudas, N.: Twittermonitor: trend detection over the twitter stream. In: Proceedings of 2010 International Conference on Management of Data, SIGMOD 2010
4. Bringay, S., Béchet, N., Bouillot, F., Poncelet, P., Roche, M., Teisseire, M.: Towards an on-line analysis of tweets processing. In: Hameurlain, A., Liddle, S.W., Schewe, K.-D., Zhou, X. (eds.) DEXA 2011, Part II. LNCS, vol. 6861, pp. 154–161. Springer, Heidelberg (2011)
5. Rehman, N., Mansmann, S., Weiler, A., Scholl, M.H.: Building a data warehouse for twitter stream exploration. In: ACM Fifteenth International Workshop on Data Warehousing and OLAP, DOLAP 2012
6. Dayal, U., Gupta, C., Castellanos, M., Wang, S., Garcia-Solaco, M.: Of cubes, DAGs and hierarchical correlations: a novel conceptual model for analyzing social media data. In: Atzeni, P., Cheung, D., Ram, S. (eds.) ER 2012 Main Conference 2012. LNCS, vol. 7532, pp. 30–49. Springer, Heidelberg (2012)
7. Gallinucci, E., Golfarelli, M., Rizzi, S.: Meta-stars: multidimensional modeling for social business intelligence. In: Proceeding of ACM International Workshop on Data Warehousing and OLAP, DOLAP, pp. 11–18 (2013)
8. Kimball, R.: The data warehouse toolkit: practical techniques for building dimensional data warehouses. John Wiley & Sons (1996)
9. AlchemyAPI, Alchemyapi:TransformingTextintoKnowledge (2008). http://www.alchemyapi.com

Data Warehouse Design Methods Review: Trends, Challenges and Future Directions for the Healthcare Domain

Christina Khnaisser[1(✉)], Luc Lavoie[1], Hassan Diab[2], and Jean-Francois Ethier[2,3,4]

[1] Département d'informatique, Université de Sherbrooke, Sherbrooke, Canada
{christina.khnaisser,luc.lavoie}@usherbrooke.ca
[2] Centre Intégré Universitaire de Santé et de Service Sociaux de l'Estrie - Centre Hospitalier de Sherbrooke, Sherbrooke, Canada
hdiab.chus@ssss.gouv.qc.ca, ethierj@gmail.com
[3] Département de Médecine, Université de Sherbrooke, Sherbrooke, Canada
[4] INSERM UMR 1138 Team 22 Centre de Recherche des Cordeliers, Université Paris Descartes - Sorbonne Paris Cité, Paris, France

Abstract. In secondary data use context, traditional data warehouse design methods don't address many of today's challenges; particularly in the healthcare domain were semantics plays an essential role to achieve an effective and implementable heterogeneous data integration while satisfying core requirements. Forty papers were selected based on seven core requirements: data integrity, sound temporal schema design, query expressiveness, heterogeneous data integration, knowledge/source evolution integration, traceability and guided automation. Proposed methods were compared based on twenty-two comparison criteria. Analysis of the results shows important trends and challenges, among them (1) a growing number of methods unify knowledge with source structure to obtain a well-defined data warehouse schema built on semantic integration; (2) none of the published methods cover all the core requirements as a whole and (3) their potential in real world is not demonstrated yet.

Keywords: Data warehouse design · Clinical data warehouse · Secondary data use · Medical informatics · Bioinformatics

1 Introduction

Large volumes of heavily fragmented healthcare domain (HD) data are generated every day from several healthcare institutions using different knowledge models and terminologies for the same episode of care. Part of this situation can be explained by the fact that patients will see different care providers in various independent organizations for the same problem. Moreover, different care processes mandate different requirements specific to the specialty and context (e.g. acute care in hospital vs. chronic care by the treating physician in clinic) resulting in heterogeneous data.

© Springer International Publishing Switzerland 2015
T. Morzy et al. (Eds): ADBIS 2015, CCIS 539, pp. 76–87, 2015.
DOI: 10.1007/978-3-319-23201-0_10

Fragmentation must thus be resolved along at least three axes: location, time, and function. The net result is that it is very difficult to have a unified and complete view of a patient's clinical state and history. While each setting may have a very efficient system from a local perspective, not having a complete picture of a patient creates difficulties in providing optimal care, conducting efficient research and managing resources. A data warehouse (DW) is needed to uniformly integrate heterogeneous data from hundreds of independent sources with minimal human resources.

Many DW issues in the HD can be found in other domains. Some of them have been deeply investigated in the DW literature although many proposed solutions are hardly implemented in commercial DW platform. Besides, many issues have been studied independently. Possible incompatibilities or negative interactions between various solutions can be present. Previous surveys [9, 12, 14, 19, 40, 46, 56] do not clearly identify the best methods that suits the HD and some of the comparison criteria used are not well-documented. Furthermore, none of the surveys compare complete methods in the context of a real-world implementation. We took this opportunity to review the scientific literature in order to identify the relevant methods in data warehouse design (DWD). While some end products like I2B2 [36] exist, it is fundamental to first examine the design methods themselves as they will have significant implications in terms of functionality and limitations of the resulting systems. Therefore, in this paper we focus on comprehensive and integrated DWD methods that can be practically implemented in the HD.

The paper is organized as follows: section 2 describes the methodology used to select and compare papers. Section 3 presents interesting points from the evaluation results. Section 4 discusses trends and remaining challenges. Finally, section 5 concludes with open questions and potential research avenues.

2 Study Methodology

The aim of this study is to help identify current DWD methods and ongoing challenges as applicable to the HD. Seven requirements have been defined from clinical data characteristics and used to evaluate methods trends and unresolved requirements. Although none of these requirements are unique to HD, they must be fully satisfied together in order to give the intended services to the HD applications.

2.1 Clinical Data Characteristics

Health care applications range from processing of very low level of data objects (e.g. mass and length) to very higher level of data objects (e.g. patient behavior, organism). Health care data must also be identified in time with multiple degrees of accuracy. Among others, these characteristics raise inevitable special issues and fundamental differences in comparison with many other domains [52].

A clinical DW must contend with three important characteristics of clinical data and its use in the context of secondary analysis of operational data. Firstly, clinical data is tightly coupled in nature and highly dependent on contextual information in order to fully derive its semantics. For example, while "diagnosis" may seem like a straightforward concept, many aspects can, and need to be taken into account to fully understand the nature of a diagnostic code present in a database. Is it: a diagnosis given when a patient was first admitted to the hospital (so it might change as more information becomes available), a discharge (final) diagnosis or a diagnosis entered to justify an investigation? Is it a diagnosis made by a medical student, a resident or an attending physician? Is it an active diagnosis (the patient has a pneumonia), a past diagnosis that is now resolved (the patient had pneumonia 2 years ago), or a diagnosis that was first identified in the past but that is chronic (the patient was first diagnosed with diabetes 10 years ago)? Etc. Many other similar of clinical data include the same level of complexity.

Secondly, as illustrated with the pneumonia/diabetes example above, temporality is a significant challenge with medical data. It covers the entire life of an individual. A bacterial infection at the age of three can have an impact on a heart valve disease identified at the age of fifty-five. There is also substantial uncertainty surrounding a significant part of temporal data. It is common to have a patient report that she or he has had diabetes for "more than ten years" (when in reality, the first diagnosis was 12 years ago but the disease has been present for 16 years). Querying and managing such data is challenging. This is compounded by the concept of "episode of care". For example, if a patient suffers from a major depression episode, she or he will likely see a physician multiple times for that episode. Clinical data will then show multiple entries for "major depression" during that time. Nevertheless, it is really only one episode. Now let's consider that the episode is resolved, but two years later, the patient has another episode and seeks medical attention again. Medical data will show again a "major depression" entry. It is very challenging to reconstruct the timeline for this patient and to decipher how many episodes are represented. Did the patient seek care as a follow-up for the previous episode that was never fully gone or is it a completely new episode? This is just one of the simpler situations. When intertwined with medication timing, investigations (process and results) and other care events, handling of temporality becomes quite complex.

Thirdly, the nature of data and its use for clinical care and research bring specific demands. As opposed to some other domains where requirements can be predefined with users and then implemented, clinical DW must support prospective analysis along axes that evolve rapidly as new knowledge arises.

2.2   Method Requirements

From these characteristics and existing requirements for management activities, we can derive a list of requirements a clinical DWD method must satisfy:

R1 - Data Integrity. The method must preserve (all available) integrity constraints to ensure data quality and correctness [45]. Data in the DW will be used to generate different kinds of reports. Results must be correct and reliable to help different end-users (e.g. managers, cardiologists or researchers). Data needs to be stored in a neutral way as not to hinder use in one context or another.

R2 - Sound Temporal Schema Design. Information variation over time is crucial for most analysis purposes. Having a well-defined temporal schema ensures correct temporal semantic and temporal constraint management. The final DW schema must be based on a sound, comprehensive and formalized temporal model to improve expressiveness and interoperability (like [8] and [11]).

R3 - Query Expressiveness. The final DW schema must simplify the expression of queries, especially temporal ones. This may be reached by automatic generation of views specific to a target problem class expressed in terms of its contributing knowledge elements. It must also be possible to define operators specific to the problem class to facilitate data manipulation (like [50] for OLAP querying).

R4 - Heterogeneous Data Integration. The method must ensure heterogeneous integration of data extracted from multiple sources in a context of high fragmentation. See [3] and [32] for interesting definitions and propositions.

R5A - Knowledge Evolution Integration. The method must provide mechanisms to minimize errors and human resources when integrating knowledge changes. Knowledge is in constant evolution and the DW must cope with it, while maintaining earlier knowledge interpretations and preserving coherent data, correctly represented.

R5B - Source Evolution Integration. The method must cope with new sources integration and structural changes in existing ones with minimal impact on the DW and no impact on the end-user view of the DW (other than the availability of new data and its supporting structure). See [48] for interesting propositions.

R6 – Traceability. The method must keep track of changes in knowledge models, source availability, source structure, schema structure, and designer choices along the DW life cycle. Using mechanisms to coordinate all DWD phases is essential [9]. Traceability helps to assess the impact of structural changes and improve reusability and maintainability [31].

R7 – Guided Automation. To account for the characteristics of clinical data and its fragmentation, DWD must support some degree of automation. The resulting DW scale inevitably calls for automated tools to minimize the resources needed. However, human involvement also remains necessary to handle ambiguous situations. Guided automation is a trade-off, balancing automation and human judgment while facilitating traceability efforts and minimizing errors.

2.3   Comparison Criteria

Twenty-two criteria are defined to compare DWD methods and evaluate the requirements. Some criteria introduced by [57] were extended, including: automation, design approach, requirement and source representation, source analysis, algorithm, conceptual data model, logical data model, physical data model and used tools. Other criteria were added to support requirements assessment [23].

We have defined four classes (to provide case study implementation's scale):

Table 1. Case study categories

Classes	Sources	Relations	Attributes	Tuples
Pedagogical example (PE)	1	3	12	1E+02
Proof of concept (PC)	3	20	100	1E+04
Scale test (ST)	8	1 000	10 000	1E+08
Realistic test (RT)	50	10 000	100 000	1E+11

The intended use is the following: PE for illustration purpose, PC for coverage demonstration, ST for evaluating practical performance at early stages, RT for benchmarking and road test before ongoing a real deployment effort.

The complete list of criteria and their definitions can be found online at http://info.usherbrooke.ca/llavoie/projets/epiiramide/DWDMR

2.4   Literature Selection Process

Throughout the entire process, we retain only papers from year 2000 and up. At first, we targeted general methods (634 papers) with Google scholar, Summon 2.0 and Engineering Village. The final group was chosen based on the inclusion of some automation (i.e. including some kind of potential automation for the creation process). A total of 40 papers were then evaluated: [1, 4–6, 10, 13, 15–18, 21, 22, 24–31, 33–35, 37–39, 41, 43, 45, 47–49, 51, 53, 55, 57–61].

3   Compilation and Results

General observations are presented, based on our results compilation available on the public share [23]. The requirements defined earlier are then reviewed and assessed.

3.1   General Observations

Many methods use a hybrid approach (19/40), 6 among them including a knowledge approach. Since 2010, most methods representing requirements and/or knowledge use ontologies (12/18). Extraction for the source representation and data integration is still mostly manual. The relational model is the most common model to represent sources (8/40), although complete information on sources structural representation is rarely available. Only three methods report significant results based on multiple sources test cases. Dimensional modeling is widely used in DWD (26/40), but relational modeling is also quite present (8/40). If we restrict to temporal DWD, (5/8) are relational, (2/8) are dimensional and (1/8) is entity-attribute-value (EAV). We also notice that most authors don't distinguish between conceptual and logical model and, when they do, they may be using different definitions form one to another. Ontology-based DW are an emerging solution to address data heterogeneity [3]. Few methods used standard data sets (6/40).

3.2   Requirements

R1 - Data Integrity. Data integrity constraints may come from knowledge models (KM) or, occasionally, from the sources themselves (see R4). Moreover, integrity constraints are often encapsulated in applications (not in the database), thereby increasing the complexity of extraction and validation (even in source-driven approach). Only 5 methods propose a hybrid approach including knowledge and source but no method proposes a dual source analysis (structure and data) with explicit integrity verification and validation. Most methods give very few indications on constraints preservation and propagation by their algorithms. As it stands, R1 is partially satisfied.

R2 - Sound Temporal Schema Design. Only 8 methods address the temporal modeling explicitly. One method provides temporal DW schema based on TRM model [11] while others use *ad hoc* models (5/40). None of these methods offer a significant automation level based on knowledge temporal constraints, source temporal structure and source temporal data. Interesting representations are given in [39] and [48]. As it stands, R2 is partially satisfied.

R3 - Query Expressiveness. No method addresses explicitly the issue of query expressiveness. Many of them seem to consider that views directly produced by DM design are adequate. In our experience, they may fulfill some of the managers' needs, but are not adequate when end-users (e.g. care provider or researcher) must be able to query the DW themselves, using multiple and complex knowledge models. As it stands, R3 is not satisfied at the design step.

R4 - Heterogeneous Data Integration. Data integration has received a large attention by the DW community over the last 30 years. Our hypothesis is that data integration must be guided by knowledge and part of the DWD design method. Only 5 methods explicitly cope with multiple sources and only 3 of them have a knowledge representation that can be used to arbitrate the heterogeneity. Only one of them addresses explicitly the ETL process, but more experiments based on a ST class case study are needed to conclude. As it stands, R4 is partially satisfied.

R5A - Knowledge Evolution Integration. No method reports support of knowledge evolution integration. As it stands, R5A is not currently satisfied.

R5B - Source Evolution Integration. Only 3 methods explicitly report support to source evolution integration. No clear indication of the ability to query retrospectively the sources based on a sound temporal model were found. As it stands, R5B is partially satisfied.

R6 – Traceability. Methods [21] and [31] report a convincing traceability approach, at different granularity level, although they don't address explicitly the knowledge representations' changes. Unfortunately, none had linked their framework with a sound temporal model. Finally, more experiments based on a ST class case study are needed. As it stands, R6 is quite fully satisfied.

R7 – Guided Automation. As expected, no methods are fully automatized, neither automatized at a level that will make our project feasible. Some methods perform

quite well on discovering dimensional concepts in sources, guided by user suggestions, others, in generating ETL. Mixing best automation results (regardless of the
compatibility of their methods) won't even be sufficient for source/knowledge evolution processes at least. Methods using model-driven architecture (MDA) approach can
be largely automated but they lack knowledge modeling. As it stands, R7 is partially
satisfied.

3.3   Compilation Summary

Within the 40 evaluated methods, no method covers all the design life cycle. When a
method shows a good level of compliance on one requirement: (1) supporting algorithms need further documentation to be independently implemented; (2) no evidence,
based on an ST class case study, is given that the proposed methods may tackle large
problems (only 2 methods report results on a PC class case study).

We conclude that current papers do not satisfy significantly *R1*, *R3* and *R5A*; partially satisfy *R2*, *R4*, *R5B* and *R7*; quite fully satisfy *R6* in an integrated method.

4   Discussion

Building a DW, taking into account clinical data characteristics and satisfying the
ensuing requirements, is a challenging issue. We will now discuss three fundamental
elements, mainly related to requirements R1 to R5.

4.1   Knowledge vs. Requirements

Secondary use of data for analysis is essential to improve the quality of care and conduct optimal research activities. DW will serve many studies for different health
fields and medical staff. Moreover, with the opportunity to easily access data, new
needs will emerge and existing needs may change. Consequently, DW must contain
all available data regardless the requirements that prevailed at initial DWD. Knowledge seems more useful than requirements to decipher source structure and isolate
interesting data elements to extract. A recent paper [20] presents a semi-automatic
guided method following hybrid requirement/source approach that covers all DWD
life cycle. Using requirements for the DWD in health domain is unfeasible regarding
the complexity and the diversity of end-users, as well as evolving needs. Moreover,
knowledge encapsulated in applications (not in the database) is hardly addressed. To
maximize reusability and extensibility, the "ideal" method should (1) take knowledge
as the basis of the initial design, (2) "easily" integrate knowledge evolution and (3) be
as "requirement neutral" as possible.

4.2   Relational vs. Dimensional

By convention, most DW schema are based on dimensional design model (DDM), although no consensus on its formalism has been established yet [14]. Also, DDM design

relies partly on non-consensual "best-known practices", some of them hardly automatable. Contrariwise, relational design theory (RDT) is algorithmically well defined [7]. DDM is based on fact/dimension dichotomy which is not universal from a problem to another [34]. Furthermore, it relies on processes identification and on requirements that are unknown at DWD time. Even if the processes were all known at design time, DW schema will depend on them, thus any change in the processes may force a change on it. RDT is based on relations and integrity constraints (functional dependencies, referential constraints, temporal constraints, etc.) relying on domain knowledge and sound axioms. DDM schema evolution will be costly and may have a large impact on the whole DW schema. DDM can be used to define known, stable problems using a requirement-driven method to address particular end user's needs. Finally, RDT can be used to define large domains using knowledge-driven approach to ensure maximum consistency and integrity of data.

Data integrity is critical when integrating a large number of data sources. Heterogeneous data integration is complicated by redundancy. Sound integration cannot be done without minimizing redundancy or adding (costly) constraints. RDT minimizes redundancy and guarantees data integrity on a sound and automatable basis. In light of the recent technology evolution, performance issues related to RDT play a much lesser role, if any.

4.3   Temporal Model

Temporal clinical data warehouses are acquiring increasing importance in the health field [2]. Temporal data is important, especially for specifying and detecting clinical phenotypes [42]. In addition, a sound temporal schema plays an essential role in minimizing data incertitude, data indeterminacy and query expressiveness. Current temporal data models [11] and [54] relies on RDT to define design guidelines and constraints regarding temporal representations and constraints. Some methods rely on *ad hoc* models that might work with requirement driven methods but carry limitations. In fact, when applied to a context where prospective operations are not predefined, it becomes essential to have a temporal model which stands on its own, provides intrinsic computability soundness, and gives (automatable) provable transformation rules.

5   Conclusion

In 2006, Rizzi et al. [44] wrote: "Though a lot has been written about how data warehouse should be designed, there is no consensus on a design method yet". This is still valid as none of the evaluated methods cover all the essential requirements, nor was tested in a large-scale implementation.

We presented here a new set of requirements and criteria that can be used to evaluate such methods in the context of clinical DW. This set may be useful in other application domain as well. We also identified certain limitations. Without public standard data sets, it is difficult to measure method efficiency and progress regarding HD.

The specification and creation of such a data set are essential to allow efficient development and evaluation of HD DWD methods.

Another key conclusion of our study is that using domain knowledge is essential to improve relevant data selection and interpretation. It also fosters users' autonomy as they can use data directly through the relevant knowledge representation instead of a requirement driven perspective. As a corollary, methods must tend to unify of source knowledge and domain knowledge, but the optimal knowledge representation method remains elusive at this point in time. In addition, the relational model and a sound temporal model are essential to simplify data queries and management (integrity and evolution).

In conclusion, this review identifies existing gaps between requirements for a fully functional HD DW and existing methods to create one. A large number of independent solutions exist for several requirements, but none of the papers propose a comprehensive and integrated method for the DWD process compliant to the requirements of HD.

References

1. Abelló, A., Martín, C.: A bitemporal storage structure for a corporate data warehouse. In: Proceedings of the 5th International Conference on Enterprise Information Systems, pp. 177–183 (2003)
2. Adlassnig, K.-P., Combi, C., Das, A.K., Keravnou, E.T., Pozzi, G.: Temporal representation and reasoning in medicine: Research directions and challenges. Artif. Intell. Med. **38**(2), 101–113 (2006)
3. Bakhtouchi, A., Bellatreche, L., Jean, S., Yamine, A.-A.: MIRSOFT: mediator for integrating and reconciling sources using ontological functional dependencies. Int. J. Web Grid Serv. **8**(1), 72–110 (2012)
4. Branson, A., Hauer, T., McClatchey, R., Rogulin, D., Shamdasani, J.: A data model for integrating heterogeneous medical data in the Health-e-Child project. Stud. Health Technol. Inform. **138**, 13–23 (2008)
5. Burney, A., Mahmood, N., Ahsan, K.: TempR-PDM: a conceptual temporal relational model for managing patient data. In: Proceedings of the 9th International Conference on Artificial Intelligence, Knowledge Engineering and Data Bases, pp. 237–243. World Scientific and Engineering Academy and Society (WSEAS), Stevens Point (2010)
6. Chute, C.G., Beck, S.A., Fisk, T.B., Mohr, D.N.: The Enterprise Data Trust at Mayo Clinic: a semantically integrated warehouse of biomedical data. J. Am. Med. Inform. Assoc. JAMIA. **17**(2), 131–135 (2010)
7. Codd, E.F.: The Relational Model for Database Management: Version 2. Addison-Wesley Longman Publishing Co., Inc., Boston (1990)
8. Combi, C., Pozzi, G.: HMAP A Temporal Data Model Managing Intervals with Different Granularities and Indeterminacy from Natural Language Sentences. VLDB J. **9**(4), 294–311 (2001)
9. Cravero, A., Sepúlveda, S.: Multidimensional design paradigms for data warehouses: a systematic mapping study. J. Softw. Eng. Appl. **2014**(7), 53–61 (2013)
10. Cravero Leal, A., Mazón, J.N., Trujillo, J.: A business-oriented approach to data warehouse development. Ing. E Investig. **33**(1), 59–65 (2013)

11. Date, C.J., Darwen, H., Lorentzos, N.A.: Time and relational theory: temporal databases in the relational model and SQL. Morgan Kaufmann, Waltham (2014)

12. Elamin, E., Feki, J. : Toward an ontology based approach fro data warehousing. Presented at the The International Arab Conference on Information Technology (ACIT2014) , University of Nizwa, Oman (2014)

13. Giorgini, P., Rizzi, S., Garzetti, M.: GRAnD: A goal-oriented approach to requirement analysis in data warehouses. Decis. Support Syst., 4–21 (2008)

14. Gosain, A., Singh, J.: Conceptual multidimensional modeling for data warehouses: a survey. In: Satapathy, S.C., Biswal, B.N., Udgata, S.K., Mandal, J.K. (eds.) Proc. of the 3rd Int. Conf. on Front. of Intell. Comput. (FICTA) 2014- Vol. 1. AISC, vol. 327, pp. 305–316. Springer, Heidelberg (2015)

15. Hachaichi, Y., Feki, J.: An Automatic Method for the Design of Multidimensional Schemas From Object Oriented Databases. Int. J. Inf. Technol. Decis. Mak. **12**(6), 1223–1259 (2013)

16. Hu, H., Correll, M., Kvecher, L., Osmond, M., Clark, J., Bekhash, A., Schwab, G., Gao, D., Gao, J., Kubatin, V., Shriver, C.D., Hooke, J.A., Maxwell, L.G., Kovatich, A.J., Sheldon, J.G., Liebman, M.N., Mural, R.J.: DW4TR: A Data Warehouse for Translational Research. J. Biomed. Inform. **44**(6), 1004–1019 (2011)

17. Husemann, B., Lechtenbörger, J., Vossen, G.: Conceptual data warehouse design. In: Proceedings of the International Workshop on Design and Management of Data Warehouses, DMDW 2000, pp. 3–9 (2000)

18. Jensen, M.R., Holmgren, T., Pedersen, T.B.: Discovering multidimensional structure in relational data. In: Kambayashi, Y., Mohania, M., Wöß, W. (eds.) DaWaK 2004. LNCS, vol. 3181, pp. 138–148. Springer, Heidelberg (2004)

19. Jindal, R., Taneja, S., et al.: Comparative study of data warehouse design approaches: a survey. Int. J. Database Manag. Syst. **4**(1), 33–45 (2012)

20. Jovanovic, P., Romero, O., Simitsis, A., Abelló, A., Candón, H., Nadal, S.: Quarry: digging up the gems of your data treasury. In: Alonso, G., Geerts, F., Popa, L., Barceló, P., Teubner, J., Ugarte, M., Bussche, J.V. den, and Paredaens, J. (eds.) Proceedings of the 18th International Conference on Extending Database Technology, EDBT 2015, Brussels, Belgium, March 23–27, 2015, pp. 549–552. OpenProceedings.org (2015)

21. Jovanovic, P., Romero, O., Simitsis, A., Abelló, A., Mayorova, D.: A requirement-driven approach to the design and evolution of data warehouses. Inf. Syst. **44**, 94–119 (2014)

22. Kerkri, E.M., Quantin, C., Allaert, F.A., Cottin, Y., Charve, P., Jouanot, F., Yétongnon, K.: An Approach for Integrating Heterogeneous Information Sources in a Medical Data Warehouse. J. Med. Syst. **25**(3), 167–176 (2001)

23. Khnaisser, C., Lavoie, L., Diab, H., Éthier, J.-F.: Data Warehouse Design Methods Review for the Healthcare Domain. http://info.usherbrooke.ca/llavoie/projets/epiiramide

24. Khouri, S., Bellatreche, L., Jean, S., Ait-Ameur, Y.: Requirements driven data warehouse design: we can go further. In: Margaria, T., Steffen, B. (eds.) ISoLA 2014, Part II. LNCS, vol. 8803, pp. 588–603. Springer, Heidelberg (2014)

25. Khouri, S., Boukhari, I., Bellatreche, L., Sardet, E., Jean, S., Baron, M.: Ontology-based structured web data warehouses for sustainable interoperability: requirement modeling, design methodology and tool. Comput. Ind. **63**(8), 799–812 (2012)

26. Krneta, D., Jovanovic, V., Marjanovic, Z.: A direct approach to physical Data Vault design. Comput. Sci. Inf. Syst. **11**(2), 569–599 (2014)

27. Lin, S.-H., Lee, Y.-C.G., Hsu, C.-Y.: Data warehouse approach to build a decision-support platform for orthopedics based on clinical and academic requirements. In: Ślęzak, D., Arslan, T., Fang, W.-C., Song, X., Kim, T.-h. (eds.) BSBT 2009. CCIS, vol. 57, pp. 89–96. Springer, Heidelberg (2009)
28. Lowe, H.J., Ferris, T.A., Hernandez, P.M., Weber, S.C.: STRIDE – an integrated standards-based translational research informatics platform. In: AMIA. Annu. Symp. Proc. 2009, pp. 391–395 (2009)
29. Lujan-Mora, S., Trujillo, J.: Applying the UML and the Unified Process to the design of Data Warehouses. J. Comput. Inf. Syst. 47(5), 30–58 (2006)
30. Malinowski, E., Zimányi, E.: A conceptual solution for representing time in data warehouse dimensions. In: Proceedings of the 3rd Asia-Pacific Conference on Conceptual Modelling. vol. 53, pp. 45–54. Australian Computer Society, Inc., Darlinghurst (2006)
31. Maté, A., Trujillo, J.: Tracing conceptual models' evolution in data warehouses by using the model driven architecture. Comput. Stand. Interfaces 36(5), 831–843 (2014)
32. Mate, S., Köpcke, F., Toddenroth, D., Martin, M., Prokosch, H.-U., Bürkle, T., Ganslandt, T.: Ontology-Based Data Integration between Clinical and Research Systems. PLoS ONE 10, 1 (2015)
33. Mazón, J.-N., Trujillo, J., Lechtenbörger, J.: Reconciling requirement-driven data warehouses with data sources via multidimensional normal forms. Data Knowl. Eng. 63(3), 725–751 (2007)
34. Moreira, J., Cordeiro, K., Campos, M.L., Borges, M.: OntoWarehousing – multidimensional design supported by a foundational ontology: a temporal perspective. In: Bellatreche, L., Mohania, M.K. (eds.) DaWaK 2014. LNCS, vol. 8646, pp. 35–44. Springer, Heidelberg (2014)
35. De Mul, M., Alons, P., van der Velde, P., Konings, I., Bakker, J., Hazelzet, J.: Development of a clinical data warehouse from an intensive care clinical information system. Comput. Methods Programs Biomed. 105(1), 22–30 (2012)
36. Murphy, S.N., Weber, G., Mendis, M., Gainer, V., Chueh, H.C., Churchill, S., Kohane, I.: Serving the enterprise and beyond with informatics for integrating biology and the bedside (i2b2). J. Am. Med. Inform. Assoc. 17(2), 124–130 (2010)
37. Nazri, M.N.M., Noah, S.A., Hamid, Z.: Using lexical ontology for semi-automatic logical data warehouse design. In: Yu, J., Greco, S., Lingras, P., Wang, G., Skowron, A. (eds.) RSKT 2010. LNCS, vol. 6401, pp. 257–264. Springer, Heidelberg (2010)
38. Nebot, V., Berlanga, R.: Building data warehouses with semantic web data. Decis. Support Syst. 52(4), 853–868 (2012)
39. Neil, C.G., De Vincenzi, M.E., Pons, C.F.: Design method for a Historical Data Warehouse, explicit valid time in multidimensional models. Ingeniare Rev. Chil. Ing. 22(2), 218–232 (2014)
40. Pardillo, J., Mazón, J.-N.: Using ontologies for the design of data warehouses. Int. J. Database Manag. Syst. 3, 2 (2011)
41. Phipps, C., Davis, K.C.: Automating Data Warehouse Conceptual Schema Design and Evaluation. Design and Management of Data Warehouses, pp. 23–32. Citeseer (2002)
42. Post, A.R., Kurc, T., Cholleti, S., Gao, J., Lin, X., Bornstein, W., Cantrell, D., Levine, D., Hohmann, S., Saltz, J.H.: The Analytic Information Warehouse (AIW): A platform for analytics using electronic health record data. J. Biomed. Inform. 46(3), 410–424 (2013)
43. Prat, N., Akoka, J., Comyn-Wattiau, I.: A UML-based data warehouse design method. Decis. Support Syst. 42(3), 1449–1473 (2006)

44. Rizzi, S., Abello, A., Lechtenborger, J., Trujillo, J.: Research in data warehouse modeling and design: dead or alive?. In: 9th ACM International Workshop on Data Warehousing and OLAP – DOLAP 2006, held in Conjunction with the ACM 15th Conference on Information and Knowledge Management, CIKM 2006, November 10, 2006–November 10, 2006, pp. 3–10. Association for Computing Machinery, New York (2006)

45. Romero, O., Abelló, A.: A framework for multidimensional design of data warehouses from ontologies. Data Knowl. Eng. **69**(11), 1138–1157 (2010)

46. Romero, O., Abelló, A.: A Survey of Multidimensional Modeling Methodologies. Int. J. Data Warehous. Min. IJDWM. **5**(2), 1–23 (2009)

47. Romero, O., Simitsis, A., Abelló, A.: GEM: requirement-driven generation of ETL and multidimensional conceptual designs. In: Cuzzocrea, A., Dayal, U. (eds.) Data Warehousing and Knowledge Discovery, pp. 80–95. Springer, Berlin Heidelberg (2011)

48. Rönnbäck, L., Regardt, O., Bergholtz, M., Johannesson, P., Wohed, P.: Anchor modeling — Agile information modeling in evolving data environments. Data Knowl. Eng. **69**(12), 1229–1253 (2010)

49. Rubin, D.L., Desser, T.S.: A Data Warehouse for Integrating Radiologic and Pathologic Data. J. Am. Coll. Radiol. **5**(3), 210–217 (2008)

50. Sabaini, A., Zimányi, E., Combi, C.: An OLAP-based approach to modeling and querying granular temporal trends. In: Bellatreche, L., Mohania, M.K. (eds.) DaWaK 2014. LNCS, vol. 8646, pp. 69–77. Springer, Heidelberg (2014)

51. Sahama, T.R., Croll, P.R.: A data warehouse architecture for clinical data warehousing. In: Proceedings of the 5th Australasian Symposium on ACSW Frontiers, pp. 227–232. Australian Computer Society, Inc., Darlinghurst (2007)

52. Shortliffe, E.H., Cimino, J.C. (eds.): Biomedical informatics: computer applications in health care and biomedicine. Springer, London (2014)

53. Sitompul, O.S., Noah, S.A.: A Transformation-oriented Methodology to Knowledge-based Conceptual Data Warehouse Design. J. Comput. Sci. **2**(5), 460–465 (2006)

54. Snodgrass, R.T.: Developing time-oriented database applications in SQL. Morgan Kaufmann Publishers, San Francisco (2000)

55. Song, I.Y., Khare, R., Dai, B.: SAMSTAR: a semi-automated lexical method for generating star schemas from an entity-relationship diagram. In: Proceedings of the ACM Tenth International Workshop on Data Warehousing and OLAP, pp. 9–16. ACM (2007)

56. Tebourski, W., Karâa, W.B.A., Ghezala, H.B.: Semi-automatic Data Warehouse Design methodologies: a survey. Int. J. Comput. Sci. Issues IJCSI. **10**(5), 48 (2013)

57. Thenmozhi, M., Vivekanandan, K.: A Tool for Data Warehouse Multidimensional Schema Design using Ontology. Int. J. Comput. Sci. Issues IJCSI. **10**(2), 161–168 (2013)

58. Di Tria, F., Lefons, E., Tangorra, F.: Hybrid methodology for data warehouse conceptual design by UML schemas. Inf. Softw. Technol. **54**(4), 360–379 (2012)

59. Wisniewski, M.F., Kieszkowski, P., Zagorski, B.M., Trick, W.E., Sommers, M., Weinstein, R.A.: Development of a Clinical Data Warehouse for Hospital Infection Control. J. Am. Med. Inform. Assoc. JAMIA. **10**(5), 454–462 (2003)

60. Zekri, M., Marsit, I., Adellatif, A.: A new data warehouse approach using graph. In: 2011 IEEE 8th International Conference on e-Business Engineering (ICEBE), pp. 65–70. IEEE Computer Society (2011)

61. Zepeda, L., Ceceña, E., Quintero, R., Zatarain, R., Vega, L., Mora, Z., Clemente, G.G.: A MDA tool for data warehouse. In: 2010 International Conference on Computational Science and Its Applications (ICCSA), pp. 261–265 (2010)

Incrementally Maintaining Materialized Temporal Views in Column-Oriented NoSQL Databases with Partial Deltas

Yong Hu$^{(\boxtimes)}$ and Stefan Dessloch

University of Kaiserslautern, Kaiserslautern, Germany
{hu,dessloch}@informatik.uni-kl.de

Abstract. Different from the relational database systems, each column in a column-oriented NoSQL database (CoNoSQLDB) maintains multiple data versions in which each data version is attached with an explicit timestamp (TS). In this paper, we study how to maintain the materialized temporal views (MTVs) in CoNoSQLDBs with partial temporal data-changes (deltas). We first review our previous work and indicate that not all change-data capture (CDC) approaches are able to provide complete deltas. Then, we propose approaches for maintaining MTVs with partial deltas.

Keywords: Column-oriented NoSQL databases · Materialized temporal view maintenance · Partial temporal delta propagation

1 Introduction

Recently, a new type of database systems called column-oriented NoSQL databases (also called column-family stores or extensible record stores in some literatures) emerge. In contrast to the relational database systems (RDBMSs), besides the concepts of table, row and column, column-oriented NoSQL databases (CoNoSQLDBs) introduce a new concept "*Column family*" which indicates all the columns which belong to the same column family are stored contiguously on disk. Each tuple in a CoNoSQLDB table is identified and distributed based on its row key and each column can store multiple data versions in which each data version is attached with an explicit timestamp (TS). Moreover, the life time of each data version can be further restricted by a so called "time-to-live (ttl)" property. Well-known examples are Google's "BigTable" [1], Apache's "HBase" and "Cassandra".

Generally, the value of TS can be either automatically assigned by CoNoSQLDBs (*transaction time* [3]) or manually specified by clients (*valid time* [3]). However, as TS may be internally used by CoNoSQLDBs in some purposes, the latter option is usually not recommended by the CoNoSQLDB community.

Although CoNoSQLDBs can store a version history for each individual column, its implicit TI representation can lead to wrong or misleading temporal query results and hence needs to be translated into the explicit TI representation

© Springer International Publishing Switzerland 2015
T. Morzy et al. (Eds): ADBIS 2015, CCIS 539, pp. 88–96, 2015.
DOI: 10.1007/978-3-319-23201-0_11

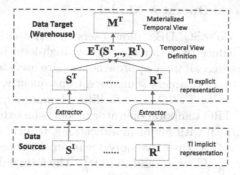

Fig. 1. Architecture of temporal query processing in CoNoSQLDBs

[2]. We utilize the *tuple-timestamping* data model [3] in which each tuple contains an explicit TI and each column contains a single data version. Corresponding examples can be found in [2].

Generally, temporal data processing in CoNoSQLDBs, i.e. temporal data transformation followed by temporal data analysis can be easily implemented as the data warehouse application [4]. The architecture of temporal query processing in CoNoSQLDBs is described in Figure 1. The data sources have the default CoNoSQLDB table representations. The extractors monitor and report the data changes made at the source data and transform the data into the explicit TI representation. Each extracted table in the data warehouse follows the tuple-timestamping model and is utilized for specifying temporal queries.

Often, temporal queries received by the data warehouse (CoNoSQLDB) are periodically repeated and source tables are modified over time. Hence, the results of temporal queries can be materialized to reduce query response time. For performance reason, materialized view maintenance uses the incremental recomputation by only propagating the deltas into materialized temporal views (MTVs) without recomputing the unchanged data.

In the temporal database context, incremental recomputation strategies are widely exploited to maintain MTVs [5,6]. Generally, the *complete delta set* is required, namely, temporal deltas are classified as insertion and deletion where update is modeled as a deletion/insertion pair. However, as already pointed out in [8], data changes which are extracted from the CoNoSQLDBs can be partial, namely, old attribute values may be missing, as the previous state of an updated tuple is unknown, or the type of deltas may be uncertain (e.g. inserted tuples cannot be distinguished from updated ones). In this paper, we address how to maintain MTV with partial deltas in the context of CoNoSQLDBs!

2   Temporal Operators and CDC Approaches

In this section, we review our previous work. We first introduce the TTRO operators [2] which are utilized to specify temporal view definitions in CoNoSQLDBs. Then, we describe feasible CDC approaches and partial temporal deltas [8].

2.1 TTRO Temporal Operator Model

For data processing, CoNoSQLDBs allow users to write either low-level pro-grams, such as MapReduce procedures or utilize high-level languages, such as Pig Latin or Hive. However, all these approaches require users to explicitly imple-ment (or specify) the temporal query semantics. Consequently, we defined a tem-poral operator model called *TTRO* [2] to facilitate temporal query processing in CoNoSQLDBs. The TTRO temporal operator model is an extension of the tem-poral relational algebra [3] with minor modifications. It includes eight temporal operators. However, in this paper, we only focus on the three most commonly used, namely, *Project* π_A^T, *Filter* σ_p^T and *Join* $\bowtie_p^T$. The formal definitions and the corresponding examples can be found in [2].

2.2 Change-Data Capture

Our previous work [8] has extensively studied the Change-data capture (CDC) issues in the context of CoNoSQLDBs and introduces 5 feasible CDC approaches, namely, *Timestamp-based approach, Audit-column approach, Log-based app-roach, Trigger-based approach* and *Snapshot-diffe-rential approach*. As already described in [8], the *Timestamp-based approach* and the *Audit-column approach* can only generate partial deltas where the *Log-based approach* and the *Snapshot-differential approach* may not be able to produce a complete change data history.

We formally defined the partial delta of source table R^I as a five-tuple of sets $(R_{ins}^I, R_{upo/upn}^I, R_{pup}^I, R_{ups}^I, R_{del}^I)$: 1) R_{ins}^I indicates a set of data values inserted into R^I; 2) $R_{upo/upn}^I$ denotes before and after-states of updated tuples; 3) R_{pup}^I denotes partial update in which the before-states of updated tuples are unkown; 4) R_{ups}^I indicates deltas are generated by either insertion or update (*upsert*) and 5) R_{del}^I indicates a set of data values deleted from R^I.

For our temporal warehouse architecture, the above delta model has to be adjusted. As base table R^T in the data warehouse maintains a version history, no current data will be ever deleted or overwritten. In consequence, $R_{upo/upn}^I$ will be represented as insertions with TI-updates. For R_{del}^I, as CoNoSQLDB supports partially delete a tuple, we also represent it as insertions (inserting the remaining values after deletion) with TI-updates. For R_{pup}^I and R_{ups}^I, we keep their original delta types.

To transform the implicit TI into the explicit TI: 1) Each tuple in R_{ins}^I, R_{pup}^I and R_{ups}^I will be attached with $[TS_{op}, \infty)$, where TS_{op} denotes when the corresponding data modifications occurred; 2) For the tuples in R_{del}^I and R_{upn}^I, TI is formed as $[TS_{gen}, TS_{op})$, where TS_{gen} indicates when that tuple is generated and 3) For R_{upo}^I, TI is represented as $[TS_{gen}, \infty)$. In consequence, the partial temporal delta set for warehousing table R^T is represented as a four-tuple of sets $(R_{ins}^T, R_{upo/upn}^T, R_{pup}^T, R_{ups}^T)$.

3   Temporal View Maintenance with Partial Deltas

Generally, traditional view maintenance approaches for both temporal and non-temporal materialized views require complete deltas [5–7], namely, inserted and deleted deltas where update is represented as a deletion/insertion pair. However, as we have already seen in Section 2, the complete delta sets cannot always be generated because of the limitations of CDC approaches or the efficiency requirements for CDC processing.

For non-temporal databases, [9] proposed a view maintenance approach with partial deltas for a specific type of view, namely, a *Dimension View*. A dimension view is a SPJ (Select-Project-Join) view in which each tuple in the view contains a unique identifier and the join predicates should at least reference the primary key of one base table. The author's general idea is to transform partial deltas into complete deltas based on additional table information and reuse traditional view maintenance approaches.

To propagate the partial deltas in the temporal context, three methodologies can be adopted:

1. Translate temporal queries and temporal data (including both temporal base relations and temporal deltas) into their non-temporal counterparts and reuse the approach proposed by [9]. Although this approach can seamlessly integrate with existing database systems (e.g. RDBMSs), transforming temporal queries and temporal data can heavily decrease query performance. Moreover, irrelevant delta propagation cannot be avoided [9].
2. Transform partial temporal deltas into complete temporal deltas by utilizing temporal base relations stored in the data warehouse or external look-up tables, and reuse the existing MTV maintenance approaches (based on temporal operators). Generally, temporal deltas are classified as deleted and inserted deltas. To deduce the MTV maintenance rules, various types of temporal queries need to be distinguished and the corresponding temporal delta contents have to be generated [5].
3. Design specific view maintenance algorithms for directly handling partial deltas (without partial delta transformations). Although this approach can be utilized in non-temporal databases, it is unsuitable for maintaining MTVs. Due to the space limitation, the corresponding examples are shown in [10].

To incrementally maintain MTVs in CoNoSQLDBs with partial deltas, we adopt the second method. However, rather than directly utilizing the existing approaches, we represent the complete temporal deltas as insertion with TI update instead of deletion and insertion. The main advantage of this change is that it can avoid the need of temporal delta transformation [5] and simplify the temporal delta propagation.

In the following sections, we first introduce how the partial temporal deltas can be logically translated into the complete temporal deltas. Then, we describe the temporal view maintenance algorithms along with the physical partial delta transformations based on various temporal operators, such as *Project* π_A^T, *Filter* σ_p^T and *Join* $\bowtie_p^T$. Note that, the auxiliary temporal data information required for physical partial delta transformation depends on both the types of temporal operators and the sorts of predicates (for σ_p^T and $\bowtie_p^T$). We use V^{old} to denote the old state of MTV in the following sections.

3.1 Logically Transforming Partial Temporal Deltas into Complete Temporal Deltas

Recall from the description of change-data capture (CDC) in Section 2.2 that an inserted tuple detected at data source will be treated as an insertion for the base relation in data warehouse. An update and a deletion will be translated into an insertion (D_{ins}) with a TI-update ($D_{upo/upn}$). Partial update means that CDC can only obtain the current state of the updated tuple whereas its previous state is unknown. Hence, we can treat the partial-updated deltas (D_{pup}) as the inserted deltas in which the TI-updated parts are unknown. To logically distinguish this from $D_{upo/upn}$, we notate this missing information as D_{mpup}. To translate upserted deltas (D_{ups}), the same method as for D_{pup} can be used.

Suppose R^T is a base relation in our warehouse and R_Δ^T and R_{UT}^T denote the *inserted* and *TI-updated* deltas of R^T. RD_{ins}, $RD_{upo/upn}$, RD_{pup} and RD_{ups} indicate the *inserted, TI-updated, partial-updated, upserted* deltas extracted by CDC approaches. RD_{mpup} and RD_{mups} represent the *missing TI-updated information* for RD_{pup} and RD_{ups}, respectively. The logical transformation from partial temporal deltas into complete temporal deltas is given as:

- $R_\Delta^T = RD_{ins} \cup RD_{ups} \cup RD_{pup}$;
- $R_{UT}^T = RD_{upo/upn} \cup RD_{mups} \cup RD_{mpup}$.

To physically transform partial deltas into complete deltas, the simplest way is to store all the tuples (T_{lsrk}) of base relation in which each tuple of T_{lsrk} represents the latest version for each distinct srk (row key of the source table) extracted from the data source. However, the main problem of this approach is the additional storage and maintenance costs incurred. In the following sections, we point out the situations in which materializing T_{lsrk} is unnecessary and can hence be avoided. For simplicity, we utilize D_m to denote $D_{mpup} \cup D_{mups}$ in the following sections.

3.2 Temporal Project π_A^T

$\pi_A^T(R)$ returns the desired columns which are denoted as projection attributes A. As the row key column is mandatory for each CoNoSQLDB table, the result of $\pi_A^T(R)$ must implicitly include the row key column even if it is not specified in A. In consequence, all the srk and TI in T_{lsrk} are included in the result table. Hence, we can derive D_{mpup} and D_{mups} by just accessing the result of V^{old}. The extractions of D_{mpup} and D_{mups} are represented as follows:

$V' = (V^{old} \bowtie_p D_{op})$ where p: $V^{old}.srk = D_{op}.srk \wedge overlap(V^{old}.TI, D_{op}.TI)$ and op is *pup* or *ups*:

1. if $V' \neq \emptyset$, $D_{mop} = \{(t'_1, t'_2)|\exists t' \in V', \exists t_i \in D_{op}|(t'_2 = t') \wedge (t'_1.Value = t'.Value) \wedge (t'_1.TI = [t'.TI.Start, t_i.TI.Start)\}$.
2. if $V' = \emptyset$, $D_{mop} = \emptyset$.

V' represents the data sets in V^{old} which are affected by partial deltas D_{ups} and D_{pup} ($\ltimes$ is left semi-join). When $V' \neq \emptyset$ (condition 1), it denotes such D_{mpup} and D_{mups} have been found in which the old state (t'_2) of TI-updated tuple is derived from $t' \in V^{old}$. Otherwise, no partial deltas exist.

The MTVs maintenance procedures of π_A^T are described in table 1. For every inserted tuple $t_{ins} \in R_\Delta^T$, the corresponding inserted tuple for V^{old} will be constructed by fetching the columns referenced in A. To process each TI-updated pair $(t_{tio}, t_{tin}) \in R_{UT}^T$ (t_{tio} and t_{tin} indicate the before and after-states of a TI-updated tuple, respectively), the temporal view maintenance approach will first locate $\pi_A^T(t_{tio})$ in V^{old} and $\pi_A^T(t_{tio}).TI$ is then updated to $t_{tin}.TI$.

Table 1. Procedures for maintaining π_A^T view

Delta_Type	View modifications
R_Δ^T	Insertion
R_{UT}^T	TI update

3.3 Temporal Filter σ_p^T

$\sigma_p^T(R)$ selects the tuples which satisfy the filter condition and discards all others. The select condition p has a form, such as $p_1 \circ ... \circ p_n$ where each p_i ($1 \leq i \leq n$) can be a non-temporal or temporal comparison and $\circ$ represents the logical connective, such as $\wedge$, $\vee$ or $\neg$ or parentheses ().

To extract D_{mpup} and D_{mups}, two different situations need to be distinguished. If the filter predicate has no temporal comparisons, V^{old} contains all the tuples from T_{lsrt} which make p true. In consequence, when the TI modifications incurred to $T_{lsrk} - \sigma_p^T(T_{lsrk})$ (indicate the tuples in T_{lsrk} which do not satisfy p), they have no effect to V^{old}. Hence, we can reuse the D_m extraction processing described for π_A^T. However, when the filter condition contains a temporal predicate, we cannot guarantee $\sigma_p^T(T_{lsrk}) \subseteq V^{old}$. So, the complete T_{lsrk} needs to be stored in the data warehouse. Corresponding examples can be found in [10].

Table 2. Procedures for maintaining σ_p^T views

Delta_Type	Predicate	Exist	View Modifications
R_Δ^T	T	F	Insertion
R_{UT}^T	T	T	TI update
R_{UT}^T	T	F	Insertion
R_{UT}^T	F	T	Deletion

We describe the view maintenance approaches for σ_p^T in table 2. Two "Status" columns, i.e. "Predicate" and "Exist" are utilized to indicate whether the newly generated tuple satisfies the filter predicate and whether the old state of a TI-modified tuple exists in V^{old}. For each inserted tuple $t_{ins} \in R_\Delta^T$, the incremental procedure will apply the filter predicate to t_{ins} and the resulting tuples will be inserted into V^{old}. For TI-updated deltas R_{UT}^T, the evaluation of each tuple pair $(t_{tio}, t_{tin}) \in R_{UT}^T$ can be distinguished in three cases:

1. Both t_{tio} and t_{tin} satisfy the filter predicate and TI of t_{tio} will be updated to TI of t_{tin}.
2. t_{tin} satisfies the filter predicate but t_{tio} not. This condition denotes the old state of updated tuple does not exist in V^{old} and t_{tin} will be inserted into V^{old}.
3. t_{tio} satisfies the filter predicate but t_{tin} not. This condition indicates t_{tio} in V^{old} should be deleted.

3.4   Temporal Join $\bowtie_p^T$

When joining two temporal tables S^T and R^T, tuples in the join result are generated if two joining tuples satisfy the join predicate and are valid during the same period of time. Similar to the discussions made for σ_p^T, the detection of D_{mpup} and D_{mups} for $\bowtie_p^T$ should be also differentiated based on whether the join condition contains temporal or non-temporal predicates.

We describe the view maintenance approaches for $R^T \bowtie_p^T S^T$ in Figure 2. To incrementally refresh the MTV, the unmodified tuples (denoted as S_0 and R_0) from two tables will be involved in the incremental recomputation. For

S^T	R^T	Predicate	Exist	View Modifications
S_0	R_Δ^T	T	F	Insertion
S_0	R_{UT}^T	T	T	TI Update
S_0	R_{UT}^T	T	F	Insertion
S_Δ^T	R_0	T	F	Insertion
S_Δ^T	R_{UT}^T	T	F	Insertion
S_{UT}^T	R_0	T	T	TI update
S_{UT}^T	R_0	T	F	Insertion
S_{UT}^T	R_0	F	T	Deletion
S_{UT}^T	R_Δ^T	T	T	Insertion
S_{UT}^T	R_{UT}^T	T	T	TI update
S_{UT}^T	R_{UT}^T	T	T	TI update
S_{UT}^T	R_{UT}^T	F	T	Deletion

Fig. 2. Procedures for maintaining $\bowtie_p^T$ view

each inserted tuple $s_{ins} \in S_\Delta^T$, s_{ins} joins with the tuples in RD where $RD = R_0 \cup R_\Delta^T \cup R_{UT}^T$. For each TI-updated pair $(s_{tio}, s_{tin}) \in S_{UT}^T$, three cases can be distinguished:

1. $s_{tin} \bowtie_p^T RD \neq \emptyset$ and $s_{tio} \bowtie_p^T RD = \emptyset$. Consequently, inserted tuples will be generated.
2. $s_{tin} \bowtie_p^T RD \neq \emptyset$ and $s_{tio} \bowtie_p^T RD \neq \emptyset$. Hence, TI of the derived tuples in V^{old} will be updated.
3. $s_{tin} \bowtie_p^T RD = \emptyset$ and $s_{tio} \bowtie_p^T RD \neq \emptyset$. In consequence, the tuples derived from s_{tio} will be deleted.

4 Related Work

In the non-temporal context, most view maintenance approaches are proposed based on complete deltas, see survey [7]. For propagating partial deltas, [9] describes the approach for maintaining a specific type of views, namely, *Dimension views* in data warehousing context. The general idea is to transform partial deltas to complete deltas and reuse the existing view maintenance approaches.

In the temporal database context, [5,6] describe the methods to maintain a MTV based on complete temporal delta sets. In [6], temporal queries and temporal data will be translated into their non-temporal counterparts and the view maintenance approaches for the non-temporal databases can be reused. [5] proposes the temporal view maintenance algorithms based on temporal operators. Temporal deltas are classified as insertion and deletion. To deduce the temporal view maintenance rules, various temporal queries need to be differentiated and the corresponding temporal delta contents have to be generated.

To our best knowledge, our work is the first intensive study which addresses the MTV maintenance with partial deltas.

5 Conclusions

In this paper, we study the problem of maintaining materialized temporal views (MTVs) with partial temporal deltas in the context of CoNoSQLDBs. Each tuple in the CoNoSQLDBs is represented by utilizing a tuple-timestamping data model. We briefly review 5 feasible change-data capture approaches and find that not all of them can produce complete delta sets. Consequently, the temporal delta sets extracted from CoNoSQLDBs are classified as *insertion, TI update, partial update* and *upsert*. For propagating partial deltas, we first describe how partial-updated and upserted deltas can be logically translated into the complete delta sets. Then, we describe our partial delta propagation algorithms along with the physical partial delta transformation. Note that, the physical partial delta transformation depends both on the types of temporal operators and the sorts of predicates (for σ_p^T and $\bowtie_p^T$).

References

1. Cange, F., et al.: Bigtable: a distributed storage system for structured data. In: OSDI (2006)
2. Hu, Y., Dessloch, S.: Defining temporal operators for column oriented NoSQL databases. In: Manolopoulos, Y., Trajcevski, G., Kon-Popovska, M. (eds.) ADBIS 2014. LNCS, vol. 8716, pp. 39–55. Springer, Heidelberg (2014)
3. Richard, S.: The TSQL2 Temporal Query Language. Kluwer (1995)
4. Kimbal, R., Kastera, J.: The Data Wareshouse ETL Toolkit. Wiley Publishing (2004)
5. Yang, J., Widom, J.: Maintaining temporal views over non-temporal information sources for data warehousing. In: Schek, H.-J., Saltor, F., Ramos, I., Alonso, G. (eds.) EDBT 1998. LNCS, vol. 1377, p. 389. Springer, Heidelberg (1998)
6. Jensen, C., et al.: Using differential techniques to efficiently support transaction time. VLDB Journal, 75–116 (1993)
7. Gupta, A., et al.: Maintenance of Materialized Views: Problems, Techniques and Applications. IEEE Data Eng. Bull. (1995)
8. Hu, Y., Dessloch, S.: Extracting deltas from column oriented NoSQL databases for different incremental applications and diverse data targets. In: Catania, B., Guerrini, G., Pokorný, J. (eds.) ADBIS 2013. LNCS, vol. 8133, pp. 372–387. Springer, Heidelberg (2013)
9. Joerg, T., Dessloch, S.: View maintenance using partial deltas. In: BTW (2009)
10. Hu, Y.: Incrementally Maintaining Materialized Temporal Views in CoNoSQLDBs with Partial Deltas. Technical report, TU Kaiserslautern (2015)

Towards Self-management in a Distributed Column-Store System

George Chernishev[(✉)]

Saint-Petersburg University, Saint-Petersburg, Russia
chernishev@gmail.com
http://www.math.spbu.ru/user/chernishev/

Abstract. In this paper, we discuss a self-managed distributed column-store system which would adapt its physical design to changing workloads. Architectural novelties of column-stores hold a great promise for construction of an efficient self-managed database. At first, we present a short survey of an existing self-managed systems. Then, we provide some views on the organization of a self-managed distributed column-store system. We discuss its three core components: alerter, reorganization controller and the set of physical design options (actions) available to such a system. We present possible approaches to each of these components and evaluate them. This study is the first step towards a creation of an adaptive distributed column-store system.

Keywords: Column-stores · Column database · Physical design · Self-management · On-line tuning

1 Introduction

The problem of an automatic database tuning is one of the oldest research problems in the database domain. It spans about forty years of evolution and can be described by the following three stages:

1. Separate attempts to automate a selection of various physical design structures: vertical partitioning [31,33,43,46], horizontal partitioning [13,16,17, 51], data allocation [22,27,28,58], indexes [12,30,53] and materialized view selection [45]. These works usually consider only one aspect of a database physical design, they typically provide an algorithm or an idea not tied to a particular database system and do not employ the "what-if" mechanism. Instead, they propose their own cost models. This stage started in the 70-es and largely ceased in the middle of 90-es, when the next approach had appeared.
2. A development of advisors or recommenders. An advisor is a tool which recommends some actions concerning database physical design using the knowledge of a data schema and a workload. This approach is characterized by the integration with some database system and the use of its query

© Springer International Publishing Switzerland 2015
T. Morzy et al. (Eds): ADBIS 2015, CCIS 539, pp. 97–107, 2015.
DOI: 10.1007/978-3-319-23201-0_12

optimizer in the "what-if" mode. Also, the advisors often consider several physical design options simultaneously. Finally, unlike the previous class, these are not only algorithms, but full-fledged tools. They are aimed for the industrial application, for the industrial end user. One of the first such tools was REDWAR [59] tool for database analysis. Later, there was AutoAdmin [18] tool for index recommendation, DB2 advisor [49,57,60,61] (index, materialized view, partitions, allocation), Oracle Advisor [23,25] (index, materialized view, partitioning support), various advisors for Microsoft's products [5,6,8,14,47] (variety of physical structures), PostgreSQL [10,26,44], Ingres [56]. This stage started in the end of 80-es and continues up to this day.

3. A self-management approach. In the majority of previous studies, it was the administrator who made the decision whether to apply or not the proposed database reorganization. Self-management approach aims for the complete elimination of a human intervention into the database tuning cycle. To the best of our knowledge, the first studies of this type appeared about ten years ago [15,52]. It is essential to note that several systems (or their parts) mentioned above can be also considered as the self-management ones.

A comprehensive survey of the automatic physical design tuning, listing a large amount of works involving all the aforementioned types, can be found in reference [21].

The self-management technology aims to automate to the largest possible extent the tasks of deploying, configuration, administration, monitoring and tuning of a database system [19]. The reason for the interest to self-managing systems is simple: it is widely accepted that the contribution of these tasks into the TCO (Total Cost of Ownership) of a database system is high. Some reports [6,19,24,38] indicate that TCO of a database system is dominated by the human-related expenses, while the others [4] claim that it is not true for very large scale systems.

In this paper, we discuss the self-management aspects of a distributed relational column-store database. A column-store database is a database which keeps each attribute in a separate file, unlike classical row-stores, which use slotted page layout [9]. The different storage model leads to different query processing schemes and subsequently to a different set of physical design options. Recently several such systems [2,39] emerged and quickly gained popularity in both academy and industry communities due to their exceptional performance on the read-only workloads. The unique properties of the column-stores hold a great promise for a construction of an efficient self-managing system. In this paper, we evaluate the prospects of a self-managed distributed column-store system development.

We propose the overall scheme of such system which includes alerter, reorganization controller and the list of possible physical design options (actions). We start with the description of the environment and the alerter component. Next we present two possible design approaches for the controller construction. Finally, we discuss the novel physical design options provided by the storage model, assess their impact on the physical design and meditate on the automating of their selection.

We assume that the reader is familiar with the basics of column-store technology. Otherwise we suggest the following surveys [2,20].

To the best of our knowledge, this is the first work to consider on-line physical design tuning in a distributed column-store database.

2 Database (Self-)Tuning Basics

A behavior of a database system can be described by the following formula [19]:

$$f : configuration \times workload \rightarrow performance$$

Here, the system configuration consists of a hardware setup (characteristics of the used hardware), software setup (boot-time parameters), database physical design and so on [19]. The workload characteristics include data schema information, query information (frequencies, attributes involved) and their arrival patterns. The performance is represented by one of the performance metrics (response time, throughput, reliability etc.) or their combination.

Having a model involving all of these components one may try to find the best configuration, the configuration maximizing a given performance metric [19]. In this paper we are interested solely in the physical design aspect of the configuration component. If we restrict configuration to a set of physical design structures we will get the general formulation of the physical database tuning problem. Finding the best configuration even for a single type of structures is usually an NP-hard problem [32,43,48,51]. Thus, a heuristic algorithm is required.

There are three popular methods for this kind of problems: the integer programming [10,33], a general heuristic algorithm [31,46] and a domain-specific heuristic algorithm [8,47,49]. The latter is usually tightly coupled with a database system and relies on a "what-if" calls.

Most of the works up to 2005 considered the physical design problem as a static problem, i.e. the workload cannot change after the selection of a configuration. However, a self-tuning system should be adaptive to the workload, thus a "dynamization" of the algorithm is needed (usually the selection algorithm is too expensive to be called multiple times at will).

Here comes into play an observe-predict-react cycle [19]. It is a general framework for construction of an adaptive database systems. The observe component monitors the specified workload characteristics, like time it takes to process a given query or a cardinality of a given relation. The predict component is used to assess performance of the current configuration in the near-future and to compare it with the possible configurations. The react component is engaged when the current configuration is found unfit and a new configuration should be selected.

There are several dimensions of a self-managed database [34]: Self-Configuration, Self-Optimization, Self-Healing and Self-Protection. Automatic physical design tuning fits into the first three of them. These dimensions were partially implemented in systems of the past, but no fully self-managed system yet exists.

Now, let's study contemporary systems featuring self-management components.

3 Related Work

In the recent years, there was a number of prototypes, which employed the self-tuning approach for the physical configuration. COLT (Continuous On-Line Tuning) [52] is a framework which adjusts the system configuration in order to maximize query performance with respect to the active query set. The proposed approach is to select the most beneficial indexes taking into account a storage budget. Authors implemented it using PostgreSQL database system.

A reference [15] describes an alerter component of a self-tuning system. This component periodically checks whether there is a configuration which will result in a performance improvement. The alerter produces the lower and upper bounds of the improvement if a tuning component is run. Alerter component is designed for indexes, but authors also describe its application for materialized views.

A reference [40] describes on-line physical design tuning in a multistore system. A multistore system is a system which encompasses several different data stores (HDFS and RDBMS) and allows for simultaneous querying of data kept in all types of the stores. Different types of data are stored in different systems, for example, HDFS can be used to keep big log files and RDBMS may contain analytical business data. The building block of the proposed physical design is a so-called opportunistic materialized view, which is the by-product of a query processing on a Hadoop-based system. The system automatically adapts to the dynamically changing workload. In order to achieve this authors tune the set of opportunistic materialized views in each store and solve the data placement problem. The same opportunistic materialized views are used for tuning of UDFs (user defined functions) in the reference [41].

COLT was used in an on-line tuning tool for PostgreSQL described in the reference [10]. This tool is capable of recommending both indexes and partitions using a unified model and is capable of adapting to changes in the workload. Additionally, the tool recommends a beneficial order of index materialization. Several other interesting algorithms and techniques are incorporated in this tool.

An on-line vertical and horizontal partitioning is considered in the reference [37]. Authors employ the idea of attribute affinity used in the works of 80-90es (e.g. [43, 46]) and "dynamize" it. The result is called AutoStore, an automatically and online partitioned database store.

H2O database system [11] proposes an on-line data reorganization with on-the-fly query compilation. The data reorganization is represented by a change of vertical partitioning schemes ranging from the row-store to the column-store.

SMOPD [42] uses closed item sets mining to perform an on-line vertical partitioning of a set of tables.

A reference [50] describes an on-line physical design tuning for in-memory database SAP HANA. This database is designed to handle both transactional and analytical workloads. The goal of the proposed physical design tuning is to select the more beneficial table storage mode: a column-store or a row-store. In order to provide such recommendations authors developed a cost-based model. The next idea is the store-aware partitioning which is as follows: split a table into different parts and keep them in different stores.

Vertica is a commercial distributed column-store database. It has an automatic physical design component [39] which helps to select a set of projections for a given storage budget. However it is not described in detail and to the best of our knowledge, it is not an on-line tool.

There are many more approaches, which involve physical self-tuning, but we are limited by the space to describe them all.

4   The Design of a Self-managed Distributed Column-Store System

The General Idea and the Alerter Component. We propose to construct a distributed adaptive column-store system which uses a single column as a minimum unit of data storage. We consider to start from the classical formulation of the physical design tuning problem [19]: given a workload, data scheme, available hardware and, possibly, a set of user constraints (e.g. storage bound of each node) find a configuration consisting of physical design structures, which maximizes the throughput of the system.

Environment. We consider the following environment. There is a set of nodes in a distributed column-store system. Each of them has its own hardware characteristics: the available disk space, processing power, network link capacity and so on. Each node stores a number of columns or their parts and is capable of performing not only scans, but complex operators like joins and aggregations.

Queries. The queries and their characteristics (frequencies, involved attributes, selectivities of their predicates) are used to control the reorganization.

Alerter. The first problem is what we should adapt to, what kind of information should use the alerter component. There are several possible approaches:

- The most common approach is to use a sliding window, which helps to keep track of the recently processed queries. These recent queries allow us to compute some aggregate characteristics, which are then used to decide whether to trigger reorganization or not.
- Another approach is to use the knowledge of some pre-defined query patterns. For example, we might know the existence of query patterns like it is done in the reference [7]. We can detect these patterns or rely on some external knowledge (e.g. we know that data loading happens only at night).
- A single query information can also be useful. Consider a query which takes a long time to complete. We can predict its performance and evaluate its plan.

Then, we can start the evaluation and simultaneously start the physical reorganization. Later, we switch its evaluation to a new plan while reusing the partial results obtained earlier. The goal is to change it in such a way that the query would benefit from the reorganization on upcoming stages of processing.

Control of the Reorganization. One of the key points of a self-managed database system is the control of the reorganization. We can propose the following approaches:

- A cost-based optimization for a particular query or a set of queries. This is the classical approach used since 70-es in the area of physical design tuning. Nowadays, it is the mainstream approach used almost in every industrial database [7,8,10,47,60] and in a majority of academic studies.
- A kind of heuristic strategy which will guide the search behavior of a system. This "forgotten" approach was also used since the earlier days of physical design tuning [21]. In this approach, no performance model is used; the process is guided by some rule set. It was employed as a separate algorithm or as a pre-filtering step in later cost-based studies [8] in order to lower computational complexity of the problem. Another prominent example is the group of affinity-based approaches [21,46].
- A combination of both.

In spite of the evident fact that the cost-based approach is superior in terms of the quality of produced recommendations, a strategy may have several strong points over cost-based optimization:

1. Firstly, a cost-based enumeration may be very expensive in terms of computation resources. Thus, a continuous re-run of the optimization routines is impossible. One may have to resort to plan caching schemes or other types of result reuse. However, the application of this approach is hindered by the next considerations.
2. Not all queries may be known in advance. A good self-tuning system should be capable to cope with such a situation. Using a cost-based approach, we may not be able to decide on any required action at all. At the same time, a strategy may offer a reasonable action before the arrival of such a query, exploiting some rational assumptions regarding the data distribution.
3. Not all queries may be run. We may optimize a workload which is not run, thus wasting a precious time and possibly harming a future workload.

Eventually, we should adopt an approach combining elements of cost-based and strategy approaches. Let's now consider what a column-store system can do to adapt to changing workloads.

Physical Design Options (Actions). There is a number of actions, which alter column-store system's physical design and provide benefits for different queries. One can regard these actions and their results as the building blocks of a physical design for a column store database. For example: replicate, relocate and set up an adaptive index structure on a some column.

Column Relocation. A relocation of a column or a set of columns from one processing node to another. This action may be taken basing on the "what-if" estimation for a particular query or according to some strategy. In this case a strategy might look like "eliminate or minimize inter-node communication for a given query". This option existed since the earliest distributed databases, relations were shipped around the network and allocated on the nodes.

Column Reorderings. A reordering of a set of columns — a physical design specialization for a some query or a group of queries. The idea of this approach is to reorder the contents of a set of columns needed for a particular query according to the ordering of a some column set (usually the most selective predicate's column). This option is already described in the Star Schema Benchmark [1] in the section 3.1 (multiple sort orders). In that paper, it was described as a query speed-up technique, while we can propose the sort order selection problem. Furthermore, we propose to consider it in a dynamic environment, where the query patterns change with the time.

Column Duplication. A creation (deletion) of a single column copy or a copy of a column set. The idea is to transparently to a user create a copy of a some column, possible modify it by other actions and then employ it to speed-up query processing. If the column is not needed anymore, e.g. the cost-based estimator shows a more beneficial configuration or a new strategy prescribes to drop it — it could be deleted. There is no direct analogue in the classical database systems. Most studies consider the replication of the whole relation or its horizontal part. The closest approach is the creation of a materialized view; however, to the best of our knowledge, there were no studies for single-columnar views. Though, the usage of materialized views was considered for emulation of the column-store database in a comparison [3].

Horizontal Partitioning. Horizontal partitioning of a column and a partition merge. Horizontal database partitioning of a relational database is a very old problem, there are a lot of approaches [21]. One may consider value-based partitioning and non value-based [55] (range or round-robin, for example), primary and derived [17]. If the partitioning is not needed anymore one can collapse a set of partitions. This action has no direct counterpart in the classical databases, but it can be emulated. Essentially, this action and the creation of a copy constitute a materialized view for a some query. However, it will require a vertical partitioning or materialized view creation, and thus it is very costly.

Column-Store Specific Actions. The first one is an adaptive indexing, a technique which constructs the index on-the-fly during the query processing. This approach is extensively used in column-store systems [2,29,35,36]. Employing it in a distributed environment looks promising, because there are a lot of potential scenarios, which may benefit from this structure. For example, we can keep two copies of a column on different nodes and perform a reorganization without any overhead for the processing, by alternating the idle and used copies. Next, there is a join index structure [54]. It is a structure used to speed-up the reconstruction of a tuple by linking parts of a single record in different projections.

As we can see, a lot of actions share the same idea with the classical relational databases. However, due to the different query processing [2] scheme and the specifics of a data layout novel cost-based models are required.

5 Conclusion

In this paper, we discussed a design of a self-managed distributed column-store system. Architectural novelties of a column-store system hold a great promise for construction of an efficient self-managed database, which would adapt its physical design to changing workloads. We surveyed state-of-the-art self-managed database systems and provided an overview of the contemporary results. Next, we presented some thoughts on the organization of a self-managed distributed column-store system. We discussed the three core components: alerter, reorganization controller and the set of physical design options (actions) available to such system. We described three approaches for the alerter design and two for the reorganization controller. Finally we discussed available physical design options — the building blocks of an adaptive distributed column-store system. We have a prototype row-store distributed query engine, which was ranked third on the ACM SIGMOD Contest 2010[1] and which we are modifying according to some of the presented ideas.

References

1. O'Neil, P.E., O'Neil, E.J., Chen, X.: The Star Schema Benchmark (SSB). http://www.cs.umb.edu/~poneil/StarSchemaB.PDF (acessed July 20, 2012)
2. Abadi, D., Boncz, P., Harizopoulos, S.: The Design and Implementation of Modern Column-Oriented Database Systems (2013)
3. Abadi, D.J., Madden, S.R., Hachem, N.: Column-stores vs. row-stores: how different are they really? In: Proc. of SIGMOD 2008, pp. 967–980 (2008)
4. Aboulnaga, A., Salem, K.: Report: 4th Int'l Workshop on Self-Managing Database Systems (SMDB 2009), pp. 2–5 (2009)
5. Agrawal, S., Chaudhuri, S., Kollar, L., Marathe, A., Narasayya, V., Syamala, M.: Database tuning advisor for microsoft SQL server 2005: demo. In: Proceedings of the SIGMOD 2005, pp. 930–932 (2005)
6. Agrawal, S., Chaudhuri, S., Kollar, L., Marathe, A., Narasayya, V., Syamala, M.: Database tuning advisor for microsoft SQL server 2005. In: Proceedings of VLDB, pp. 1110–1121 (2004)
7. Agrawal, S., Chu, E., Narasayya, V.: Automatic physical design tuning: workload as a sequence. In: Proceedings of SIGMOD 2006, pp. 683–694 (2006)
8. Agrawal, S., Narasayya, V., Yang, B.: Integrating vertical and horizontal partitioning into automated physical database design. In: Proceedings of SIGMOD 2004 (2004)

---

[1] http://dbweb.enst.fr/events/sigmod10contest/results/#winner

9. Ailamaki, A., DeWitt, D.J., Hill, M.D., Skounakis, M.: Weaving relations for cache performance. In: Proceedings of VLDB 2001, pp. 169–180 (2001)
10. Alagiannis, I., Dash, D., Schnaitter, K., Ailamaki, A., Polyzotis, N.: An automated, yet interactive and portable DB designer. In: Proceedings of SIGMOD 2010 (2010)
11. Alagiannis, I., Idreos, S., Ailamaki, A.: H2O: a hands-free adaptive store. In: Proceedings of SIGMOD 2014, pp. 1103–1114 (2014)
12. Bellatreche, L., Boukhalfa, K.: Yet another algorithms for selecting bitmap join indexes. In: Bach Pedersen, T., Mohania, M.K., Tjoa, A.M. (eds.) DAWAK 2010. LNCS, vol. 6263, pp. 105–116. Springer, Heidelberg (2010)
13. Bellatreche, L., Woameno, K.Y.: Dimension table driven approach to referential partition relational data warehouses. In: Proc. of the DOLAP 2009, pp. 9–16 (2009)
14. Bruno, N., Chaudhuri, S.: Automatic physical database tuning: a relaxation-based approach. In: Proceedings of SIGMOD 2005, pp. 227–238 (2005)
15. Bruno, N., Chaudhuri, S.: To tune or not to tune?: a lightweight physical design alerter. In: Proceedings of VLDB 2006, pp. 499–510 (2006)
16. Ceri, S., Navathe, S., Wiederhold, G.: Distribution design of logical database schemas. IEEE Transactions on Software Engineering 9, 487–504 (1983)
17. Ceri, S., Negri, M., Pelagatti, G.: Horizontal data partitioning in database design. In: Proceedings of SIGMOD 1982, pp. 128–136 (1982)
18. Chaudhuri, S., Narasayya, V.: Self-tuning database systems: a decade of progress. In: Proceedings of VLDB 2007, pp. 3–14 (2007)
19. Chaudhuri, S., Weikum, G.: Self-management technology in databases. In: Liu, L., Özsu, M. (eds.) Encyclopedia of Database Systems, pp. 2550–2555 (2009)
20. Chernishev, G.: Physical design approaches for column-stores. SPIIRAS Proceedings 30, 204–222 (2013). www.mathnet.ru/trspy682
21. Chernishev, G.: A survey of dbms physical design approaches. SPIIRAS Proceedings 24, 222–276 (2013). www.mathnet.ru/trspy580
22. Copeland, G., Alexander, W., Boughter, E., Keller, T.: Data placement in Bubba. In: Proceedings of SIGMOD 1988, pp. 99–108 (1988)
23. Dageville, B., Das, D., Dias, K., Yagoub, K., Zait, M., Ziauddin, M.: Automatic SQL tuning in oracle 10g. In: Proceedings of VLDB 2004, pp. 1098–1109 (2004)
24. Dageville, B., Dias, K.: Oracle's self-tuning architecture and solutions. IEEE Data Eng. Bull. 29(3), 24–31 (2006)
25. Eadon, G., Chong, E.I., Shankar, S., Raghavan, A., Srinivasan, J., Das, S.: Supporting table partitioning by reference in oracle. In: Proceedings of SIGMOD 2008 (2008)
26. Gebaly, K.E., Aboulnaga, A.: Robustness in automatic physical database design. In: Proceedings EDBT 2008, pp. 145–156 (2008)
27. Ghandeharizadeh, S., DeWitt, D.J.: Hybrid-range partitioning strategy: a new declustering strategy for multiprocessor database machines. In: Proceedings of VLDB 1990, pp. 481–492 (1990)
28. Ghandeharizadeh, S., DeWitt, D.J., Qureshi, W.: A performance analysis of alternative multi-attribute declustering strategies. SIGMOD Rec. 21(2), 29–38 (1992)
29. Graefe, G., Kuno, H.: Self-selecting, self-tuning, incrementally optimized indexes. In: Proceedings of ICDE 2010, pp. 371–381 (2010)
30. Hammer, M., Chan, A.: Index selection in a self-adaptive data base management system. In: Proceedings of SIGMOD 1976, pp. 1–8 (1976)
31. Hammer, M., Niamir, B.: A heuristic approach to attribute partitioning. In: Proceedings of SIGMOD 1979, pp. 93–101 (1979)
32. Harinarayan, V., Rajaraman, A., Ullman, J.D.: Implementing data cubes efficiently. SIGMOD Rec. 25(2), 205–216 (1996)

33. Hoffer, J.A.: An integer programming formulation of computer database design problems. Inf. Sci. **11**, 29–48 (1976)
34. Holze, M., Ritter, N.: Towards workload shift detection and prediction for autonomic databases. In: Proceedings of PIKM 2007, pp. 109–116 (2007)
35. Idreos, S., Kersten, M.L., Manegold, S.: Self-organizing tuple reconstruction in column-stores. In: Proceedings of SIGMOD 2009, pp. 297–308 (2009)
36. Idreos, S., Kersten, M.L., Manegold, S.: Database cracking. In: CIDR, pp. 68–78 (2007). www.cidrdb.org
37. Jindal, A., Dittrich, J.: Relax and let the database do the partitioning online. In: Castellanos, M., Dayal, U., Lehner, W. (eds.) BIRTE 2011. LNBIP, vol. 126, pp. 65–80. Springer, Heidelberg (2012)
38. Kwan, E., Lightstone, S., Storm, A., Wu, L.: Automatic configuration for ibm db2 universal database: Compressing years of performance tuning experience into seconds of execution. Tech. rep., Performance technical report, IBM (2002)
39. Lamb, A., Fuller, M., Varadarajan, R., Tran, N., Vandiver, B., Doshi, L., Bear, C.: The vertica analytic database: C-store 7 years later. Proc. VLDB Endow **5**(12)
40. LeFevre, J., Sankaranarayanan, J., Hacigumus, H., Tatemura, J., Polyzotis, N., Carey, M.J.: MISO: souping up big data query processing with a multistore system. In: Proceedings of SIGMOD 2014, pp. 1591–1602 (2014)
41. LeFevre, J., Sankaranarayanan, J., Hacigumus, H., Tatemura, J., Polyzotis, N., Carey, M.J.: Opportunistic physical design for big data analytics. In: Proceedings of SIGMOD 2014, pp. 851–862 (2014)
42. Li, L., Gruenwald, L.: Self-managing online partitioner for databases (smopd): a vertical database partitioning system with a fully automatic online approach. In: Proceedings of IDEAS 2013, pp. 168–173 (2013)
43. Lin, X., Orlowska, M., Zhang, Y.: A graph based cluster approach for vertical partitioning in database design. Data & Knowl. Eng. **11**(2), 151–169 (1993)
44. Maier, C., Dash, D., Alagiannis, I., Ailamaki, A., Heinis, T.: PARINDA: an interactive physical designer for Postgre SQL. In: Proceedings of EDBT 2010, pp. 701–704 (2010)
45. Mami, I., Bellahsene, Z.: A survey of view selection methods. SIGMOD Rec. **41**(1), 20–29 (2012)
46. Navathe, S., Ceri, S., Wiederhold, G., Dou, J.: Vertical partitioning algorithms for database design. ACM Trans. Database Syst. **9**, 680–710 (1984)
47. Nehme, R., Bruno, N.: Automated partitioning design in parallel database systems. In: Proceedings of SIGMOD 2011, pp. 1137–1148 (2011)
48. Piatetsky-Shapiro, G.: The optimal selection of secondary indices is np-complete. SIGMOD Rec. **13**(2), 72–75 (1983)
49. Rao, J., Zhang, C., Megiddo, N., Lohman, G.: Automating physical database design in a parallel database. In: Proceedings of SIGMOD 2002, pp. 558–569 (2002)
50. Rösch, P., Dannecker, L., Färber, F., Hackenbroich, G.: A storage advisor for hybrid-store databases. Proc. VLDB Endow. **5**(12), 1748–1758 (2012)
51. Sacca, D., Wiederhold, G.: Database partitioning in a cluster of processors. ACM Trans. Database Syst. **10**, 29–56 (1985)
52. Schnaitter, K., Abiteboul, S., Milo, T., Polyzotis, N.: COLT: continuous on-line tuning. In: Proceedings of SIGMOD 2006, pp. 793–795 (2006)
53. Stonebraker, M.: The choice of partial inversions and combined indices. International Journal of Computer & Information Sciences **3**(2), 167–188 (1974)

54. Stonebraker, M., Abadi, D.J., Batkin, A., Chen, X., Cherniack, M., Ferreira, M., Lau, E., Lin, A., Madden, S., O'Neil, E., O'Neil, P., Rasin, A., Tran, N., Zdonik, S.: C-store: a column-oriented DBMS. In: Proceedings of VLDB 2005, pp. 553–564 (2005)
55. Taniar, D., Leung, C.H.C., Rahayu, W., Goel, S.: High Performance Parallel Database Processing and Grid Databases. Wiley Publishing (2008)
56. Thiem, A., Sattler, K.U.: An integrated approach to performance monitoring for autonomous tuning. In: Proceedings of ICDE 2009, pp. 1671–1678 (2009)
57. Valentin, G., Zuliani, M., Zilio, D., Lohman, G., Skelley, A.: DB2 advisor: an optimizer smart enough to recommend its own indexes. In: Proc. of ICDE 2000 (2000)
58. Wong, E., Katz, R.H.: Distributing a database for parallelism. SIGMOD Rec. 13(4), 23–29 (1983)
59. Yu, P.S., Chen, M.S., Heiss, H.U., Lee, S.: On workload characterization of relational database environments. IEEE Trans. Softw. Eng. 18(4), 347–355 (1992)
60. Zilio, D.C., Rao, J., Lightstone, S., Lohman, G., Storm, A., Garcia-Arellano, C., Fadden, S.: DB2 design advisor: integrated automatic physical database design. In: Proceedings of VLDB 2004, pp. 1087–1097 (2004)
61. Zilio, D., Zuzarte, C., Lightstone, S., Ma, W., Lohman, G., Cochrane, R., Pirahesh, H., Colby, L., Gryz, J., Alton, E., Valentin, G.: Recommending materialized views and indexes with the IBM DB2 design advisor. In: Proceedings of ICAC 2004, pp. 180–187 (2004)

Relational-Based Sensor Data Cleansing

Nadeem Iftikhar[1], Xiufeng Liu[2]($\boxtimes$), and Finn Ebertsen Nordbjerg[1]

[1] University College of Northern Denmark, Aalborg, Denmark
{naif,fen}@ucn.dk
[2] Technical University of Denmark, Kgs. Lyngby, Denmark
xiuli@dtu.dk

Abstract. Today sensors are widely used in many monitoring applications. Due to some random environmental effects and/or sensing failures, the collected sensor data is typically noisy. Thus, it is critical to cleanse the data before using it for answering queries or for data analysis. Popular data cleansing approaches, such as classification, prediction and moving average, are not suited for embedded sensor devices, due to their limit storage and processing capabilities. In this paper, we propose a sensor data cleansing approach using the relational-based technologies, including constraints, triggers and granularity-based data aggregation. The proposed approach is simple but effective to cleanse different types of dirty data, including *delayed data, incomplete data, incorrect data, duplicate data* and *missing data*. We evaluate the proposed strategy to verify its efficiency and effectiveness.

Keywords: Data cleansing · Sensor data · Relational-based · Dirty data

1 Introduction

In recent years, sensor data processing is becoming increasingly important, due to the widely use of sensory devices. On the other hand, more and more sensing capabilities are developed to sensors [1]. Sensors generate large amounts of data which are used to generate high-level of accurate information for applications. Also, the reporting within sensor devices is an integral functionality for many embedded systems for monitoring the activities of the systems in a timely manner. It is challenging to process data on the sensor devices locally since sensor devices typically have limited processing and storage capabilities [2]. Furthermore, sensor devices often have a high probability of generating dirty data, e.g., due to sensing failure, instability of power supply or wireless communication. In this regard, the existing work of detecting and cleaning dirty data, such as reasoning engine based approaches, classification and prediction based techniques, and fuzzy/approximate based matching algorithms, are not suitable for sensor data processing. This might be due to the need of intense programming effort, the constraint of processing and storage capacity of sensors, and other domain specific problems. Relational-based technologies have matured, and using relational

© Springer International Publishing Switzerland 2015
T. Morzy et al. (Eds): ADBIS 2015, CCIS 539, pp. 108–118, 2015.
DOI: 10.1007/978-3-319-23201-0_13

technologies becomes a common practice for sensor devices, such as RFID technology [3]. Inspired by this fact, this paper proposes a sensor-side data cleansing solution based on the relational technologies. The proposed solution greatly simplifies sensor data cleansing compared with using traditional cleansing tools. The proposed solution does not require significant processing capabilities and storage capacity, and is able to cleanse dirty data without using complex logic. Thus, it is well-suited for deploying into resource-constrained systems, such as sensor devices, for doing data cleansing locally.

In sensor devices, the detailed sensor data might lose its value or may not have the same value as before when growing older. Thus, the detailed data can be aggregated into a higher granularity to save space, and to support efficient queries. Various granularity-based data aggregation techniques have been proposed [4–9], aiming at reducing the size of the data stored in sensors. In our previous work [6,7], we discussed how to store multi-granular data efficiently in resource-constrained systems. In this paper we will focus on how to improve the quality of the data at different granularity levels. We will adopt the conservative grouping principle [10], which suggests that the data to be grouped should be correct; and, incorrect data and the data not belonging the group should be discarded.

We organize the rest of the paper as follows. Section 2 presents a problem statement. Section 3 introduces data validation and interval-based aggregation techniques. Section 4 evaluates the proposed solution, followed by presenting the related work in Section 5. We conclude the paper and point out the future research directions in Section 6.

2 Problem Statement

Data generated by sensor devices are notoriously dirty, in which readings are frequently missed, delayed and individual readings are unreliable. In this section, we will present a real-world example of dirty data issues from a farm project, LandIT [11]. The sensor data is logged as the detailed records into a dimensionless fact table (see Table 1). The fact table consists of Task, Parameter, Timestamp, Granularity, and Value attributes. A task represents the particular work carried out by a farm contractor in a field of a farm. A parameter represents a variable code for what type of data is recorded, e.g., the codes of 1, 41, 247 and 248 represent the speed of a tractor in km/h, the area sprayed in hectares, the amount of chemical sprayed in liters, and the distance covered in km, respectively. The timestamp is the point of the time of reading a sensor value, which is in UTC format [12]. The granularity represents the frequency of reading a sensor value in seconds, e.g., 20, 30 or 60 seconds. A value is the sensor reading, which is a numeric measure. In this example, we highlight the dirty data using the bold font, which includes incomplete data, duplicate data, incorrect data, missing data, and delayed (or late-arriving) data. For example, the row ($r = 4$) contains the incomplete data in Timestamp attribute. The row ($r = 6$) are the duplicate to the row ($r = 5$). The row ($r = 8$) contains the incorrect

Table 1. Data_Log (with multiple data granularities)

r	Task	Parameter	Timestamp	Granularity	Value
1	8	41	1420102800	20	19.12
2	8	248	1420102810	30	31.44
3	8	247	1420102810	30	84.20
4	8	41	142010	20	19.12
5	8	1	1420102840	60	9.00
6	8	1	1420102840	60	9.00
7	8	41	1420102840	20	19.13
8	8	248	1420102840	30	31450000
9	8	247	1420102840	30	84.23
10	8	41	1420102860	20	19.15
11	8	248	1420102870	30	31.46
12	8	247	1420102870	30	84.25
13	8	41	null	null	null
14	8	1	1420102900	60	9.50
15	8	41	1420102900	20	19.16
16	8	248	1420102925	30	31.47
17	8	247	1420102935	30	84.27
...	...	...	...	...	...

sensor reading in the `Value` attribute. The row ($r = 13$) has the missing values in `Timestamp`, `Granularity` and `Value` attributes. The rows ($r = 16$ and 17) are the late-arriving data whose timestamp values both should be 1420102900. From this log table, we could see the raw sensor data is quite "dirty", which is indefeasible to use the data for analysis and reporting.

3 Data Cleansing

In this section, we will present the approach for data cleansing using relational technologies on sensor devices. The cleansing approach consists of two components, *data validation* and *interval-based aggregation*. Data validation is the process of removing incorrect, incomplete and duplicate rows, while interval-based

---

Algorithm 1. Data Cleansing

– Read the rows from `Data_Log` table and insert into `Detailed_Fact` table after validation
  • Validation process removes incorrect, incomplete and duplicate rows
– Delete the rows from the `Data_Log` table after moving to `Detailed_Fact` table
– Aggregate the detailed data to a higher granularity if its age is older than n months
  • Interval-based aggregation process takes care of missing and delayed values
– Insert the newly aggregated rows (with a higher granularity) into `Aggregated_Fact` table.
– Delete the old/detailed data from `Detailed_Fact` table after the aggregation.

---

aggregation is the process of managing missing and delayed values. We outline the complete data cleansing procedure in Algorithm 1.

The algorithm first reads the rows from Data_Log table, and inserts into Detailed_Fact table. During the insertion, the rows are validated by the predefined conditions, including KEY constraint, NULL value constraint, CHECK constraint and triggers (see Section 3.1). Second, the rows are deleted from Data_Log table when they are moved to Detailed_Fact table. Third, the rows are aggregated. The aggregation is conducted based on time intervals (see Section 3.2). Finally, the aggregated rows are inserted into Aggregated_Fact table, and the rows are deleted from Detailed_Fact table after the aggregation. Fig. 1 illustrates the process of relational-based sensor data cleansing (taking the first eight rows from Table 1). Sensor readings are logged into Data_Log. No validations are applied to the data during logging process in order to obtain real-time performance, instead, the validations are done in a batch mode to cleanse the logged data afterwards. In addition, since the sensor readings are captured for hundreds of parameters at different granularities, e.g., in field spray sensors about 350 parameters are captured with the granularities ranging from 1 to 60 seconds, it is reasonable to use a staging Data_Log table to store the data temporarily before applying data transformation and validations. It is also possible to bypass Data_Log table, i.e., directly insert the sensor readings from the spray equipment into Detailed_Fact table, if there are only a few parameters and/or the logging frequencies of the parameters are slightly coarse. Four types of data issues can be found for the rows ($r = 1$–8) in Data_log table (see Fig. 1), including incomplete, duplicate, incorrect, and missing values. The dirty data is cleansed by validation process when it is extracted from Data_Log table to Detailed_Fact table. After the validation, only the rows that satisfy the constraints and conditions are moved to Detailed_Fact table, while the incorrect, incomplete, or duplicate rows are ignored. The final step is to aggregate older data from a fine granularity to a coarse granularity, and insert the data into the Aggregated_Fact table.

3.1   Data Validation

Since sensors are the resource-constrained devices, it is appropriate to select a light-weight DBMS to manage sensor data, e.g., SQLite[13], a server-less database engine. In our example, the validation is triggered when a row is inserted into Detailed_Fact table. As mentioned earlier, the data from the spray equipment sensors is first saved into Data_Log table which no constraints are applied, in order to achieve better efficiency. On the other hand, Detailed_Fact table (see line 1–7 in the script bellow) is created with KEY, NULL value and CHECK constraints. These constraints can automatically take care of duplicate, missing and to some extent incorrect data values. In Table 1, the row ($r = 6$) will be discarded when it is added into the Detailed_Fact table, due to the primary key constraint on Task, Parameter and Timestamp attributes. Similarly, the null value constraint takes care of managing the rows with missing values,

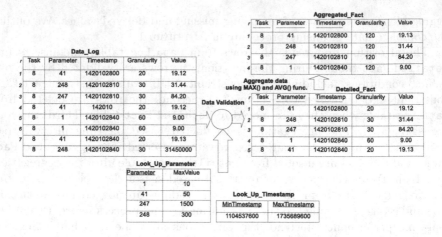

Fig. 1. The process of relational-based sensor data cleansing

```
1 CREATE TABLE Detailed_Fact(
2  Task INT UNSIGNED NOT NULL,
3  Parameter INT NOT NULL,
4  Timestamp INT NOT NULL,
5  Granularity INT NOT NULL CHECK(Granularity IN (20, 30, 60)),
6  Value DECIMAL(6,2) NOT NULL CHECK(Value > 0),
7  PRIMARY KEY(Task, Parameter, Timestamp));
8 CREATE TRIGGER Validate
9 BEFORE INSERT ON Detailed_Fact
10 FOR EACH ROW
11  WHEN (NOT EXISTS(
12       SELECT *
13       FROM Look_Up_Parameter
14        WHERE NEW.Value <= Look_Up_Parameter.MaxValue
15       AND NEW.Parameter = Look_Up_Parameter.Parameter)
16  OR (NOT EXISTS(
17       SELECT *
18       FROM Look_Up_Timestamp
19       WHERE (NEW.Timestamp > Look_Up_Timestamp.MinTimestamp
20       AND NEW.Timestamp < Look_Up_Timestamp.MaxTimestamp))))
21 BEGIN
22  SELECT RAISE(IGNORE);
23 END;
```

e.g., the rows ($r = 13$) will be discarded since the value is checked by the null constraint. Furthermore, more complex validations have to done by creating triggers. In this example we create the trigger to validate the rows with incorrect and incomplete values (see line 8–23 in the script bellow). This trigger uses two lookup tables, Look_Up_Parameter and Look_Up_Timestamp (see Fig. 1), which are to validate the values of parameter and timestamp, respectively.

3.2   Interval-Based Data Aggregation

We now describe the *interval-based* method for aggregating detailed data from fine granular levels to a higher granular level. We illustrate the aggregation process by the example of using MAX function in the script shown in the line 24–31. This aggregation statement can aggregate the data with multiple fine granularities into a single higher granularity for different parameters in a single SQL statement. The time interval-based aggregation process only considers the rows that belong to the same time interval, while late-arriving data and the missing data will not be considered for that particular time interval. Instead, the late-arriving data will be considered by the next time interval. The time-interval based aggregation is achieved by the GROUP BY clause in the SQL statement (see line 30). The GROUP BY clause calculates the beginning of each time interval based on the time granularity, and aggregates the data with several different granularity levels into their corresponding higher granularity levels. The value of Timestamp attribute is first divided by TimeGranularity. The value of the TimeGranularity variable can be manually inserted by users or automatically set by a rule. The *ROUND* function is used to round a decimal result to the nearest integer. Further, the rounded value is then multiplied by the same denominator to get the higher level of granularity.

```
24 INSERT INTO Aggregated_Fact (Task, Parameter, Timestamp,
25                                 Granularity, Value)
26 SELECT Task, Parameter, Timestamp, TimeGranularity, MAX(Value)
27 FROM Detailed_Fact
28 WHERE Parameter IN (41, 247, 248)
29 AND Timestamp < strftime('%s', 'now', '-1 month')
30 GROUP BY Task, Parameter, ROUND((Timestamp/TimeGranularity),0)
31                                 *(TimeGranularity);
```

4   Evaluation

We conduct the tests on an embedded hardware platform with the configuration of a compact flash card of size 2 GB, 512 MB RAM, CPU 600 MHz Intel Centrino M 100 Processor; and with SQLite 3.8 RDBMS installed on Linux. In the embedded database, Data_Log table is created without any constraints and indexes, whereas Detailed_Fact and Aggregated_Fact tables are created

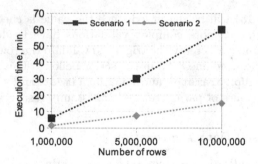

Fig. 2. The efficiency of data cleansing

with KEY, NULL and CHECK constraints as well as UNIQUE indexes. The data used in the experiment are the real-world data sets collected from sensors installed on farm tractors. The data are the readings for 350 sensor parameters, with the granularity of 20-, 30- and 60-second levels. In the following tests, we will measure data cleansing efficiency (in terms of time) and evaluate its effectiveness (in terms of data accuracy). We evaluate the efficiency using the following two scenarios. The first scenario is given that sensor data has already been logged into Data_Log table, we measure the efficiency of validating the rows inserted into Detailed_Fact table, and aggregating the rows inserted into Aggregated_Fact table. In the second scenario, the data is validated during the logging process in real-time, therefore, the time of validating and inserting the rows into the Detailed_Fact table is not counted, while only aggregation time is recorded. The time we measure for each scenario is the total time used by all its corresponding steps. Fig. 2 shows the times used by the scenario 1 and 2, respectively, when the number of rows is scaled from one to five million. As shown, the two scenarios both scale nearly linearly when more rows are added. In scenario 1, the validation and aggregation is done in a bulk mode within a single transaction, whereas, in scenario 2, only the aggregation is done in the bulk mode. This experiment aggregates the rows from 20-, 30- and 60-second granular levels to 120-second granular level. According to the results, we could conclude that scenario 2 is a better option, when the data granularity is coarse or only a few parameters are collected, otherwise, scenario 1 is better when the data granularity is fine or many parameters are collected.

We now evaluate the effectiveness. We consider the sensor data shown in Table 1. After the validation process, Detailed_Fact table contains the rows with the correct values (see Table 2), i.e., the rows with incorrect, incomplete and missing values have been identified and removed. From the aggregation results, we can see that the late-arriving data are aggregated according to their actual timestamp values (see Table 3), instead of the original time interval that they should belong to. For example, the row ($r = 11$) in Detailed_Fact table belongs to the time interval of $[1420102800, 1420102800 + 120)$ during the aggregation; therefore, it is in the same group with the other three rows with the same

Table 2. Detailed_Fact

r	Task	Parameter	Timestamp	Granularity	Value
1	8	41	1420102800	20	19.12
2	8	248	1420102810	30	31.44
3	8	247	1420102810	30	84.20
4	8	1	1420102840	60	9.00
5	8	41	1420102840	20	19.13
6	8	247	1420102840	30	84.23
7	8	41	1420102860	20	19.15
8	8	248	1420102870	30	31.46
9	8	247	1420108870	30	84.25
10	8	1	1420102900	60	9.50
11	8	41	1420102900	20	19.16
12	8	248	1420102925	30	31.47
13	8	247	1420102935	30	84.27

Table 3. Aggregated_Fact

r	Task	Parameter	Timestamp	Granularity	Value
1	8	41	1420102800	120	19.16
2	8	248	1420102810	120	31.47
3	8	247	1420102810	120	84.25
4	8	1	1420102840	120	9.25
5	8	247	1420102930	120	84.27

parameter value ($r = 1, 5$ and 7), resulting in the row ($r = 1$) in
Aggregated_Fact table. On the contrary, the late-arriving data ($r = 13$) falls
into the time interval of $[1420102930, 1420102930+120)$ in the aggregation, which
is different to the time interval of the rows ($r = 3, 6$ and 9). Thus, it results in
the row ($r = 5$ in Aggregated_Fact) in the next time interval, i.e., with the
timestamp of 1420102930.

5   Related Work

Previously, other studies on dirty data have been done. Several promising techniques for correcting missing data, duplicate data, uncertain data and dirty
data have been proposed in [14], [15], [16], [17], [18] and [19]. A reasoning engine
based approach to rectify the missing data is proposed by [14]. Similarly, classification and prediction based techniques to predict unknown or missing values
are described in [15]. In order to detect duplicate data, fuzzy matching algorithms are presented in [16] and [17]. Another promising approach to manage
uncertain data is proposed by [18]. In this approach, semantic uncertainties such
as duplicate data are detected using probabilities. A comprehensive taxonomy
of dirty data and the impact of dirty data on data mining is presented in [19].

The authors have also examined the technologies for preventing and repairing dirty data, and found that modern technologies address only some of the dirty data issues. Furthermore, issues such as substantial manual effort, limited interoperability and domain dependence with various state-of-the-art data cleansing tools are also discussed in [20] and [21]. In addition, [22] and [23] have emphasized that existing data management solutions are not designed for embedded sensor devices, hence tailor-made data management solutions should be built. As obvious, most of the data cleansing approaches require considerable software/hardware resources to operate, as a result, they may not be suitable for systems that are low on resources. In the context of gradual granular data aggregation, work has also been reported. A number of time granularity-based, time interval-based and ratio-based aggregation methods with implementation strategies and real-life examples have been proposed in [4], [5], [6], [7] and [8]. Time granularity-based aggregation methods are explained in [4] and [5]. The granularity-based methods perform aggregation using time hierarchies. Time interval-based and row-based aggregation methods are presented in [6] and [7]. The interval-based method aggregates time intervals and the row-based method aggregate rows. Finally, ratio-based aggregation methods are discussed in [8]. The ratio-based methods are capable of aggregating data with user defined ratio. The main purpose of these approaches is to aggregate multi-granular data without much consideration to the consequences of dirty data. In comparison to all the above mentioned approaches, the current paper provides the data cleansing solution based on relational technology for embedded sensor devices.

6   Conclusions and Future Work

Data cleansing is a non-trivial task on sensors, due to the complexity of data cleansing process and the resource constraints of sensors. This paper proposed a simple but effective solution for cleansing sensor data. The proposed solution does not require substantial memory and processing capacity. It uses a lightweight RDBMS to manage sensor data locally, and uses standard relational technologies, including constraints, triggers and aggregation, to cleanse data. To the best of our knowledge, this is the first work to present relational-based data cleansing solution on embedded systems. The paper evaluated the proposed solution using farming sensor data. The results showed that the proposed solution is efficient and effective. In the future, we will develop a GUI-based tool for the interactive data cleansing and make the proposed solution adaptable to different business areas.

Acknowledgment. We thank Razvan Dan Bucioaca and Jelizaveta Kuznecova for assistance with testing the code.

References

1. Qin, Z., Han, Q., Mehrotra, S., Venkatasubramanian, N.: Quality-aware sensor data management. In: Ammari, H.M. (ed.) The Art of Wireless Sensor Networks, pp. 429–464. Springer, Heidelberg (2014)
2. Zimmerman, A.T., Lynch, J.P., Ferrese, F.T.: Market-based Resource Allocation for Distributed Data Processing in Wireless Sensor Networks. ACM Transactions on Embedded Computing Systems 12(3), Article 84 (2013)
3. Jeffery, S.R., Alonso, G., Franklin, M.J., Hong, W., Widom, J.: Declarative support for sensor data cleaning. In: Fishkin, K.P., Schiele, B., Nixon, P., Quigley, A. (eds.) PERVASIVE 2006. LNCS, vol. 3968, pp. 83–100. Springer, Heidelberg (2006)
4. Iftikhar, N., Pedersen, T.B.: Using a Time Granularity Table for Gradual Granular Data Aggregation. Fundamenta Informaticae 132(2), 153–176 (2014)
5. Iftikhar, N., Pedersen, T.B.: A rule-based tool for gradual granular data aggregation. In: 14th ACM Int. Workshop on DW and OLAP, pp. 1–8. ACM Press, NY (2011)
6. Iftikhar, N., Pedersen T.B.: An embedded database application for the aggregation of farming device data. In: 16th European Conference on Information Systems in Agriculture and Forestry, pp 51–59. Czech University of Life Sciences (2010)
7. Iftikhar, N., Pedersen, T.B.: Gradual data aggregation in multi-granular fact tables on resource-constrained systems. In: Setchi, R., Jordanov, I., Howlett, R.J., Jain, L.C. (eds.) KES 2010, Part III. LNCS, vol. 6278, pp. 349–358. Springer, Heidelberg (2010)
8. Iftikhar, N.: Ratio-based gradual aggregation of data. In: Benlamri, R. (ed.) NDT 2012, Part I. CCIS, vol. 293, pp. 316–329. Springer, Heidelberg (2012)
9. Iftikhar, N.: Integration, aggregation and exchange of farming device data: a high level perspective. In: 2nd International Conference on the Applications of Digital Information and Web Technologies, pp. 14–19. IEEE (2009)
10. Pedersen, T.B., Jensen, C.S., Dyreson, C.E.: Supporting imprecision in multidimensional databases using granularities. In: 11th International Conference on Scientific and Statistical Database Management, pp. 90–101. IEEE (1999)
11. LandIT. http://daisy.aau.dk/education/proposals/farmingdevicedata.php
12. UTC. http://en.wikipedia.org/wiki/Coordinated_Universal_Time
13. SQLite. https://sqlite.org
14. Darcy, P., Stantic, B., Sattar, A.: Correcting missing data anomalies with clausal defeasible logic. In: Catania, B., Ivanović, M., Thalheim, B. (eds.) ADBIS 2010. LNCS, vol. 6295, pp. 149–163. Springer, Heidelberg (2010)
15. Han, J., Kamber, M., Pei, J.: Data Mining: Concepts and Techniques. Morgan Kaufmann (2006)
16. Naumann, F., Herschel, M.: An Introduction to Duplicate Detection. Morgan & Claypool Publishers (2010)
17. Rajaraman, A., Ullman, J.D.: Mining of Massive Datasets. Cambridge University Press (2012)
18. Maurice, V.K.: Managing Uncertainty: The Road Towards Better Data Interoperability. IT - Information Technology 54(3), 138–146 (2012)
19. Kim, W., Choi, B.J., Hong, E.K., Kim, S.K., Lee, D.: A Taxonomy of Dirty Data. Data Mining and Knowledge Discovery 7(1), 81–99 (2003)
20. Rahm, E., Do, H.H.: Data Cleaning: Problems and Current Approaches. IEEE Data Engineering Bulletin 23(4), 3–13 (2000)

21. Barateiro, J., Galhardas, H.: A Survey of Data Quality Tools. IEEE Data Engineering Bulletin, Datenbank-Spektrum **14**, 15–21 (2005)
22. Rosenmuller, M., Siegmund, N., Schirmeier, H., Sincero, J., Apel, S., Leich, T., Spinczyk, O., Saake, G.: FAME-DBMS: tailor-made data management solutions for embedded systems. In: EDBT Workshop on Software Engineering for Tailor-made Data Management, pp. 1–6. ACM Press, NY (2008)
23. Kim, G.J., Baek, S.C., Lee, H.S., Lee, H.D., Joe, M.J.: LGeDBMS: a small DBMS for embedded system with flash memory. In: 32nd International Conference on Very Large Data Bases, pp. 1255–1258 (2006)

Avoiding Ontology Confusion in ETL Processes

Selma Khouri[1,2]([✉]), Sabrina Abdellaoui[2], and Fahima Nader[2]

[1] LIAS ISAE/ENSMA, Futuroscope, Chasseneuil-du-poitou, France
selma.khouri@ensma.fr
[2] Ecole Nationale Suprieure D'Informatique (ESI), El Harrach, Algeria
{s_abdellaoui,f_nader}@esi.dz

Abstract. Extract-Transform-Load ($\mathcal{ETL}$) is a crucial phase in Data Warehouse ($\mathcal{DW}$) design life-cycle that copes with many issues: data provenance, data heterogeneity, process automation, data refreshment, execution time, etc. Ontologies and Semantic Web technologies have been largely used in the $\mathcal{ETL}$ phase. Ontologies are a *buzzword* used by many research communities such as: *Databases*, *Artificial Intelligence* (AI), *Natural Language Processing* (NLP), where each community has its type of ontologies: conceptual canonical ontologies (for databases), conceptual non-canonical ontologies (for AI), and linguistic ontologies (for NLP). In $\mathcal{ETL}$ approaches, these three types of ontologies are considered. However, these studies do not consider the types of the used ontologies which usually affect the quality of the managed data. We propose in this paper a semantic $\mathcal{ETL}$ approach which considers both canonical and non-canonical layers. To evaluate the effectiveness of our approach, experiments are conducted using Oracle semantic databases referencing LUBM benchmark ontology.

Keywords: Ontology layers · $\mathcal{ETL}$ · Data quality · Consistency

1 Introduction

The rationale behind data warehouse systems ($\mathcal{DW}$) is to integrate data stored in various heterogeneous data sources. The Extract-Transform-Load ($\mathcal{ETL}$) phase plays a crucial role in designing a $\mathcal{DW}$, because the quality of the warehouse strongly depends on $\mathcal{ETL}$ (garbage in garbage out principle). The $\mathcal{DW}$ design cycle is composed of five main phases: requirements definition, conceptual design, logical design, $\mathcal{ETL}$ phase and physical design [6]. The $\mathcal{ETL}$ phase has been placed right after the logical phase, it is thus executed at the logical level. Different studies have recently proposed to migrate some $\mathcal{ETL}$ tasks (e.g., the global schema, the sources schemes and/or the mappings) to the conceptual level for different purposes like the automation or the documentation of the $\mathcal{ETL}$ process. To ensure this migration, several research efforts have been proposed and recommended the use of ontologies. Ontologies are defined as an explicit specification of a conceptualization [7].

Our work is motivated by two different types of know-how, that we have gathered over the years: (i) in our laboratory[1], we have significant expertise in

---

[1] http://www.lias-lab.fr/

© Springer International Publishing Switzerland 2015
T. Morzy et al. (Eds): ADBIS 2015, CCIS 539, pp. 119–126, 2015.
DOI: 10.1007/978-3-319-23201-0_14

constructing ontologies in Engineering domains such as Avionics, Energy, etc. and we are involved in many projects dedicated to the construction and publication of normalized ontologies. (ii) The second fact concerns our experience in constructing and exploiting materialized data integration systems, where semantic solutions were proposed [3]. The fruit of the first experience is the proposition of the onion model, proposing an ontological taxonomy where three ontological layers are identified [8]: (1) *Linguistic Ontologies (LOs)* (e.g. Wordnet [2]): define the terms that appear in a given domain. Integration solutions based on these approaches offer approximate results. (2) *Conceptual Canonical Ontologies (CCOs)*: define ontologies which describe primitive or canonical concepts (CC) of a domain. (3) *Non Conceptual Canonical Ontologies (NCCOs)*: define primitive concepts, and also defined or non-canonical concepts (NC), i.e. concepts that can be defined from other concepts by the use of expression languages. The fruit of the second experience is our ability to propose solutions for constructing data integration systems by considering the types of the used ontologies. In this work, we mainly concentrate on ontology-based $\mathcal{ETL}$ approaches. After exploring these approaches, we realized that they do not distinguish between canonical and non-canonical layers. This situation affects the quality of managed data. Data quality is defined as a multidimensional concept including many concepts like consistency, timeliness, accuracy, recoverability and traceability. The *consistency* dimension is ensured by the satisfaction of semantic or integrity constraints expressed on a set of data items [2]. Ignoring the distinction between canonical and non-canonical layers of the ontologies involved in the $\mathcal{ETL}$ process violates these constraints. To illustrate these issues, let us take the following example from LUBM[3] ontology schema. Let *Student* be a canonical concept, defined as the domain of the following properties: Id, Grade, Birthdate and Age. The non-canonical concept *PhdStudent* can be defined as a *Student* having a *Phd grade* (Grade=Phd) by using the following OWL restriction: (PhdStudent $\equiv$ Student $\cap$ Grade:Phd).

Figure 1 illustrates the storage of *Student* and *PhdStudent* concepts in the triple table (the storage model usually used in semantic $\mathcal{ETL}$ studies). The representation of non-canonical concepts in this table introduces the following anomaly: if the Grade of Student1 is modified (from Phd to Postdoc), then the triple (Student1, type, PhdStudent) becomes incorrect. Same anomaly is noticed for Birthdate property. These inconsistencies are due to the dependencies between CC and NC layers that introduce redundancies. Different studies pointed out the presence of these inconsistencies in semantic sources. For instance, [5] identified such inconsistencies when implementing Jena and Oracle semantic databases. Park et al. [12] analyzed 27 major semantic biological databases referencing the Gene Ontology and identified different types of inconsistencies including high rates of redundant and taxonomy inconsistent instances. One can easily imagine how difficult it is for users and designers to detect such inconsistencies in big amounts of instances, stored in different sources. If the

---

[2] http://wordnet.princeton.edu/
[3] http://swat.cse.lehigh.edu/projects/lubm/

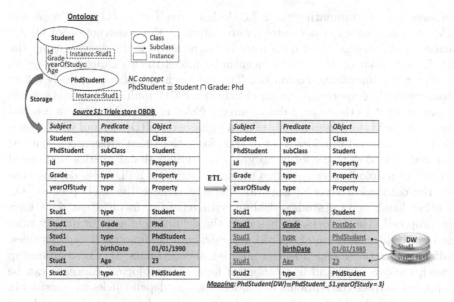

Fig. 1. Anomalies of storing NC concepts

following mapping is defined ($PhdStudent$(DW)$\equiv PhdStudent\_S1.yearOfStudy = 3$), then the inconsistency would be loaded in the $\mathcal{DW}$. Extracting such anomalies from sources and loading them in target data stores is critical for the $\mathcal{ETL}$ process management. To handle the above problems, two directions are possible: (1) using data stores that fully manage the non-canonical layer (like in IBM SOR) for source and target data stores. This solution greatly limits the design autonomy. (2) Consider an $\mathcal{ETL}$ process that is defined at the ontological level which allows the identification of the canonical and non-canonical layers. Our proposal provides a specific treatment for each class type. The rest of the paper is organized as follows: section 2 presents the related work of ontology-based $\mathcal{ETL}$ approaches. Section 3 formalizes the framework defining the proposed $\mathcal{ETL}$ approach. Section 4 presents a case tool and experiments evaluating the effectiveness of our finding. Section 5 concludes the paper and exposes some future works.

2   Related Work

Various studies used ontologies to facilitate the design and construction of the $\mathcal{ETL}$ process. Some studies have used the linguistic layer of ontologies like [14] who proposed to use the Wordnet linguistic ontology in order to provide a textual description of a conceptual $\mathcal{ETL}$ design. Other methods used conceptual ontologies composed of canonical and non-canonical concepts. For instance, [15,16] annotate the schemata of the source and target data stores by complex expressions built from concepts of a corresponding domain ontology. The $\mathcal{ETL}$ design is defined based on the semantic relationships between these expressions. The approach aims at automating the $\mathcal{ETL}$ design. *Niemi et al.* [11] used a generic OLAP ontology as an upper ontology to automate the construction of OLAP schemes. The integration of data from different sources is achieved using the RDF format. *Romero et al.* [13] conceptually model the $\mathcal{ETL}$ process using an ontology which integrates the sources. The approach identifies iteratively a flow of conceptual $\mathcal{ETL}$ operations required to feed the $\mathcal{DW}$ multidimensional schema by connecting it to the related data sources. *Nebot et al.* [10] propose a method to generate fact tables from semantic data published as RDF(S) and OWL annotations. The method provides a set of transformations that guarantees that the generated factual data conforms to given multidimensional queries (expressed in terms of ontological concepts). Otherwise, existing ETL tools do not consider the ontological layer of semantic data to achieve the ETL process. On the other hand, several studies dealing with repairing inconsistent data have been proposed. These works are based on the definition of a set of quality rules based on the identification of defined *dependencies* like functional dependencies, conditional functional dependencies and denial constraints [1]. The relationship between the canonical and non canonical layers of ontological concepts can be assimilated to these conventional dependencies. The dependencies we consider in this study are defined between concepts based on their definitions. A dependency between two concepts C_i and C_j exists if the definition of C_j can be derived from C_i. Contrary to existing studies cited, we propose in this paper an approach that: (1) leverages the $\mathcal{ETL}$ process entirely at the ontological level, in order to clearly identify the ontological dependencies. We note that such dependencies cannot be clearly identified at the logical level (eg. relational storage schema) which usually involves implementation constraints. (2) Reviews the $\mathcal{ETL}$ process in order to manage efficiently the two ontological layers.

3   Formalization and Proposition

$\mathcal{DW}$s are considered as Data Integration Systems (DIS) defined by the framework $< G, S, M >$ [9], where $\mathcal{G}$ is the global schema, $\mathcal{S}$ is a set of local schemes that describe the structure of each source participating in the integration process, and $\mathcal{M}$ is a set of mapping assertions relating elements of $\mathcal{G}$ with elements of schemas $\mathcal{S}$. We define in what follows this framework at the ontological level. The model-based formalization defining the details of framework is fully provided in [4]. We focus in what follows on the process-based formalization and define the $\mathcal{ETL}$ process alimenting $\mathcal{DW}$ classes by instances from sources using ontological mappings, which is based on: (1) a set of $\mathcal{ETL}$ operators and (2) the $\mathcal{ETL}$ algorithm, that will be detailed in what follows. Note that the canonicity algorithm we consider has been presented in [5] and aims at identifying canonical and non canonical concepts in a given ontology.

$\mathcal{ETL}$ **operators:** Skoutas et al. [15] defined ten generic operators typically encountered in an $\mathcal{ETL}$ process, which are: Extract, Retrieve, Merge, Union, Join, Store, DD, Filter, Convert, Aggregate. In our approach, each $\mathcal{ETL}$ operator is translated into an ontological query, using *Sparql* query language. For

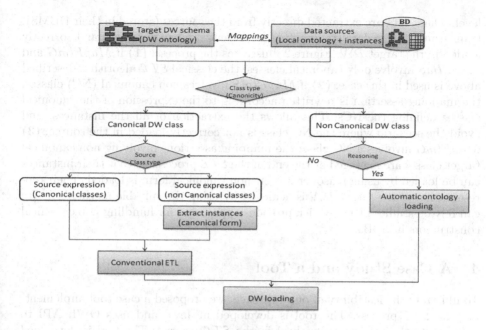

Fig. 2. Proposed $\mathcal{ETL}$ process

example, Retrieve operator is translated as follows: Select ?Instances# Where ?Instances# rdf :type NameSpaceED:Class

The $\mathcal{ETL}$ algorithm: Our algorithm takes as inputs the framework $< G, S, M >$. The mappings follows the GaV approach (Global as View), where the target classes are defined in terms of the sources classes. The integration process depends on the semantics of mappings (*SemanticRelation*): equivalent and sound mappings require the extraction and the merging of instances. Complete and overlap mappings require the transformation of instances before their merging. Several studies, including ours [4], have proposed ontology-based $\mathcal{ETL}$ algorithms. Consequently, and because of lack of space, we focus on the management of CC and NC layers. The canonicity algorithm is first used to identify the CC and NC layers. Two scenarios are expected in our approach: (1) The $\mathcal{ETL}$ process can be processed on the ontologies files. In this case, the ontologies are extracted from sources (extracted from the source if an SDB is used, or defined if a conventional source is used). The target ontology is already available (defined from G). A reasoner is used in order to reconstitute the missing instances for each source class. The target canonical classes are loaded using the mapping assertions as defined by the $\mathcal{ETL}$ algorithm described above. The instances of non canonical (NC) classes are loaded by the reasoner if possible, otherwise they are loaded using the defined mappings. For this first scenario, the use of ontology files can degrade the performances of the $\mathcal{ETL}$ process. The second scenario is preferable. (2) The $\mathcal{ETL}$ process is processed at the ontological level, where data are extracted directly from the sources (stored in their DBMS), transformed in the data staging area (represented by an ontology) and correctly loaded in the target $\mathcal{DW}$. Figure 2 illustrates the process: **(1)** if $MapElmG$ and $MapElmS$ involve only canonical classes: the classical $\mathcal{ETL}$ algorithm described above is used in this case. **(2)** if $MapElmS$ involves non canonical (NC) classes, the mapping assertion is rewritten according to the expression of the canonical classes defining the NCs. This allows the extraction of all the instances, and avoid the scenario where the NC class is not correctly loaded in the source. **(3)** $MapElmG$ involves a NC class: the mapping assertions involving non canonical target classes are managed at the end of the $\mathcal{ETL}$ process, so that their instances can be loaded by using reasoner if their non canonical form is loaded. Their representation in the target $\mathcal{DW}$ is achieved by using the suitable mechanism. We can rely on studies like [5] which provides approaches for handling non canonical constructors in SDBs.

4   A Case Study and a Tool

To illustrate the feasibility of our approach, we proposed a case tool implementing the $\mathcal{ETL}$ process. The tool is developed in Java, and uses OWL API to manipulate the ontologies involved in the $\mathcal{ETL}$ process. The tool is developed following the MVC (model-view-controller) architecture. The tool offers a main interface allowing the designer to: load and browse the involved ontologies and execute the $\mathcal{ETL}$ algorithm. To show the effectiveness of our approach, we conducted an experiment using LUBM benchmark[4] (*Lehigh University BenchMark*). LUBM is an ontological benchmark describing university domain. The adopted scenario for evaluating our proposal consists in creating 4 Oracle SDBs (S_1, S_2, S_3 and S_4), where each source references LUBM ontology using simple and complex mappings. The ontology defines 43 classes and 32 properties (including 25 object properties and 7 datatype properties). 40% of the classes are non canonical. Consistency can be calculated using different metrics like the amount of percentage of tuples that violate semantic constraints or business constraints [2]. Based on these metrics, we illustrate in Figure 3 the amount of instances loaded in the two layers : canonical and non canonical layers. This graph shows for each mapping, the important amount of *possible* inconsistent instances. The figure also illustrates the feasibility of the proposed $\mathcal{ETL}$ algorithm (in term of execution time) which linearly evolves according to the amount of loaded instances. We also calculated the precision of our $\mathcal{ETL}$ process compared to a conventional $\mathcal{ETL}$ process using the following formula: the ratio of inconsistent instances loaded to the number of all inconsistent instances. We have introduced different inconsistencies in the SDBs: referential integrity constraints inconsistencies (like key violations) and class dependencies inconsistencies in the following classes: *University, Institute, MasterStudent, StudentEmployee* and *MasterCourse*. The experiment is achieved using Protege ontology editor. We then proceeded to the execution of a conventional $\mathcal{ETL}$ process, and then the execution of our

---

[4] http://swat.cse.lehigh.edu/projects/lubm/

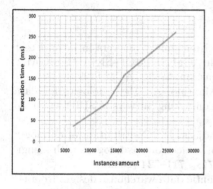
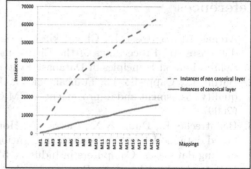

Fig. 3. Amount of instances (NC layer) that can cause inconsistency & ETL Process Execution time

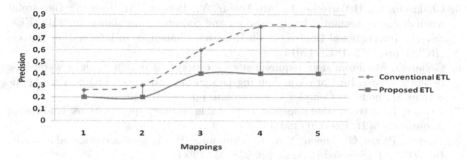

Fig. 4. Accuracy of the proposed $\mathcal{ETL}$ process

proposed $\mathcal{ETL}$ process (Figure 4). This comparison showed us that class dependencies inconsistencies introduced have not been loaded in the DW data store, since the mappings are rewritten in their canonical form. However, the referential integrity constraints inconsistencies have been loaded in both $\mathcal{ETL}$ processes.

5   Conclusion

Semantic-aware $\mathcal{ETL}$ approaches use ontologies for different purposes: the automation of the $\mathcal{ETL}$ process, the documentation of the $\mathcal{ETL}$ process, the management of target data store deployment. Ontologies are composed of three layers: canonical, non-canonical and linguistic. We noticed that the canonical and non-canonical layers are treated without a real distinction in existing studies. We showed that this situation may cause redundancies and inconsistencies. In this paper, we proposed and implemented in this paper a new $\mathcal{ETL}$ approach that analyzes class dependencies for efficiently managing the two ontological layers. This study opens diverse perspectives, among them: (i) studying the complexity of our approach and the impact of reasoning tasks on the quality of the ETL process and (ii) integrating the linguistic layer to cover all ontology types.

References

1. Arenas, M., Bertossi, L., Chomicki, J.: Consistent query answers in inconsistent databases. In: Proceedings of the Eighteenth ACM SIGMOD-SIGACT-SIGART Symposium on Principles of Database Systems. pp. 68–79. ACM (1999)
2. Batini, C., Cappiello, C., Francalanci, C., Maurino, A.: Methodologies for data quality assessment and improvement. ACM Computing Surveys (CSUR) 41(3), 16 (2009)
3. Bellatreche, L., Dung, N.X., Pierra, G., Hondjack, D.: Contribution of ontology-based data modeling to automatic integration of electronic catalogues within engineering databases. Computers in Industry 57(8), 711–724 (2006)
4. Bellatreche, L., Khouri, S., Berkani, N.: Semantic data warehouse design: from etl to deployment à la carte. In: Meng, W., Feng, L., Bressan, S., Winiwarter, W., Song, W. (eds.) DASFAA 2013, Part II. LNCS, vol. 7826, pp. 64–83. Springer, Heidelberg (2013)
5. Chakroun, C., Bellatreche, L., Ait-Ameur, Y., Berkani, N., Jean, S.: Be careful when designing semantic databases: data and concepts redundancy. In: 2013 IEEE Seventh International Conference on Research Challenges in Information Science (RCIS), pp. 1–12. IEEE (2013)
6. Golfarelli, M.: From user requirements to conceptual design in data warehouse design a survey. In: Data Warehousing Design and Advanced Engineering Applications Methods for Complex Construction, pp. 1–16 (2010)
7. Gruber, T.: A translation approach to portable ontology specifications. Knowledge Acquisition 5(2), 199–220 (1993)
8. Jean, S., Pierra, G., Ameur, Y.A.: Domain ontologies: a database-oriented analysis. In: WEBIST (Selected Papers), pp. 238–254 (2006)
9. Lenzerini, M.: Data integration: a theoretical perspective. In: PODS, pp. 233–246 (2002)
10. Nebot, V., Berlanga, R.: Building data warehouses with semantic web data. Decision Support Systems (2012)
11. Niinimäki, M., Niemi, T.: An ETL process for OLAP using RDF/OWL ontologies. In: Spaccapietra, S., Zimányi, E., Song, I.-Y. (eds.) Journal on Data Semantics XIII. LNCS, vol. 5530, pp. 97–119. Springer, Heidelberg (2009)
12. Park, Y.R., Kim, J., Lee, H.W., Yoon, Y.J., Kim, J.H.: Gochase-ii: correcting semantic inconsistencies from gene ontology-based annotations for gene products. BMC Bioinformatics 12(1), 1–7 (2011)
13. Romero, O., Simitsis, A., Abelló, A.: GEM: requirement-driven generation of ETL and multidimensional conceptual designs. In: Cuzzocrea, A., Dayal, U. (eds.) DaWaK 2011. LNCS, vol. 6862, pp. 80–95. Springer, Heidelberg (2011)
14. Simitsis, A., Skoutas, D., Castellanos, M.: Representation of conceptual etl designs in natural language using semantic web technology. Data & Knowledge Engineering 69(1), 96–115 (2010)
15. Skoutas, D., Simitsis, A.: Ontology-based conceptual design of etl processes for both structured and semi-structured data. International Journal on Semantic Web and Information Systems (IJSWIS) 3(4), 1–24 (2007)
16. Skoutas, D., Simitsis, A., Sellis, T.: Ontology-driven conceptual design of ETL processes using graph transformations. In: Spaccapietra, S., Zimányi, E., Song, I.-Y. (eds.) Journal on Data Semantics XIII. LNCS, vol. 5530, pp. 120–146. Springer, Heidelberg (2009)

Towards a Generic Approach for the Management and the Assessment of Cooperative Work

Amina Cherouana[1(✉)], Amina Aouine[1], Abdelaziz Khadraoui[2], and Latifa Mahdaoui[1]

[1] RIIMA Laboratory, University of Sciences and Technology Houari Boumediene (USTHB), Algiers, Algeria
{acherouana,aaouine,lmahdaoui}@usthb.dz
[2] Institute of Services Science (ISS), University of Geneva, (CUI), Geneva, Switzerland
abdelaziz.khadraoui@unige.ch

Abstract. Cooperative work, teamwork or networking of individuals and collective-labors have become the key elements in modern organizations. These new forms of work organization have favored the birth of Computer-Supported Cooperative Work tools (CSCW), who study the individual mechanisms as well as the group-work collectives and then research how to assemble and cooperate many actors with various skills and different prerequisites. However, despite the enormous benefits of CSCW, few of these ones focus on the assessment aspect. Thus, this paper proposes a generic approach which combines Workflow Man-agement Systems (WfMS) with generic design patterns in order to generate a model for the management and specifically for the assessment of cooperative processes and their final rendering.

Keywords: Information and Communication Technologies (ICT) · Computer Supported Cooperative Work (CSCW) · Activity theory · Workflow Management System (WfMS) · Design patterns

1 Introduction

Cooperative work, teamwork or networking of individuals and collective-labors have become the key elements in modern organizations. Indeed, they are increasingly based on work organizations that bring together a large number of professional skills around the same processes or activities. In such conditions, define and coordinate the development teams and the project activities are proving to be a difficult task. Consequently, it is necessary to adopt methods and supporting-tools for managing the complex activities of this cooperative work.

Therefore, these new forms of work organization have spread widely by exploiting the development of new Information and Communication Technologies (ICT). These have favored the birth of Computer Supported Cooperative Work (CSCW), who study the individual mechanisms as well as the group-work collectives, and then research how to assemble and to cooperate many actors with various skills and different prerequisites.

© Springer International Publishing Switzerland 2015
T. Morzy et al. (Eds): ADBIS 2015, CCIS 539, pp. 127–134, 2015.
DOI: 10.1007/978-3-319-23201-0_15

However, if we assume that the modern technologies provide a set of tools in order to communicate, coordinate, as well as collaborate, then what strategy (i-e means) to follow, in order to measure the effectiveness and the quality of added value in a professional context. Thus, the arising question in this context is to say: *"how to assess, firstly, the collaborative process (or even its components activities), and secondly, their final rendering?"*. Let's specify that the fields concerned by these solutions revolve around one key dimension namely the "Activity Theory".

This paper proposes a generic approach which combines CSCW tools with generic design patterns in order to reach a model for the management and the assessment of cooperative processes and their final rendering. The current proposed solution makes specifically recourse to the Workflow Management System (WfMS) in order to ensure the management aspect and the assessment strategies in order to ensure the assessment one. The WfMS are used because they allow the description of structured and cooperative activities, while Learning Management Systems (LMS) don't.

In addition, the generic feature comes to the use of a textual pattern permitting the construction of an assessment device which will be used to establish the cooperative assessment models. These are transformed to cooperative process models using graphical patterns and then to executable processes. This model considers also the involvement degree of the "3C" on which a given cooperation is based, namely: Collaboration, Coordination and Communication.

The remainder of this paper is organized as follow:

- Several research works may be related to this work. An overview of some of these works is made in the second section.

- As mentioned above, the foundations of the proposed solution are based on those of the "Activity Theory". A brief description of this key dimension is made in the third section.

- Computer-Supported Cooperative Work tools and design patterns are the main used technologies. These are presented in the fourth and the fifth sections.

- The description of the proposed approach and its component levels are presented in the sixth section.

2   Related Works

In a quest for a better and effective cooperation, more research works have addressed the issue of cooperative work management and especially the adoption of CSCW technologies. A thorough analysis of a set of these works has shown that the proposed solutions are generally focused on the management of complex cooperative activities using these tools (e.g. [9][2][11][10]). Indeed, the common question that brings these solutions is *"how to ensure and provide an efficient cooperation of many actors, each with different prerequisites and skills?"*

In this context, we cite as example the works of [9] that tempted to build a computer-based technology in order to support all forms of group work. Therefore, they defined a method for group work analysis and in particularly for the analysis and the design of Workflow applications using as a basis OSSAD methodology. These authors proposed the use of group-oriented models in order to represent the cooperative work phases that cannot be considered by a workflow tool. Another example consolidating this same issue is the works of [2] who proposed a methodological approach allowing to structure cooperative work and reduce organizational complexity. This approach is an extension of the methodological framework of ISO SPICE.

In summary, the proposed solutions under this issue are usually deployed as Groupware or Workflow. In contrast, few studies focus on the assessment of these cooperative activities (or even of cooperative processes), and their final rendering. Indeed, the assessment aspect is mainly considered in e-learning field (e.g. [1][4]).

A reflection has allowed us to find that all these fields (whatever is its nature) are built around the same dimension, namely the "Activity Theory". Consequently, in this work we propose a generic approach allowing the generation of specific software model for the management and specifically for the assessment of cooperative processes (i-e their component activities) as well as their final rendering for a given field.

3   Activity Theory

As mentioned above, the proposed generic approach in this work is based on the foundations of the theory of activity.

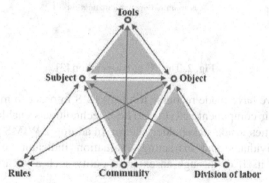

Fig. 1. Activity theory foundations[7].

In this theory, an activity is divided into a hierarchical structure consisting of three levels: "Activity", "Action", and "Operation". Each activity has a directed "Object" and an active "Subject" (actor) including its main objective. The existing mediation between the subject and the object is presented by the supporting "Tool".

Moreover, this theory shows that an actor is not isolated and he is part of a "Community". The community represents the set of subjects that share the same object of activity. When new subjects arrive within a community, they must appropriate a distributed knowledge and be located relatively of why, what and how. Thus the two new elements are considered: the concept of "Rules" that mainly mediates the subject-community relationship and the "Division of labor" that mediates the object-community relationship.

For this work, we focus on three main areas (colored surfaces in Fig.1). Indeed, in our context an actor is part of a team and holds a set of activities to reach a determined objective by means of tools. In addition, a set of strategies for division of labor should be imposed within a team.

4   Computer-Supported Cooperative Work

Computer-Supported Cooperative Work is a vast domain in continuous expands. According to [12]: "CSCW is an identifiable area of research that seeks to understand the nature and the characteristics of cooperative work whose objective is to develop appropriate technologies to assist this type of work". The supporting tools of this work are called Groupware [6]. The Groupware covers areas as vast as cooperation, human-computer interaction and interpersonal interaction [5]. These tools are generally in compliance with the "3C" standard which defines cooperation as an activity involving the Collaboration, the Coordination and the Communication dimension [3].

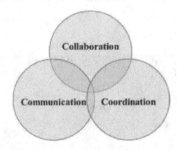

Fig. 2. The 3C of cooperation [3].

In this work, we have made recourse to the WfMS in order to manage cooperative processes and their component activities. These technologies enable parallel working and sharing activities inside stakeholder's team. In addition, WfMS provide the means for tracking individuals and extracting evaluation indicators either through the workflow system itself or through an external entity such as Learning Management System (LMS).

5   Design Patterns

The patterns were initially proposed for the architectural design field by Alexander & al. [1]. Their use was, then, widely held to cover several contexts such as CSCW [7] or software engineering [10]. Whatever the target area, patterns share a common language structured into three levels allowing the representation of the problem, the solution and the relation between some features [8]. These levels can be extended in order to give birth to new patterns for specific fields. In addition, patterns can be also used as support encapsulating the knowledge and the experience of actors within the same team or community.

In this work, patterns are used for the assessment purposes. Thus, we have defined a generic pattern based on assessment strategies arising from the CSCW field. These strategies can be improved progressively with the accumulated knowledge and experience extracted from the real execution environment.

6   Approach Description

As mentioned above, this paper proposes a generic approach for the management and the assessment of cooperative processes and their final rendering. It makes recourse to WfMS in order to ensure the management aspect and generic patterns based on assessment strategies in order to ensure the assessment aspect. It also considers the involvement degree of the 3C around which any cooperation is articulated.

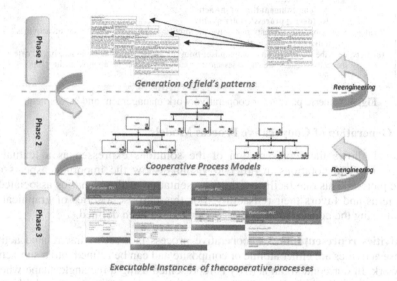

Fig. 3. The levels of the proposed approach.

As shown in the Fig.3, the approach is a three-level stages according to the levels of Model Driven Engineering (MDE) [13]. The main advantage of this approach comes to the transformation of design patterns, as textual form, to graphical patterns adapted to the automation and the execution. In addition, the experience accumulated from these last phases contributes to the enrichment of the generic patterns through reengineering links.

6.1   Generation of Field's Patterns

The objective of this phase is the generation of the field's patterns from the generic pattern. This notion of 'Generic Pattern' allows the reuse of this approach regardless the target field. Fig.4 shows the developed generic pattern and illustrates its parts.

Title: set of words used to identify a pattern.
Context : represents the situation where the pattern can be applied.
Problem : describes the problem that the pattern tries to solve. In other words how to solve the
    mounting problems of the cooperation around the 3C.
Objective: the assessment of the cooperation process (rendering process) or the final output
    (render). The assessment of the both can also be performed.
Type of cooperation: communication, collaboration and coordination. A hybrid
      cooperation between these three classes may also be considered.
Criteria to assess
        Score : indicates the value that describes the individual/collective effort of
        stakeholders.
        Accomplishment time of the activity
        Rendering process/render quality
Strategy : explains the elements that solve the problem, that is to say the set of activities,
    relations and communications, etc.
Impacts: describe the results and compromises from the application of the solution. In our case
    the consequences are the assessment strategies that can be applied in the proposed
    solution.

Fig. 4. Generic pattern for cooperative work management and assessment.

6.2   Generation of Cooperative Process Models

This level allows the representation of the solutions expressed as a textual form
through graphical notations in order to visually show the important aspects of the
generic pattern. This one facilitates the representation of the scenarios associated with
the patterns and favors their reusability. For this purpose a series of graphical notations allowing the construction of these models have been defined.

1. **Activities representation:** a cooperative process is a set of interrelated activities.
   These activities are either atomic or composite and can be refined into a sub-activities
   network. In our context, an activity is represented using a rectangle shape where the
   composite activity includes a 'C' Character in the lower-left corner. In addition, a rectangle with a single small symbol (one individual) in the lower-right corner
   represents an individual atomic activity, while a small icon with two symbols (several
   individuals) in the same place represents a collective atomic activity.

Fig. 5. Graphical representation of activities.

2. **Flow-control representation:** flow control is used to express the relationships between activities. For the modeling of cooperative situations, we have proposed an
   extension (adaptations) for the notations/patterns of the workflow-nets formalism.

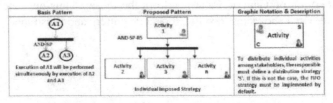

Fig. 6. Example of the proposed AND-split pattern.

On other hand, the concept of activities distribution between the actors must be considered. This one is expressed using a given strategy that represents the way in which activities will be distributed. For this purpose, a new type called "Routing Activity" is defined. The graphical notation of this one is presented as a rectangle shape where the character 'S' in the upper-right corner represents the distribution strategy of activities (e.g. F:FIFO, C:by competence, L:by level, etc.) and the character 'C' in the lower-left corner represents the entity that imposes this strategy.

3. **Data-flow representation:** for this component, we use comments to model and express the resources used and manipulated by the actors. On other hand, the comments can also be used to highlight the final rendering.

4. **Roles representation:** the actors must play a specific role within the cooperative processes. Therefore, in order to represent these roles as well as the activities organization, this approach exploits the Swim lanes partitions of activity diagram defined by the Unified Modeling Language (UML).

6.3   Generation of Executable Instances for Cooperative Processes

This is the level where the resulting cooperative models are executed as instances. However, in order to support the execution of the aspects described through the previous levels, a hybrid software architecture allowing the generation of a dedicated platform is designed.

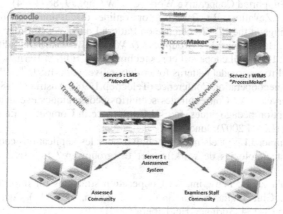

Fig. 7. Software architecture for the execution of cooperative process instances.

This architecture is composed of several software components. It mainly implies the Workflow Management System 'Process Maker' and the Learning Management System 'Moodle'. The interaction between these two systems is ensured through our developed system which makes recourse to database transactions and web-services invocation.

7   Conclusion

The objective of this work is to propose unifying approach allowing the generation of a specific model for the management and the assessment of activities under computer supported cooperative work. The main particularity of this solution comes to possibility of its application regardless the application field.

Our main objective is to put in cooperation several actors with different skills, and to assess the quality of their cooperative activities and their resulting rendering. Let's recall that the activity theory is the common base that regroups the fields that can be considered by this solution.

This approach is applied to the legal ontology construction process. This ontology is used to support the development of an e-government system and it construction is a strongly cooperative process which puts on interaction three types of roles, each with different skills and prerequisites: (1) legislation experts, (2) IT experts, and (3) public institution managers.

References

1. Alexander, C., Ishikawa, S., Silverstein, M., Jacobson, M., Fiksdahl-King, I., Angel, S.: A pattern language: town, buildings, construction. Oxford University Press (1977)
2. David, M., Idelmerfaa, Z., Richard, J.: Organizing cooperative work for the product development process. In: Information Control Problems in Manufacturing (INCOM 2004) (2004)
3. Ellis, C.A., Wainer, J.: A conceptual model of groupware. In: Malone, T. (ed.) Conference on Computer-Supported Cooperative Work (CSCW), pp. 79–88 (1994)
4. O'sullivan, J., Zevallos, R.: Patterns for online discussions, published in: PLOP. In: 10 Proceedings of the 17th Conference on Pattern Languages of Programs (2010)
5. Khoshafian, S., Buckiewics, M.: Groupware & Workflow, MASSON Editions (1998)
6. Longchamp, J.: Le travail coopératif et ses technologies. Edition Hermes, Paris (2003)
7. Lonchamp, J.: Process model patterns for collaborative work. In: 15th IFIP World Computer Congress, Telecooperation Conference (Telecoop 1998), Austria (1998)
8. Moura, C., Derycke, A.: Concevoir des scénarios pédagogiques exécutables avec des patrons de conception pédagogiques. In: 3rd Conference in Computing Environment for Humain Learning (EIAH 2007), June 2007
9. Nurcan, S., Chirac, J.L.: Quels modèles choisir pour les applications coopératives mettant en œuvre les technologies de workflow et de groupware? In: Actes Congrès AFCET (1995)
10. Panti, M., Spalazzi, L., Penserini, L.: Cooperation strategies for information integration. In: Batini, C., Giunchiglia, F., Giorgini, P., Mecella, M. (eds.) CoopIS 2001. LNCS, vol. 2172, pp. 123–134. Springer, Heidelberg (2001)
11. Perrin, O., Wynen, F., Bitcheva, J., Godart, C.: A model to support collaborative work in virtual enterprises. In: van der Aalst, W.M., ter Hofstede, A.H., Weske, M. (eds.) BPM 2003. LNCS, vol. 2678, pp. 104–119. Springer, Heidelberg (2003)
12. Schael, T.: Théorie et pratique du Workflow: des processus métier renouvelés. Springer Edition (2000)
13. Schmidt, D.C.: Model-Driven Engineering. IEEE Computer Society (2006)
14. Schmidt, D.C.: Using design patterns to develop reusable object-oriented communication software. In: Published in Magazine: Communications of the ACM-Special Issue on Object–Oriented Experiences And Future Trends (1995)

MLES: Multilayer Exploration Structure for Multimedia Exploration

Juraj Moško[1]([✉]), Jakub Lokoč[1],
Tomáš Grošup[1], Přemysl Čech[1], Tomáš Skopal[1], and Jan Lánský[2]

[1] SIRET Research Group, Department of Software Engineering, Faculty of
Mathematics and Physics, Charles University in Prague, Prague, Czech Republic
{mosko,lokoc,grosup,cech,skopal}@ksi.mff.cuni.cz
http://www.siret.cz
[2] Department of Computer Science and Mathematics,
University of Finance and Administration, Prague, Czech Republic
lansky@mail.vsfs.cz

Abstract. The traditional content-based retrieval approaches usually use flat querying, where whole multimedia database is searched for a result of some similarity query with a user specified query object. However, there are retrieval scenarios (e.g., multimedia exploration), where users may not have a clear search intents in their minds, they just want to inspect a content of the multimedia collection. In such scenarios, flat querying is not suitable for the first phases of browsing, because it retrieves the most similar objects and does not consider a view on part of a multimedia space from different perspectives. Therefore, we defined a new Multilayer Exploration Structure (MLES), that enables exploration of a multimedia collection in different levels of details. Using the MLES, we formally defined popular exploration operations (zoom-in/out, pan) to enable horizontal and vertical browsing in explored space and we discussed several problems related to the area of multimedia exploration.

Keywords: Similarity search · Multimedia exploration · Content-based retrieval · Exploration operation · Multimedia browsing

1 Introduction

Besides traditional similarity retrieval tasks, there are emerging novel means of retrieval that focus on more complex retrieval scenarios, considering advanced system features like restricted GUI (e.g., smart phones), interactivity and entertainment. The *multimedia exploration* [2] is a typical example of such retrieval scenario where users want to explore and get idea of an unknown multimedia collection. In such cases, a typical *query-by-example* similarity search is not possible as the user does not have any example to provide or to choose from. Often the only thing the user has, is an idea or a picture of the result in her/his mind. The system has to give some insight into the data and guide the user through the collection to satisfy her/his information needs. The tasks integrating iterative

© Springer International Publishing Switzerland 2015
T. Morzy et al. (Eds): ADBIS 2015, CCIS 539, pp. 135–144, 2015.
DOI: 10.1007/978-3-319-23201-0_16

navigation, browsing and visualization of the collection are commonly summarized under the term multimedia exploration. Various exploration operations serve the user as an analogy of navigating in a map, while two typical scenarios for navigating in a map are either horizontal (panning) or vertical (zooming in, zooming out) browsing.

An area of multimedia exploration has been highly examined during the last decade. Various proposals differ from each other in many details, but when we look at them from a bigger perspective, differences are generally in a way how an access method/index is designed or what kind of user interface is provided to a user for intuitive navigation in an explored space. For example according to our previous work [9,11], the exploration structures can be divided into two categories – structures for *iterative querying* that represents a sequence of consecutive similarity queries and *iterative browsing* that represents moving in some hierarchical precomputed structure. Another work, by Schaefer [17], focuses on solution that allows exploration with different level on granularity, while the work by Chen et al. [6] follows a similar idea with hierarchical 2D grids and so-called similarity pyramids. The approach by Beecks et al. [1] focuses on efficient evaluation of similarity queries when their implementation of retrieval is supported with a tree-based structure.

When it comes to operations that multimedia exploration system intuitively should provide, we can get to analogy with web mapping applications, where the operations like zooming and panning are already used over decades. According to the work by Schaefer [17], when a user "zooms in" an image (being a cluster representative), the images from the lower level of the clustered hierarchy are retrieved. Such scenario can be used with a use of different hierarchical structures as a supporting index, like M-tree [7] or PM-Tree [18]. A direct analogy to the panning operation from the web mapping systems was used in [5], where the panning operation is used as a way how a user can cross the boundary of a current view. A quite different but still a useful operation is magnifying, which enables a user to display some objects in more detail [10,16].

In this paper, we introduce a new multilayer multimedia exploration structure that natively supports zoom in, zoom out and pan exploration operations. Such operations enable users to decide at any time at which level of detail he/she would like to browse the multimedia collection.

2   Multilayer Multimedia Exploration

In this section, we present our new structure for multimedia exploration that combines the approaches mentioned in section 1. The structure reflects requirements of a scalable multimedia exploration system presented in [12].

2.1   Exploration Structure

The multimedia exploration usually starts in a so-called *page zero view*, where users start browsing of the multimedia collection using a limited number of representative objects. Then the users consecutively zoom in to specific parts of the view, pan to other groups of objects, or zoom out if the actual view is filled with undesired objects, all the time seeing the same (small) number of objects.

From the above exploration use case, we assume the exploration structure should be able to provide representative distinct objects for earlier phases of the exploration, while more related (similar) objects should be retrieved in later stages, i.e. it should provide some mechanism to search in different level of details. At the same time, whenever users select an object to zoom in details of its neighborhood, this object should be visualized also in a query result to preserve fluency of the exploration. Based on these assumptions, we define a Multilayer Exploration Structure (1):

Definition 1. *(Multilayer Exploration Structure). Given a dataset $\mathbb{S}$ of objects (descriptors of the respective multimedia objects), the Multilayer Exploration Structure, $MLES(\mathbb{S}, m, v, \phi)$, is a system of $m + 1$ subsets $\mathbb{L}_i$ (layers), where the subset condition holds:*

$$\mathbb{L}_i \subset \mathbb{L}_{i+1}, \forall i = \{0, .., m - 1\}$$

The smallest subset $\mathbb{L}_0$ represents the first depicted v objects (i.e., page zero view) and the proper subset $\mathbb{L}_m = \mathbb{S}$ represents the whole database. Selection of objects for each layer is determined by a selection function $\phi : \mathbb{N} \to 2^{\mathbb{S}}$ that has to comply with the subset condition.

With such definition, each layer could be indexed by the most suitable similarity index that supports the k-NN similarity query processing. For example, the upper layers containing a small number of objects can use an efficient memory based index (e.g., pivot table [13]), while lower layers in MLES hierarchy containing a lot of objects can use efficient disk-based indexes [7,18] and/or GPU computing [19].

In order to select objects for particular layers in the MLES, various techniques can be utilized (e.g., random sampling, (hierarchical) clustering techniques, or their combination). In this work, we consider just simple random sampling of the subsets[1] which is applicable for huge datasets. The dataset has to be just randomly mixed and then the subsets corresponding to layers can be simply defined using prefixes of such dataset.

Since the number of displayed objects v depends on the size of a client device and on user preferences, we need to answer the question of a different zero layer configuration. However, as we use prefixes of a mixed dataset to define layers, the page zero view corresponds just to the first prefix of the dataset. Therefore, the page zero view can be changed dynamically according to given device needs.

2.2   Exploration Operations

The typical querying task in similarity search, the k-nearest neighbor (k-NN) query [20] (Definition 2), can be used also as a supportive operation for exploration operations. For lower layers of the MLES with huge number of objects,

---

[1] Techniques guaranteeing representativeness of the objects in top layers are out of the scope of this paper and we leave them for future work.

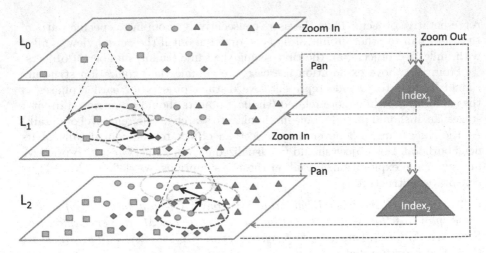

Fig. 1. MLES with indexes $Index_1$ for layer L_1 and $Index_2$ for layer L_2.

also approximate k-NN search techniques [4,13–15] can be utilized for higher efficiency of exploration. We believe that multimedia exploration belongs among retrieval tasks where the maximal precision is not required and thus can profit from approximate search.

Definition 2. *(k-nearest neighbor query). Given a descriptor universe* $\mathbb{U}$*, a distance function* $\delta : \mathbb{U} \times \mathbb{U} \to \mathbb{R}_0^+$*, a dataset* $\mathbb{S} \subseteq \mathbb{U}$ *of object descriptors, a query object* $q \in \mathbb{U}$ *and a query parameter* k*, the k-nearest neighbor query is defined as:*

$$kNN(q) = \{\mathbb{T} \subseteq \mathbb{S}, |\mathbb{T}| = k \wedge \forall o \in \mathbb{T}, o' \in \mathbb{S} - \mathbb{T} : \delta(o,q) \leq \delta(o',q)\}$$

In the following, the *current user view* represents a collection of objects currently visualized on the user screen, while the *next user view* represents the collection of objects resulting from some exploration operation. We assume that the *current user view* of k objects is visualized as a static network (see Section 1), where only the most similar objects are connected to each other and positions of objects are influenced by the similarities between them (we employed force-directed placement introduced in [8]). In this work, we do not focus on precise positioning of particular objects in a user view.

We start with operations *Zoom-In* and *Zoom-Out*, allowing to narrow or broaden the concept being actually explored in the collection.

Zoom-In. In operation Zoom-In, the user selects a query object from objects available in the *current user view*. Using query object, the k-NN query is performed on the layer immediately below the layer of the *current user view*.

Definition 3. *Given a Multilayer Exploration Structure* $E = MLES$ $(\mathbb{S}, m, v, \phi)$*, a query object* $q \in \mathbb{L}_i$ *and a parameters* k, i*, the operation Zoom-In(E, q, k, i) on a layer* $\mathbb{L}_i, \forall i = \{0, .., m-2\}$ *returns a set of objects being the k nearest neighbors to the query object* q *from objects in layer* $\mathbb{L}_{i+1}$*.*

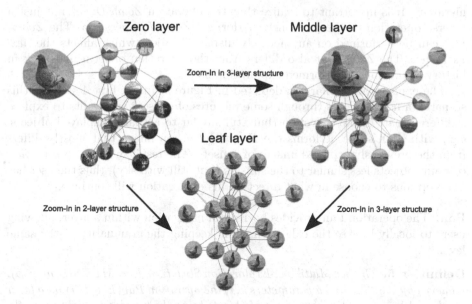

Fig. 2. Operation Zoom-in on two MLES structures with different number of layers (the enlarged image represents the query object for the following operation).

In Figure 2, differences between two MLES structures with a different configuration are depicted. On the left part, the operation *Zoom-In* is performed in the MLES that has only two layers in total, the zero layer and the leaf layer. In such case, performing the *Zoom-In* operation is equivalent to the iterative querying mentioned in Section 1. On the contrary, in the 3-layer MLES the *Zoom-In* operation from the same *zero page view* returns objects from the first (in this case labeled as middle) layer. The results represent similar objects, but in a lower level of details than in the leaf layer and so users can select from more diverse set of objects (unlike the case with the 2-layer MLES). If the user selects the *Zoom-In* operation with the same query object, the same *next user view* as in the case of the 2-layer MLES scenario is returned.

Zoom-Out. The operation *Zoom-Out* is similar to the operation *Zoom-In* except that the kNN query is performed on the layer immediately above the layer of the *current user view*. As the layer L_i does not contain all objects from the layer L_{i+1}, the query object does not necessary have to be in the *next user view*, because of the *the subset condition* in Definition 1 of the MLES.

Definition 4. *Given a Multilayer Exploration Structure, $E = MLES(\mathbb{S}, m, v, \phi)$, a query object $q \in \mathbb{L}_i$ and a parameters k, i, the operation Zoom-Out(E, q, k, i) on a layer $\mathbb{L}_i, \forall i = \{1, .., m - 1\}$ returns a set of objects being the k nearest neighbors to the query object q from objects in layer $\mathbb{L}_{i-1}$.*

The motivation for the operation *Zoom-Out* is to lower the level of details in the *current user view* by navigating vertically to the upper layers of the hierarchy. It is important to realize that the operation *Zoom-Out* is not just a reverse operation to the previously performed *Zoom-In* operation. The *Zoom-Out* can be performed on any actually displayed object which affects the *next user view*. The Zoom-Out also differs from the operation of returning back in history of previously performed exploration steps.

The explanatory scenario is depicted in Figure 3. In the first 4 steps of this scenario, a user explores through some pictures of sunsets and wants to explore a different part of the collection (but still similar to actually displayed objects, e.g., with sky), so he performs the *Zoom-Out* on an image that mostly differs from the previously explored images of sunsets. The result in the *next user view* contains objects less similar to the sunsets (but still with sky), thus the user has more options to choose in what direction his exploration will continue.

Pan. The operation Pan provides horizontal navigation within a layer, allowing users to locally browse the collection while keeping the granularity at the same level.

Definition 5. *Given a Multilayer Exploration Structure, $E = MLES(\mathbb{S}, m, v, \phi)$, a query object $q \in \mathbb{L}_i$ and a parameters k, i, the operation $Pan(E, q, k, i)$ on a layer $\mathbb{L}_i, \forall i = \{0, .., m-1\}$ returns a set of objects being the k nearest neighbors to the query object q from objects in layer $\mathbb{L}_i$.*

This simple definition of the operation *Pan* follows the definitions of operations *Zoom-In* and *Zoom-Out*, where k-NN query is evaluated for one selected query object in the same layer. The operation *Pan* allows a user to investigate a neighborhood of some object in the *current user view*, while a current level of details is not changed. In Figure 3, the operation *Pan* is used for two purposes. In the leaf layer, the operation is used to retrieve more similar images of a searched concept (e.g., sunsets) and in the middle layer as a method for navigation to other parts of the explored collection.

There is one problem in our definitions of exploration operations, that comes from the non-symmetry of the k-NN queries. Lets assume that k is fixed and that some object B is in the result of the k-NN query for some object A. Then, object A does not have to be in the result of the k-NN query for object B. As a consequence, some object does not have to be reachable in the MLES using defined exploration operations. For completion of the definitions, we need to ensure that each object can be reachable by some sequence of exploration operations starting from *the page zero view*. As the definitions of the zooming operations seems to be quite intuitive, we simply extend the definition of the panning by adding some random objects from the same layer to the result of the k-NN query. Although we add some level of non-determinism into the exploration process, we ensure that each object is with non-zero probability reachable from *the page-zero view*. The proportion between the number of objects that come from the result of the k-NN query and the number of randomly selected objects can be handled by some parameter of the operation Pan. More complex solution, involving detection of outliers is out of scope of this paper, we suggest it for a future work. The randomly selected objects added into the result of the operation

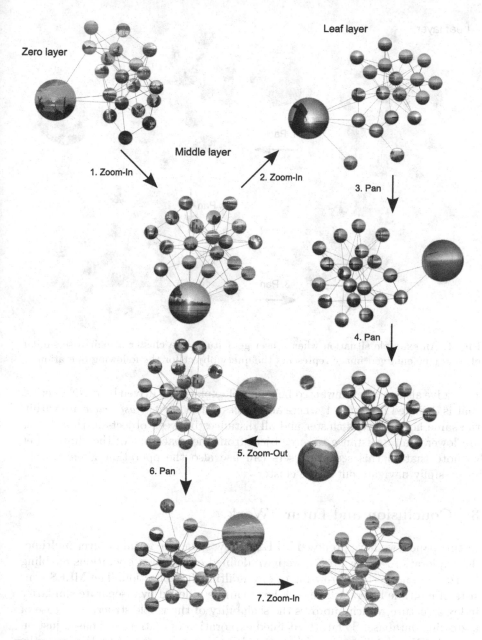

Fig. 3. Exploration scenario in 3-layer MLES involving all exploration operation for navigating horizontally and vertically through all layers. A user start exploring images of sunsets and he consecutively navigate to images of nature in daylight (the enlarged image represents the query object for the following operation).

Leaf layer

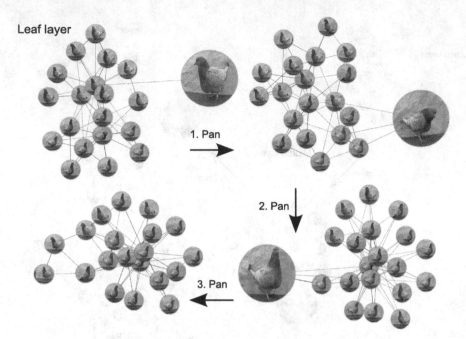

Fig. 4. An example situation when a user gets stuck in a cluster of pair-wise similar objects (the enlarged image represents the query object for the following operation).

Pan solve also another unwanted behavior of exploration driven by k-NN queries, that is depicted in Figure 4, where all objects in the *current user view* are within the same homogeneous cluster and all distances between objects in this cluster are lower than a distance to any object from the space out of the cluster. Let us note that in this scenario users can use also the operation *Zoom-Out* to successfully navigate out of the cluster.

3   Conclusion and Future Work

In this paper, we have proposed MLES, a new structure for multilayer multimedia exploration. Using MLES, we have defined exploration operations enabling horizontal and vertical browsing of any multimedia collection. The MLES consists of multiple layers, where each layer can be indexed by a separate similarity index structure, which improves the scalability of the whole structure in case of a growing database. So far, the defined exploration operations are based just on exact k-NN queries, however, further enhancements are possible. For example, approximate k-NN queries could be used for more efficient retrieval in lower layers of MLES, or multiple k-NN queries or similarity models could be combined in one exploration operation.

In order to demonstrate benefits of more layers in multimedia exploration, many user studies focusing on various retrieval scenarios have to be performed. According to our preliminary experience, multilayer structures can help with a database inspection phase. More specifically, we have asked several tens of users to find as much objects of a selected class as possible, using our *Find the image* web application [11]. The application uses the PROFIMEDIA test collection [3] comprising 21993 small thumbnail images divided into 100 classes and a similarity model based on position-color-texture feature signatures. The searched classes are described just by a keyword and the users can use only 15 exploration operations. Given either two or three layer variant of the MLES structure, the users start in the predefined initial view showing selected 50 objects from the dataset, where the searched class is not present. Then the users can browse to different parts of the dataset just by selecting objects with similar color distribution as the searched images. The results of the browsing show that when using three layer MLES, the users are able to find a first object of the searched class sooner. Also a number of unsuccessful searches seems to be lower for three layer MLES. On the other hand, when users find a first object of the searched class, a number of found objects is often similar for both two and three layer MLES. According to our observations, we conclude that multilayer exploration helps especially with the inspection phase of the multimedia exploration. In the future we plan to perform several extensive user studies to compare our multilayer exploration structure with traditional flat retrieval approaches.

Acknowledgments. This research has been supported in part by Czech Science Foundation project 15-08916S, by Grant Agency of Charles University projects 201515 and 910913 and by project SVV-2015-260222.

References

1. Beecks, C., Uysal, M., Driessen, P., Seidl, T.: Content-based exploration of multimedia databases. In: 2013 11th International Workshop on Content-Based Multimedia Indexing (CBMI), pp. 59–64, June 2013
2. Beecks, C., Driessen, P., Seidl, T.: Index support for content-based multimedia exploration. In: Proceedings of the International Conference on Multimedia, MM 2010, pp. 999–1002. ACM, New York (2010)
3. Budikova, P., Batko, M., Zezula, P.: Evaluation platform for content-based image retrieval systems. In: Gradmann, S., Borri, F., Meghini, C., Schuldt, H. (eds.) TPDL 2011. LNCS, vol. 6966, pp. 130–142. Springer, Heidelberg (2011)
4. Chavez, G.E., Figueroa, K., Navarro, G.: Effective proximity retrieval by ordering permutations. IEEE Trans. Pattern Anal. Mach. Intell. **30**(9), 1647–1658 (2008)
5. Chen, C., Gagaudakis, G., Rosin, P.: Similarity-based image browsing (2000)
6. Chen, J.Y., Bouman, C., Dalton, J.: Hierarchical browsing and search of large image databases. IEEE Transact. on Image Processing **9**(3), 442–455 (2000)
7. Ciaccia, P., Patella, M., Zezula, P.: M-tree: an efficient access method for similarity search in metric spaces. In: VLDB 197, pp. 426–435. Morgan Kaufmann Publishers Inc. (1997)
8. Fruchterman, T.M.J., Reingold, E.M.: Graph drawing by force-directed placement. Softw. Pract. Exper. **21**(11), 1129–1164 (1991)
9. Grošup, T., Čech, P., Lokoč, J., Skopal, T.: A web portal for effective multi-model exploration. In: He, X., Luo, S., Tao, D., Xu, C., Yang, J., Hasan, M.A. (eds.) MMM 2015, Part II. LNCS, vol. 8936, pp. 315–318. Springer, Heidelberg (2015)

10. Liu, H., Xie, X., Tang, X., Li, Z.W., Ma, W.Y.: Effective browsing of web image search results. In: Proceedings of the 6th ACM SIGMM International Workshop on Multimedia Information Retrieval, MIR 2004, pp. 84–90. ACM (2004)
11. Lokoč, J., Grošup, T., Čech, P., Skopal, T.: Towards efficient multimedia exploration using the metric space approach. In: 2014 12th International Workshop on Content-Based Multimedia Indexing (CBMI), pp. 1–4, June 2014
12. Moško, J., Skopal, T., Bartoš, T., Lokoč, J.: Real-time exploration of multimedia collections. In: Wang, H., Sharaf, M.A. (eds.) ADC 2014. LNCS, vol. 8506, pp. 198–205. Springer, Heidelberg (2014)
13. Navarro, G.: Searching in metric spaces by spatial approximation. The VLDB Journal 11(1), 28–46 (2002)
14. Novak, D., Batko, M., Zezula, P.: Metric index: An efficient and scalable solution for precise and approximate similarity search. Inf. Syst. 36(4), 721–733 (2011)
15. Patella, M., Ciaccia, P.: Approximate similarity search: A multi-faceted problem. J. of Discrete Algorithms 7(1), 36–48 (2009)
16. Pecenovic, Z., Do, M.N., Vetterli, M., Pu, P.: Integrated browsing and searching of large image collections. In: Laurini, R. (ed.) VISUAL 2000. LNCS, vol. 1929, pp. 279–289. Springer, Heidelberg (2000)
17. Schaefer, G.: A next generation browsing environment for large image repositories. Multimedia Tools and Applications 47, 105–120 (2010). doi:10.1007/s11042-009-0409-2
18. Skopal, T., Pokorný, J., Snášel, V.: PM-tree: pivoting metric tree for similarity search in multimedia databases. In: Advances in Databases and Information Systems (2004)
19. Strong, G., Gong, M.: Browsing a large collection of community photos based on similarity on GPU. In: Bebis, G., Boyle, R., Parvin, B., Koracin, D., Remagnino, P., Porikli, F., Peters, J., Klosowski, J., Arns, L., Chun, Y.K., Rhyne, T.-M., Monroe, L. (eds.) ISVC 2008, Part II. LNCS, vol. 5359, pp. 390–399. Springer, Heidelberg (2008)
20. Zezula, P., Amato, G., Dohnal, V., Batko, M.: Similarity Search: The Metric Space Approach. Springer, Heidelberg (2005)

SLA Ontology-Based Elasticity in Cloud Computing

Taher Labidi[1,2(✉)], Achraf Mtibaa[1,2], and Faiez Gargouri[1,3]

[1] MIRACL Laboratory, University of Sfax, Sfax, Tunisia
{taherlabidi,achrafmtibaa}@gmail.com
[2] National School of Electronic and Telecommunications, University of Sfax, Sfax, Tunisia
faiez.gargouri@gmail.com
[3] Higher Institute of Computing and Multimedia, University of Sfax, Sfax, Tunisia

Abstract. Service Level Agreements (SLA) is the principal means of control which defines the Quality of Service (QoS) requirements in cloud computing. These requirements have to be guaranteed in order to avoid costly SLA violations. However, elasticity strategies, which have not been deeply considered yet in SLA documents, may significantly ameliorate the QoS. Therefore, in this paper, our aim is to guarantee the QoS by introducing the semantic meaning of the elasticity strategies in SLA. In this regard, we propose an ontology-based elasticity approach which allows getting an elastic cloud service by dynamically apply corrective actions. These corrective actions present the elasticity strategies applied following a violation or in prediction of a violation. Our proposed approach allows getting an interactive and flexible SLA document in order to maintain a reliable QoS and respect the SLA parameters.

Keywords: Cloud computing · Service level agreement · Quality of service · Ontology · Elasticity

1 Introduction

The National Institute of Standards and Technology (NIST) identified cloud SLAs as an important gap that needs further clarification in cloud computing [1]. A primary question is why the classic SLAs are not sufficient for the cloud? For example, if we want to develop a contract for a house; this contract remains valid for a wide duration because it does not change the dimension. It is the same thing for the old SLAs. Hence, if we use these contracts in the cloud, due to the dynamic nature of this latter, we shall elaborate SLA whenever we have a change. In this case, we shall develop, perhaps, an SLA every second. So, the dynamic change of cloud resources and services makes old SLAs insufficient [2], [3], [4]. Therefore, we require developing flexible SLA which takes into account the dynamic nature of the service as we have a house with a dimension that can change at any moment. These changes must be managed and planned in the SLA document.

In addition, some providers in cloud computing are compelled to reject consumers' requests of resources when the SLA is not respected [5]. This is due to the inability to allocate the various cloud services to their available resources and the ambiguity of

© Springer International Publishing Switzerland 2015
T. Morzy et al. (Eds): ADBIS 2015, CCIS 539, pp. 145–152, 2015.
DOI: 10.1007/978-3-319-23201-0_17

elasticity strategies offered. This can negatively affect the QoS and cause inconsistencies with the agreements negotiated in the SLA. For this reason, we will provide the semantic meaning of the different elasticity strategies. In fact, these strategies are applied in a given context and situation, which can significantly influence how a service can be executed, and thus affect the actual performance of the service (i.e., choose an implementation that conforms to the current context of the cloud resources). We opt to improve the SLA by introducing elasticity strategies and their semantic meaning to dynamically apply guarantee actions applied following a violation or prediction of a violation in order to get an elastic cloud service and ensure a reliable QoS.

The remaining of this paper is organized as follows: In the next section, we will be describing our motivation, where the importance of elasticity and ontology in SLA is shown. Next, we will present the related works. Then, in section 4, we intend to provide our ontology-based elasticity approach. In addition, we detail our generic ontological model Cloud SLA Elasticity Ontology (CSLAEOnto) while showing its ability in modeling elasticity of cloud computing. Finally, section 5 will be devoted to the conclusion.

2   Requirements for an Ontological Representation of Elasticity Strategies in Cloud Computing SLA

According to NIST, elasticity means that capabilities can be rapidly and elastically provisioned, in some cases automatically, to quickly scale out and rapidly released to quickly scale in [1]. We can distinguish different strategies employed in the implementation of elasticity solutions as replication, resizing and migration [6]. The lack of coordination between these concepts is due to the inexistence of a universally accepted elasticity representation. In addition, for those who are not specialist in cloud computing, there is an ambiguity between elasticity, scalability, resources management, etc. In order to clarify this vagueness and unify discourse languages between various actors in the cloud, ontological representation is required. It allows projecting the elasticity concepts in an ontology to make them understandable. The validity of the elasticity concepts is ensured by the reasoning techniques which avoid the contradictory information modeling. Thus, ontologies allow deducing new information through inference. Inference rules are rules set by the developer who assists in deducing or inferring new knowledge from a set of captured raw data. They provide a formal support to express beliefs and knowledge in a domain.

Moreover, the dynamicity of cloud computing pushes us to the necessity of managing the variability and the configurability at the conceptual level. Variability means the possibility to cover a very large area which can be varied, as needed, while keeping the semantic concepts. Hence, configurability allows substituting element by element in this area or navigate in hierarchy of concepts according to our needs. In this regard, practical techniques are required to satisfy the variability and the configurability of elasticity concepts. Obtaining a formalized elasticity concept helps set up an enforcement process to be automated and hence cover the dynamic aspect of cloud.

Expressing elasticity concepts in a machine-readable format using ontologies could be a production step to ameliorate the SLA document and to support automated interaction following a violation. In fact, ontology can cover a very large area which can be varied while keeping the semantic concepts. For example, we can inherit only the concepts which interest us. In addition, in a hierarchy of concepts, we can substitute element by element or navigate according to our needs, which guarantee configurability. We will detail more these aspects in our ontology by instances to better understand them.

With these clearly identified requirements, we proceed to improve SLAs in the cloud. To do this, we prefer to use elasticity concept to adapt cloud services depending on the QoS value in order to avoid costly SLA violations. However, the use of the ontology is essential for two reasons. On the one hand, it allows the modeling of elasticity concepts to be readable and automatically selected, thus improving the understanding of elasticity strategies. On the other hand, it allows the variability and the configurability which manage the dynamic nature of the cloud and improve the interactivity and the flexibility of SLA. Indeed, using the benefits of inference in ontology, we can deduce, from elasticity concepts, new knowledge about corrective actions to be performed in order to obtain an elastic cloud service.

3   Related Works

The research on SLA and QoS metrics has been considered by many researchers in many domains, such as E-commerce and web services. In these domains, many efforts have been undertaken concerning the representation of QoS concepts [2]. However, due to the dynamic nature of the cloud, those representations are insufficient [2], [3], [4]. Therefore, few works try to develop flexible SLA which takes into account the different characteristics of cloud services.

García García et al. [4] propose the "Cloudcompaas" platform to dynamically and completely manage the life-cycle of resources cloud PaaS satisfying the multiple requirements of the users. To do this, a new architecture headed by SLA for automatic provision, planning, allocation and dynamic resource management for Cloud is presented. This architecture is based on an extension of the specification WS-agreement. It is adapted to the specific needs of Cloud Computing and allows the definition of QoS rules and automatic actions for arbitrary and corrective preventions. Thus, in order to automate the process of managing SLAs in the cloud, Dastjerdi et al. [7] present architecture for the deployment of cloud services. This architecture is able to describe, deploy (discover, organize and coordinate) and execute monitoring services in an automatic way. For this reason, the WSMO ontology was used for the modeling and description of SLAs monitoring capabilities. This ontology built an inter-Cloud language that allows the semantic correspondence between the SLAs of different cloud layers. In addition, the effects of QoS dependencies between services that generate SLA violations were treated. To overcome these violations, the deployment of appropriate service monitoring and filtering report violation is performed using dependency knowledge. Kouki [8] proposes an SLA-driven cloud elasticity management approach. He addresses the trade-offs between the benefit of SaaS vendor and customer's satisfaction. Thereby, his work aims to provide a solution to implement the elasticity of cloud computing to meet this compromise. The main occupations of his work are the definition of SLA, the SLA dependencies between XaaS layers and the management of resource capacity.

SLA models must provide a flexible, scalable and dynamic approach to suit the dynamic nature of the cloud. However, we noted that there is a lack of representation for the elasticity strategies and their semantic meaning in cloud computing SLA using ontology.

4   Ontology-Based Elasticity Approach

SLA management covers three steps described in our previous work [9]. First of all, the SLA establishment phase is accomplished and it consists of the definition and followed by the negotiation steps. These steps allow the specification of an SLA. Then, the SLA validation step where the corrective actions are made in order to make the cloud service elastic. This phase must be performed throughout the service execution and in an automatic way to ensure that the QoS is always reliable. Finally, SLA monitoring phase is achieved by detecting violations and verifying the reliability of the QoS. In the scope of this paper, we focus on SLA validation and especially corrective actions that must be performed following a violation or in prediction of a violation.

4.1   Ontology-Based Elasticity Approach Overview

Our proposal uses ontology to represent the knowledge. The elasticity strategies will be modelled in the ontology and processed with SLA concepts presented in [10]. Next step, using inference mechanisms and custom rules, we determine the best configuration that will be applied after a violation or prediction of a violation. This configuration is automatically chosen, throughout the service execution, depending on the violation degree and the resources availability in order to offer an elastic cloud service. This guarantees the QoS defined and negotiated in SLA document. In the whole process, the restrictions required by the providers or consumers are respected which enables a customized configuration.

Modeling elasticity strategies in cloud computing SLA is performed by ontology. As a consequence, we take advantage of benefits in sharing knowledge and maintaining semantic information. Ontology contains different elasticity strategies, their properties, their relationships and axioms to choose the most suitable resource depending on the condition and situation of the other concepts. Hence, we take another advantage of ontology which is the inference. Inference is considered as one of the most powerful features of the web semantic. It makes each data item more valuable, because it can have an effect on the creation of new information. Each new piece of information has the capacity to add a great deal of new information via inference [11]. Hence, semantic inferences allow us to detect indirect correspondences that are not detectable by the different languages.

In addition, our approach provides both consumers and providers needs. Firstly, it assists the provider to implement corrective actions when violations are detected or predicted. In this case, the provider can respond quickly and avoid the consequences of such violation in order to minimize penalties risk. In addition, it offers to the provider an advanced mechanism for controlling its available resources in such a way that they serve exactly the current customer demand in order to avoid excessive use of resources which maximizes its revenue. Secondly, it provides to the customer profitable services with acceptable QoS and reasonable usage cost. Moreover, it offers software, hardware and QoS requirement to the consumers in times. In this way, we guarantee a high profit provider while satisfying to the utmost the expectations of its customers. The proposed approach is shown in figure 1.

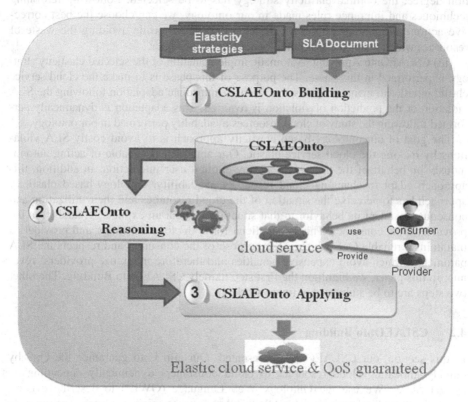

Fig. 1. Ontology-Based Elasticity Approach

This approach is performed during the execution of a cloud service. Once a violation is detected or predicted, elasticity strategies depending in the available resources have to be configured properly to be able to deliver the SLO in SLA contract. For this end, we present our ontology-based elasticity approach to enable self-elasticity in cloud computing. To do so, cloud resources adaptation depending on the service execution is a very important step that must be studied. We propose three main parts:

(i) CSLAEOnto Building: This step allows the construction of a generic ontology containing the main elasticity strategies. Ontological model is used for two reasons. On the one hand, it allows the modeling of elasticity concepts to be readable and automatically selected, thus improving understanding in elasticity strategies. On the other hand, it allows the variability and the configurability which enable the selection of the best elasticity strategies and improve the SLA interactivity and flexibility. The CSLAEOnto will be added to our generic SLA ontology presented in [10]. In this latter work, we have developed an automated monitoring system in order to detect SLA violations. Following this violation, this CSLAEOnto will be used to automatically triggering corrective actions.

(ii) CSLAEOnto Reasoning: Depending in the available resources and QoS violation degree, the optimal elasticity strategy has to be selected. Following reasoning techniques and inference rules made in our ontology, we can choose the best corrective action that guarantee QoS for the cloud consumer while avoiding the waste of resources which keeps the maximum benefit to the provider.

(iii) CSLAEOnto Applying: Automatic implementation of the selected elasticity strategy is performed in this phase. The purpose of this phase is to make the cloud service elastic in order to maintain QoS stable. Hence, immediate adaptation following the SLA violation or the prediction of violation is required. This adaptation is dynamically performed following the study of cloud resources availability performed in our ontology.

The goal of our ontology-based elasticity approach is to avoid costly SLA violations by making the cloud service elastic. Our approach is capable of acting autonomously on behalf of the users without any explicit user interaction. In addition, this approach adapt to changing cloud resources availability. Ontology-based elasticity approach has to perceive the situation of the cloud resources and their utilization and consequently adapt its behavior to that situation without an explicit demand from the provider or the consumer. This is beneficial for both cloud consumer and provider. It maintains a reliable QoS which is the purpose of the consumer and respects the SLA parameters which avoid expensive penalties and therefore maximize providers' revenue. In this paper, we highlight the first step namely CSLAEOnto Building. The other two steps are to be addressed in future work.

4.2   CSLAEOnto Building

In this section, our CSLAEOnto is presented. Our aim is to guarantee the QoS by quickly allocating cloud resources to cloud consumers dynamically depending on agreed SLAs. We use the Ontology Web Language (OWL[1]) to describe classes, constraints, and properties. The OWL allows us to give semantic meaning to SLA elasticity strategies as well as combine and connect them in a manner understandable by machines. Figure 2 presents the generic structure of our CSLAEOnto. The elasticity strategies presented covers the Application 'ApplicationElasticity' the Resource 'ResourceElasticity' and the Price 'PricingElasticity'.

- ApplicationElasticity: it controls the elasticity at the SaaS application. This strategy is based on two methods managing the application: the degradation of service and the application migration. The first method involves applying a deterioration of Ser-

---

[1] http://www.w3.org/TR/owl-ref/

vice (QoS or functionality) to absorb the load after adding a horizontal dimensioning VMs (Virtual Machines) or PMs (Physical Machines) (AddVM, AddPM) or a vertical dimensioning of adding resources (AddCPU , AddRAM) until what the resources requested will be adjusted. However, the second method is to migrate the application to another VM, PM, or to another cloud (private or public). Each operation is aimed at a particular situation. For example, due to lack of VM resources hosting the SaaS application, we believe that the application migration to another VM or PM can solve this problem. However, migrating to a private cloud can be seen as a solution for the security challenges in the public cloud during the increase of workload.

- ResourceElasticity: it ensures the resources elasticity through automatic dimensioning. Our model considers both the horizontal and the vertical dimensions to optimize the management of cloud resources. It can be applied to any type of resource including compute, storage and network.

- PricingElasticity: it consists in changing the service or resource price based on current QoS provided. For example, we can increase with 10% the service price if its availability is greater than 99.5% or decrease it in the opposite case. These operations are discussed in the SLA negotiation step between the consumers and the provider.

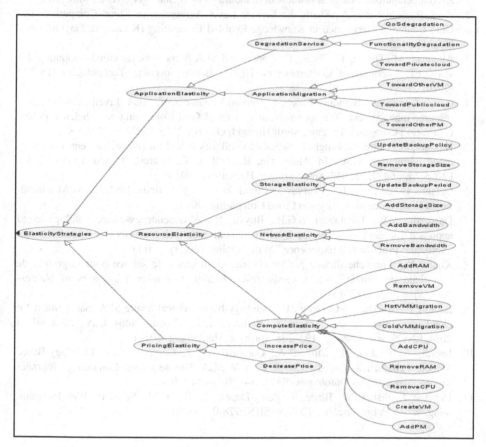

Fig. 2. CSLAEOnto

5   Conclusion

Elasticity is considered as a key feature allowing to maintain a guaranteed QoS in Cloud Computing. In this paper, we present an SLA Ontology-Based Elasticity in Cloud Computing. Using the benefits of representation and inference of ontology, we can choose the appropriate elasticity strategies and corrective actions to be used following an SLA violation or violation prediction in order to make the cloud service elastic and ensure reliable QoS. In the ongoing work, we are completing the two steps of our approach in order to test its efficiency in a real cloud computing platform. For future work, we opt to predict the SLA violation by considering the contextual parameters of the cloud consumers.

References

1. Badger, M.L., Grance, T., Patt-Corner, R., Voas, J.M.: Cloud Computing Synopsis and Recommendations. National Institute of Standards and Technology (NIST), May 2012
2. Patel, P., Ranabahu, A., Sheth, A.: Service Level Agreement in Cloud Computing. The Ohio Center of Excellence in Knowledge-Enabled Computing (Kno.e.sis), Dayton, Ohio U.S. state (2009)
3. Alhamad, M., Dillon, T., Chang, E.: Conceptual SLA framework for cloud computing. In: 4th IEEE International Conference on Digital Ecosystems and Technologies (DEST) (2010)
4. García García, A., Blanquer Espert, I., Hernández García, V.: SLA-driven dynamic cloud resource management. The International Journal of Grid Computing and eScience Future Generation Computer Systems, North-Holland, October 2013. ISSN: 0167-739X
5. Hussain, O., Dong, H., Singh, J.: Semantic similarity model for risk assessment in forming cloud computing SLAs. In: Meersman, R., Dillon, T., Herrero, P. (eds.) OTM 2010. LNCS, vol. 6427, pp. 843–860. Springer, Heidelberg (2010)
6. Galante, G., de Bona, L.: A survey on Cloud computing elasticity. In: IEEE/ACM International Conference on Utility and Cloud Computing (2012)
7. Dastjerdi, A.V., Tabatabaei, S.G.H., Buyya, R.: A dependency-aware ontology-based approach for deploying service level agreement monitoring services in Cloud. Journal "Software: Practice and Experience". Wiley Online Library (2011)
8. Kouki, Y.: Approche dirigée par les contrats de niveaux de service pour la gestion de l'élasticité du "nuage", Ph.D. thesis, research unit: Computer Laboratory of Nantes-Atlantique (LINA) (2013)
9. Labidi, T., Mtibaa, A., Gargouri, F.: Ontology-based context-aware SLA management for cloud computing. In: Ait Ameur, Y., Bellatreche, L., Papadopoulos, G.A. (eds.) MEDI 2014. LNCS, vol. 8748, pp. 193–208. Springer, Heidelberg (2014)
10. Labidi, T., Brabra, H., Mtibaa, A., Gargouri, F.: A Comprehensive Ontology-Based Modèle exhaustif à base d'ontologie pour le SLA dans le Cloud Computing. Journées Francophones sur les Ontologies (JFO 2014), 101–114 (2014)
11. Hebeler, J., Fisher, M., Blace, R., Perez-Lopez, A., Dean, M.: Semantic Web Programming. Wiley Publishing, Inc. (2009). ISBN: 978-0-470-41801-7

Bi-objective Optimization for Approximate Query Evaluation

Anna Yarygina[✉] and Boris Novikov

St. Petersburg University, St. Petersburg, Russia
anya_safonova@mail.ru, b.novikov@spbu.ru

Abstract. A problem of effective and efficient approximate query evaluation is addressed as a special case of multi-objective optimization with 2 criteria: the computational resources and the quality of result. The proposed optimization and execution model provides for interactive trade of quality for speed.

Keywords: Multi-objective query optimization · Parametric query optimization · Approximate query evaluation

1 Introduction

Query evaluation is approximate if the output is, in a certain sense, incomplete or imprecise. In contrast with OLAP, the approximate analytical processing might be unavoidable, e.g. for ranking or similarity queries, or desirable to speed-up execution based on approximate algorithms. A user query submitted for approximate evaluation may be accompanied with certain constraints, e.g. at most specified response time or at least specified quality of the output.

An approximate execution is called controllable if it admits a specification of amount of computational resources and yields the output of different quality depending on the provided constraint. In this research we address the optimization problem for approximate query evaluation based on controllable implementations of operations used to express a query.

As soon as both resources and quality are taken into consideration, the optimization problem becomes bi-objective. We solve this problem as parametric and construct a compact representation of a Pareto set as a data structure that for any amount of resources keeps a plan that yields the maximum possible expected result quality. A user may then choose the actual values for the constraints to trade quality for speed.

2 Preliminaries

Any optimization problem is to find a value of certain arguments that minimize an objective function subject to given constraints. An optimization problem

© Springer International Publishing Switzerland 2015
T. Morzy et al. (Eds): ADBIS 2015, CCIS 539, pp. 153–161, 2015.
DOI: 10.1007/978-3-319-23201-0_18

is called multi-objective if several objectives (that is a vector-valued objective function) are considered.

Possible argument value X of an objective function belongs to the Pareto set if there is no argument Y such that in all dimensions the objective function value for Y is better than it for X. Any solution of the multi-objective optimization problem is in the Pareto set.

An optimization problem is called parametric if the objective function has certain parameters and the problem should be solved for any combination of parameter values. A multi-objective problem may be transformed into an equivalent parametric one if some of objectives are converted into parameterized constraints.

In the context of query optimization the multi-objective problem was considered with several different types of criteria (such as CPU, I/O) in [1,5,13,14]. We consider two kinds of objectives: computational resources and result quality metrics. We provide here a brief summary of concepts related to quality estimations defined and discussed in [15]. To make our algorithms efficient, we assume that all quality and resource-related objectives are aggregated into single scalar objectives $q \in \mathfrak{Q}$ and $t \in \mathfrak{R}$, respectively.

Operations in our model are approximate and may produce different quality of the output depending on the quality of the input and the amount of allocated resources. For example, processing of a larger sample to calculate an average value is more expensive but, in general, provides better quality. The controllable behavior of operations is represented by cost/quality models which describe the dependence of the amount of resources for operation execution and the relative operation quality. The quality of a query plan can be calculated as a combination of operation qualities. In this research we estimate a plan quality as the product of the quality achieved by operations in the plan.

The dependency of a relative quality of a plan on the amount of allocated resources is expressed with a plan *quality function* $Q_P : \mathfrak{R} \to \mathfrak{Q}$, where P is a query plan. We assume that the relative quality of a (sub)plan is a non-decreasing continuous bounded function of the allocated amount of resources. Further we work with piece-wise linear approximations of quality functions. A plan quality function Q_P is defined on a closed interval $[t_{min}, t_{max}]$. To simplify notation, we extend the argument range to $[0, \infty)$ as follows: for $t < t_{min}$ $Q_P(t) = 0$ and for $t > t_{max}$ $Q_P(t) = Q_P(t_{max})$.

Let us consider a query with two controllable plans A and B having the quality functions depicted in figure 1(a). Implementation B is optimal if available amount of resources is greater than 5, otherwise A provides better quality. An optimal solution of bi-objective query optimization problem is based on the quality function labeled as OCS and where different plans may be chosen as optimal depending on run-time constraints.

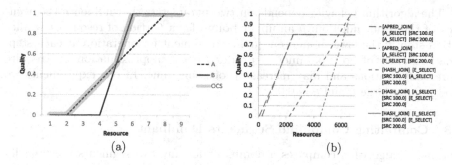

Fig. 1. Artificial and Real Non-Trivial OCSs

3 Model and Algorithms

3.1 Model

Since an optimal plan depends on the amount of resources, hence the concept of quality function is generalized so that it may be associated with several plans. We define a *compound segment* S for a query as a non-decreasing piecewise linear function $\bar{Q}_S : \mathfrak{R} \to \mathfrak{Q}$ each interval of which has an associated plan (not necessarily the same for different intervals) and $\bar{Q}_S(t) = Q_P(t)$, where Q_P is a quality function for the plan P associated with an interval containing t. Any plan quality function can be represented as a compound segment with the same plan associated with all its intervals.

A compound segment S can be represented as a finite set $I_S = \{\langle i, P_i \rangle : i = [t_l, t_u)\}$ of non-intersecting resource intervals $\{[t_l, t_u)\}$, covering the whole range $[0, \infty)$, with associated plans. For query from figure 1(a) we can construct, for example, a compound segment: $I_S = \{\langle [0,2), A \rangle, \langle [2,5), B \rangle, \langle [5,8), A \rangle, \langle [5, \infty), B \rangle\}$.

A compound segment S *dominates* over T if, for any $t \in [0, \infty)$ $\bar{Q}_S(t) \geq \bar{Q}_T(t)$. For any segments S and T a *dominant* $D = \mathrm{dom}(S, T)$ is a compound segment such that D dominates over S and T and any other compound segment dominating over both S and T dominates over D.

An *optimal compound segment* (OCS) dominates over all compound segments for the query. We use OCS as a compact representation of the Pareto set for bi-objective query optimization problem. For query from figure 1(a) a compound segment: $I_S = \{\langle [0,2), A \rangle, \langle [2,5), A \rangle, \langle [5,6), B \rangle, \langle [5, \infty), B \rangle\}$ is optimal.

3.2 Plan Quality Function Computation

To construct a plan quality function we apply a modified polynomial resource allocation algorithm introduced in [15]. The incremental allocation of resources defined there is repeated until the maximum possible quality for a plan is reached, and the output is organized into a compound segment.

The algorithm behavior depends on two parameters that influence the accuracy of the result function: μ an upper bound for a portion of resources at allocation step; ν the lower bound on quality increment for operation at each step. Smaller values of both parameters produce better approximation but require more computations and may increase processing time. In our experiments we used infinite large μ and $\nu = 0.02$.

3.3   Computing Compound Segments Dominant

The *merge* algorithm computes a dominant for any two segments. Technically the intervals for a segment are stored in ascending order hence a dominant can be constructed in a single pass over input segments.

To estimate the number of intervals in a dominant, we observe that if incoming segments contain m and n intervals respectively, the number of intersections is at most $m + n$. Further, if quality functions are piecewise linear, each of the intersections can be split into at most two intervals and hence the total number of intervals cannot exceed $2(m+n)$. Consequently, the complexity of the merge is linear in a number of segment intervals. To prevent rapid growth of the number of intervals, the merge procedure is followed by a compression that replaces the result compound segment with its approximation.

3.4   Enumerators: Traversing the Plan Space

Algorithms for an OCS construction can be derived from any known query optimization approach and an enumerator based on a set of transformation rules. For each plan listed by the enumerator, a quality function represented as a compound segment is built, and these segments are (iteratively) merged into a candidate segment which becomes approximately optimal when the enumerator terminates. The algorithm is more expensive than single-objective optimizer because quality function construction and merge are more expensive than just cost estimation and the number of plans to be stored in solution is larger.

Three different enumerators were explored: *full* (exhaustive scan of the plan space), *iterative improvement*, and *recursive descent*. The full scan is impractical due to unacceptable performance but we need it to obtain exact solution for comparison with other techniques.

The iterative improvement algorithm starts from an arbitrary query plan and merges its quality function with segments for all plans accessible from the current one with a single transformation. If a new plan participates in the merged segment, it is selected as a new current plan and the algorithm iterates. Plans that do not participate in the merge result are discarded.

The recursive descent algorithm applies all transformations to the root operation in the plan, then recursively processes distinct sub-plans obtaining OCSs for corresponding sub-queries. The output is used to construct promising top-level plans from sub-plans participating in OCSs for sub-queries. Finally, quality functions for them are generated and these segments are merged, producing the OCS for the root. This procedure avoids generation of sub-plans that cannot be included into the complete plan and therefore is efficient for extended algebra with reduced number of algebraic equivalences. However, our experiments suggest that cost of OCS computation for sub-queries might be too high.

4 Experiments

We generated synthetic queries from a representative set of parameterized operations (algorithms):

- joins (JOIN) and non-associative joins (NA-JOIN):
  - nested-loops join (NL-JOIN);
  - hash-join (HASH-JOIN);
  - approximate join based on nested-loops with sampling (A-JOIN);
  - nested-loops join with approximate predicate (APRED-JOIN);
- filters: operations considering one object at a time (FILTER);
- selection (SELECT) (extract data from data sources (SRC)): exact selection (E-SELECT) and approximate selection based on sampling (A-SELECT).

A set of transformation rules includes:

- algorithm replacements, e.g. exact selection into approximate selection;
- associativity for joins;
- commutativity for (non-associative) joins;
- distributivity for (non-associative) joins and filters;
- linear associativity for filters.

Operations and transformations listed above model all essential features of extended query languages that affect the optimization algorithms and their performance.

4.1 Non-trivial Optimal Compound Segment

In this experiment a join of data from two sources demonstrates that optimal plan may depend on the amount of allocated resources even for small queries:

[JOIN 100 400]
   [SELECT][SRC 100]
   [SELECT][SRC 400].

Here and further we specify the number of objects in primary source and selectivity of operations in constants corresponding to operation or algorithm names. Estimations for cardinality of the join output are based on harmonic mean of cardinalities of sources used in the predicate.

In exact case the optimizer selects a plan containing hash-join and exact selection algorithms. In approximate case the space of possible plans is extended by plans containing approximate algorithms for operations. Figure 1(b) demonstrates quality functions for several query plans that are optimal in different constraint intervals; thus, the query has a non-trivial OCS.

4.2   Iterative Improvement and Recursive Descent

Traditional high-performance join algorithms may be unavailable in an extended query evaluation environment based on similarity. We assume that the above applies to our environment; therefore the hash-join algorithm is excluded from further experiments.

We compare our algorithms in terms of performance (time) and quality (how close is the constructed compound segment to the optimal one) and started from a quite simple query:

[JOIN 1000 4000]
    [SELECT][SRC 1000]
    [JOIN 4000 16000]
        [SELECT][SRC 4000]
        [SELECT][SRC 16000].

The experiment ended with the following outcome: all 3 algorithms produced similar dependence of the query result quality on the allocated amount of resources; the approximate join algorithm based on sampling participated in plans for the majority of query evaluation constraints.

We use short notation to refer to queries with selections and joins: S and J stand for SELECT and JOIN, respectively. Table 1 presents (see SSSJJ) an average absolute time spent for construction of the OCSs.

Table 1. Performance of OCS Construction for Simple Queries

Query	Algorithm	Full Time (ms)	Quality Function Time (ms)	Visited plans
	Full	562	271	576
SSSJJ	Recursive	246	119	441
	Iterative	171	56	259
	Full	36719	22673	17280
SSSSJJJ	Recursive	2623	1428	2952
	Iterative	752	325	911

All 3 algorithms also included the same set of optimal plans into the OCS for more complex query with 3 join operations. Time measurements for this query are shown in table 1 (see SSSSJJJ). One may see that the iterative improvement technique takes less time. The reason is significantly reduced set of visited plans.

To measure the accuracy of the proposed algorithms we use the area of zone above an OCS as its quality measure: lower the area means better quality. The experiments have shown that although recursive descent produces slightly better OCSs, the quality of the result OCSs does not differ significantly.

4.3   Performance of Optimal Compound Segment Construction

The problem of OCS construction is more complex compared to the single-objective optimization. A comparison of our implementation with industrial optimizers would not be fair as the algebraic properties of operations may differ from those of relational operations. Instead, we replace the cost-quality models of all operations with trivial ones. This effectively degenerates the OCS construction into the single-objective (exact) optimization.

The purpose of the experiment was to compare performance of the single-objective optimization with the OCS construction for queries with growing number of non-associative joins and filters above selections from primary sources. Starting from a query with two filters and one join at each step join and filter operations were added. We generated 5 queries for each number of joins varying filters selectivity and increasing number of objects in primary sources .

Figure 2(a) demonstrates the relative performance of our algorithms compared to the corresponding single-objective optimization implementations. The figure shows the time spent for the OCS construction, the time of quality functions construction, and the number of enumerated plans.

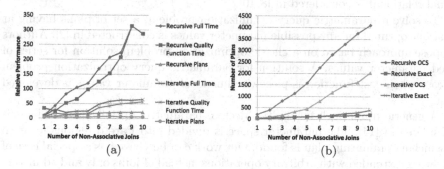

(a)   (b)

Fig. 2. Iterative Improvement and Recursive Construction Performance

Figure 2(b) demonstrates how the number of visited plans increases with the growing size of a query. As expected, the number of visited plans for OCS construction is significantly larger than that for single-objective optimization, but the growth of the ratio is moderate, indicating feasibility of acceptable performance for OCS construction.

5   Related Work

Query optimization for RDBMS was studied for decades [9,11]. The underlying algebras in analytical query processing systems are extended with additional operations, such as similarity joins, for which some of nice algebraic equivalences of relational algebra do not hold. Authors of [4] proposed efficient enumerators suitable for extended algebras.

Our work relies on firmly approximate algorithms, such as an algorithm for aggregation operation proposed in [12], approximate join algorithms discussed in [3], or any algorithm based on sampling.

Multi-objective query optimization is considered in [1,7,13]. Authors of [5] use multi-criteria optimization for parallel execution of SPJ queries, minimizing evaluation time and meeting constraints on capacity and performance.

Authors of [13] developed two approximate algorithms for multi-objective optimization based on principle of near-optimality. Multi-objective optimization problem is described by vector of criteria; cost function calculated as weighted sum on criteria values; and set of constraints on optimization criteria values.

Framework for parallel approximate query evaluation based on sampling is developed in [1]. Proposed system supports real time approximate evaluation of SPJA queries providing statistic guaranties on result quality. A specific multi-objective optimization problem on evaluation time and result quality is addressed in [1]: at the first step optimal set of table samples is selected; then optimal join order is constructed.

Yet another multi-objective optimization problem is solved in [7]: additive utility metric is constructed based on cost and capacity of a query.

The problem of multi-objective optimization for queries different from traditional relational is considered in [8,10].

To solve a parametric query optimization problem, a set of plans having at least one optimal for any possible parameter values is constructed in [2]. Authors propose approach based on a classic optimization problem solution for series of fixed parameter values. A solution for parametric query optimization problem in case when cost function is piecewise linear on parameter values is developed in [6] .

A generic framework for multi-objective parametric optimization of joins is considered in [14]. The parameter space is divided into Pareto regions for each of which a dominating plan is found. Our work can be viewed as a special case of this work, extended with arbitrary operations instead of joins only and admitting efficient algorithms for calculation of regions.

6   Conclusion

Declarative querying facilities both require and enable sophisticated query optimization. In the context of approximate query evaluation need in controlled trade-offs between quality and speed makes query optimization multi-objective.

We introduced techniques for the bi-objective optimization problem with resources and result quality as objectives. The solution is based on a compact representation of Pareto set. Experimental evaluation of the proposed algorithms based on different plan enumeration strategies demonstrated their feasibility.

References

1. Agarwal, S., Iyer, A.P., Panda, A., Madden, S., Mozafari, B., Stoica, I.: Blink and it's done: interactive queries on very large data. Proceedings of the VLDB Endowment 5(12), 1902–1905 (2012)
2. Bizarro, P., Bruno, N., DeWitt, D.J.: Progressive parametric query optimization. IEEE Transactions on Knowledge and Data Engineering 21(4), 582–594 (2009)
3. Braga, D., Campi, A., Ceri, S., Raffio, A.: Joining the results of heterogeneous search engines. Inf. Syst. 33(7–8), 658–680 (2008)

 4. Fender, P., Moerkotte, G.: Counter strike: generic top-down join enumeration for hypergraphs. Proceedings of the VLDB Endowment **6**(14), 1822–1833 (2013)
 5. Ganguly, S., Hasan, W., Krishnamurthy, R.: Query optimization for parallel execution. In: Proceedings of the 1992 ACM SIGMOD International Conference on Management of Data, SIGMOD 1992, pp. 9–18. ACM, New York (1992)
 6. Hulgeri, A., Sudarshan, S.: Parametric query optimization for linear and piecewise linear cost functions. In: Proceedings of the 28th International Conference on Very Large Data Bases, pp. 167–178. VLDB Endowment (2002)
 7. Kambhampati, S., Nambiar, U., Nie, Z., Vaddi, S.: Havasu: a multi-objective, adaptive query processing framework for web data integration. In: ASU CSE. Citeseer (2002)
 8. Kllapi, H., Sitaridi, E., Tsangaris, M.M., Ioannidis, Y.: Schedule optimization for data processing flows on the cloud. In: Proceedings of the 2011 ACM SIGMOD International Conference on Management of Data, pp. 289–300. ACM (2011)
 9. Kossmann, D., Stocker, K.: Iterative dynamic programming: a new class of query optimization algorithms. ACM Trans. Database Syst. **25**(1), 43–82 (2000)
10. Simitsis, A., Wilkinson, K., Castellanos, M., Dayal, U.: Optimizing analytic data flows for multiple execution engines. In: Candan, K.S., Chen, Y., Snodgrass, R.T., Gravano, L., Fuxman, A. (eds.) SIGMOD Conference, pp. 829–840. ACM (2012)
11. Steinbrunn, M., Moerkotte, G., Kemper, A.: Heuristic and randomized optimization for the join ordering problem. VLDB J. **6**(3), 191–208 (1997)
12. Theobald, M., Weikum, G., Schenkel, R.: Top-k query evaluation with probabilistic guarantees. In: Nascimento, M.A., Özsu, M.T., Kossmann, D., Miller, R.J., Blakeley, J.A., Schiefer, K.B. (eds.) VLDB, pp. 648–659. Morgan Kaufmann (2004)
13. Trummer, I., Koch, C.: Approximation schemes for many-objective query optimization. In: Dyreson, C.E., Li, F., Özsu, M.T. (eds.) SIGMOD Conference, pp. 1299–1310. ACM (2014)
14. Trummer, I., Koch, C.: Multi-objective parametric query optimization. Proceedings of the VLDB Endowment **8**(3) (2014)
15. Yarygina, A., Novikov, B.: Optimizing resource allocation for approximate real-time query processing. Computer Science and Information Systems **11**(1), 69–88 (2014)

Second International Workshop on Big Data Applications and Principles (BigDap 2015)

Cross-Checking Data Sources in MapReduce

Foto Afrati, Zaid Momani, and Nikos Stasinopoulos(✉)

National Technical University of Athens, Kesariani, Greece
afrati@softlab.ece.ntua.gr, {zed,stasino}@central.ntua.gr

Abstract. Fact checking from multiple sources is investigated from different and diverse angles and the complexity and diversity of the problem calls for a wide range of methods and techniques [1]. Fact checking tasks are not easy to perform and, most importantly, it is not clear what kind of computations they involve. Fact checking usually involves a large number of data sources that talk about the same thing but we are not sure which holds the correct information, or which has any information at all about the query we care for [2]. A join among all or some data sources can guide us through a fact checking process. However, when we want to perform this join on a distributed computational environment such as MapReduce, it is not obvious how to distribute efficiently the records in the data sources to the reduce tasks in order to join any subset of them in a single MapReduce job. In this paper, we show that the nature of such sources (i.e., since they talk about similar things) offers this opportunity, i.e., to distribute the records with low replication. We also show that the multiway algorithm in [3] can be implemented efficiently in MapReduce when the relations in the join have large overlaps in their schemas (i.e., they share a large number of attributes).

Keywords: Fact checking · MapReduce · Multiway join

1 Introduction

In many applications, we need to extract meaningful conclusions from multiple data sources that provide information about seemingly the same attributes[1]. Such different data sources provide pieces of information that may be slightly different, contradicting each other or even incomplete. E.g., in the latter case, the missing data problem is not at all uncommon as there exist abundant examples in the social sciences [4], field and clinical research [5], [6], as well as disciplines such as knowledge discovery, web personalization and fact-checking [7], [8], [9], [10]. In order to make sense of the data, we must address the missing data problem

---

This work was supported by the project Handling Uncertainty in Data Intensive Applications, co-financed by the European Union (European Social Fund - ESF) and Greek national funds, through the Operational Program "Education and Lifelong Learning", under the program THALES.

[1] Although the attributes may not have the same name, we have good reason to believe that they refer to the same entity/info.

T. Morzy et al. (Eds): ADBIS 2015, CCIS 539, pp. 165–174, 2015.
DOI: 10.1007/978-3-319-23201-0_19

while at the same time coping with the sheer amount of data presented to us. The typical approaches to dealing with the nuisance of missing data involve employing statistical methods to attack the problem beginning with seminal work such as [11] moving forward to data mining [9] and data integration techniques [2].

The view that is presented in this paper gives priority to incorporating as many data sources as possible to our knowledge. Since often not a single data source offers good information about all the potential queries, it would be valuable to combine and compare the information the sources are offering and find answers to our queries that are based on such a combination of many data sources that are partially reliable. It is not uncommon that some queries are answered reliably from certain data sources and some other for a different set of data sources.

In the following is an example for a problem that can be viewed as a part of an entity resolution problem (it is somewhat contrived for the sake of simplicity).

Example 1. Suppose we have four data sources that offer information about movies and actors participating in them and demographic information about the actors such as telephone numbers, addresses, etc. In many modern applications (e.g. Dremel [12], [13]) all this information is collected in a single relation with many attributes rather than in many relations (as would be the case, e.g., in a star schema with a big fact table and several smaller dimension tables). It might be the case that we have many data sources that offer similar such information, which therefore are over almost the same attributes but not quite. In our actors example, let us assume that we have four relations, R_1, R_2, R_3, R_4 where R_1 is over attributes $movieTitle, actorName, telNumber, producer$, R_2 is over attributes $movie - title, actor, address, tel$, R_3 is over attributes $actorName, address, producerCollaborating, telNumber$ and R_4 is over attributes $movieTitle, producer, producerTelNumber, prodAddress$ (remember some actors may also be producers). Since the sources are incomplete and not reliable, if we want the telephone number of Tom Cruise, we would inquire all the available sources and either find a single number or find the number that appears more often.

In this paper, we offer a solution to the problem demonstrated in the example above that can be implemented in the MapReduce [14] computational environment. In particular, we show that we can distribute many relations with large overlaps over their attributes in an efficient way across the reducers of a single MapReduce job by using the multiway join algorithm of [3]. We distribute our input relation records as if we intended to join all relations. However, having efficiently distributed our relation tuples, we may choose to join in the reducers any subset of the relations we find suitable for our particular query. We refer again to the example above to show how we may benefit from that.

Example 2. Suppose we want to find the telephone number of Tom Cruise. Then in each reducer we join all relations that contain telephone numbers (observe that relation R_4 may also contain a reference to Tom Cruise if he had been a producer) and see what we find.

```
SELECT tel FROM R1 JOIN R2 JOIN R3 JOIN R4
WHERE actor_name=Tom Cruise OR prod_name=Tom Cruise
```

If the output is not satisfying (e.g., empty, which means that some tel. numbers are wrong or there are multiple *tel* numbers and therefore they didn't join), we can join fewer relations in the reducers and come up with answers.

We may use the same distribution, of course, to resolve a different query, e.g., find the address of Angelina Jolie. Thus we have used the same distribution to the reducers for multiple queries as well as to resolve queries by choosing which relations to join (without the extra overhead of re-distributing the relations).

All of the above is feasible because of a very important property that the multiway algorithm of [3] has and which we will explain in subsection 2.

The rest of the paper is organized as follows. In **Section 2**, we review the multiway algorithm described in [3] for the case of the n-way join. In **Section 3**, we examine closely the communication cost of the multiway join in the case of "fat" relations and how this compares to binary multiway joins. In **Section 4**, we provide representative experimental results. **Section 5** refers to related work and **Section 6** discusses future directions.

2   Multiway Join Algorithm for MapReduce

MapReduce is a parallel computation framework for processing over large amounts of distributed data. The user provides an implementation of the *map* function which transforms its input to `<key, value>` pairs. Those pairs are shuffled along the distributed system and assigned to *reduce tasks*, each one distinctly identified by the `key` value. Each *reduce* task applies a reduce function to the `values` associated with each `key` to produce a final result.

In the multiway join algorithm [3], the map phase provides an efficient way to allocate tuples of relations to reduce tasks which in turn perform the actual join. We demonstrate the algorithm using two examples, a simple one and a more complicated one.

Example 3. Assume we have the following 3-way chain join:

$$A_1(x_1, x_2) \bowtie A_2(x_2, x_3) \bowtie A_3(x_3, x_4)$$

To perform a multiway join, the *map task* assigns tuples coming from different relations to different keys with the goal of merging tuples with matching attribute values at the reduce phase. In this example, at the reduce phase two separate joins must occur: one between A_1 and A_2 (based on a matching x_2 attribute), the other between A_2 and A_3 (based on a matching x_3 attribute).

In order to perform both joins inside the same reduce task all three tuples that have a match either in x_2 or x_3 must be passed to the same reducer. In our example, the middle relation A_2 joins on x_2 with A_1, so all tuples within A_1 with $A_1.x_2 = A_2.x_2$ must be in the same reducer, likewise for $A_2.x_3 = A_3.x_3$. To accomplish this, the middle relation acts as a key consisting of (x_2, x_3) to indicate which reducer is to receive the emitted `<key,value>` by the mapper. To restrict the arbitrary size of the key produced by (x_2, x_3) to a fixed size, the actual key is composed of hash values reflecting a set of buckets corresponding to a number of reducers (see figure 1).

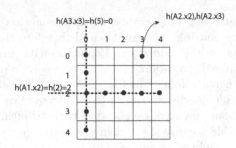

Fig. 1. Buckets corresponding to the number of reducers.

Let h be the hash function with range $0...m\text{-}1$ and each bucket is identified by a pair (i, j) where each i, j ranges between 0 and $m\text{-}1$. Hence the number of buckets is $m \times m$. All tuples received from the middle relation A_2 will be hashed to bucket $(h(x_2), h(x_3))$ whereas tuples coming from A_1 and A_3 will be replicated over buckets $(h(x_2), j)$ for all values of j and buckets $(i, h(x3))$ for all values of i respectively. As mentioned earlier the *mapper* emits the pairs `<key,value>`, the value part is the actual tuple including the name of the relation; for example (A_1, x_1, x_2). Then comes the role of the reducer that loops through the tuples received and merges the tuples with successful matches on x_2 and x_3.

Example 4. Assume now we want to perform the following 4-way chain join.

$$A_1(x_1, x_2) \bowtie A_2(x_2, x_3) \bowtie A_3(x_3, x_4) \bowtie A_4(x_4, x_5)$$

We extend the same multiway join method to join tuples in a single round. As before, all middle relations will act as keys. The key will consist of three attributes (x_2, x_3, x_4) hashed to $m \times m \times m$ reduce tasks. Notice that figure 1 now becomes a 3-dimensional array of buckets. Since all three attributes do not lie in the same tuple , the *map tasks* must hash according to what attributes are available and replicate the rest. For example, the key for a tuple $A_2(x_2, x_3)$ is $(h(x_2), h(x_3), l)$, where l is a third dimension or index for the third hashed value. Likewise, the key for tuple $A_1(x_1, x_2)$ is $(i, h(x_2), l)$, tuple $A_3(x_3, x_4)$ is $(i, h(x_3), h(x_4))$, and tuple $A_4(x_4, x_5)$ is $(i, j, h(x_4))$. The value part of the `<key, value>` pair emitted by the *map task* will be like before, the actual tuple and the relation name.

In general, the hash function applied on the value of attributes is different for each attribute (selected appropriately to balance the skewness introduced by the different sizes of relations). The number of buckets into which the values of a certain attribute are hashed is call the *share* of this attribute. The product of all shares should be equal to the number of reducers k.

So, the communication cost is a sum of terms, one term for each relation. Each term is the product of the size of the relation multiplied by the shares of the attributes that are missing from the relation.

In example 3, e.g.

$$A_1(x_1, x_2) \bowtie A_2(x_2, x_3) \bowtie A_3(x_3, x_4)$$

the communication cost expression is

$$a_1 \cdot s_3 + a_2 \cdot 1 + a_3 \cdot s_2$$

where a_i is the size of relation A_i and s_i is the share of attribute x_i.
In example 4, e.g.

$$A_1(x_1, x_2) \bowtie A_2(x_2, x_3) \bowtie A_3(x_3, x_4) \bowtie A_4(x_4, x_5)$$

the communication cost expression is

$$a_1 \cdot s_3 s_4 + a_2 \cdot s_4 + a_3 \cdot s_2 + a_4 \cdot s_2 s_3$$

Dominance Rule

> We say that an attribute Y is "dominated" by another attribute X if every
> relation R_i that contains Y also contains X. In other words, Y always appears
> with X in our join schema, while the contrary may not be true.

According to the dominance rule, if an attribute is dominates then the number
of shares it gets is equal to 1.

3   Optimizing Shares for the Fat Relation Join

We call relations with many joining attributes with each other "fat" relations; to
contrast a small, binary tuple of only two attributes. Computing joins over "fat"
relations is an application of the n-way chain join.

3.1   Cross-Checking *fat* Relations in MapReduce

We will use the multiway join for fat relations. In the next example, we construct
the map-key and calculate the communication cost expression for the join.

Example 5. We assume the following fat relations that form a 4-way fat join:

$$A_1(x_3, x_4, x_5) \bowtie A_2(x_1, x_4, x_5) \bowtie A_3(x_1, x_2, x_5) \bowtie A_4(x_1, x_2, x_3)$$

Using dominance rule for this query, we take the attributes in the middle
relations to act as keys. We do not need to take all the attributes in the middle
relations considering that some attributes are dominated by other attributes.
Notice that x_2 is dominated by x_1, because x_2 is always showing up along with
x_1 (in relations A_3 and A_4). Similarly, x_4 is dominated by x_5. However, this is
not the case, for example, with attribute x_3. Attribute x_3 is accompanied by x_4
and x_5 in A_1 and with x_1 and x_2 in A_4. Thus, x_3 must be part of the map-key
since it does not dominate nor is it dominated. Finally the key is composed of
$(x_1; x_3; x_5)$.

The communication cost expression for this join is[2] as follows:

$$a_1 \cdot s_1 + a_2 \cdot s_3 + a_3 \cdot s_3 + a_4 \cdot s_5$$

---

[2] Remember we used the convention that a_i is the size of relation A_i and s_i is the
share of attribute x_i.

Observe that, if all relations are of the same size, that is $a_1 = a_2 = a_3 = a_4 = a$, then because of symmetry all the shares are equal, $s_1 = s_3 = s_5$, since we also have the constraint $s_1 \cdot s_2 \cdot s_3 \cdot s_4 = k$.

Also, notice that going back to example 4, under the same assumption of equal relation size, the communication cost is minimized when the shares are the following: $s_2 = s_4 = \sqrt{k}$ and $s_3 = 1$.

This is not a coincidence, because in the join of example 5, all relations share two attributes with the map-key, whereas in example 4 relations A_1 and A_4 share only one attribute with the map-key. This is an indication that records from those two relations will have high replication. We make this point more formal in the next subsection.

In fact checking, as we established in the introduction, we often have to extract information from collections of relations with large overlaps over their attributes. This can be done by taking the join of all or some of those relations. We have just demonstrated that we can achieve that in MapReduce with a low communication cost. Thus, in this paper, we present an efficient way to perform fact checking tasks.

3.2   Examining the Cost of Fat Relations Join

In this section, we examine the cost of using the multiway approach in the case of "fat" relations. Such relations overlap over many attributes in their schemas and each one is missing the same small number of attributes.

Suppose we have a multiway join query Q that is symmetric in the sense that it doesn't matter which relation we choose to begin with when constructing the map-key for computing query Q.

Let d be the number of all attributes present in the join query Q and k the number of reducers.

Because of symmetry each share is the same: $k^{\frac{1}{d}}$ (so it holds that $(k^{\frac{1}{d}})^d = k$)

Suppose we have n relations and r tuples on each relation. If each relation participating in the join is missing 2 attributes from the join key, then each relation's tuple must be must shared across $(k^{1/d})^2 = k^{2/d}$ reducers. Since, there are n relations the total communication cost is $nk^{2/d}$.

The communication cost per reducer is:

$$q = (rnk^{2/d})/k = \frac{rn}{k^{\frac{d-2}{d}}}$$

Similarly, if there are i missing attributes the cost is: $q_i = \frac{rn}{k^{\frac{d-i}{d}}}$

Notice that rn is the input size of our query which we denote with IN, so that the per reducer communication cost is:

$$q_i = \frac{IN}{k^{\frac{d-i}{d}}}$$

Remember that d is the number of all attributes and i is the number of missing attributes at each relation.

We make the following observations:

- If i is 1, $q_1 = \frac{IN}{k^{\frac{d-1}{d}}}$, that is the cost for each reducer is almost IN/k, which is the cost in the case of embarrassing parallelization.
- If i is close to d, e.g. $d-1$, that is the relation contains very few of the attributes present in the map-key, then the cost on each reducer is almost all the input IN. In this case, the tuples need to be replicated to many reducers, which is undesirable.
- The above extreme points made mean that this approach is more effective for relations that join on many attributes, that is, for $i \ll d$, the decreasing q_i expression moves towards an optimal parallelization.

4   Experimental Evaluation

For the experimental evaluation we computed 2 joins, the 4-way binary join

$$A_1(x_1, x_2) \bowtie A_2(x_2, x_3) \bowtie A_3(x_3, x_4) \bowtie A_4(x_4, x_5)$$

and the 4-way fat relation join

$$A_1(x_3, x_4, x_5) \bowtie A_2(x_1, x_4, x_5) \bowtie A_3(x_1, x_2, x_5) \bowtie A_4(x_1, x_2, x_3)$$

We designed our experiments to demonstrate our claim that the distribution of the relations to the reducers is "cheaper" in the case of fat relations as compared to the case of not fat relations. Thus, our first experiment computes the two queries but without doing the actual join in the reducers (we just piped the tuples through from the mappers to the reducers). In this experiment, we did in fact (see table 1) demonstrate our claim that the communication cost is lower in the case of the fat query.

In our second experiment, we did actually compute the whole join in the reducers and, as expected (see table 2), the fat relation join took longer to run than the binary relation join. This difference is attributed not only to the difference in communication cost mentioned before, but also to a lower per reducer cost as shown in 3.2.

For the experiments we used datasets that are randomly generated. The experiments were conducted on a Hadoop cluster of 4 computation nodes with Intel Core i5 processors (2.7GHz) and 4GB of RAM each over a 1Gbps Ethernet.

Table 1. Processing time in Piped through 4-way binary and fat relation join

Extended Binary Multiway Join: Pipe through			
	27 Reducers	64 Reducers	125 Reducers
20000 records	11504 ms	19804 ms	47096 ms
24000 records	12628 ms	21656 ms	51664 ms
Fat Relations Multiway Join: Pipe through			
	27 Reducers	64 Reducers	125 Reducers
20000 records	11432 ms	18876 ms	42916 ms
24000 records	10840 ms	20924 ms	44496 ms

Table 2. Processing time in 4-way binary and fat relation join

Extended Binary Multiway Join			
	27 Reducers	64 Reducers	125 Reducers
20000 records	335836 ms	313912 ms	358444 ms
24000 records	623148 ms	580612 ms	635772 ms
Fat Relations Multiway Join			
	27 Reducers	64 Reducers	125 Reducers
20000 records	239310 ms	195910 ms	309420 ms
24000 records	175808 ms	136476 ms	166960 ms

5   Related Work

Much work and thought has been put into performing the basic challenges of *Big Data Integration*: Schema Alignment and Record Linkage. Especially the Record Linkage steps of Blocking and Pairwise Matching for heterogeneous data have been explored in the distributed setting, particularly MapReduce. In [15], multiple MapReduce rounds are used to perform a set similarity join of input records. In [16], blocking key techniques are transferred to the MapReduce setting with a view to balance the load across computation nodes. Within MapReduce, a

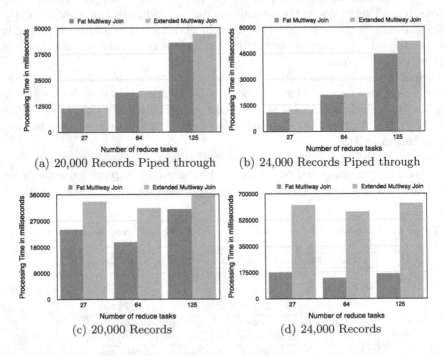

Fig. 2. Processing time comparison of Binary Multiway vs. Fat Multiway Join

dynamic approach for record blocking is proposed in [17] and a dynamic Sorted Neighborhood blocking approach in [18]. Dedoop[19] is an Entity Resolution framework based on MapReduce. Pair-wise similarity computation with MapReduce is refined for redundant pair comparisons in [20].

Data integration has also been explored in the setting of computational fact-checking [1],[2], where heterogeneous data are considered as sources of information that need to be cross-checked. Fusing together fat relations provides a stepping stone for further similarity research and source trustworthiness characterization.

6   Conclusions and Future Work

We have demonstrated an efficient way to match information coming from different data sources. Building on top of that we intend to refine cross-checking of data. Here are possible ways to achieve that:

- Dangling records that didn't join completely on the reducer can be either partially matching or referring to completely different entities. For the former, we may decide to lower our join threshold by using smaller map-keys. However, this comes with the inefficiencies we discussed earlier in Section 3.2.
- By counting the number of records matching from each origin data source, we can measure the authority of the data source. The authority of a data source is an indicator of how much we trust it. Knowing about the trustworthiness of source, we will then characterize facts coming from it.
- Records that join at the reducers match on their critical (map-key) attributes. We can assume that they should match also in other attributes. For example, actors matching on name and age may come with slightly different telephone numbers which is not a prerequisite for entity matching. We can correct errors this way at the reducer by employing similarity techniques.
- Data.gov [21] claims to have up to 400,000 data sets. Our method can be very useful since many of these data sets have missing values and characters.
- We can cross-check our results with RDF data, but since HDFS only stores and manipulates flat files an RDF triples will have to be transformed into tabular form.

References

1. Dong, X.L., Gabrilovich, E., Heitz, G., Horn, W., Murphy, K., Sun, S., Zhang, W.: From data fusion to knowledge fusion. Proceedings of the VLDB Endowment **7**(10) (2014)
2. Dong, X.L., Berti-Equille, L., Srivastava, D.: Integrating conflicting data: the role of source dependence. Proceedings of the VLDB Endowment **2**(1), 550–561 (2009)
3. Afrati, F.N., Ullman, J.D.: Optimizing joins in a map-reduce environment. In: Proceedings of the 13th International Conference on Extending Database Technology, pp. 99–110. ACM (2010)

4. Juster, F.T., Smith, J.P.: Improving the quality of economic data: Lessons from the hrs and ahead. Journal of the American Statistical Association **92**(440), 1268–1278 (1997)
5. Graham, J.W.: Missing data analysis: Making it work in the real world. Annual Review of Psychology **60**, 549–576 (2009)
6. Acock, A.C.: Working with missing values. Journal of Marriage and Family **67**(4), 1012–1028 (2005)
7. Grzymała-Busse, J.W., Hu, M.: A comparison of several approaches to missing attribute values in data mining. In: Ziarko, W.P., Yao, Y. (eds.) RSCTC 2000. LNCS (LNAI), vol. 2005, pp. 378–385. Springer, Heidelberg (2001)
8. Padmanabhan, B., Zheng, Z., Kimbrough, S.O.: Personalization from incomplete data: what you don't know can hurt. In: Proceedings of the Seventh ACM SIGKDD International Conference on Knowledge Discovery and Data Mining, pp. 154–163. ACM (2001)
9. Magnani, M.: Techniques for dealing with missing data in knowledge discovery tasks. Obtido **15**(01), 2007 (2004). http://magnanim.web.cs.unibo.it/index.html
10. Li, X., Dong, X.L., Lyons, K., Meng, W., Srivastava, D.: Truth finding on the deep web: is the problem solved? Proceedings of the VLDB Endowment **6**(2), 97–108 (2012)
11. Dempster, A.P., Laird, N.M., Rubin, D.B.: Maximum likelihood from incomplete data via the em algorithm. Journal of the Royal Statistical Society. Series B (methodological), 1–38 (1977)
12. Afrati, F.N., Delorey, D., Pasumansky, M., Ullman, J.D.: Storing and querying tree-structured records in dremel. PVLDB **7**(12), 1131–1142 (2014)
13. Melnik, S., Gubarev, A., Long, J.J., Romer, G., Shivakumar, S., Tolton, M., Vassilakis, T.: Dremel: interactive analysis of web-scale datasets. Commun. ACM **54**(6), 114–123 (2011)
14. Dean, J., Ghemawat, S.: Mapreduce: simplified data processing on large clusters. Communications of the ACM **51**(1), 107–113 (2008)
15. Vernica, R., Carey, M.J., Li, C.: Efficient parallel set-similarity joins using mapreduce. In: Proceedings of the 2010 ACM SIGMOD International Conference on Management of Data, pp. 495–506. ACM (2010)
16. Kolb, L., Thor, A., Rahm, E.: Load balancing for mapreduce-based entity resolution. In: 2012 IEEE 28th International Conference on Data Engineering (ICDE), pp. 618–629. IEEE (2012)
17. McNeill, N., Kardes, H., Borthwick, A.: Dynamic record blocking: efficient linking of massive databases in mapreduce. In: Proceedings of the 10th International Workshop on Quality in Databases (QDB) (2012)
18. Mestre, D.G., Pires, C.E.: An adaptive blocking approach for entity matching with mapreduce
19. Kolb, L., Rahm, E.: Parallel entity resolution with dedoop. Datenbank-Spektrum **13**(1), 23–32 (2013)
20. Kolb, L., Thor, A., Rahm, E.: Don't match twice: redundancy-free similarity computation with mapreduce. In: Proceedings of the Second Workshop on Data Analytics in the Cloud, pp. 1–5. ACM (2013)
21. U.S. General Services Administration: U.S. government's open data (2013). http://www.data.gov/ (accessed June 19, 2015)

CLUS: Parallel Subspace Clustering Algorithm on Spark

Bo Zhu$^{(\boxtimes)}$, Alexandru Mara, and Alberto Mozo

Universidad Politécnica de Madrid, Madrid, Spain
bozhumatias@ict-ontic.eu, alexandru.mara@alumnos.upm.es, a.mozo@upm.es

Abstract. Subspace clustering techniques were proposed to discover hidden clusters that only exist in certain subsets of the full feature spaces. However, the time complexity of such algorithms is at most exponential with respect to the dimensionality of the dataset. In addition, datasets are generally too large to fit in a single machine under the current big data scenarios. The extremely high computational complexity, which results in poor scalability with respect to both size and dimensionality of these datasets, give us strong motivations to propose a parallelized subspace clustering algorithm able to handle large high dimensional data. To the best of our knowledge, there are no other parallel subspace clustering algorithms that run on top of new generation big data distributed platforms such as MapReduce and Spark. In this paper we introduce CLUS: a novel parallel solution of subspace clustering based on SUB-CLU algorithm. CLUS uses a new dynamic data partitioning method specifically designed to continuously optimize the varying size and content of required data for each iteration in order to fully take advantage of Spark's in-memory primitives. This method minimizes communication cost between nodes, maximizes their CPU usage, and balances the load among them. Consequently the execution time is significantly reduced. Finally, we conduct several experiments with a series of real and synthetic datasets to demonstrate the scalability, accuracy and the nearly linear speedup with respect to number of nodes of the implementation.

Keywords: Subspace · Parallel · Clustering · Spark · Big data

1 Introduction

Clustering is one of the main techniques for unsupervised knowledge discovering out of unlabeled datasets. This technique uses the notion of similarity to group data points in entities known as clusters. Traditional clustering algorithms such as partitioning based approaches (e.g. K-Means[1]), density based approaches (e.g. DBSCAN[2]) and hierarchical approaches (e.g. DIANA[3]), take the full

The research leading to these results has been developed within the ONTIC project,which has received funding from the European Union's Seventh Framework Programme (FP7/2007-2011) under grant agreement no. 619633.

© Springer International Publishing Switzerland 2015
T. Morzy et al. (Eds): ADBIS 2015, CCIS 539, pp. 175–185, 2015.
DOI: 10.1007/978-3-319-23201-0_20

feature space into consideration. However, as current datasets become larger and higher-dimensional, these algorithms fail to uncover meaningful clusters due to the existence of irrelevant features and the curse of dimensionality[4].

In many application domains, such as sensor networks, bioinformatics, and network traffic, objects are normally described by hundreds of features. Since data collection and storage become cheaper and more convenient, bigger datasets are generated without an analysis of relevance. Consequently, a number of techniques were intensively studied to address the clustering task for high dimensional datasets. Dimensionality reduction techniques like Principle Component Analysis[5] and feature selection techniques like mRMR feature selection algorithm[6], generate an optimal subset of features containing the most relevant information. Given the fact that clusters can be found in different subsets of features, such techniques fail to detect locally relevant features for each cluster[4].

A special family of algorithms, which derived from the frequent pattern mining field[7], rapidly constituted a novel field named subspace clustering. These algorithms aimed to find clusters hidden in subsets of the original feature space and therefore avoid the curse of dimensionality. Nevertheless, up to our knowledge none of these techniques scales well with respect to the size of datasets. They are generally under the assumption that the whole dataset can fit in a single machine. SUBCLU[8], an Apriori based subspaces clustering algorithm uses a traditional DBSCAN implementation to cluster the data in each subspace resulting in an inefficient and non-scalable solution. In this context, parallelization comes out as a natural solution to improve the efficiency and scalability of existing subspace clustering algorithms. However, during our research we have detected a lack of scalable and parallel subspace clustering approaches for high dimensional data on top of current Big Data distributed platforms. Recently, MapReduce[9] paradigm jointly with the Hadoop framework[10] have become the most popular Big Data distributed framework. Nevertheless, the limitations of MapReduce and its lack of suitability for iterative Machine Learning algorithms have motivated researchers to propose different alternatives, being Spark [11] one of the most representative. Therefore, designing scalable subspace clustering algorithms on top of Spark is an interesting challenge.

Contributions. In this paper, we present CLUS, a novel parallel algorithm on top of Spark based on SUBCLU that overcomes the latter's limitations using a dynamic data partitioning method. CLUS reduces the dimensionality and time complexity of SUBCLU by parallelizing the clustering tasks across different nodes. CLUS also eliminates the dataset size limitation of the centralized algorithm to the available RAM in one machine by distributing it across nodes and spilling it to disk when needed.

To develop CLUS, specific Spark primitives were used as they have the potential to provide higher performance if employed in an appropriate manner. Unlike MapReduce, Spark gives more flexibility allowing users to manage memory, disk and network usage in order to obtain more efficient algorithms. Moreover, instead of writing intermediate results to disk like MapReduce, Spark intends to maintain them in memory. Due to this in-memory computing Spark outperforms by an order of magnitude other Big Data platforms [24].

In summary, the main contributions of CLUS algorithm are:

1. A dynamic data partitioning method is carefully designed using Spark's specific operations in order to induce data locality. This optimization step reduces the cost of communications by avoiding unnecessary slow data shuffling, which is common in most MapReduce and Spark applications.
2. CLUS avoids replications of the whole dataset in each node in order to process it in parallel. Based on Spark's specific operations, CLUS can be deployed on a cluster of commodity machines and efficiently process huge datasets.
3. I/O cost is minimized by using indexing so as to access and move to specific nodes only the necessary data in each iteration. This results in faster and more efficient data management and better use of available RAM.
4. We report experimental results on real and synthetic datasets to show the scalability and accuracy of our algorithm. The results show a dramatic decrease in CLUS execution time w.r.t. centralized SUBCLU.

The reminder of this paper is organized as follows: Section 2 presents an overview of the related work. Section 3 gives a brief introduction to SUBCLU algorithm and presents the implementation of CLUS. In Section 4 we show experimental results on various datasets. Finally, Section 5 outlines our conclusions and future work.

2   Related Work

2.1   Subspace Clustering

Many research works have tried to address the subspace clustering task in the last two decades. There are some excellent surveys[4,12,13] that conducted either theoretical or experimental comparisons among different subspace clustering algorithms. These approaches were categorized into several classes considering different algorithmic aspects. In [12] a number of classical algorithms were divided into bottom-up and top-down groups based on the applied search method.

The bottom-up group first generates 1-D histogram information for each dimension, and then tries to perform a bottom-up traversal on the subspace lattice. Since the time complexity of naive traversal on the lattice is exponential w.r.t. the dimensionality of the dataset, most of these approaches conduct a pruning step for selecting candidates. This filtering procedure is based on the anti-monotonicity property.

Anti-Monotonicity Property: if no cluster exists in subspace S_k, then there is no cluster in any higher dimensional subspaces S_{k+1} either. i.e.

$$\exists S_k, C_{S_k} = \varnothing \Rightarrow \forall S_{k+1} \supset S_k, C_{S_{k+1}} = \varnothing.$$

By starting from one dimension and adding another dimension each time, bottom-up approaches tend to work efficiently in relatively small subspaces. Consequently, they generally show better scalability when uncovering hidden subspace clusters in lower dimensions. However, the performance decreases dramatically with the size of candidate subspaces containing clusters[12]. Examples of bottom-up approaches are CLIQUE [14], ENCLUS[15] etc.

In contrast with bottom-up approaches, top-down methods start from the equally-weighted full feature space and generate an approximation of the set of clusters. After the initialization, updates of the weight for each dimension in each cluster and the regeneration of clusters are iteratively conducted. Finally, a refinement of the clustering result is carried out to achieve a better quality of clusters. Since multiple iterations of clustering process are conducted in the full feature space, sampling techniques are generally used to increase efficiency by reducing the accuracy of the results. Clusters generated by this kind of approaches are non-overlapping and of similar dimensionality due to the mandatory input parameters. Algorithms such as PROCLUS[16] and ORCLUS[17] are typical examples of top-down approaches.

In some recent summary research works[4,13] subspace clustering algorithms are classified into three paradigms with regards to the underlying cluster definition and parametrization. Grid-based approaches, e.g. SCHISM[18], MaxnCluster[19], try to find sets of grid bins which contain more data than a density threshold for different subspaces. Density-based approaches, e.g. SUBCLU[8], INSCY[20], search for dense regions separated by sparse regions by calculating the distance of relevant dimensions. Clustering-oriented approaches, e.g. STATPC[21], Gamer[22], assume global properties of the whole cluster set similar to those of top-down approaches[12]. As Emmanuel Müller summarized in [4], "Depending on the underlying clustering model, algorithms always have to tackle the trade-off between quality and runtime. Typically, high quality results have to be paid with high runtime." Our effort in this paper is to increase efficiency by means of parallelization while maintaining high quality clustering results.

2.2   Parallel Subspace Clustering

During a thorough survey of the state-of-the-art, we detected a lack of parallel subspace clustering algorithms despite of the potential performance improvements that could be obtained from parallelization. Up to our knowledge, only two parallel implementations have been proposed by now. They used specific architectures like Message Passing Interface (MPI) and Parallel Random Access Machine (PRAM). Compared with the novel Spark framework, these models suffer from a non-negligibly high communication cost and a complex failure recovery mechanism. Furthermore, they neither scale as well as Spark nor provide as much flexibility to programmers.

In [23] the grid based parallel MAFIA algorithm was proposed as an extension of CLIQUE. MAFIA partitions each dimension into several discrete adaptive bins with distinct densities. In order to run the algorithm in parallel, the original dataset was randomly partitioned into several parts and read into different nodes. Data parallelization based on a shared-nothing architecture, which assembles a naive version of MapReduce, can bring a significant reduction in execution times. However, the generated partitions can be highly skewed and greatly affect the quality of the clustering results.

The other parallel algorithm was based on the Locally Adaptive Clustering (LAC) algorithm. LAC was proposed in [25] as a top-down approach by assigning values to a weight vector based on the relevance of dimensions within the corresponding subspace cluster. Parallel LAC proposed in [26], transforms subspace clustering task to the problem of finding K centroids of clusters in an N-dimensional space. Given K x N processors that share a global memory, PLAC managed to distribute the whole dataset into a grid of K centroids and N nodes for each centroid. In the experiments, they used one machine as the global shared memory, assuming that the whole dataset could fit in a single node. This architectural design severely limited the scalability of PLAC, as the maximum size of the dataset shown in [26] was no larger than 10,000 points.

3 CLUS: Parallel Subspace Clustering Algorithm on Spark

3.1 SUBCLU Algorithm

SUBCLU follows a bottom-up, greedy strategy intending to detect all density-based clusters in all subspaces by using DBSCAN algorithm. First, a clustering process is performed over each dimension to generate the set of 1-Dim subspaces containing clusters. Then, SUBCLU recursively generates the set of candidates for the (K+1)-Dim by combining pairs of K-Dim candidates with clusters sharing K-1 dimensions. The Anti Monotonicity property is used to prune irrelevant candidates. To increase the efficiency of the subsequent clustering process, the K-Dim subspace with minimum amount of clustered data is chosen as the "best subspace" for running DBSCAN. The recurrence terminates when no more clusters are detected. The algorithm takes the same two parameters: epsilon and minpts as DBSCAN. A more detailed description of SUBCLU, the process of generating candidates as well as experimental results can be found in [8].

3.2 CLUS Algorithm

CLUS is a parallel algorithm on top of the Spark that overcomes the severe limitations of SUBCLU. While SUBCLU sequentially runs DBSCAN clustering for each subspace and has to wait for the termination of all the instances before generating the candidate set for next dimensionality, CLUS is able to execute multiple DBSCANs in parallel. The candidate generation and pruning steps are also performed in a distributed manner without iterating over the dataset. CLUS takes advantage of specific Spark primitives such as ReduceByKey, AggregateByKey etc. to induce data locality and improve the overall performance.

It should be noted that Spark is able to autonomously place the data in the nodes that require it by means of shuffle and repartition operations. However, these primitives imply data movements resulting in slow disk and RAM I/O operations and even slower network I/O operations. Moreover, Spark's attempts to assure data locality using the previously mentioned autonomous mechanisms might result in inefficiencies as they could further trigger other unnecessary shuffles. These dispensable time consuming operations are a very common problem in most MapReduce based platforms and should be avoided as much as possible. The data management strategy to achieve this has to be carefully considered in order to provide the precise partitions at any moment that will achieve load balance in each node and boost the performance.

The main challenges of CLUS are to assure that: 1) each node has the data required to run subspace clustering; 2) partial data replication might be necessary to assure simultaneous subspace clustering; 3) the minimum number of shuffles and repartitions are performed; 4) nodes are well balanced and only specific data is moved across them; 5) there are no idle nodes; 4) the information is always stored and accessed using efficient (key,value) structures.

```
Input: data matrix M ∈ ℝ^{m×n}

1.  Compute CRV RDD({Column, Row, Value}) aplying transformations to M
2.  RDD DBS1 := Assign one feature per node, compute DBSCAN and remove features with no clusters
3.  maxSubs := DBS1.count()
4.  if maxSubs > 1
5.     currentSubs := 0
6.     while (currentSubs < maxSubs-1) and (not stop)
7.        currentSubs := currentSubs + 1
9.        candidates := Generate (K+1)-D candidates
8.        if currentSubs != 1 and candToPrune != 0
13.          candidates := candidates.substractByKey(candToPrune)
14.       numCandidates := candidates.count()
15.       if numCandidates != 0
16.          Get real data => candidates.flatMap().leftOuterJoin(CRV).rreduceByKey(CandidatePerNode)
17.          Compute DBSCAN for each candidate
18.          candToPrune := candidates with no cluster
19.          if currentSubs < 2
20.             stop := True
21.       else
22.             stop := True
```

Fig. 1. Pseudocode of CLUS algorithm

CLUS pseudocode is shown in Figure 1. The algorithm follows a similar concept to SUBCLU, but designed to reduce data reallocation and improve the efficiency of parallel density-based clustering tasks. CLUS manages the input information by columns/features so that different columns are stored in different machines as (key,value) pairs. The algorithm starts by running parallel DBSCANs on each of these features. The set of dimensions with clustered points are further used to generate (K+1)-Dim candidates. These candidates are generated by adding to each K-Dim subspace a disjoint dimension containing at least one cluster. The resulting (K+1)-Dim subspaces might require the same column to be present in different machines (e.g. to compute the subspaces f1-f2 and f1-f3 at the same time), so an efficient flatMap-leftOuterJoin-reduceByKey schema is

used for data replication. This mechanism allows CLUS to simultaneously execute independent DBSCANs on each subspace. The density based clustering is
performed again on each of these higher dimensional subspaces, but using only
the points that were not marked as noise in lower subspaces. A pruning step is
conducted over the resulting set of (K+1)-Dim subspaces in order to remove the
overlapping candidates (e.g. subspace f1-f2 plus subspace f3 generates the same
candidate as f2-f3 plus f1). In addition, based on the Anti-Monotonicity property
all the (K+1)-Dim subspaces containing a K-Dim subspace without any cluster
are also removed. To this end, the subspaces eliminated because of their lack
of clusters are used to generate (K+1)-Dim subspaces. These subspaces to be
removed are substracted from the pool of valid (K+1)-Dim subspaces of the next
iteration. CLUS ends when no higher subspace can be generated.

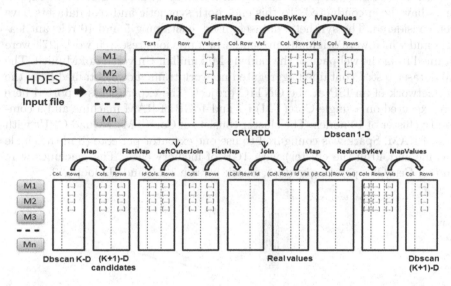

Fig. 2. Workflow of CLUS

The data workflow of CLUS is illustrated in Figure 2. For the sake of clarity some operations have been omitted, thus showing only the most important
steps. The top part of the figure shows how clustering is performed on each
dimension and how the work is distributed across Spark nodes. As mentioned,
data is obtained form a text file stored in the HDFS and reorganized by columns
and rows using a (key,value) associative data structure. Each real value has a
column and row index and they constitute an RDD named CRV. This CRV
RDD is stored across nodes so that each of them has one column. DBSCAN is
performed in a distributed fashion and a new RDD with the indexes of the clustered points is obtained. This RDD is further used to generate the (K+1)-Dim
candidates. The bottom part of Figure 2 shows the process of generating the
set of candidates from prior K-Dim subspaces. By managing only the indexes of
clustered points memory and network usage are optimized. The real data values
are only accessed when DBSCAN needs to be performed trough a join operation

with the CRV RDD. The special structures of CRV and the index RDDs allows each node to access the precise information it requires for each subspace analysis and assures concurrent access to these data. The keys of the index RDD constitute the set of subspaces with clusters and therefore are used to generate new candidates. For each subspaces dimensionality a single data shuffle and repartition is strictly required. The results generated in each iteration of the algorithm are stored in a specific RDD through a union operation and wrote to HDFS at the and of the execution. This provides a clear and easy way to manage outputs.

4   Preliminary Experiments

In order to evaluate the performance and accuracy of CLUS a series of experiments have been conducted. To this end, both synthetic and real datasets have been considered. The synthetic datasets used, containing 5 and 10 relevant features and which were originally presented in a previous research work [27], were obtained from the webpage of the Ludwig Maximilian University of Munich. The real dataset tested contained aggregated network traffic flows obtained from the core network of an ISP at the ONTIC project. The experiments reported here were executed on a single Core(TM) i-7 and 16GB of RAM machine and a commodity cluster of 10 nodes. The cluster consists of Core(TM) 2 Quad CPUs with 4GB of RAM. Spark was configured to use one executor per core (8 in the single machine and 40 in the cluster) with 1GB of memory each. The machines were running CentOS operating system, Spark version 1.2.2 and Hadoop 2.0.

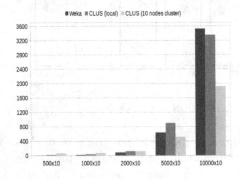 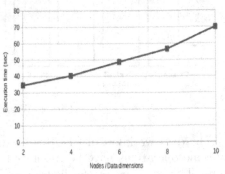

Fig. 3. SUBCLU and CLUS execution time w.r.t. number of nodes

Fig. 4. Execution time with one node per dimension

4.1   Scalability

First we have evaluated the scalability of the algorithm comparing it to a centralized version of SUBCLU available in the OpenSubspace v3.31 project [4]. OpenSubspace extends with subspace clustering algorithms the well known Weka ML library. Figure 3 shows the execution times of SUBCLU and CLUS on a single machine and the cluster for different dataset sizes.

During the initial tests, Weka, without any additional requirement for cluster evaluation or visualization ran out of memory with 1GB of Java heap. CLUS running on a single node and with the same dataset used less than 3,5 MB. The total amount of heap for Weka was further increased to 8GB in order to be able to run all the experiments.

Additional experiments show, as expected, that CLUS's execution time grows with a quadratic factor w.r.t. dataset size and number of features on a constant number of cluster nodes (Figure 5 and Figure 6). However, as the number of machines in the cluster increases, CLUS achieves a nearly linear speedup by taking full advantage of the data partitioning strategy and simultaneous clustering executions. With a number of machines equal to the dimensions of the dataset a nearly linear speedup is achieved, as shown in Figure 4.

We have also tested CLUS with real network traffic data of up to 10000 flows and 10 features obtaining interesting insights. The algorithm discovered clusters in up to 8 dimensions and others in lower ones aggregating flows of the same type. The results were consistent in time and dimensionality with the ones using synthetic data.

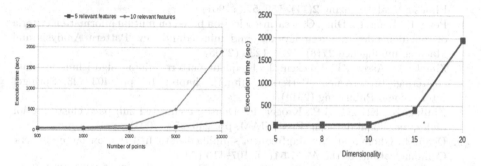

Fig. 5. Execution time w.r.t. dataset size.   **Fig. 6.** Execution time w.r.t. dimensionality

4.2   Accuracy

In order to evaluate the accuracy of CLUS the algorithm was compared to the Weka implementation of SUBCLU providing the same clusters. Nevertheless, Weka implementation misclassified some points, so a Java version of the algorithm was implemented from the original paper [8]. Results showed a perfect match between this version of SUBCLU and CLUS on different datasets.

5   Conclusions and Future Work

In this paper we present CLUS: a novel scalable and parallel subspace clustering algorithm on top of Spark. CLUS is highly inspired by the well-known SUB-CLU algorithm. Relying on specific Spark primitives, CLUS is able to execute multiple DBSCAN tasks in parallel achieving a significant speedup. In addition, unnecessary data shuffle is avoided by using a carefully designed data partitioning strategy to induce data locality. We carried out some preliminary tests on a modest cluster showing promising results with respect to scalability. The obtained results show that CLUS efficiently finds correct subspace clusters. In the future, we plan to extend our experiments with respect to scalability both on the number of dimensions and the size of datasets.

References

1. Forgy, E.W.: Cluster analysis of multivariate data: efficiency versus interpretability of classifications. Biometrics **21**, 768–769 (1965)
2. Ester, M., Kriegel, H.P., Sander, J., Xu, X.: A density-based algorithm for discovering clusters in large spatial databases with noise, pp. 226–231. AAAI Press (1996)
3. Kaufman, L., Rousseeuw, P.J.: Finding Groups in Data. John Wiley & Sons (1990)
4. Müller, E., Günnemann, S., Assent, I., Seidl, T.: Evaluating clustering in subspace projections of high dimensional data. In: Proc. VLDB, vol. 2(1) (2009)
5. Pearson, K.: On Lines and Planes of Closest Fit to Systems of Points in Space. Philosophical Magazine **2**(11), 559–572 (1901)
6. Peng, H., Long, F., Ding, C.: Feature selection based on mutual information criteria of max-dependency, max-relevance, and min-redundancy. Pattern Analysis and Machine Intelligence **27**(8), 1226–1238 (2005)
7. Zimek, A., Assent, I., Vreeken, J.: Frequent pattern mining algorithms for data clustering. In: Frequent Pattering Mining, chapter 16, pp. 403–423. Springer International Publishing (2014)
8. Kailing, K., Kriegel, H.P., Kröger, P.: Density-connected subspace clustering for high-dimensional data. In: Proc. SIAM, pp. 246–257 (2004)
9. Dean, J., Ghemawat, S.: MapReduce: simplified data In Proc. on large clusters. Communications of the ACM **51**(1), 107–113 (2008)
10. Shvachko, K., et al.: The hadoop distributed file system. In: 2010 IEEE 26th Symposium on Mass Storage Systems and Technologies (MSST). IEEE (2010)
11. Zaharia, M., et al.: Resilient distributed datasets: a fault-tolerant abstraction for in-memory cluster computing. In: Proc. USENIX (2012)
12. Parsons, L., Haque, E., Liu, H.: Subspace clustering for high dimensional data: a review. ACM SIGKDD Explorations Newsletter **6**(1), 90–105 (2004)
13. Sim, K., Gopalkrishnan, V., Zimek, A., Cong, G.: A survey on enhanced subspace clustering. Data Mining and Knowledge Discovery **26**(2) (2013)
14. Agrawal, R., Gehrke, J., Gunopulos, D., Raghavan, P.: Automatic subspace clustering of high dimensional data for data mining applications. In: Proc. ACM SIGMOD, pp. 94–105 (1998)
15. Cheng, C., Fu, A., Zhang, Y.: Entropy-based subspace clustering for mining numerical data. In: Proc. SIGKDD, pp. 84–93 (1999)
16. Aggarwal, C.C., Wolf, J.L., Yu, P.S., Procopiuc, C., Park, J.S.: Fast algorithms for projected clustering. In: Proc. ACM SIGMOD, pp. 61–72 (1999)
17. Aggarwal, C.C., Yu, P.S.: Finding generalized projected clusters in high dimensional spaces. In: Proc. ACM SIGMOD, pp. 70–81 (2000)
18. Sequeira, K., Zaki, M.: SCHISM: a new approach for interesting subspace mining. In: Proc. ICDM, pp. 186–193 (2004)

19. Liu, G., Sim, K., Li, J., Wong, L.: Efficient mining of distance-based subspace clusters. Statistical Analysis and Data Mining **2**(5–6), 427–444 (2010)
20. Assent, I., Krieger, R., Müller, E., Seidl, T.: INSCY: indexing subspace clusters with in-process-removal of redundancy. In: Proc. ICDM, pp. 719–724 (2008)
21. Moise, G., Sander, J.: Finding non-redundant, statistically significant regions in high dimensional data: a novel approach to projected and subspace clustering. In: Proc. SIGKDD, pp. 533–541 (2008)
22. Gunnemann, S., Farber, I., Boden, B., Seidl, T.: Subspace clustering meets dense subgraph mining: a synthesis of two paradigms. In: Proc. ICDM (2010)
23. Goil, S., Nagesh, H., Choudhary, A.: MAFIA: efficient and scalable subspace clustering for very large data sets. In: Proc. SIGKDD (1999)
24. Spark. https://spark.apache.org/
25. Domenoconi, C., Papadopoulos, D., Gunopulos, D., Ma, S.: Subspace clustering of high dimensional data. In: Proc. SIAM (2004)
26. Nazerzadeh, H., Ghodsi, M., Sadjadian, S.: Parallel subspace clustering. In: Proc. the 10th Annual Conference of Computer Society of Iran (2005)
27. Achtert, E., Kriegel, H.-P., Zimek, A.: ELKI: a software system for evaluation of subspace clustering algorithms. In: Ludäscher, B., Mamoulis, N. (eds.) SSDBM 2008. LNCS, vol. 5069, pp. 580–585. Springer, Heidelberg (2008)

Massively Parallel Unsupervised Feature Selection on Spark

Bruno Ordozgoiti[✉], Sandra Gómez Canaval, and Alberto Mozo

Department of Computer Systems, University College of Computer Science,
Universidad Politécnica de Madrid, Crta. de Valencia Km. 7, 28031 Madrid, Spain
{bruno.ordozgoiti,sgomez}@etsisi.upm.es, a.mozo@upm.es

Abstract. High dimensional data sets pose important challenges such as the curse of dimensionality and increased computational costs. Dimensionality reduction is therefore a crucial step for most data mining applications. Feature selection techniques allow us to achieve said reduction. However, it is nowadays common to deal with huge data sets, and most existing feature selection algorithms are designed to function in a centralized fashion, which makes them non scalable. Moreover, some of them require the selection process to be validated according to some target, which constrains their applicability to the supervised learning setting. In this paper we propose as novelty a parallel, scalable, exact implementation of an existing centralized, unsupervised feature selection algorithm on Spark, an efficient big data framework for large-scale distributed computation that outperforms MapReduce when applied to multi-pass algorithms. We validate the efficiency of the implementation using 1GB of real Internet traffic captured at a medium-sized ISP.

Keywords: Feature selection · Unsupervised learning · Data mining

1 Introduction

Over the last few years, machine learning has proved to be a useful discipline for many applications such as automated classification, forecasting and anomaly detection. However, the data are often high dimensional, which poses certain challenges. First, this entails what is known as the curse of dimensionality. The amount of data required to train a system increases exponentially in the number of dimensions, and the sparsity of the data grows dramatically in settings in which distance measures are used. In addition, some of the present features might be redundant or noisy, which might hamper the accuracy of the model.

There exist techniques to fight these problems, some of which being based on Principal Components Analysis or the Singular Value Decomposition. These

The research leading to these results has been developed within the ONTIC project, which has received funding from the European Union's Seventh Framework Programme (FP7/2007-2011) under grant agreement no. 619633.

T. Morzy et al. (Eds): ADBIS 2015, CCIS 539, pp. 186–196, 2015.
DOI: 10.1007/978-3-319-23201-0_21

transform the data into a new, lower dimensional subspace which captures the majority of the information of the original records.

Another family of dimensionality reduction algorithms is the one known as feature selection. These techniques eliminate redundant or uninformative features in favor of the most relevant ones, which results in a lower dimensional and hence more manageable data set. Feature selection algorithms can be classified into filter methods, which are model-agnostic; wrapper methods, based on the performance of the model with the chosen features; and embedded methods, which are built into the training algorithm of the specific model being used.

The algorithm presented in this paper, which we will refer to as PPCSS (Parallel Partition-based Column Subset Selection) can be thought of as a filter method. It is based on the centralized algorithm described in [1], described in section 5. Our method presents several advantages over other proposals: (1) it does not rely on arbitrary partitions or concise representations to distribute the data set, which would impact the final result, (2) it is unsupervised and can be used in conjunction with any learning algorithm and (3) it provides a provable theoretical performance metric.

The rest of the paper is structured as follows: Section 2 describes the related work on parallel feature selection algorithms and Section 3 details the proposed contribution. In Section 4, a summary of big data distributed computing frameworks is shown. Sections 5 and 6 present the centralized version of the algorithm and our parallel version proposal respectively. In Section 7 we present experimental results and we conclude in Section 8.

2   Related Work

In recent years, some parallel feature selection approaches have been proposed to deal with the enormous size of today's data sets. In [3] a framework for choosing a column subset that is close to the best rank-k approximation is proposed. It relies on partitioning the data matrix $A \in \mathbb{R}^{m \times n}$ into submatrices of at least $k = \text{rank}(A)$ columns, which limits scalability to the ratio n/k. In [4] a parallel method based on mutual information is described. The data, however, are assumed to be fully discrete, and the method works separately on arbitrary subsets of the data set, making the final result a heuristic approximation. In [5] a MapReduce implementation of the minimum redundancy maximum relevance algorithm is described. This method, though, is only suitable for supervised learning models. In [7] a parallel feature selection algorithm is proposed for logistic regression. This method too is only suitable for the supervised setting, plus it requires to retrain the model at each iteration. In [6] a distributed parallel feature selection is proposed for microarray data classification. This method suffers from the same drawbacks as the previous one, and it imposes a completely random choice to be part of the final feature subset. In [8] a parallel algorithm for unsupervised feature selection is described. Their technique, however, relies on the distribution of the feature covariance matrix in full-column partitions. If the number of features is not very large, the algorithm cannot benefit from parallelization. If it is, workers might not be able to handle the partitions. Other data structures present in the algorithm present the same issue. In [9] a MapReduce implementation of a greedy method for solving the column subset selection problem is proposed. This solution relies on a random projection to represent the data in the different computing nodes. Finally, in [10] a MapReduce implementation of a method based on rough set theory is described.

3   Contribution

In this paper we describe a parallel, scalable version of the algorithm described in [1], which we have designed to be implemented on Spark. This platform is much more efficient than the widely used MapReduce paradigm when the data need to be accessed accross iterations.

Unlike other algorithms, the method we propose does not rely on concise approximations to distribute the data set or on heuristic approaches executed on arbitrary submatrices. Therefore, except for the variations inherent to its probabilistic nature, the proposed method yields the same results as the centralized version of the algorithm would if run on the same data. Moreover, the authors of the original algorithm provide a proven bound on the residual norms of the difference between the data matrix and its projection onto the span of the selected columns. To the best of our knowledge, none of the existing truly scalable versions of feature selection algorithms provide a theoretical performance metric of that nature.

The proposed algorithm functions by performing a judicious random column sampling and then computing the residual norm. In order to achieve a desirable result with a probability close to 1 it is necessary to build various samples and choose the one that provides the best residual norm. When dealing with very large data sets, the computations involved become prohibitively expensive and entail further difficulties. The norms to be computed consist of very large scale summations, which can cause numerical errors and overflow to arise. In this work we deal with some of these issues and provide guidelines for further control.

4   Parallel Computing Platforms

The increasing availability of enormous data sets has contributed to a significant surge of the popularity of large-scale data processing platforms during the last few years. One of the first frameworks to become widely popular was MapReduce [15]. The MapReduce framework is a programming model for processing large amounts of data based on two steps, *map* and *reduce*. The *map* step takes the input data and generates a set of key-value pairs according to a user-chosen criterion. The *reduce* step takes all the elements with the same key and performs an aggregation function, also implemented by the user. Each of the *map* and *reduce* operations can be executed independently on a separate machine. Therefore, together with the distributed, fault-tolerant storage capabilities of Hadoop [16], MapReduce allows for extreme scalability, and enjoys widespread use in research and industrial applications. Other systems that expose a programming API hiding distribution details are Ciel [17] and Dryad [18], the latter of which has been discontinued.

4.1 Spark

The algorithm described in this paper has been implemented on Apache Spark, a novel parallel computation platform designed to provide distribution and fault tolerance efficiently. The key idea distinguishing Spark is its in-memory computation capabilities, allowing data chunks to be reliably cached in memory across iterations. This results in significant performance improvements.

Spark revolves around an abstraction called Resilient Distributed Dataset (RDD) [12]. RDDs enable efficient data reuse by caching in main memory, which is extremely beneficial for certain types of algorithms. To efficiently enable resiliency, each RDD stores its *lineage*, i.e. the sequence of transformations necessary to build it. In fact, RDDs are loaded *lazily*, i.e. their computation is deferred until the data they contain are needed. Spark also permits different RDD *persistence* and *partitioning* strategies, which allow for substantial optimizations, and provides the *Broadcast* and *Accumulator* tools for efficient manipulation of shared variables.

5   Centralized Algorithm Description

The algorithm described in [1] provides a sub-optimal solution to the column subset selection problem (CSSP), which is formulated as follows.

Definition 1. *Given a matrix $A \in \mathbb{R}^{m \times n}$ and a positive integer k, pick k columns of A forming a matrix $C \in \mathbb{R}^{m \times k}$ such that the residual*

$$\|A - P_C A\|_\xi$$

is minimized over all possible $\binom{n}{k}$ choices for the matrix C. $P_C = CC^+$ denotes the projection onto the k-dimensional space spanned by the columns of C and $\xi = 2$ or F denotes the spectral norm or the Frobenius norm.

The challenge of the CSSP is to find a subset of columns that contains as much information of the original matrix as possible. All the possible subsets can be generated in $O(n^k)$, making it unfeasible to find an optimal solution even for relatively small choices of k when the data are represented in hundreds or thousands of dimensions.

The algorithm proposed by Boutsidis et al. in [1] works in two stages, a randomized and a deterministic one. In the randomized phase, a judicious random column sample is taken from the right singular vectors (V_k^T) of the data matrix, using probability distribution $\{p_i | i \in \{1, \ldots, n\}\}$ obtained from (1). This phase outputs a matrix $(V_k^T S_1 D_1) \in \mathbb{R}^{k \times c}$, where S_1 is a sampling matrix to keep only the chosen columns and D_1 scales these columns according to the factors previously described.

$$p_i = \frac{\|(V_k)_{(i)}\|_2^2}{2k} + \frac{\|(A)^{(i)}\|_2^2 - \|(AV_kV_k^T)^{(i)}\|_2^2}{2\left(\|A\|_F^2 - \|AV_kV_k^T\|_F^2\right)} \tag{1}$$

In the second phase, Algorithm 1 of [11] is run on matrix $V_k^T S_1 D_1$ to obtain a permutation revealing exactly the k columns to be kept.

In [1] it is proved with a probability of at least 0.7 this algorithm provides a matrix C (comprised of a column subset of A) such that

$$\|A - P_C A\|_2 \leq O(k^{3/4} \log^{1/2}(k)(\rho - k)^{1/4})\|A - A_k\|_2$$

$$\|A - P_C A\|_F \leq O(k \log^{1/2} k)\|A - A_k\|_F$$

where A_k is the best rank-k approximation of A and ρ is the rank of A.

In order to attain these bounds with sufficient certainty it is necessary to retrieve numerous different column samples in the randomized phase. In [2], where this algorithm is used to apply PCA, 40 samples are drawn in order to achieve the desired result with a probability of almost 1. The computations required to project the original matrix onto the span of the chosen subsets and obtain the residual norm are computationally intense and may become extremely demanding and problematic when dealing when large datasets. The main goal of the work presented in this paper is to dramatically improve the efficiency of this phase by means of distributed computing.

6 Distributed Algorithm Description

The algorithm described in this paper has been implemented on Apache Spark, based on RDDs (refer to section 4). When a map operation is run on an RDD, the function is applied in parallel to all of its elements.

6.1 Notation

- By RDD(S) for some set S we denote an RDD whose elements are the elements of set S.
- By $A \odot B$ we denote the element-wise product of matrices A and B.
- $A_{i:}$ is the i-th row of matrix A.
- Lowercase bold letters (e.g. d) represent vectors.

6.2 The PPCSS Algorithm

The PPCSS algorithm (see figure 1) is implemented in two separate modules, representing the column sampling and the residual norm computation respectively. The first step of the column sampling phase is to load the data set into an RDD -rows- whose elements are the indexed rows of the data matrix A, and to compute its top-k right singular vectors V_k. To compute the set of sampling probabilities $\{p_i\}$ we need to use formula 1, whose factors are obtained in parallel calling mapper *mapSamplingPartitions* then reducer *reduceSamplingPartitions* (figure 2) on RDD *rows*. This will produce a, containing the squared ℓ^2-norms of the column vectors of A (note that a equals the diagonal of the gramian matrix of A, $A^T A$), and b, containing the squared ℓ^2-norms of the column vectors of $AV_k V_k^T$. The corresponding squared Frobenius norms can be obtained adding up the elements of each of these vectors[1]. The sampling probabilities can now be easily computed on the master.

Afterwards, we must take 40 random samples of $c = O(k \log k)$ column indices following the obtained probability distribution, forming matrices S_i and D_i for every $i \in [1, 40]$ as described in [1], and then perform a rank-revealing QR factorization on each of the matrices $V_k^T S_i D_i \in \mathbb{R}^{k \times c}$. To obtain exactly k columns, we will keep the first k elements of the resulting permutation. For this task, we use LAPACK's DGEQP3 routine[2].

The next phase of the algorithm consists in finding the column choice that minimizes the residual norm $\|A - P_C A\|_F$. We compute the pseudo-inverse of each matrix C_i, C_i^+ using the singular value decomposition, and the product $C_i^+ A \in \mathbb{R}^{k \times n}$ using a simple MapReduce operation for large matrix multiplication that we will not describe for the sake of space. It then remains to compute $A - CC_i^+ A$, done simultaneously for all matrices C_i using a single MapReduce operation. First, we set up an RDD -*allMats* - that can be described as follows: the jth element of *allMats* is a list containing the jth row of each matrix C_i as well as the jth row of matrix A. Then, we call *mapNormAddends* (figure 3) on this RDD. Each map operation will compute one row of $C_i C_i^+ A$ for each matrix C_i (the matrices $C_i^+ A$ are transmitted to each worker via Broadcast), their differences with the corresponding row of A and the contribution to the residual norm of this row. This will result in a vector d containing the addends of the residual norm for each column choice. Reducer *reduceNormAddends* (figure 3) can then add up all of the vectors produced by the mappers to obtain the final residual norms.

The fact that the norms are computed simultaneously for all matrices C_i allows us to control overflow by subtracting the minimum vector component from all the elements of said vector in the reduce step. This way we can significantly decrease the chance of overflow without incurring additional numerical error. It should also be noted that since the reduce operations are performed in a pairwise fashion, the expected numerical error is $O(\log n)$, where n is the number of addends [13]. A slower alternative using Kahan's algorithm and Accumulators could be implemented to achieve $O(1)$ error [14].

---

[1] Note that we are building a probability vector. Therefore, these elements can be rescaled to avoid overflow if the data set is truly huge. With respect to the numerical error of the sum, the pairwise nature of reduce operations constrains it to $O(\log n)$ [13]

[2] We consider the case where $n \ll m$. Matrix $V_k^T S_i D_i$ is therefore of a manageable size and this operation can thus be performed locally.

Column sampling phase

Input: data matrix $A \in \mathbb{R}^{m \times n}$, number of columns k

1. $rows := \text{RDD}(\{A_{i:} \mid i \in [1, m]\})$
2. Compute the top k right singular vectors of A, V_k.
3. $(colNorms, prodNorms) :=$
   $rows.\text{mapSamplingPartitions}(\text{broadcast}(V_k V_k^T)).\text{reduceSamplingPartitions}()$
4. Compute the sampling probabilities $\{p_i\}$ for $1 \leq i \leq n$
5. Obtain 40 random samples of $c = O(k \log k)$ columns, forming matrices S_i and D_i for all i in $\{1, \ldots, 40\}$.
6. for $1 \leq i \leq 40$
   - Run LAPACK's DGEQP3 on $V_k^T S_i D_i$ and keep the k first columns, obtaining the column indices to form matrix C_i.
7. output $\{C_i \mid i \in [1, 40]\}$

Norm minimization phase

Input: $\{C_i \mid 1 \leq i \leq 40\}$

1. $P := \emptyset$
2. for $1 \leq i \leq 40$
   - $P := P \cup \{C_i^+ A\}$
3. $allMats := \text{RDD}(\{F_j = \{(C_i)_{j:} \mid i \in [1, 40]\} \cup \{A_{j:}\} \mid j \in [1, m]\})$
4. $\boldsymbol{d} := allMats.\text{mapNormAddends}(\text{broadcast}(P)).\text{reduceNormAddends}()$
5. output $\underset{C}{\arg\min} \|A - P_C A\|_F$, i.e. C_i with $i = \underset{i}{\arg\min} \, \boldsymbol{d_i}$

Fig. 1. PPCSS algorithm

mapSamplingPartitions
$(A_{i:}, \text{broadcast}(V_k V_k^T))$

1. $\boldsymbol{a} := A_{i:} \odot A_{i:}$
2. $\boldsymbol{b} := (A_{i:} V_k V_k^T) \odot (A_{i:} V_k V_k^T)$

reduceSamplingPartitions
$((\boldsymbol{a}_i, \boldsymbol{b}_i), (\boldsymbol{a}_j, \boldsymbol{b}_j))$

1. $\boldsymbol{a}_{out} := \boldsymbol{a}_i + \boldsymbol{a}_j$
2. $\boldsymbol{b}_{out} := \boldsymbol{b}_i + \boldsymbol{b}_j$
3. output $(\boldsymbol{a}_{out}, \boldsymbol{b}_{out})$

mapNormAddends
$(F_j, \text{broadcast}(P))$

1. $D_i := (C_i)_{j:} C_i^+ A - A_{j:} \forall (C_i)_{j:} \in F_j \backslash \{A_{j:}\}$
2. output vector
   $\boldsymbol{d} = (D_1 \boldsymbol{u}, \ldots, D_{40} \boldsymbol{u})$, where $\boldsymbol{u}$ is a vector of ones

reduceNormAddends $(\boldsymbol{d}_i, \boldsymbol{d}_j)$

1. output $\boldsymbol{d}_i + \boldsymbol{d}_j$

Fig. 2. MapReduce operations for the addends of the sampling probability distribution

Fig. 3. MapReduce operations for computing the final residual norms for each column choice

6.3   The Complexity of PPCSS

The complexity of the algorithm on which PPCSS is based is $O(mn\min(m,n))$ for the computation of the top k right singular vectors, $O(k^3 \log k)$ for the deterministic phase and $O(mnk)$ for the final norm computation. Since we are considering the case that $m \gg n \gg k$, the running time of the algorithm is dominated by the computation of V_k and the norms, which are the operations that we parallelize. In general, the complexity of these calculations will scale linearly with the inverse of the number of nodes in the cluster.

With respect to network usage, the algorithm requires to move $O(mn + n^2 + kn)$ between the master and the workers: the data set, matrix $V_k V_k^T \in \mathbb{R}^{n \times n}$ and set $P = \{C_i^+ \in \mathbb{R}^{k \times n} | i \in [1, 40]\}$. Therefore, a large number of features could severely impact the total running time. However, we have observed that in certain domains, e.g. network traffic management, this number tends to be below a few hundred, which our algorithm seems to be able to handle adequately.

7   Preliminary Experimental Results

We have tested the algorithm on a homemade cluster of 10 nodes equipped with a Quad-core processor and 4Gb of RAM each. The dataset used was a 1GB real Internet traffic capture from a Spanish medium-sized ISP consisting of 2,500,000 flow instances and 105 features.

The purpose of the first of the experiments was to test the scalability of the algorithm with respect to the number of workers. We launched the algorithm on the full 2,500,000 × 105 matrix using 1, 2, 5, and all 10 nodes. This set of executions shows that our algorithm benefits greatly from parallel computation (figure 4). As expected, the speedup decreases as the number of workers grows, and the graph suggests that a point of asymptotic lack of improvement will be reached. However, this point is expected to shift to the right as the size of the input data set grows.

In the second set of experiments we executed the algorithm on datasets of different sizes. We trimmed the original data to obtain sets of 50,000, 100,000, 200,000, 400,000, 800,000 and 1.6 million instances, and tested the algorithm on these subsets. As seen on figure 5, our method yields the best results when the data set is truly huge. For instance, the difference in execution times with 200,000 and 400,000 instances is not that large. This is to be expected, given the fact that the total time is a function of Spark's cluster management tasks, network usage and I/O operations and the complexity of the algorithm. When the size of the data set is small, cluster management time (which is more or less constant for a fixed-size cluster) and the overhead of I/O operations are significant with respect to the total time. Since the running time of the algorithm increases linearly with m (the number of flows), and both n and k (the dimensionality of the data and the number of selected features respectively) are fixed in this experiment, it is evident that data seem to be handled more efficiently by Spark when the input is large.

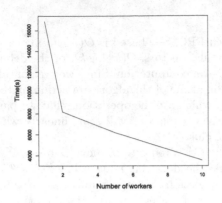

Fig. 4. Experiments with different numbers of workers

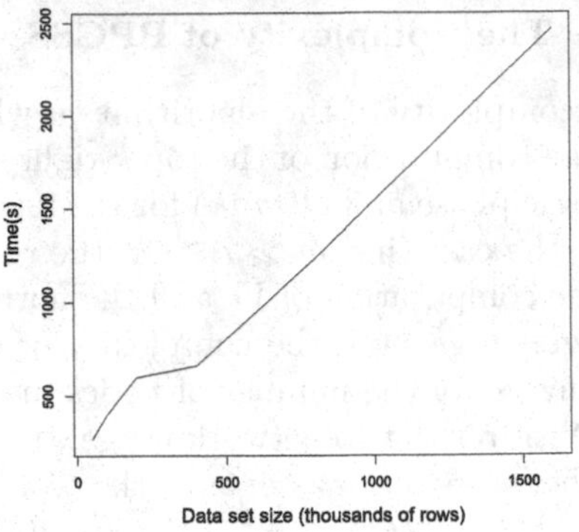

Fig. 5. Experiments with data sets of different sizes

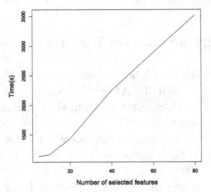

Fig. 6. Experiments with different choices of k

The third experiment was carried out to determine the impact of the parameter k (i.e. the number of selected features) on the running time. As explained in section 6.3, the algorithm depends cubically on k. Since in general $k \ll n$, this will not have a significant effect on the performance of the algorithm provided that n is not extremely large. This fact is shown by the results of the experiments reflected in figure 6, in which the value of k does not dramatically impact the total time. Nevertheless, in settings where the dimensionality is extremely high, the cubic dependence on k could be an important drawback.

8   Conclusions and Future Work

We have presented a truly scalable feature selection algorithm able to cope with enormous data sets. Our proposal is a parallel, scalable, exact implementation of an existing centralized, unsupervised feature selection algorithm on Spark. As opposed to other algorithms of this kind already present in the literature, our technique does not rely on concise representations of the original data or arbitrary partitioning that might mask certain properties and relevant charac-teristics. We have carried out some preliminary tests on a modest cluster and the results with regards to scalability and execution times are very promising. However, the performance of the algorithm is sensitive to the number of fea-tures, which impose penalties on both running time and network usage. Hence, extremely high-dimensional data sets pose significant performance concerns. It is therefore an open challenge to address scalability in that sense. The experiments have shown, nevertheless, that the algorithm can comfortably handle big data sets of over 100 features, which are frequent in domains such as network traffic management.

We plan to perform additional experiments on larger data sets, as well as to evaluate the possibilities of large-scale feature selection on certain domains that could immensely benefit from a truly scalable and efficient algorithm of this kind. We also intend to run implementations of other existing proposals in order to assess the performance of our method compared to those. In addition, we will continue our work on parallel unsupervised feature selection to address the problem of scalability with respect the number of dimensions.

References

1. Boutsidis, C., Mahoney, M.W., Drineas, P.: An improved approximation algorithm for the column subset selection problem. In: Proceedings of the Twentieth Annual ACM-SIAM Symposium on Discrete Algorithms, pp. 968–977. Society for Indus-trial and Applied Mathematics, January 2009
2. Boutsidis, C., Mahoney, M.W., Drineas, P.: Unsupervised feature selection for prin-cipal components analysis. In: Proceedings of the 14th ACM SIGKDD Interna-tional Conference on Knowledge Discovery and Data Mining. ACM (2008)
3. Pi, Y., et al.: A scalable approach to column-based low-rank matrix approximation. In: Proceedings of the Twenty-Third International Joint Conference on Artificial Intelligence. AAAI Press (2013)
4. Sun, Z., Li, Z.: Data intensive parallel feature selection method study. In: 2014 International Joint Conference on Neural Networks (IJCNN). IEEE (2014)
5. Reggiani, C., et al.: Minimum redundancy maximum relevance: mapreduce imple-mentation using apache hadoop. In: BENELEARN 2014, p. 2 (2014)
6. Bolón-Canedo, V., Sánchez-Maroño, N., Alonso-Betanzos, A.: Distributed feature selection: An application to microarray data classification. Applied Soft Computing 30, 136–150 (2015)
7. Singh, S., et al.: Parallel large scale feature selection for logistic regression. In: SDM (2009)
8. Zhao, Z., et al.: Massively parallel feature selection: an approach based on variance preservation. Machine Learning 92(1), 195–220 (2013)
9. Farahat, A.K., et al.: Distributed column subset selection on MapReduce. In: 2013 IEEE 13th International Conference on Data Mining (ICDM). IEEE (2013)
10. He, Q., et al.: Parallel feature selection using positive approximation based on MapReduce. In: 2014 11th International Conference on Fuzzy Systems and Knowl-edge Discovery (FSKD). IEEE (2014)
11. Pan, C.-T.: On the existence and computation of rank-revealing LU factorizations. Linear Algebra and its Applications 316(1), 199–222 (2000)

12. Zaharia, M., et al.: Resilient distributed datasets: a fault-tolerant abstraction for in-memory cluster computing. In: Proceedings of the 9th USENIX Conference on Networked Systems Design and Implementation. USENIX Association (2012)
13. Higham, N.J.: The accuracy of floating point summation. SIAM Journal on Scientific Computing **14**(4), 783–799 (1993)
14. Kahan, W.: 1965. Pracniques: further remarks on reducing truncation errors. Commun. ACM **8**(1), January 1965
15. Dean, J., Ghemawat, S.: MapReduce: simplified data processing on large clusters. Communications of the ACM **51**(1), 107–113 (2008)
16. Shvachko, K., et al.: The hadoop distributed file system. In: 2010 IEEE 26th Symposium on Mass Storage Systems and Technologies (MSST). IEEE (2010)
17. Murray, D.G., et al.: CIEL: a universal execution engine for distributed data-flow computing. In: NSDI, vol. 11 (2011)
18. Isard, M., et al.: Dryad: distributed data-parallel programs from sequential building blocks. ACM SIGOPS Operating Systems Review **41**(3) (2007)

Unsupervised Network Anomaly Detection in Real-Time on Big Data

Juliette Dromard[✉], Gilles Roudière, and Philippe Owezarski

CNRS, LAAS, 7, Avenue du Colonel Roche, 31031 Toulouse Cedex 4, France
{juliette.dromard,gilles.roudiere,philippe.owezarski}@laas.fr

Abstract. Network anomaly detection relies on intrusion detection systems based on knowledge databases. However, building this knowledge may take time as it requires manual inspection of experts. Actual detection systems are unable to deal with 0-day attack or new user's behavior and in consequence they may fail in correctly detecting intrusions. Unsupervised network anomaly detectors overcome this issue as no previous knowledge is required. In counterpart, these systems may be very slow as they need to learn traffic's pattern in order to acquire the necessary knowledge to detect anomalous flows. To improve speed, these systems are often only exposed to sampled traffic, harmful traffic may then avoid the detector examination. In this paper, we propose to take advantage of new distributed computing framework in order to speed up an Unsupervised Network Anomaly Detector Algorithm, UNADA. The evaluation shows that the execution time can be improved by a factor of 13 allowing UNADA to process large traces of traffic in real time.

1 Introduction

Nowadays, networks suffer from many different failures and attacks like, for example DOS, DDOS, network scanning, port scanning, spreading of worms and viruses which can damage them. In order to prevent these damages, network administrators rely on intrusion detection systems (IDS). Two main types of IDS are largely dominant, signature-based IDSs and behavior-based IDSs.

Behavior-based IDSs detect all the operations which deviate from a known normal behavior whereas signature-based IDSs detect only known attacks for which they possess a signature. Therefore these two techniques rely on previous knowledges. Building new signatures and new models of normal behaviors take time as it requires manual inspection of human experts. It implies that signatures-based IDSs may not be aware of new attacks and behavior-based IDSs of new users' behaviors, as a result they may launch many false negatives or false positives.

In order to overcome these knowledge-based IDSs' issues, researchers focuse their attention on detectors which rely on no previous knowledge: the unsupervised network anomaly detectors. These systems aim at detecting anomalous flows, i.e rare flows that possess different patterns from normal flows. They rely on the two following hypothesis:

© Springer International Publishing Switzerland 2015
T. Morzy et al. (Eds): ADBIS 2015, CCIS 539, pp. 197–206, 2015.
DOI: 10.1007/978-3-319-23201-0_22

Hypothesis 1. *Intrusive activities represent a minority of the whole traffic [18]*

Hypothesis 2. *Intrusive activities's pattern is different from normal activities' patterns [17]*

Therefore the central premise of anomaly detection is that intrusive activity is a large subset of anomalous activity [17]. To detect anomalous flows,these detectors often rely on machine learning techniques (MLTs) which can be either "supervised", label data are then required to identify patterns, or "unsupervised", no previous knowledge is required to discover data's patterns.

Unsupervised network anomaly detector exploits unsupervised MLTs to identify flows which have rare patterns and are thus anomalous. However, detecting anomalies may then be time-consuming as these systems need to dive deeply in the flows' features to identify their patterns, therefore they can hardly meet IDS's real time requirements. To solve this issue, existing detectors may process only sampled data which implies that harmful traffic may be not processed and so not detected [4]. To overcome this limitation,we propose in this paper to take advantage of a new distributed computing system to deploy an unsupervised network anomaly detector on a large cluster of servers.

The remainder of this paper is organized as follows. Section 2 presents related works. Section 3 presents UNADA, an unsupervised network anomaly detector which has been previously proposed by our team. Section 4 presents the implementation and validation of UNADA over a large cluster of servers in terms of computational time and scalability. The possible future works are then discussed and we conclude in Section 6.

2   Related Works

The problem of unsupervised network anomaly detection has been extensively studied in the last decade. Existing systems generally detect anomalies others. For that purpose, many techniques can be used such as multi resolution Gamma modeling [7], Hough transform [12], the Kullback-Leibler (KL) distance [13], however, two main techniques are largely dominant in this area: principal component analysis [16] [14] (PCA) and clustering methods [6] [19].

PCA identifies the main components of the normal traffic, the flows distant from these components are considered anomalous. The pioneering contribution in this area was published by Lakhina et al. [16]. PCA based detectors' main drawback is that they they don't allow to retrieve the original traffic features of the anomalous traffic. This difficulty is overcame in [14] by using random projection technique (sketches): the source IP addresses of the anomalies can then be identified.

Most of unsupervised network anomaly detector rely on clustering techniques, [6] [19] are some relevant examples. Clustering algorithm group similar flows in clusters, to determine similarities between flows distance function like lie outside the clusters are rare flows and are then flagged as anomalous. UNADA falls within the clustering-based detectors and presents several advantages w.r.t. existing network anomaly detectors. First, it works in a complete unsupervised way, so it can be plugged to any monitoring system and works from scratch. Secondly, it combines the notions of subspace clustering and evidence accumulation (EA) and can so overcome the curse of dimensionality [10]. Finally, UNADA is highly parallelizable and can so be easily implemented over a large cluster of nodes to process large amount of traffic in a small amount of time. These unsupervised detectors have to dive deeply in the collected traffic in order to identify patterns. As a result, they often rely on techniques which time complexity is not linear which prevents them from being scalable and real-time.

Closely related to our work is Hashdoop [11]. Hashdoop is a generic framework, based on the MapReduce paradigm, to distribute the computing of any unsupervised network anomaly detector in order to speedup the detection. In Hashdoop, the traffic is collected in time-bins which are then cut in horizontal slices: all the traffic from or to a same IP address must lie in the same slice. Many detectors are then launched in parallel, each processes the traffic of one slice, the obtained results are then aggregated. The authors claim that their framework is generic, but they only validate it on very simple detectors based on change detection. Furthermore, in Hashdoop, as all the traffic, from and to a same IP address must lie in a same slice, this latter may possess a majority of intruder's traffic and hypothesis 1 may no longer apply. Therefore no detector which relies on the hypothesis 1 and so on unsupervised MLTs like UNADA can be distributed with Hashdoop.

3   The Network Anomaly Detector

UNADA is a network anomaly detector which has already been proposed by our team in [5]. UNADA works on single-link packet-level traffic captured in consecutive time-slots of fixed length ΔT and runs in three main steps: the change detection, the clustering and the EA step.

The first step aims at detecting anomalous slots by detecting change in the traffic. To this end, multi-resolution flow aggregation is applied on the traffic at each time slot. It consists in aggregating packets at different level from finer to coarser-grained resolution in "flows". There is 8 aggregation levels, each level l_i ($i \in [1,8]$) is defined by a mask (/32, /24, /16, /8) and the IP address (IPx) on which the mask is applied which can be be either the IP source (IPsrc) or the IP destination (IPdst). For each level, the traffic is represented by a set $Y = \{y_1, ..., y_f, ...y_F\}$ where each element y_f is considered as a "flow" and F is the total number of flows. In this context, a flow y_f is a subset of the original traffic having the same IPx/mask. For each level l_i, multiple time series $Z_t^{l_i}$ are then computed, each time series considers a simple metric t such as number of bytes, number of packets, number of IP flows. One point for each time series is built by slot. Then, a generic change detection algorithm is applied on each time series at each new time slot. If the detection algorithm launches an alarm, it implies that there is a change in the traffic pattern probably due to anomalous activities: this slot requires a deeper inspection and is then flagged.

The second step aims at extracting anomalies from the flagged slots thanks to a subspace clustering algorithm. It takes as input the traffic $Y = \{y_1, ..., y_F\}$ extracted from the flagged slots and aggregated according to the level which has raised the alarm. Each flow y_f can be described by a set of A features in a vector $x_f \in \mathbb{R}^A$. The set of vectors of every flow is denoted by a normalized matrix $X = \{x_1, , ...x_F\}$ representing the features space. Numerous features can be computed over a flow y_f such as: nDsts (# of different IPdst), nSrcs (# of different IPsrc), nPkts (# of pkts), nSYN/nPkts, nICMP/nPkts, etc. Any other attribute sensitive to anomalies can be chosen making UNADA a very flexible tool. To detect anomalous flows in Y, a subspace clustering algorithm is applied to the features space. Each vector of features is considered as a point in the clustering algorithm and each point isolated from the others and which does not belong to any cluster is identified an an outlier. In order to identify any form of cluster, UNADA is based on a density based grid algorithm DBSCAN (Density-Based Spatial Clustering of Applications with Noise) [8] which takes two parameters, n_{min} which represents the minimum number of neighbors a point must have to be a core point in a cluster and $\varepsilon_{neighbor}$ which defines the distance of a point's neighborhood.In DBSCAN, a cluster is defined as a area in the data space of higher density. However, DBSCAN suffers from the curse of dimensionality, i.e. when the dimensionality increases the points become increasingly sparse and distance between pairs of points becomes meaningless: DBSCAN can then hardly identify clusters. To overcome this issue, UNADA is based on subspace clustering [15] and EA. The subspace clustering consists in dividing the features space into N many subspaces and performing DBSCAN on each independently. At the end of this step a clustering ensemble $P = \{p_1, ...p_N\}$ is formed where each each element p_i represents the partition obtained on the ith subspace. As DBSCAN provides better results in low-dimensional spaces, the dimension of each subspace is set to 2, which gives $N = m(m\ 1)/2$ partitions.

The third step of DBSCAN is the EA for outliers identification (EA4O) where the N different partitions are combined to identify the anomalies. This step accumulates for each point the level of abnormality it gets in each subspace. In a subspace, if a point belongs to a cluster its level of abnormality is null, otherwise its level of abnormality is proportional to its distance with the centroid of the biggest cluster. A dissimilarity vector $D = \{d_1, .., d_F\}$ is built where each element d_f reflects the accumulated level of abnormality of the flow y_f. In order not to overwhelm the network administrator with anomalies, only the most pertinent anomalies are selected. To identify them, the dissimilarity vector is sorted and an anomaly detection threshold t_h is defined. t_h is set at the value for which the slope of the sorted dissimilarity presented a major change. Finally, every flow y_f with a dissimilarity score d_f above this threshold is considered as anomalous. UNADA's pseudo-code is presented in algorithm 1.

4   Performances Evaluation of UNADA

In this paper, we aim at detecting the anomalies of a Spanish Internet service provider's core network which is crossed by around 1.2Gbit/s of data.

Algorithm 1. Steps 2 and 3 of UNADA

1: **Initialize:**
2: Set n_{min} and ε
3: Set D the dissimilarity vector to a null F*1 vector
4: **for** i=1:N **do**
5: $\quad P_i = DBSCAN(X, n_{min}, \varepsilon_{neighbor})$
6: $\quad UpdateD, \forall f \in F$
7: $\quad$ **if** $o_i^f \in P_i$ **then**
8: $\quad\quad w_i \leftarrow n/(n - n_{max_i})$
9: $\quad\quad D(f) \leftarrow D(f) + d(o_i^f, C_i^{max}) * w_i$
10: $\quad$ **end if**
11: **end for**
12: $t_h = computeTresh(D)$
13: **Retrieve anomalous flows** $F' = \{f, D(f) > t_h\}$

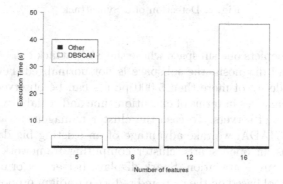

Fig. 1. UNADA's mean execution time for a 15s time slot

As UNADA requires only the packets' header to detect anomalies, only 64 bits of each packet is stored. To deal with such a traffic, UNADA has to process 19,2 Mbit/s, so 1.6 TeraBytes per day, which makes a huge amount of data.

For UNADA's evaluation, the aggregation's level is set to IPsrc/16 and the time slot ΔT to 15 seconds. A slot is made up of around 2.000.000 packets. Furthermore, the evaluation does not consider the flows' features computation time, we assume that a dedicated hardware processes this task upstream. Figure 1 displays UNADA's mean execution time of one slot on a single machine with 16 Gbit of RAM and an Intel Core i5-4310U CPU 2.00GHz. For the local mode, Spark is not used, as it would slow down UNADA's execution and it takes advantage of two cores; one is dedicated to the Java's garbage collection and the second to UNADA . It shows that UNADA's execution time is mainly due to the clustering step and increases with the number of features.

We analyze 60 slots of traffic and found many anomalies, most of them were induced by large point to multipoint traffic or flashcrowds which could represent from 25 to 10% of the total core network traffic. Furthermore UNADA has detected alpha flows, accounting for 20% of the whole traffic, unknown anomalous flows detected due to a hight number of ICMP packets and finally a SYN

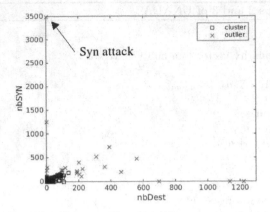

Fig. 2. Detection of a Syn attack

attack. Figure 2 depicts one subspace where the SYN attack can be observed, for easier viewing and diagnosis, the subspace is not normalized. Few outliers and a big cluster made up of more than 5.000 points can be observed. To improve UNADA's performances in terms of execution time and scalability, it is deployed on a large cluster of servers. To ease the cluster management and computing distribution of UNADA, we take advantage of an existing big data tool Spark 1.2.0 [1] which is an open source cluster computing framework developed by the Apache Software Foundation. Spark displays better performance than the famous Hadoop tool based on the map and reduce paradigm proposed by Google. Indeed, for certain computing tasks, Spark can fasten ×100 the execution time compared to Hadoop [20]. Furthermore, Spark offers over 80 high-level operators that make it easy to build parallel applications. The validation has been performed on the Grid5000 platform [2], a large-scale and versatile testbed which provides access to a large amount of resources: 1000 nodes, 8000 cores, grouped

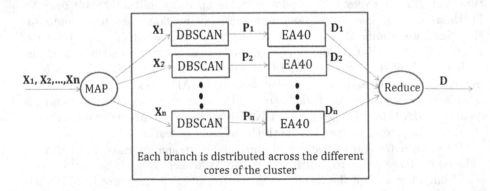

Fig. 3. UNADA's deployment over a cluster of servers with Spark

in homogeneous clusters. In Spark, a master node runs the main() function of the application, creates the SparkContext, acquires executors on nodes in the cluster and then sends tasks to run to the executors. The memory the executors use, as well as the total number of cores across all executors can be finely tuned. UNADA's evaluation is performed on nodes with 8 GB of RAM, two CPUs at a frequency of 2.26GHz, each with 4 cores. To deploy UNADA over a cluster of servers, two Spark's high-level operators are used (see Figure 3):

- the map operator which sends across the different cores of the cluster the processing of the N subspaces. The clustering and the EA of each subspace are thus parallelized. The map function returns a dissimilarity vector for each subspace.
- the reduce operator which aggregates the dissimilarity vector obtained in each subspace. It simply sums the dissimilarity vector of each subspace to obtain the final dissimilarity vector.

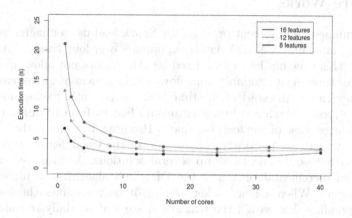

Fig. 4. Execution Time of UNADA

UNADA's deployment has been validated in terms of scalability and execution time. Figure 4 displays UNADA's execution time according to the number of features and cores considered. UNADA's execution time for each feature decreases until reaching a threshold. Furthermore, the difference in execution time of UNADA with different number of features tends to decrease while adding new cores, which implies that a high number of features does not prevent from using the detector. Figure 5 depicts the improvement in UNADA's execution time with different number of cores compared with a local execution of the detector. The gain in time increases with the number of cores till reaching a threshold . However, for UNADA with 8 and 12 features, from a certain number of cores this gain slightly decreases. This decrease may be due to the fact that serializing the data and sending them to the servers of the cluster may take longer than processing them. Finally, the validation shows that distributing UNADA's computing can improve significantly the processing time till reaching a speedup factor of 13.

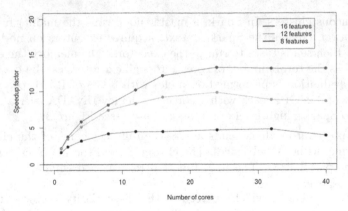

Fig. 5. Speedup of UNADA

5   Future Works

The performance improvement provided by Spark lead us to imagine new ways of detecting anomalies. UNADA detects anomalies over long fixed-length period of time. As this time-bin has a long fixed length, it does not allow an accurate analysis over time. As an example some flows might appear as anomalous simply because they end at the start of the time-bin, consequently having not enough packets analyzed. Therefore, our system would benefit from having a more frequently updated view of the features space. However, because of the algorithm complexity, we can not perform a clustering over the whole features space more often. Nevertheless, we can think up several solutions. As proposed in [9], we could use an incremental version of the DBSCAN algorithm to incrementally update clusters. When a new packet arrives, it only modifies the features of the corresponding flow within the features space and partially re-compute the cluters. As a per-packet computation is not realistic, aggregating several packets before updating the corresponding flow within the DBSCAN algorithm can be considered. Furthermore, as incremental DBSCAN does not provide an efficient way of updating already clustered points, we also consider updating only points that are enough different from their previous value. This Incremental version of DBSCAN, could allow to detect earlier and more efficiently anomalies. [3] proposes a GPU based implementation of the DBSCAN algorithm which allows to gain a x100 speed-up factor from a typical CPU based implementation. Therefore, as UNADA is based on DBSCAN and is, at least, per-subspace parallelizable, we can expect good results from a GPU based implementation. This speed-up factor could let us perform more frequent updates of UNADA results, moving UNADA closer to real-time efficiency.

6   Conclusion

Unsupervised network anomaly detectors mainly rely on machine learning techniques which complexity are often hight and which can thus hardly be real time or deal with large traffic traces. In this paper, we have proposed to take advantage of new distributed computing systems in order to speed up an unsupervised network anomaly algorithm. Dividing the features space in subspaces allows UNADA to run in parallel multiple DBSCANs and EA algorithms. Spark's high operators distributes the processing of the subspaces over the servers of the cluster and aggregates then the result. UNADA's deployment with Spark can improve the execution time by a factor of 13. This paper is a step forward for detecting network anomalies in real time on large non sampled traffic such as the traffic of an Internet provider's core network.

Acknowledgments. This work has been done in the framework of the FP7 ONTIC project (see http://ict-ontic.eu) funded by European Commission under the Seventh Framework Programme. Experiments presented in this paper were carried out using the Grid'5000 testbed, supported by a scientific interest group hosted by Inria and including CNRS, RENATER and several Universities as well as other organizations (see https://www.grid5000.fr).

References

1. Apache spark - lightning-fast cluster computing. https://spark.apache.org/ (accessed April 29, 2015)
2. Grid5000. https://www.grid5000.fr (accessed April 29, 2015)
3. Andrade, G., Ramos, G., Madeira, D., Sachetto, R., Ferreira, R., Rocha, L.: G-dbscan: A GPU accelerated algorithm for density-based clustering. Procedia Computer Science, 369–378 (2013)
4. Brauckhoff, D., Tellenbach, B., Wagner, A., May, M., Lakhina, A.: Impact of packet sampling on anomaly detection metrics. In: Proc. of the 6th ACM SIGCOMM Conference on Internet Measurement, pp. 159–164 (2006)
5. Casas, P., Mazel, J., Owezarski, P.: Unsupervised network intrusion detection systems: Detecting the unknown without knowledge. Computer Communications, 772–783 (2012)
6. Celenk, M., Conley, T., Willis, J., Graham, J.: Anomaly detection and visualization using fisher discriminant clustering of network entropy. In: Third International Conference on Digital Information Management, pp. 216–220, November 2008
7. Dewaele, G., Fukuda, K., Borgnat, P., Abry, P., Cho, K.: Extracting hidden anomalies using sketch and non gaussian multiresolution statistical detection procedures. In: Proc. of the 2007 Workshop on Large Scale Attack Defense, pp. 145–152. ACM (2007)
8. Ester, M., peter Kriegel, H., S, J., Xu, X.: A density-based algorithm for discovering clusters in large spatial databases with noise, pp. 226–231. AAAI Press (1996)
9. Ester, M., Kriegel, H.P., Sander, J., Wimmer, M., Xu, X.: Incremental clustering for mining in a data warehousing environment. In: Proc. of the 24rd International Conference on Very Large Data Bases, pp. 323–333 (1998)
10. Fahad, A., Alshatri, N., Tari, Z., Alamri, A., Khalil, I., Zomaya, A., Foufou, S., Bouras, A.: A survey of clustering algorithms for big data: Taxonomy and empirical analysis. IEEE Transactions on Emerging Topics in Computing, 267–279, September 2014
11. Fontugne, R., Mazel, J., Fukuda, K.: Hashdoop: a mapreduce framework for network anomaly detection. In: INFOCOM WKSHPS, pp. 494–499, April 2014

12. Fontugne, R., Fukuda, K.: A hough-transform-based anomaly detector with an adaptive time interval. SIGAPP Appl. Comput. Rev., 41–51 (2011)
13. Gu, Y., McCallum, A., Towsley, D.: Detecting anomalies in network traffic using maximum entropy estimation. In: Proc. of the 5th ACM SIGCOMM Conference on Internet Measurement, pp. 32–32 (2005)
14. Kanda, Y., Fukuda, K., Sugawara, T.: Evaluation of anomaly detection based on sketch and pca. In: GLOBECOM 2010, pp. 1–5. IEEE (2010)
15. Kriegel, H.P., Kroger, P., Zimek, A.: Clustering high-dimensional data: A survey on subspace clustering, pattern-based clustering, and correlation clustering. ACM Trans. Knowl. Discov. Data (2009)
16. Lakhina, A., Crovella, M., Diot, C.: Diagnosing network-wide traffic anomalies. In: Proc. of ACM SIGCOMM 2004, pp. 219–230, Auguest 2004
17. Patcha, A., Park, J.M.: An overview of anomaly detection techniques: Existing solutions and latest technological trends. Comput. Netw., 3448–3470 (2007)
18. Portnoy, L., Eskin, E., Stolfo, S.: Intrusion detection with unlabeled data using clustering. In: Proc. of ACM CSS Workshop on Data Mining Applied to Security, pp. 5–8 (2001)
19. Wei, X., Huang, H., Tian, S.: A grid-based clustering algorithm for network anomaly detection. In: The First International Symposium on Data, Privacy, and E-Commerce, ISDPE 2007, pp. 104–106, November 2007
20. Xin, R.S., Rosen, J., Zaharia, M., Franklin, M.J., Shenker, S., Stoica, I.: Shark: SQL and rich analytics at scale. In: Proc. of the 2013 ACM SIGMOD International Conference on Management of Data, pp. 13–24 (2013)

NPEPE: Massive Natural Computing Engine for Optimally Solving NP-complete Problems in Big Data Scenarios

Sandra Gómez Canaval[✉], Bruno Ordozgoiti Rubio, and Alberto Mozo

Department of Computer Systems, University College of Computer Science,
Universidad Politécnica de Madrid, Crta. de Valencia Km. 7, 28031 Madrid, Spain
{sgomez,bruno.ordozgoiti}@etsisi.upm.es, a.mozo@upm.es

Abstract. *Networks of Evolutionary Processors* (NEP) is a bio-inspired computational model defining theoretical computing devices able to solve NP-complete problems in an efficient manner. *Networks of Polarized Evolutionary Processors* (NPEP) is an evolution of the NEP model that presents a simpler and more natural filtering strategy to simulate the communication between cells. Up to now, it has not been possible to have implementations neither in vivo nor in vitro of these models. Therefore, the only way to analyze and execute NPEP devices is by means of ultra-scalable simulators able to encapsulate the inherent parallelism in their computations. Nowadays, there is a lack of such simulators able to handle the size of non trivial problems in a massively distributed computing environment. We propose as novelty NPEPE, a high scalability engine that runs NPEP descriptions using Apache Giraph on top of Hadoop platforms. Giraph is the open source counterpart of Google Pregel, an iterative graph processing system built for high scalability. NPEPE takes advantage of the inherent Giraph and Hadoop parallelism and scalablity to be able to deploy and run massive networks of NPEPs. We show several experiments to demonstrate that NPEP descriptions can be easily deployed and run using a NPEPE engine on a Giraph+Hadoop platform. To this end, the well known 3-colorability NP complete problem is described as a network of NPEPs and run on a 10 nodes cluster.

Keywords: Big data · Distributed architectures · Parallel computation · Bio-inspired computational models · Networks of evolutionary processors

1 Introduction

Bio-inspired computational models can solve NP-complete problems efficiently by mimicking the way in which the nature computes. Networks of Evolutionary

The research leading to these results has been developed within the ONTIC project, which has received funding from the European Union's Seventh Framework Programme (FP7/2007-2011) under grant agreement no. 619633.

T. Morzy et al. (Eds): ADBIS 2015, CCIS 539, pp. 207–217, 2015.
DOI: 10.1007/978-3-319-23201-0_23

Processors (NEP) [1] is one of the most well known bio-inspired models based on the Formal Language Theory. Networks of Polarized Evolutionary Processors (NPEP) [2] is an evolution of the NEP model that presents a simpler and more natural filtering strategy to simulate the communication between cells. The NEP family has been widely investigated from a theoretical point of view as (i) devices accepting and generating languages and problem solvers [3], (ii) computationally complete devices [4], and (iii) universal and efficient devices [5]. From a computational perspective, a NEP can be defined as a graph whose nodes are processors performing very simple operations over strings that can be sent to other nodes. These operations consist in erasing, adding or substituting symbols. Every node has filters that block some strings from being sent and/or received. As computing devices, NEPs alternate evolutionary and communication steps, until a predefined stopping condition is fulfilled. All processors change their contents at the same time in each evolutionary step. In the communication step, the strings that pass the corresponding filters are interchanged between connected nodes.

Up until now, it has not been possible to have implementations neither in vivo nor in vitro of many bio-inspired models including NEP model and its variants. Therefore, the only way to analyze and execute NEP devices is by means of ultra-scalable simulators able to encapsulate the inherent parallelism in their computations. When these devices have to be simulated on conventional computers, the total amount of space needed to simulate the model and to actually run the algorithm usually becomes exponential [6]. This is one of the main reasons why natural computing models are not yet widely used to solve real problems. To the best of our knowledge, there is a lack of simulators able to handle the size of non trivial problems in massively distributed computing environments. A specific proposal in this area was made by Navarrete et al [6] but unfortunately, the authors only propose an ad-hoc solution (based on java threads) for a specific NP-complete problem (Hamiltonian Path) by using a MPI based cluster. In fact, the MPI paradigm is not well suited to support ultra-scalable distributed computing because it does not have built-in resilience and fault tolerance capabilities that are mandatory in this context.

Nowadays, the emergence of massively distributed platforms for big data scenarios open a new line of interest for the development of ultra-scalable simulators able to execute non-trivial bio-inspired computational models. Due to their inherently parallel and distributed nature, the NEP and its variants fit nicely in these distributed computing platforms. Moreover, actual applicability of the NEP family is possible considering the scalability these platforms offer.

Contribution. For that reason, we propose as novelty NPEPE, a high scalability engine that runs NPEP descriptions using Apache Giraph on top of the Hadoop platform. Giraph is the open source counterpart of Google Pregel, an iterative graph processing system built for high scalability. NPEPE takes advantage of the inherent Giraph and Hadoop parallelism and scalablity to be able to deploy and run massive networks of NPEPs. With respect to other existing solutions our engine provides (1) generality (i.e. it can execute any NPEP description) and (2) the high scalability, fault tolerance and resilience that are built-in by design in the Giraph and Hadoop frameworks. In addition, we perform several experiments to demonstrate that NPEP descriptions can be easily deployed and run using a NPEPE engine on a Giraph+Hadoop platform. We run several NPEP descriptions of the *3-colorability* NP-complete problem with different graph sizes and topologies, using NPEPE on top of a Hadoop platform with 10 commodity PC nodes. In these experiments, we show that NPEPE engine can correctly deploy and execute NPEP descriptions.

The rest of the paper is organized as follows. In Section 2, we introduce NEP and NPEP natural computing models. Section 3 describes big data distributed computing frameworks, and graph oriented ones in particular. Section 4 describes the NPEPE engine architecture, its main components and how it interacts with Giraph. In Section 5, we show the obtained results when executing different NPEP descriptions of the 3-colorability NP complete problem on NPEPE engine. Finally, in section 6 the conclusions and future work are discussed.

2   Network of Evolutionary Processors

Network of Evolutionary Processors (NEP) [1] is a bio-inspired computational model that resembles a common architecture for massively parallel and distributed symbolic processing. NEP and their variants have extensively proved that NP-complete problems can be solved in an efficient way, in linear time and with a linear amount of resources [3, 4].

A NEP description consists of several processors placed in nodes of a virtual graph. Each processor node acts on local data in accordance with some predefined rules. Local data become mobile agents which can navigate in the network following a given protocol. This navigation is defined by communication processes that indicate the conditions imposed by the processors when sending or receiving data (even simultaneously). Conditions are defined by using a variety of filtering strategies (filters or polarizations) characterizing each variant in the model. When the processors use filters, the model is named Network of Evolutionary Processors (NEP). If the processors use polarizations, the model is called Network of Polarized Evolutionary Processors (NPEP).

From a bio-inspired point of view, these processors can be viewed as cells that contain genetic information encoded in DNA sequences which may evolve by point mutations. These mutations are represented as grammar rules representing the insertion, deletion or substitution of a pair of nucleotides. In addition, the polarization resembles an electrical charge that allows to filter words according to their sign. To evaluate words, polarized processors contain a valuation mapping that extracts the sign of each word. Therefore, the string migration from one node to another simulates a transport channel between two cells.

A brief informal introduction to NPEP model is done below. A formal description and details of the NPEP model can be found in [7]. We summarize some basic notation to introduce the model. An *alphabet* is a finite and non-empty set of symbols. Any sequence of symbols from an alphabet A is called *word* over A. The set of all words over A is denoted by A^* and the empty word is denoted by ε. The set of evolutionary processors over A are denoted by EP_A. A NPEP can be formally defined as a 7-tuple $\Gamma = (V, U, G, N, \varphi, x_I, x_O)$:

- V and U are the input and network alphabets respectively, such that $V \subseteq U$.
- $G = (X_G, E_G)$ is an undirected graph with vertices X_G and edges E_G.
- $N : X_G \to EP_U$ is a mapping that associates each node with the corresponding processor in a given NPEP. Each node $N(x)$ contains:
    - M_x is a finite set of evolutionary rules (insertion, deletion, substitution).
    - α_x defines how the rules will be applied: on the left, on the right or anywhere. Substitution rules are always applied in an arbitrary position.
    - $\pi_x \in \{-, 0, +\}$ defines the polarization of the node.
- $\varphi : U^* \to \mathbb{Z}$ is the valuation function that reveals the polarity of a word.
- $x_I, x_O \in X_G$ are the *input* and the *output* nodes, respectively.

The dynamic of the NPEP model is determined by *evolutionary* and *communication* steps performed sequentially on the configuration of each network node. A node configuration can be understood as the set of words that are present in any node at a given moment. Firstly, the input word is injected to x_I (input node) and the rest of the nodes are empty. Then, a sequence of evolutionary and communication steps are performed until a *halt* condition is reached. In each iteration an evolutionary and a communication step are executed in order. In each evolutionary step, nodes apply their predefined rules on their configuration and so, new node configurations can be obtained. In communication steps, each node outputs a copy of every word not matching its polarity, and receive words matching it. The *halt* condition of a Γ computation is defined by one of the following events: 1) There exists a configuration in which the set of words contained in the output node is non-empty, and 2) No further change can be produced in the configuration of the output node.

3   Distributed Computing Platforms for Big Data

The increasing availability of enormous data sets has contributed to a significant surge of popularity for large-scale data mining and data processing platforms during the last few years. One of the first frameworks to become widely popular was MapReduce [8]. The MapReduce framework is a programming model for processing large amounts of data based on two steps, *map* and *reduce*. The *map* step takes the input data and generates a set of key-value pairs according to a user-chosen criterion. The *reduce* step takes all the elements with the same key and performs an aggregation function, also implemented by the user. Each of the *map* and *reduce* operations can be executed independently on a separate machine. Therefore, together with the distributed, fault-tolerant storage capabilities of Hadoop [9], MapReduce allows for extreme scalability, and enjoys widespread use in research and industrial applications.

However, the limitations of MapReduce and its lack of suitability for certain classes of applications have motivated researchers to propose different alternatives, mainly in the context of iterative algorithms (e.g. Spark [10]), real-time analytics (e.g. Storm [11]), and graph processing. Specifically, despite its capabilities for large-scale data processing, MapReduce is ill-suited for graph-based algorithms. This shortfall, along with the pervasiveness of graph-like structures in today's data (e.g. social networks, search engine indices), has stimulated the design of new parallel computing platforms able to tackle the realization of high scalable graph computations. In this context, inspired by the Bulk Synchronous Parallel model [12], researchers at Google proposed Pregel [13] for processing social and other graphs. The main motivation for Pregel was the lack of high scalable and fault tolerant graph-processing frameworks. Pregel computations consist of a sequence of iterations, called supersteps, that are applied to the vertices of a graph. Each vertex in the graph is associated with a user-defined compute function that defines the behaviour of a single vertex. During a superstep the Pregel framework invokes this user-defined function for each vertex in parallel. In addition, the vertices can send messages through the edges and exchange values with other vertices. As far as the vertices are distributed on a cluster of machines, the computations can be done massively in parallel. Finally, a global check procedure determines if all compute functions are finished. Other notable graph-based large-scale processing platforms include Dato [14] (formerly GraphLab), Spark GraphX [15] and PowerGraph [16]. Recently, Giraph [17] has been proposed as the open source counterpart of Google Pregel. Due to its open source nature and how well its model fits the problem at hand, Giraph was chosen as the basic building block of NPEPE engine. In section 4 we provide a detailed description of how the NPEPE design on top of Giraph has been done.

4 NPEPE. A Massively Parallel NPEP Computation Engine

As previously explained, a Network of Polarized Evolutionary Processors (NPEP) works by running evolutionary steps in network nodes which communicate synchronously. We propose to model this parallel behavior using Giraph. Evolutionary steps fit nicely into Giraph's supersteps and huge NPEPs can take advantage of the scalability and fault tolerance of Giraph. We have designed NPEPE, a NPEP engine that can deploy and execute NPEP descriptions containing as many nodes as needed by using the Giraph framework on top of Hadoop.

NPEP descriptions can be represented both graphically and with text. Due to the lack of a widely accepted textual description, we propose one containing a line per NPEP node. The following fields describe the node behavior:

- "id": The node identifier.
- "input": If equal to 1, it identifies the input node
- "output": If equal to 1, it identifies the output node
- "polarity": The node polarity (i.e. -1, 0 or 1).
- "type". Which kind of rule the node will execute (insert, delete or susbtitute)
- "action": How the rules will be applied (on the left, on the right or anywhere)
- "rules": The set of rules to be applied to the word in each evolutionary step.
- "edges": The nodes that are directly connected to this node.

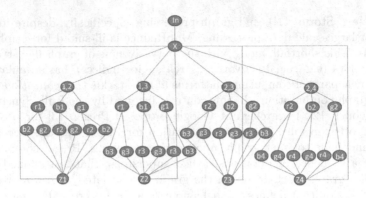

Fig. 1. A NPEP graphical description.

{"id": 0, "value": {"input": 1, "output": 0, "polarity": 1, "type": 0,
"action": 1, "rules": []}, "edges": [1]}

{"id": 1, "value": {"input": 0, "output": 0, "polarity": 0, "type": 0,
"action": 1, "rules": [["T1","T'0"], ["T2","T'1"], ["T3","T'2"],
["T4","T'3"], ["T0","T''0"]]}, "edges": [0, 2, 12, 13, 23, 24, 34, 35,
45, 46]}

{"id": 2, "value": {"input": 0, "output": 0, "polarity": 1, "type": 0,
"action": 1, "rules": [["e12","e'12"]]}, "edges": [1, 3, 6, 9]}

{"id": 3, "value": {"input": 0, "output": 0, "polarity": -1, "type": 0,
"action": 1, "rules": [["r1","r'1"]]}, "edges": [2, 4, 5]}

Fig. 2. A NPEP textual description (only nodes 0 to 3 are shown).

Figures 1 and 2 show a sample of a graphical NPEP description and its corresponding textual description (only nodes 0 to 3 are shown).

The NPEPE engine can directly execute any input NPEP, i.e. it can solve any NP-complete problem described in NPEP form without the need to implement additional features. NPEPE takes as input (1) an NPEP textual description and (2) a configuration file in which parameters such as alphabets, symbol weights and the initial input word are described. After parsing said description, the engine generates and deploys a set of Giraph components representing NPEP elements (e.g. nodes or messages) that can be run on top of a Hadoop platform.

The Giraph execution model, the two principal components of NPEPE engine architecture (*Computation* and *I/O* modules), and the way they are mapped to the Giraph execution model are described below.

Giraph Execution Model. Giraph requires the user to implement the actions to be done by each graph vertex in a specialization of class *BasicComputation*. During each superstep, Giraph invokes all vertices in parallel so that they can process their received messages and send out the ones they must. Once they are all done, the next superstep takes place. This process goes on until all vertices decide to stop. Since vertices function in a distributed manner, they cannot access the state of the rest of the graph. During a superstep, a vertex can only access its internal state and the messages it received. Therefore, termination is achieved through a voting process. When a vertex has nothing to do it *votes* to halt by invoking a specific method. This vote will remain active unless the vertex receives a message, in which case the vote will be called off until recast. The algorithm terminates once all vertices have voted to halt and no messages remain in flight.

NPEPE Architecture. Computation module: Each processor of an NPEP performs a set of distinct operations on incoming messages. These operations are implemented and managed by the following elements:

- *NpepComputation*: This class is a specialization of *BasicComputation*, i.e. it contains the method that Giraph invokes for every vertex on each superstep. For each NPEP processor (i.e. each node in the NPEP graph), Giraph will instantiate a *Vertex*. Therefore, on each superstep all NPEP processors will receive their messages and process them according to their type and their set of rules. After processing a message, they will discard it, store it for further processing or send it to all connected vertices.
- *Processor*: Every *Vertex* instance contains an object of this class, which encode the transformations to be applied to incoming messages. They can be of one of the following three types:
  - *InputProcessor*: Represents the *In* processor. To mimic the effect of this processor's rules, this vertex generates all possible solution candidates.
  - *OutputProcessor*: Represents the *Out* processor. When this vertex receives a message, it votes to halt and propagates a termination message.
  - *RegularProcessor*: Represents the rest of the processors. These vertices process incoming strings (but those of a different polarity) according to the rules specified in the NPEP input file. Afterwards, they will keep or send those strings to other vertices depending on their new polarity.

NPEPE architecture. I/O module: Given the large scale of the graphs that Giraph can handle, vertex instances support serialization. Giraph employs some of the Hadoop input and output abstractions, and provides some predefined formats for vertex representation, although the characteristics of the NPEP model required that we wrote our own. The *I/O* components are described below.

- *NpepInputFormat*: This class reads an NPEP text representation and converts it to a Giraph graph. To represent NPEPs, we have designed a JSON format stored in plain text, where each line describes exactly one vertex.
- *NpepOutputFormat*: When a Giraph execution ends, each vertex dumps its data to an external file. In the case of NPEPs, we have defined class *NpepOutputFormat* to dump the content of the output vertex.
- *NpepWritable*: This class implements the serialization logic necessary to keep vertex state across supersteps, as well as the static characteristics of each vertex. NPEPs store the strings that processors keep for further processing.

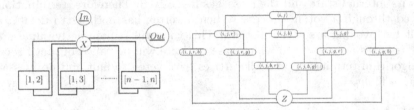

Fig. 3. Generic NPEP description of the 3-colorability problem as shown in [2]. Left side: Top level view of the NPEP description. Right side: Detail of a subnetwork that deals with coloring two vertices.

5   Experimental Results

In order to demonstrate that NPEPE can execute NPEP descriptions we attack the well known NP-complete "3-colorability" decision problem. The challenge of this problem is to decide whether each vertex in a connected undirected graph can be colored using three colors such that no two connected vertices have the same color. Graph coloring numerous applications: scheduling, register allocation, pattern matching, frequency assignment for mobile radio stations, etc.

Let a graph $\mathfrak{G}$ (a set of vertices and edges) be the input to the *3-colorability* problem. In [2] a generic NPEP description Γ is proposed to decide whether a graph $\mathfrak{G}$ can be colored with three colors, namely red, blue and green. If $\mathfrak{G}$ can be colored, Γ will generate as output the set of words representing the different ways $\mathfrak{G}$ can be colored. Figure 3 shows a generic graphical description of the proposed solution Γ and Table 1 describes the rules (M_x column) and the polarity (π_x column) of nodes in Γ. It must be noted that specific textual NPEP descriptions must be derived from the former generic solution to be able to execute the *3-colorability* problem for different graphs. To this end, we have developed a generator of NPEP textual descriptions that (1) takes as input a graph to be colored and a table describing the rules and polarity of nodes, and (2) generates a runable NPEP textual description of the *3-colorability* problem for this graph. For example, Figure 2 shows part of the NPEP textual description that our generator produced as output taking as input the 4 nodes graph shown in Figure 4 and the NPEP node behaviour described in Table 1.

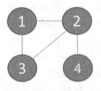

Fig. 4. A sample 4 node graph to be used as input to the *3-colorability* problem.

Table 1. Definition of rules and polarity of nodes in Γ as shown in [2]

Node	M_x	π_x
$\underline{In}$	$\bigcup_{i=1}^{n}\{a \to r_i, a \to b_i, a \to g_i\}$	$+$
X	$\{T_k \to T_{k-1} \mid 1 \le k \le n(n-1)/2\} \cup \{T_0 \to T_0''\}$	0
$[i,j]$	$\{e_k \to e_k' \mid e_k = \{i,j\}\}$	$+$
$[i,j,z], z \in \{r,g,b\}$	$z_i \to z_i'$	$-$
$[i,j,z,y], z,y \in \{r,g,b\}, z \ne y$	$y_i \to y_j''$	0
Z	$\{T_k' \to T_k \mid 0 \le k \le n(n-1)/2\} \cup$ $\{z_i' \to z_i, g_j'' \to g_j'\} \cup \{e_k' \to \bar{e}_k \; e_k = \{i,j\}\}$	$+$
$\underline{Out}$	$\emptyset$	$-$

We run several NPEP descriptions of the *3-colorability* problem with different graph sizes and topologies using NPEPE on top of a Hadoop cluster with 10 commodity PC nodes. Initially, we produced NPEP textual descriptions for several small size graphs in order to test that all components of NPEPE behaved adequately. Table 2 shows a summary of the execution results for each graph size, namely the total execution time, the number of NPEP nodes contained in the corresponding NPEP description, the number of all of the processed words and the number of results obtained at the end of the process. It must be noted that, in all cases, the obtained results matched the solutions to the problem. The main goal of these experiments is to show that NPEPE can deploy and run NPEP descriptions taking full advantage of the inherent Giraph+Hadoop cluster parallelism. In summary, we observe in these preliminary experiments that NPEPE engine can correctly deploy and execute NPEP descriptions.

Currently, we are deploying and running a new collection of NPEP textual descriptions for non trivial graphs to assess how well the proposed framework scales when the number of NPEP nodes and messages grow exponentially.

Table 2. Results of the execution of the 3-colorability problem with different graph sizes

Graph size (nodes)	NPEP size (nodes)	Total words	Solutions	execution time (secs)
4	43	81	12	36
6	68	729	66	66
7	68	2187	192	128
8	79	6561	384	356

6   Conclusions and Final Remarks

NP-complete problems are often attacked with ad-hoc algorithms. However, the NEP bio-inspired model and its variants can solve NP-complete problems efficiently. Up until now, it has not been possible to implement these models in vivo or in vitro. Therefore, it is extremely useful to provide generic, ultra-scalable simulators able to run them. Currently, there are no such simulators. Thus, our paper addresses this challenge proposing as novelty NPEPE, a highly scalable engine that can directly execute NPEP textual descriptions (i.e. it can solve any NP-complete problem described in the form of the corresponding NPEP without the need to implement additional features).

NPEPE is deployed on top of the Giraph and Hadoop frameworks, so it provides nice big data characteristics not present in previous proposals such as fault tolerance, resilience and ultra-scalability. We run initial experiments to demonstrate that NPEPE can correctly deploy and execute NPEP descriptions. Specifically, we executed several NPEP descriptions of the well-known NP-complete 3-colorability problem with different graph sizes and topologies and using NPEPE on top of a Giraph+Hadoop platform with 10 commodity PC nodes. The obtained results show that NPEP descriptions are correctly executed and their outputs match the expected results. We plan to run more complex NPEP descriptions of the 3-colorablity problem in order to test how well NPEPE scales when the number of NPEP nodes grows. In addition, we plan to generate NPEP descriptions and execute other well known NP-complete problems such as the "hamiltonian path" and the "knapsack" problems in order to test NPEPE performance when applying different node rules, actions and graph topologies.

We would like to thank Stanislav Vakaruv for helping us to debug and run the experiments and to configure the Giraph+Hadoop cluster.

References

1. Castellanos, J., Martín-Vide, C., Mitrana, V., Sempere, J.M.: Networks of evolutionary processors. Acta Informática **39**, 517–529 (2003)
2. Alarcón, P., Arroyo, F., Mitrana, V.: Networks of polarized evolutionary processors. Information Sciences **265**, 189–197 (2014)
3. Manea, F., Martín-Vide, C., Mitrana, V.: All NP-problems can be solved in polynomial time by accepting networks of splicing processors of constant size. In: Mao, C., Yokomori, T. (eds.) DNA12. LNCS, vol. 4287, pp. 47–57. Springer, Heidelberg (2006)
4. Manea, F., Margenstern, M., Mitrana, V., Pérez-Jiménez, M.J.: A new characterization of NP, P, and PSPACE with accepting hybrid networks of evolutionary processors. Theory Comput. Syst. **46**, 174–192 (2010)
5. Manea, F., Martín-Vide, C., Mitrana, V.: On the size complexity of universal accepting hybrid networks of evolutionary processors. Mathematical Structures in Computer Science **17**, 753–771 (2007)
6. Navarrete, C., Echeandia, M., Anguiano, E., Ortega, A., Rojas, J.: Parallel simulation of NEPs on clusters. In: Proceedings of the 2011 IEEE/WIC/ACM International Conferences on Web Intelligence and Intelligent Agent Technology, vol. 3, pp. 171–174. IEEE Computer Society (2011)
7. Arroyo, F., Gómez Canaval, S., Mitrana, V., Popescu, Ş.: Networks of polarized evolutionary processors are computationally complete. In: Dediu, A.-H., Martín-Vide, C., Sierra-Rodríguez, J.-L., Truthe, B. (eds.) LATA 2014. LNCS, vol. 8370, pp. 101–112. Springer, Heidelberg (2014)

8. Dean, J., Ghemawat, S.: MapReduce: simplified data processing on large clusters. Communications of the ACM **51**(1), 107–113 (2008)
9. Shvachko, K., et al.: The hadoop distributed file system. In: 2010 IEEE 26th Symposium on Mass Storage Systems and Technologies (MSST). IEEE (2010)
10. Zaharia, M., et al.: Resilient distributed datasets: a fault-tolerant abstraction for in-memory cluster computing. In: Proceedings of the 9th USENIX conference on Networked Systems Design and Implementation. USENIX Association (2012)
11. Apache Storm. http://storm.apache.org
12. Valiant, L.G.: A bridging model for parallel computation. Communications of the ACM **33**(8), 103–111 (1990)
13. Malewicz, G., et al.: Pregel: a system for large-scale graph processing. In: Proceedings of the 2010 ACM SIGMOD International Conference on Management of data. ACM (2010)
14. Yucheng, L., et al.: Distributed GraphLab: a framework for machine learning and data mining in the cloud. Proceedings of the VLDB Endowment **5**(8), 716–727 (2012)
15. Xin, R.S., et al.: Graphx: a resilient distributed graph system on spark. In: First International Workshop on Graph Data Management Experiences and Systems. ACM (2013)
16. Gonzalez, J.E., et al.: Powergraph: distributed graph-parallel computation on natural graphs. In: OSDI, vol. 12, No. 1 (2012)
17. Apache Giraph. http://giraph.apache.org/

Andromeda: A System for Processing Queries and Updates on Big XML Documents

Nicole Bidoit[1], Dario Colazzo[2], Carlo Sartiani[3]([⊠]), Alessandro Solimando[4], and Federico Ulliana[5]

[1] BD&OAK Team, Université Paris Sud - INRIA, Orsay, France
[2] LAMSADE, Université Paris Dauphine, Paris, France
[3] DIMIE, Università Della Basilicata, Potenza, Italy
`sartiani@gmail.com`
[4] DIBRIS, Università di Genova, Genova, Italy
[5] LIRMM, Université Montpellier 2, Montpellier, France

Abstract. In this paper we present Andromeda, a system for processing queries and updates on large XML documents. The system is based on the idea of statically and dynamically partitioning the input document, so to distribute the computing load among the machines of a Map/Reduce cluster.

1 Introduction

In the last few years cloud computing has attracted much attention from the database community. Indeed, cloud computing architectures like Google Map/Reduce [1] and Amazon EC2 proved to be very scalable and elastic, while allowing the programmer to write her own data analytics applications without worrying about interprocess communication, recovery from machine failures, and load balancing. Therefore, it is not surprising that cloud platforms are used by large companies like Yahoo!, Facebook, and Google to process and analyze huge amounts of data on a daily basis.

The advent of this novel paradigm is posing new challenges to the database community. Indeed, cloud computing applications might also be built upon *parallel databases*, that were introduced nearly two decades ago to manage huge amounts of data in a very scalable way. These systems are very robust and very efficient, but for the following reasons their adoption is still very limited: (i) they are very expensive; (ii) their installation, set up, and maintenance are very complex; and, (iii) they require clusters of high-end servers, which are more expensive than cloud computing clusters.

Our Contribution. In this paper we present Andromeda, a system that is able to process both queries and updates on very large XML documents, usually generated and processed in contexts involving scientific data and logs [2].

Our system supports a large fragment of XQuery [3] and XUF (XQuery Update Facility) [4]. The system exploits dynamic and static partitioning to distribute the processing load among the machines of a Map/Reduce cluster.

© Springer International Publishing Switzerland 2015
T. Morzy et al. (Eds): ADBIS 2015, CCIS 539, pp. 218–228, 2015.
DOI: 10.1007/978-3-319-23201-0_24

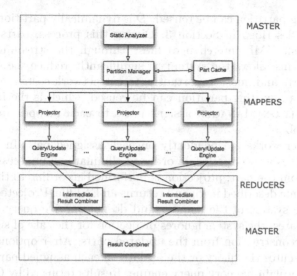

Fig. 1. System architecture.

The proposed technique applies when queries and updates are *iterative*, *i.e.*, they iterate the same query/update operations on a sequence of subtrees of the input document. From our experience many real world queries and updates actually meet this property.

Paper Outline. The rest of the paper is structured as follows. Section 2 introduces the general architecture of Andromeda, while Section 3 details the query and update processing technique used by the system. In Section 4, then, we present an experimental evaluation validating the scalability properties of the system. Finally, in Sections 5 and 6, we discuss some related work and draw our conclusions.

2   System Architecture

The basic idea of our system is to dynamically and/or statically partition the input data to leverage on the parallelism of a Map/Reduce cluster and to increase the scalability. The architecture of our system is shown in Figure 1.

When a user submits a query or an update to the system, the STATIC ANA-LYZER parses the input query or update, and extracts relevant information for partitioning the input document D. This information is passed to the PARTITION MANAGER, which verifies if D has already been partitioned; in that case, as a single document can be partitioned in multiple ways, the PARTITION MANAGER checks if there exists a partition that is still valid (*i.e.*, D has not been updated or externally modified after partitioning), and that it is compatible with the submitted query or update. Parts are stored in the distributed file system, so to be globally available.

If no existing partition can be reused, D is dynamically partitioned according to the scheme described in Section 3. During this process, parts are encoded as EXI (Efficient XML Interchange) files[1] through the streaming encoder of EXIficient [5]; this allows the system to significantly reduce the storage space required for parts and, most importantly, to cut network costs.

If, instead, an existing partition can be reused, which is the most common case, the PARTITION MANAGER assigns parts to each mapper and launches a Map/Reduce job.

Each mapper works independently on each assigned part. In the case of a query, each part is also *projected*, in order to eliminate all unnecessary elements or attributes from the part; projection is performed according to the path-based projection scheme described in [6] and returns an EXI file. Projected parts reside in the local file system of the mapper and do not survive query execution. In the case of updates, the system ignores projection for the sake of simplifying the global result reconstruction from the updated parts. After optional projection, the mapper executes the query or the update on each assigned part by invoking Qizx-open [7], a main-memory query engine. Results returned by Qizx-open are stored in the distributed file system.

Query/update results produced by mappers are combined into a single file in two phases. In the first phase, reducers perform a preliminary result combination, which is then refined by the RESULT COMBINER.

3   Processing Queries and Updates

3.1   Iterative Queries and Updates

Our system supports the execution of *iterative* XQuery queries and updates, *i.e.*, queries and updates that i) use forward XPath axes, and ii) first select a sequence of subtrees of the input document, and then iterate some operation on each of the subtrees. Iterative queries and updates are widely used in practice, and a static analysis technique has been proposed to recognize them [6].

As an example of iterative query, consider the following query on XMark documents [8] (assume $auction is bound to the document node $doc("xmark.xml")$).

```
for $i in $auction/site//description
where contains(string(exactly-one($i)), "gold")
return $i/node()
```

The query iterates the same operation on each subtree selected by $auction/site //description and, hence, is iterative.

---

[1] EXI is a binary format, proposed by the W3C, for compressing and storing XML documents.

This property is enjoyed by many real world queries: for instance, in the XMark benchmark 11 out of the 20 predefined queries are iterative[2]. Non iterative queries are typically those performing join operations on two independent sequences of nodes of the input documents. Notice that, however, iterative queries may perform join operations, as in the following case:

```
for $i in $auction/site//description
    $x in $i//keyword
    $y in $i//listitem
where $x = $y
return $x
```

Iterative updates include the wide class of updates that modify a sequence of subtrees, and such that each delete/rename/insert/replace operation does not need data outside the current subtree. As an example of iterative update, consider the following one:

```
for $x in $auction/site/regions//item/location
where $x/text() = "United  States"
return (replace value of node $x with "USA")
```

This update iterates over *location* elements and replaces each occurrence of "United States" with "USA". As no information outside the subtrees rooted by *location* elements is required for processing the replace operation, the update is iterative.

3.2 Data Partitioning

As described in Section 2, the STATIC ANALYZER parses the input query/update to extract the information required for checking the property of Section 3.1 and for partitioning the input data. This information, which is passed to the PARTITION MANAGER, is essentially the set of paths used in the query/update, enriched with details about bound variables, and it guides the partitioning process.

To illustrate, consider the following iterative query:

```
for $x in /a,
    $y in $x/b
where $y/c/d
return < res > $y/c/e < /res >
```

For this query the STATIC ANALYZER extracts the following set of paths:

$$\{ \ /a\{for\ x\}, /a\{for\ x\}/b\{for\ y\},$$
$$/a\{for\ x\}/b\{for\ y\}/c/d, /a\{for\ x\}/b\{for\ y\}/c/e\}$$

---

[2] Queries from Q_1 to Q_5, Q_{14}, and Q_{16} to Q_{20} are iterative.

By analyzing this set of paths, the STATIC ANALYZER derives that $/a/b$ is the path on which the query iterates; this path is called *partitioning path* and is used during the partitioning process to identify *indivisible* subtrees, *i.e.*, subtrees that cannot be split among multiple parts. In particular, if a node matches this path, then the whole subtree is kept in the current part; subtrees rooted at nodes outside subtrees selected by the partitioning path can be split across consecutive parts. This indivisibility property is necessary to ensure that query result on the input document is equal to the ordered concatenation of query results on each part.

In the case of updates, the system must distinguish between *simple updates*, *i.e.*, updates consisting of a single `delete/rename/insert/replace` operation without `for`-iterations, and update containing iterations. In the first case, the STATIC ANALYZER extracts paths selecting target nodes of the update operations, and considers these paths as partitioning paths. In the second case, the partitioning path is computed as for queries. Composite updates are treated by summing the partitioning paths of each update. As happens for queries, partitioning paths are used to recognise subtrees that should not be split. Again, this indivisibility property is necessary in order to ensure semantics preservation once the update is distributed over the partition.

When a document is partitioned for the first time, the PARTITION MANAGER uses the partitioning paths to perform the actual partitioning. The PARTITION MANAGER also computes a DataGuide [9] for an input document D. The DataGuide is later used to verify the compatibility of a newly issued query/update with an existing partition, by verifying that the indivisible subtrees identified by the partition paths of the new query/update are already indivisible in an existing partition.

For both queries and updates, the PARTITION MANAGER ensures that each part in the partition does not exceed the memory capacity of the main-memory query engine by ending the current part and creating a new one when the size of the current part exceeds a given threshold (if this happens during the visit of an indivisible subtree, then the part is terminated only after the subtree has been totally parsed). Also, for both queries and updates, artificial tags are added during partitioning to ensure each generated part is well-formed and rooted (so that the query/update engine can process it).

3.3 Query/Update Processing

Once the STATIC ANALYZER has extracted path information from the input query/update, and the PARTITION MANAGER has found an existing partition or created a new one for processing the query/update, parts are assigned to mappers for query/update processing.

When processing a query, each mapper receives not only the address on the distributed file system of each assigned part, but also the path set extracted by the STATIC ANALYZER. This set is used to project the parts, *i.e.*, to remove elements and attributes not necessary for the query. While original parts are stored in the distributed file system, projected parts are stored in the local file system of the mapper and do not survive query execution. The input query is executed on each projected part by a local instance of Qizx-open, which exports the results, encoded in XML format, to the distributed file system.

When processing an update, instead, projection cannot be applied, as each fragment of a given input part is necessary. As a consequence, the local instance of Qizx-open just executes the update on the original part, and stores its updated version, encoded in EXI format, in the distributed file system.

3.4   Result Combination

Result combination works a bit differently for queries and updates. Indeed, partial results of a query can be simply concatenated together, while partial results of an update must be merged.

The combination of partial query results is performed in two steps. In the first step, each reducer receives a set of consecutive part results, which are then combined through high-speed Java NIO channels; the RESULT COMBINER, finally, links together the combined part results produced by the reducers. In the case of updates, untouched parts must be merged with updated parts; this process requires the system to read all parts and drop artificial tags introduced by the data partitioning technique.

4   Experimental Evaluation

Our experiments aim at i) proving the efficiency of the system in processing queries and updates on large documents, and ii) showing how the system scales with the document size and the number of nodes in the cluster.

4.1   Experimental Setup

We performed our experiments on a *multitenant* cluster running Hadoop 2.2 on RHEL Linux and Java 1.7. The cluster comprises 1 master node and 100 slave nodes connected through an InfiniBand network. To reduce issues related to independent system activities and other jobs in the cluster, we ran each experiment five times, discarded both the highest (worst) and the lowest (best) processing times, and reported the average processing time of the remaining runs.

4.2   Test Sets

We performed our experiments on two distinct datasets. The first dataset is dedicated to query experiments, and comprises five XMark [8] XML documents obtained by running the XMark data generator with factors 100, 150, 200, 250, and 300, respectively; the resulting documents have approximate sizes ranging from 10GB to 32GB. The second dataset is used for update tests and contains ten XMark documents whose size ranges approximately from 1GB to 10GB.

4.3 Evaluating Queries

In our first battery of experiments we tested the performance and the scalability of our system when processing queries. In the first test we selected the iterative fragment of the XMark benchmark query set (*i.e.*, queries Q_1, Q_2, Q_3, Q_4, Q_5, Q_{14}, Q_{15}, Q_{17}, Q_{18}, Q_{19}, and Q_{20}) and processed each query individually on the documents of the first data set; in this experiment we used parts of size 100000000 bytes. The results we obtained are shown in Figure 2(a). This graph indicates that the evaluation time is only partially affected by the size of the input document: indeed, given an input query Q, Andromeda filters out parts that do not structurally match Q, and processes Q only on those parts that may give a contribution to the result. Hence, even for large documents, the number of machines actually used by the system is below the cluster size.

Partitioning time for exemplifying queries Q_1, Q_2, Q_5, and Q_{14} is reported in Table 1, together with the number of *generated* parts and the percentage of *used* parts. As we mentioned before, unused parts are discarded. As it can be easily observed, the partitioning time grows linearly with the size of the input document and the number of used parts is only a small fraction of the total number of parts, with the only notable exception of query Q_{14}, which is not very selective. This explains why the processing time of queries Q_{14} and Q_{19}, that uses exactly the same partitioning scheme of query Q_{14}, is bigger than that of the remaining queries.

Table 1. Partitioning time (sec.), generated parts, and used parts (%).

Size	Q_1			Q_2			Q_5			Q_{14}		
	Time	Gen.	Used	Time	Gen.	Used	Time	Gen.	Used	Time	Gen.	Used)
10GB	851.686	142	9.1%	706.45	138	22.4%	813.448	144	11.8%	810.014	138	42.7%
15GB	1148	214	8.8%	1060	207	22.7%	1243	217	11.9%	1250	208	42.7%
20GB	1564	285	9.1%	1461	277	22.3%	1666	290	12%	1700	277	42.5%
25GB	2007	357	8.9%	1808	347	22.1%	2215	363	11.8%	2299	347	42.6%
30GB	2391	429	8.8%	2147	417	22.3%	2526	436	11.9%	2534	417	42.4%

Processing Workloads. In our third experiment we evaluated the performance of our system when processing a workload comprising all the queries of the iterative fragment of XMark. The results we collected are shown in Figure 2(b) and Table 2.

In Figure 2(b) we reported the total workload processing time. It is worthy to note that workload processing time grows linearly with the size of the input document. This is implied by the fact that, even on smaller documents, the parallel execution of the queries in the workload involves the use of all the machines in the cluster, as confirmed by Table 2, which reports the partitioning time, the number of generated parts, and the number of map input records (parts to process) for each input document: as shown in this table, even on the 10GB document the cluster is fully exploited.

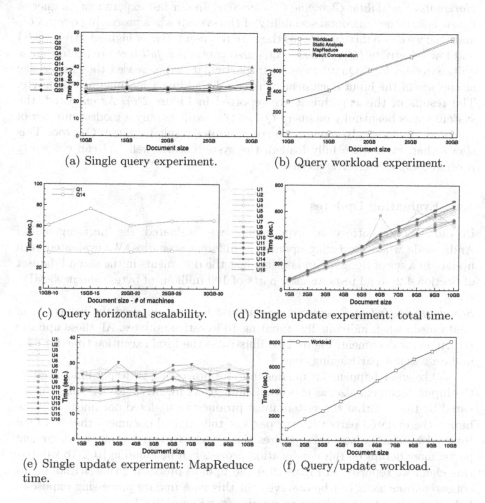

(a) Single query experiment.

(b) Query workload experiment.

(c) Query horizontal scalability.

(d) Single update experiment: total time.

(e) Single update experiment: MapReduce time.

(f) Query/update workload.

Fig. 2. Experiments.

Table 2. Workload: partitioning time (sec.), generated parts, and map input records.

Size	Time	Gen.	Map input records
10GB	694.342	120	1309
15GB	1070106	181	1980
20GB	1424.379	241	2618
25GB	1876.539	302	3289
30GB	2138.638	362	3938

Horizontal Scalability: Changing Cluster Size. In our last experiment on queries we evaluated the horizontal scalability of the system when processing queries Q_1 and Q_{14}: we chose these queries as they are representative of high selectivity (Q_1) and low selectivity (Q_{14}) queries; Q_{14} also contains a *full-text* predicate that is quite stressful for XQuery engines. In particular, we increased the cluster size as the size of the input document increases, by adding 1 machine per Gigabyte. The results of this experiment are reported in Figure 2(c). As expected, the system scales beautifully on query Q_1, as this exploits only a modest number of machines. Surprisingly enough, we got a similar result for query Q_{14} too. This shows that, even when fully loaded, the system scales well and can efficiently process complex iterative queries.

4.4   Evaluating Updates

In our second battery of experiments we evaluated the performance of Andromeda when processing updates in different scenarios. We evaluated each update in a set of iterative updates against the documents in the second dataset of Section 4.2; in all tests we used parts of 100 millions of bytes (about 95 MB).

Scalability of Update Processing. In our first test we analysed the behaviour of Andromeda when individually executing 16 iterative updates. All these updates return a new document. Figure 2(d) illustrates the total execution time for each update without partitioning time.

Unlike what happens for queries, update processing is deeply influenced by the input document size, as execution time grows linearly with it. This is motivated by the fact that the system must produce an updated document by combining the updated parts with the parts of the original document that were not touched by the update: this requires the system to traverse all the document parts; more details on this combination process can be found in [10]. To validate this claim we reported in Figure 2(e) the update processing time without part concatenation; as it can be observed, in this case update processing exposes a behaviour close to that shown on queries (see Figure 2(a)).

Processing Mixed Workloads. In our second test we created a random query/update workload and analyzed the behaviour of the system when processing the workload on documents of increasing size. The workload comprises 20 expressions randomly chosen by an initialization script, that also chooses the execution order: queries and updates are executed according to the *reader/writer* semantics, hence queries can be evaluated simultaneously, while updates have to be processed individually. Queries and updates are selected by respecting a 80:20 ratio, hence the workload contains 16 queries and 4 updates. The composition of the workload we considered is reported below:

$$W = (U_2, U_{12}, [Q_{18}, Q_{17}, Q_3, Q_1, Q_{18}],$$
$$U_4, U_{14}, [Q_{15}, Q_5, Q_2, Q_{17}, Q_{15}, Q_{15}, Q_{20}, Q_{10}, Q_1, Q_5, Q_{18}])$$

Figure 2(f) describes the behaviour of the system when processing the workload. As it can be observed, the workload execution time grows linearly with the input size, despite the fact that 16 tasks out of 20 are queries. This is caused by the presence of updates, which not only require result concatenation, but also force the system to partition the updated document for processing the next task, hence making partition reuse much less effective.

5 Related Works

There exist only a few systems able to process queries on XML data in distributed and cloud environments, e.g., ChuQL [11], MRQL [12], HadoopXML [2], PAXQuery [13], and VXQuery [14]. Among them, HadoopXML is the system that most closely resembles Andromeda as it can transparently process XPath queries on an Hadoop cluster. HadoopXML requires a preliminary document indexing phase, close to Andromeda partitioning phase. Despite these similarities, HadoopXML only supports XPath queries, and, unlike Andromeda, cannot process XQuery queries or XUF updates.

PAXQuery and VXQuery are systems for processing XQuery queries on collections of (relatively) small XML documents scattered across a cloud computing cluster. While very efficient even on small clusters, they were not designed to evaluate queries on big documents. MRQL is a query processing system that supports an SQL-like query language that can be used to query XML and JSON data; MRQL directly translates queries into Java code that can be executed on top of Hadoop or Spark. While more powerful than PigLatin, MRQL cannot process complex XQuery queries and does not support updates. ChuQL, finally, is a language embedding XQuery that allows the programmer to distribute XQuery queries over Map/Reduce clusters. The programmer has the duty to manage low-level details about query parallelization, while Andromeda completely hides the underlying processing environment.

To the best of our knowledge, there is no system supporting XUF updates on big XML documents.

6 Conclusions and Future Work

In this paper we described the architecture of Andromeda, and analyzed its performance and scalability. This analysis confirms that Andromeda scales with the document size and the number of nodes in the cluster, and that it can efficiently process queries and updates on very large XML documents.

In the near future we want to extend Andromeda in several ways. First of all, we want to improve the partitioning technique, so to obtain new partitions from existing ones without the need of reading and parsing again the whole input document. Second, we want to explore new techniques for result fusion in order to lower its cost. Finally, we want to understand if and how Hadoop can be replaced with Apache Spark.

References

1. Dean, J., Ghemawat, S.: MapReduce: simplified data processing on large clusters. In: OSDI, pp. 137–150. USENIX Association (2004)
2. Choi, H., Lee, K.H., Kim, S.H., Lee, Y.J., Moon, B.: HadoopXML: a suite for parallel processing of massive XML data with multiple twig pattern queries. In: wen Chen, X., Lebanon, G., Wang, H., Zaki, M.J. (eds.) CIKM, pp. 2737–2739. ACM (2012)
3. Boag, S., Chamberlin, D., Fernández, M.F., Florescu, D., Robie, J., Siméon, J.: XQuery 1.0: An XML Query Language (Second Edition). Technical report, World Wide Web Consortium. W3C Recommendation (2010)
4. Robie, J., Chamberlin, D., Dyck, M., Florescu, D., Melton, J., Siméon, J.: XQuery Update Facility 1.0. Technical report, World Wide Web Consortium. W3C Recommendation (2011)
5. (Exificient). http://exificient.sourceforge.net
6. Bidoit, N., Colazzo, D., Malla, N., Sartiani, C.: Partitioning XML documents for iterative queries. In: Desai, B.C., Pokorný, J., Bernardino, J. (eds.) IDEAS, pp. 51–60. ACM (2012)
7. (Qizx-open). http://www.xmlmind.com/qizxopen/
8. Schmidt, A., Waas, F., Kersten, M.L., Carey, M.J., Manolescu, I., Busse, R.: XMark: a benchmark for XML data management. In: VLDB, pp. 974–985. Morgan Kaufmann (2002)
9. Goldman, R., Widom, J.: DataGuides: enabling query formulation and optimization in semistructured databases. In: VLDB (1997)
10. Malla, N.: Partitioning XML data, towards distributed and parallel management. PhD thesis, Université Paris Sud (2012)
11. Khatchadourian, S., Consens, M.P., Siméon, J.: Having a ChuQL at XML on the cloud. In: Barceló, P., Tannen, V. (eds.) Proceedings of the 5th Alberto Mendelzon International Workshop on Foundations of Data Management, Santiago, Chile, May 9–12, 2011. CEUR Workshop Proceedings, vol. 749. CEUR-WS.org (2011)
12. Fegaras, L., Li, C., Gupta, U., Philip, J.: XML query optimization in map-reduce. In: WebDB (2011)
13. Camacho-Rodríguez, J., Colazzo, D., Manolescu, I.: PAXQuery: a massively parallel XQuery processor. In: Katsifodimos, A., Tzoumas, K., Babu, S. (eds.) Proceedings of the Third Workshop on Data analytics in the Cloud, DanaC 2014, June 22, 2014, Snowbird, Utah, USA. In conjunction with ACM SIGMOD/PODS Conference, pp. 1–4 ACM (2014)
14. Preston Carman Jr., E., Westmann, T., Borkar, V.R., Carey, M.J., Tsotras, V.J.: Apache VXQuery: A scalable XQuery implementation (2015). CoRR abs/1504.00331

Fast and Effective Decision Support for Crisis Management by the Analysis of People's Reactions Collected from Twitter

Antonio Attanasio[1,2]([✉]), Louis Jallet[3], Antonio Lotito[1], Michele Osella[1], and Francesco Ruà[1]

[1] Istituto Superiore Mario Boella, Torino, Italy
{attanasio,lotito,osella,rua}@ismb.it
[2] Politecnico di Torino, Torino, Italy
[3] Ineo Digital, Vélizy Villacoublay, France
louis.jallet@cofelyineo-gdfsuez.com

Abstract. The impact of human behavior during a crisis is a crucial factor that should always be taken into account by emergency managers. An early estimation of people's reaction can be performed through information posted on social networks. This paper proposes a platform for the extraction of real time information about an ongoing crisis from social networks, to understand the main concerns issued by users involved in the crisis. Such information is combined with other contextual data, in order to estimate the impacts of different alternative actions that can be undertaken by decision makers.

Keywords: Crisis management · Social networks · Event detection · Human behavoir · Sentiment analysis · Decision support system

1 Introduction

During disasters and emergencies, the success of rescue operations can be determined by many factors. The human behavior is among the most valuable ones and sometimes it may aggravate (or diminish) hazards, hindering also the most carefully prepared rescue plans. Therefore, the opportunity to monitor people's behavior in real-time during a crisis would be a strong advantage. In the last years, lots of people directly involved in large-scale emergencies have reported on social networks the happenings they have witnessed, sharing millions of posts and huge volumes of data, especially during the very first stages [1].

Contents posted on social networks can provide valuable information that is unlikely to be discovered through other media [2]. Aggregated information about people's perception of the crisis can be also deduced, through appropriate data analysis and data mining algorithms [3,4]. Furthermore, in many cases real-time messages posted on social networks allow to detect hazardous events, anticipating announcements from official sources [5].

© Springer International Publishing Switzerland 2015
T. Morzy et al. (Eds): ADBIS 2015, CCIS 539, pp. 229–234, 2015.
DOI: 10.1007/978-3-319-23201-0_25

This paper proposes a solution to provide a timely notification of hazardous events and then to evaluate people's reaction, through the real-time analysis of messages on Twitter. Contextual data about an ongoing crisis will be combined with historical data about past crises and processed by a Decision Support System (DSS), to provide decision makers with a ranking of alternative actions and an estimated measure of their impacts. The work is part of Snowball, an ongoing research project funded under the European Seventh Framework Programme. The main purpose of the project is to provide a deep analysis of cascading effects during a crisis and to develop methods to anticipate them. Three different scenarios will be simulated to validate the solution, using data collected during past crises: a storm in Finland and Poland, a volcanic eruption in Santorini and a flooding of Danube in Hungary. This paper is focused on the cascading effects caused by human behavior and exploits the knowledge from social networks to evaluate how people react to a crisis.

The rest of the paper is organized as follows. Section 2 illustrates the overall architecture of the proposed solution. Section 3 describes how information is extracted from Twitter. Section 4 introduces the adopted database and data model. Section 5 explains the role of the DSS within the platform.

2   Overall Architecture

The schema in Fig. 1 illustrates the architecture of the proposed solution (a reduced view of the overall Snowball architecture).

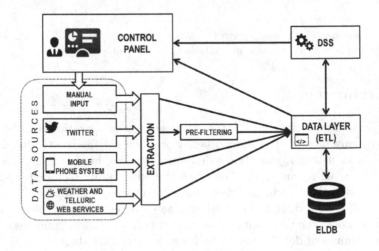

Fig. 1. High-level architecture of the proposed solution

A control panel will display reports about an ongoing crisis, allowing decision makers to control and coordinate the rescue operations through an interface for manual inputs, which represents one of the data sources of the platform.

The others are: the Twitter social network; the mobile phones in the hand of first responders and rescue managers; and meteorological and telluric Web services.

Based on statistics on past crises and on other related works [1], Twitter can be considered as the only source potentially providing high bursts of data to be analyzed, especially during the first stages of a crisis. For this reason, in the rest of the paper the other data sources will not be explicitly addressed, since they are not directly involved in the big data processing issue.

Data extracted from their own sources are sent to the Extraction Transformation and Loading (ETL) module, that shall reformat and store them in the Event Log DataBase (ELDB).

Finally, the Decision Support System (DSS) takes its input data from the ELDB and runs an algorithm on the modeled crisis scenario, to assess the impacts of different alternative actions.

3   Twitter Data Mining

To extract relevant information useful along the crisis management process, three phases have been identified: Collection, Filtering and Mining.

During the collection phase, data coming from Twitter will be extracted using the public Streaming API, in order to allow a continuous research based on some filtering parameters, such as keywords, hashtags, geolocation, languages.

The purpose of the filtering phase is to discard non relevant data using a list of forbidden keywords and to filter on the geolocation, using some specific geographical dictionaries that will help to understand the place where the tweet is referring to.

The mining phase will include both event detection analysis and sentiment analysis. The event detection analysis will be done using specific dictionaries for every crisis type, including keywords arranged in 3 levels, respectively: keywords to filter the search on Twitter; keywords to refine the mining, identifying words which characterize the crisis; and adjectives to define the dangerousness of the crisis. A score is associated to each keyword, representing its relative dangerousness. The sentiment analysis will be done using a specific dictionary for every crisis that the system includes in the analysis.

Beside these two analyses, a graph data model of all the relations between tweets is populated in near real time, in order to understand and identify the roles played by users in Twitter (influencers, followers, bridges) and to consider different weights for the related tweets. In this direction, in a precedent work we analyzed two past crises happened in 2014 - the flood of river Secchia (Italy) and the torrential flood in the Bastan Valley (France) - modeling the extracted tweets with a graph. The open source NodeXL tool [7] was used to cluster the tweets, using various techniques and to extract some relevant parameters which characterize the relations, such as *betweenness centrality* and *closeness centrality*. The same approach will be applied within the Snowball platform. Afterwards, another tool [6] will be used to classify the content, using human and machine intelligence to collect and automatically tag a big quantity of messages in near real time. The classification needs a training period, using a set of items that can be tagged by a human operator or with other methods. The data will be tagged using the aforementioned dictionaries, then the system automatically will apply a classifier [8] to tag the collected data in real-time.

4   Events Log Database (ELDB)

The ELDB is the big data repository for the whole Snowball platform. It is deployed on a cluster of 4 servers running the Apache Hadoop Distributed File System (HDFS) [9].

The designed data model is inspired by the EDXL-DE specification, defined by the OASIS Emergency Management Technical Committee [10].Data will be stored in JSON messages containing the objects and attributes defined by EDXL-DE. The "ContentObject" list is the main element of the whole message and it is basically structured in two parts: "ContentData" and "ContentDescriptor".

The "ContentData" element is in turn a list and includes the real payloads carried by the message. The structure of its elements varies with the repository. For Twitter data, each element includes the content of the tweet, the number of followers of the corresponding account, the number of retweets, a reliability measure for the account and its location and the score derived from sentiment analysis.

The "ContentDescriptor" element contains some metadata that will be used for data indexing.

5   Decision Support System

The DSS aims at improving the preparedness and the effectiveness of actions that decision makers may adopt in the near future to answer upcoming crisis situations. Such a challenging evaluation of responses meant to mitigate the aftermath of a crisis yet to come is performed through the appraisal of social, economic and environmental impacts of cascading effects underlying the potential crisis. This is made possible by world-class decision algorithms leveraging available data. Indeed, the distinctive trait of the DSS at hand in comparison with the real-time services previously described apropos of on-going crises lies in the data sources that are selected as inputs. In fact, given the breadth and the depth of decision services offered at the fingertips of public decision makers in relation to upcoming crises, a cohort of additional data sources is leveraged to meaningfully complement Big Data streams stemming from social media. Prominent data sources that are combined with social media data are as follows:

- Contextual data (e.g., profile of the population, workforce composition, industrial base, emergency centers, weather data, road network, telecommunication grid).
- Historical data on past crises (e.g., extent of the damage, crisis management, reactions of authorities, population behavior).

- Outcomes of previous researches (e.g., already-known mathematical laws that relate emergencies to disaster management and impacts).
- Primary data (e.g., data obtained from discussions with experts of the crisis domain).

The DSS, drawing on aforementioned data sources and on simulations of scenarios realized through Agent-Based computational models [11] in a separate module of SnowBall suite, adopts Multi-Criteria Decision Modeling (MCDM) [12] as methodology allowing decision makers to compare and rank the different alternatives at stake reaching an acceptable compromise among the several goals pursued, either when performances are pre-determined or inputted by the user. Taking into account the manifold techniques falling under the banner of MCDM [13], Snowball DSS opts for Electre [14] due to the needs of leaving many degrees of personalization to end-users (i.e., using weights as decision levers) and keeping moderate the cognitive effort required to end-users (i.e., unraveling the inherent complexity of cascading effects). In view of the recourse to Electre, the type of output data that the DSS component returns is the ranking of alternative actions in light of impacts exerted and of priorities assigned to impacts. Such a ranking could be used as-is or as basis for selection (i.e., choice of the most effective option) of sorting (i.e., assignment of each action to a pre-defined category). The resulting outputs are made available both in a tabular format (i.e., estimated impact metrics for each alternative measure are stored into the ELDB) and in a graphical way (statistical distributions of expected impacts are displayed in the control panel).

6   Conclusion and Future Work

This paper proposed a solution to include the impact of human behavior during the evaluation of a crisis scenario and of the possible decisions concerning rescue operations. Social networks were chosen as a primary source of information directly provided by people involved in the crisis. A technique to extract useful information from Twitter was also described. Finally, a MCDM methodology was proposed to estimate the impact of several alternative decisions that can be considered by emergency managers.

The platform will be tested within the Snowball project, by simulating three possible scenarios, using historical data - including tweets - related to the considered real world past crises.

Acknowledgments. The authors wish to thank all the partners involved in the Snowball EU FP7 project for their contribution to the work presented in this paper.

References

1. Federal Emergency Management Agency: 2013 National Preparedness Report. Report (2013)
2. Birregah, B., Top. T., Perez, C., Chatelet, E., Matta, N., Lemercier, M., Snoussi, H.: Multi-layer crisis mapping: a social media-based approach. In: 22nd IEEE International WETICE Conference (WETICE 2013), pp. 379–384. Hammameth (2012)
3. Terpstra, T., de Vries, A., Stronkman, R., Paradies, G. L.: Towards a realtime twitter analysis during crises for operational crisis management. In: Proceedings of the 9th International ISCRAM Conference, Vancouver, Canada, pp. 1–9 (2012)
4. Ritter, A., Mausam, Etzioni, O., Clark, S.: Open domain event extraction from twitter. In: Proceedings of the 18th ACM SIGKDD International Conference on Knowledge Discovery and Data Mining (KDD 2012), pp. 1104–1112. ACM, New York (2012)
5. Sakaki, T., Okazaki, M., Matsuo, Y.: Earthquake shakes twitter users: real-time event detection by social sensors. In: Proceedings of the 19th International Conference on World Wide Web (WWW 2010), pp. 851–860. ACM, New York (2010)
6. Artificial Intelligence for Disaster Response. http://aidr.qcri.org
7. NodeXL. http://nodexl.codeplex.com
8. Imran, M., Castillo, C., Lucas, J., Meier, P., Vieweg, S.: AIDR: artificial intelligence for disaster response. In: Proceedings of the Companion Publication of the 23rd International Conference on World Wide Web Companion (WWW Companion 2014), pp. 159–162. International World Wide Web Conferences Steering Committee, Republic and Canton of Geneva, Switzerland (2014)
9. Hadoop Distributed File System. http://hadoop.apache.org/
10. OASIS Emergency Management TC. http://www.oasis-open.org/committees/emergency/
11. Gilbert, N.: Agent-Based Models (No. 153). Sage, Thousand Oaks (2008)
12. Köksalan, M., Wallenius, J., Zionts, S.: Multiple Criteria Decision Making: From Early History to the 21st Century. World Scientific, Singapore (2011)
13. Triantaphyllou, E.: Multi-Criteria Decision Making Methods: a Comparative Study. Kluwer Academic Publishers, Dordrecht (2000)
14. Roy, B.: The Outranking Approach and the Foundations of ELECTRE Methods. Theory and Decision **31**(1), 49–73 (1991)

Adaptive Quality of Experience: A Novel Approach to Real-Time Big Data Analysis in Core Networks

Alejandro Bascuñana, Manuel Lorenzo,
Miguel-Ángel Monjas[✉], and Patricia Sánchez

Technology and Innovation, Madrid R&D Center, Ericsson Spain, Madrid, Spain
{alejandro.bascunana,manuel.lorenzo,
miguel-angel.monjas,patricia.sanchez.canton}@ericsson.com

Abstract. Mobile networks are more and more a key element in how users access services. Users' perception of access to services is difficult to be grasped and enhanced if degraded. Adaptive Quality of Experience (AQoE) aims to provide a comprehensive framework for predicting and preventing QoE degradation situations. This paper provides a high-level architectural model for managing communication service providers' (CSP) mobile networks. AQoE relies on real-time big data analysis techniques to measure and anticipate QoE to allow CSP's to perform actions to maintain a certain level of QoE. Thus, CSP's can adapt users' QoE dynamically and pro-actively in order to deal with service experience degradations in mobile networks.

1 Introduction

Propelled by the dramatic growth in the number of smart phones and mobile Internet connections [1], the networks of mobile communication service providers (CSP) have become a key element in the delivery of high-quality services to users. Therefore, they have to face a conceptual shift from the management of network-based quality of service parameters to putting emphasis on the users' Quality of Experience (QoE). Quality of Experience control in mobile CSP's has to consider, at least, three different aspects: (a) a QoE model; (b) the way QoE is measured and anticipated; and (c) QoE actuation (that is, which corrective actions have to be enforced in order to enhance QoE or to mitigate its degradation). We name this overall concept Adaptive Quality of Experience (AQoE) and relies on the availability of real-time big data analysis techniques able to measure and anticipate QoE so that proper actuation can be carried out. This application paper focuses on said aspect, while providing also a high-level description of the remaining items.

2 Adaptive Quality of Experience

Adaptive Quality of Experience (AQoE) is all about predicting and preventing QoE degradation scenarios by optimizing, in real-time, network control policies and enforcing them. AQoE aims to provide the required flexibility in order to secure that

T. Morzy et al. (Eds): ADBIS 2015, CCIS 539, pp. 235–242, 2015.
DOI: 10.1007/978-3-319-23201-0_26

QoE targets, established for every customer segment, are met, despite the rate and level of changes in the ways that users access and consume the communications services delivered through mobile CSP's networks.

The Adaptive Quality of Experience concept builds on top of the regular network management and model:

- **Network Monitoring:** The CSP network is continuously supervised. Whenever a degradation or failure is detected, it is notified to a network administrator (either a human or a computer).
- **Network Policies Selection.** Proper decisions are made. They usually depend on a set of predefined rules or, simply, on an expert experience. As a result, a set of network policies that could cope with the problematic situation is decided to be applied.
- **Network Policies Deployment.** Deployment and enforcement of network policies follow the regular policy decision / enforcement pattern [4]. A Policy Decision Point evaluates network policies according to the events generated in the network and already existing policies and decides how to enforce them.
- **Network Policies Execution**. Finally, network policies are enforced in a Policy Enforcement Point. The new policies are expected to have an effect on the network conditions and hopefully enhance or mitigate the problematic situation.

Adaptive Quality of Experience enhances that model by introducing a real-time big data analytics functionality that we name Analytics Function. This functionality is the responsible of analyzing the data items extracted from the network, compute QoE indicators from the network traffic and detecting potential QoE degradation situations. A Policy Governance Function is also introduced to complement existing Policy Decision Points and deciding, according to the predicted degradation parameters, which and how mitigation policies are to be applied.

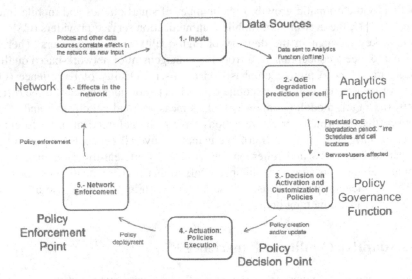

Fig. 1. Adaptive Quality of Experience high-level architecture

3 Quality of Experience Model

Quality of Experience (QoE) has been defined as the overall performance of a system from the point of view of the users [2]. QoE measures end-to-end performance at the services level from the user perspective and an indication of how well the system meets the user's needs [3]. Although there is a definition of QoE, no standard measures of QoE have been established or are widely accepted by the industry. The approach taken by Ericsson is based in its Ericsson Expert Analytics (EEA) [6]. It has defined a number of service level indices in order to predict customer satisfaction. These indices are named Key Performance Indicators (KPI) and can be computed in near real-time by continuously analyzing the traffic networks and other data sources, such as the Business Support System (BSS). How relevant said KPI's are for the users' QoE is defined by a KPI-to-QoE model that has been learned from field tests that involve the deployment of an application in the users' terminals and users' surveys. That way, it is possible to assess QoE for three main types of services, each of them with an associated set of KPI's: Video, File Transfer and Web Browsing. Examples of indicators for video services are:

- **Video Freeze Rate** (measured by comparing the media timestamp to the actual transport packet timestamp, so that it is possible to estimate the amount of media buffered and missing from the client playback buffer; usual critical threshold: 50%),
- **Video Accessibility** (ratio between the number of HTTP request and successful responses when the user starts downloading a video; usual critical threshold: 90%).

The KPI-to-QoE model is one of the aspects of the overall picture. The model contains which KPI's are the most relevant of the users' QoE, which critical thresholds are defined for each of them; and which KPI combinations or paths are the ones that better meet the users' expectations. However, although the general principles are similar regardless of the mobile CSP, in practice each CSP (or even each business segment within a same CSP) have a distinctive KPI-to-QoE model.

4 Quality of Experience Detection and Prediction

The AQoE Analytics Function uses probes to continuously monitor user traffic that goes in and out the CSP network towards the Internet. It also is able to collect events from the Radio Access Network (RAN) nodes. Typically, the AQoE AF must be able to access the following nodes and interfaces in real time:

- Radio Network Controller (RNC)/eNodeB[1] – they provide location information, radio environment and radio management or control information.

---

[1] eNodeB is the equivalent of GSM base stations in LTE.

- SGSN-MME[2] – it provides user identification and bearer information.
- Gn/S1-U/S11 Interfaces[3] – by using deep packet inspection techniques, it provides information on the service being used by the user.

The key element here is to be able to predict when a QoE degradation situation will happen. A simple mechanism consists in detecting when a KPI threshold crossing is reached, persists, and the rate of worsening exceeds some defined value. When such QoE degradation situation is predicted, the Analytics Function sends a notification to the Policy Governance Function (PGF). Upon reception, depending on the predefined business requirements and the parameters of the prediction, the PGF decides which mitigation policies to activate, which Policy Decision Points to contact and with which conditions.

5   Quality of Experience Actuation

In mobile CSP's there are mechanisms to cope with congestion at radio level. Upon a congestion situation, the Radio Access Network (RAN) can apply admission control measures and even tear down established bearers based on the bearers' priority. However, said procedures are applied to all users in the same way, regardless of their profile or the way they are consuming services. In addition to the native RAN procedures, AQoE is able to provide preventive mitigation actions before or in addition to RAN actions.

The AQoE actuation features rely on the existing capabilities provided already existing in mobile CSP's. The main component is the Policy and Charging Control (PCC), which provides operators with the means to enforce service-aware QoS and charging control. The architecture that supports PCC functionality is depicted in Fig. 2. It has been taken from TS 23.203 [7], which specifies the PCC functionality for Evolved Packet Core (EPC) [8]. It comprises three main entities: the Application Function (AF), the Policy and Charging Rules Function (PCRF), which plays the role of Policy Decision Point, and the Policy and Charging Enforcement Function (PCEF), as a Policy Enforcement Point. In EPC, the PCEF is always located in the PDN Gateway (PGW).[4]

---

2   The Serving GPRS Support Node (SGSN) is the 3GPP entity responsible for the delivery of data packets from and to the mobile stations within a given geographical area. Its functionality moves in LTE to the MME (Mobility Management Entity).

3   The Gn reference point is used to support mobility between the SGSN and GGSN.
   In LTE, the S1-U reference point between E-UTRAN and Serving GW is used for per bearer user plane tunneling and inter eNodeB path switching during handover.
   In LTE, S11 is the reference point between the MME and the Serving GW.

4   The PDN Gateway provides connectivity from the user terminal to external packet data networks, such as the Internet, by being the point of exit and entry of traffic for the UE.

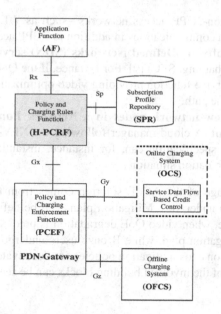

Fig. 2. PCC (Policy and Charging Control architecture)

Therefore, the main anchor for AQoE actuation features is the PCRF, but not only that. Mitigation actions may belong to any of the following categories:

- **Bandwidth limitation**: this is one of the standard network policy handled by PCC through the PGW. It enables the limitation of the bandwidth available for a given user. Bandwidth limitation will impact negatively on the user QoE, but may save bandwidth that can be used by other users.
- **Traffic gating,** so that access to specific types of services (video, file transfer, web browsing) is disabled. This is also one of the standard network policy handled by PCC through the PGW.
- **Radio Access Technology Steering**. One of the possibilities to cope with QoE degradation situations in specific areas is the selection of radio access technology (for instance, from LTE to WCDMA or GSM) or even the frequency selection within the same radio technology (for instance, from LTE FDD to LTE TDD).[5] There are several ways to enforce this type of mitigation policies. One of them is the Sx reference point [5], a proprietary interface implemented by Ericsson products that directly connects the PCRF with the MME[6] and enables direct interaction without going through the PGW.
- **Offload to Wi-Fi**, through an Access Network Discovery and Selection Function (ANDSF) server. That is an entity introduced by 3GPP to assist User Equipment

---

[5] FDD: Frequency-Division Duplex. TDD: Time-Division Duplex
[6] The Mobility Management Entity (MME) is the key control-node for the LTE access-network. It is involved in the bearer activation/deactivation process and is also responsible for choosing the SGW for a UE at the initial attach.

(UE) to discover non-3GPP access networks –such as WLAN or WIMAX– that can be used for data communications in addition to 3GPP access networks [10].

- **Introduction of Software-Defined Networks (SDN) service chains**[7] (through a Service Function Chaining, SFC) [9]. For instance, if the QoE degradation refers to video services, a service function providing video optimization can be added to the affected users' traffic path.
- **Instantiation of new network nodes in a Network Functions Virtualization (NFV) environment**. A cloud manager following the NFV-MANO (Management and Orchestration) specification can, for instance, instantiate cloud resources to cope with QoE degradation situations.

An additional advantage of the AQoE solution comes from the different treatment each subscriber group undergoes. Mitigation plans can be defined for each customer segments (for instance, when video QoE degradation is predicted, Gold users may be applied a lenient mitigation plan, while Bronze users undergo a stricter plan). When the degradation situation ends, mitigation policies are deactivated.

A high-level view of the involved building blocks can be seen in the figure below:

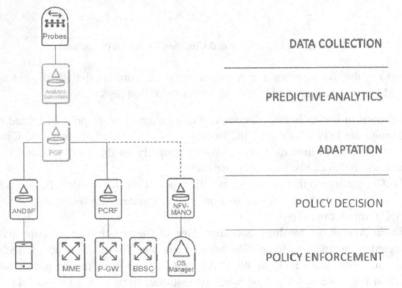

Fig. 3. Adaptive Quality of Experience high-level architecture

6   Related Works

3GPP has recently started the standardization work to create specifications to deal with radio network congestion management. Radio Access Network (RAN) user plane congestion occurs when the demand for RAN resources exceeds the available

---

[7] Service Chaining allows dynamic steering of traffic coming out of a PGW through a bunch of Value Added Services (VAS) before it hits the final destination.

capacity to deliver the user data for a period of time. RAN user plane congestion leads, for example, to packet drops or delays, and may or may not result in degraded end-user experience. 3GPP has discussed several options to cope with user plane congestion and has come out with the User Plane Congestion (UPCON) architecture.[0] Said architecture is depicted in the figure below. In fact, UPCON architecture extends the role of the Policy and Charging Control architecture by providing new inputs to the PCRF to trigger the enforcement of alleviation measures.

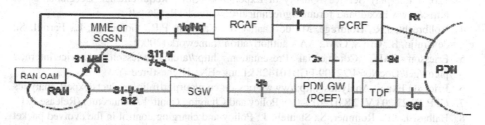

Fig. 4. User Plane Congestion (UPCON) architecture

That approach, although novel in the 3GPP context, poses a number of drawbacks that make UPCON just a starting point towards what AQoE proposes:

- It does not consider fine-grained QoE but only degradation as a result of radio resources shortage. Therefore, cause analysis cannot be performed and actuation is not as efficient as when dealing with specific KPI's.
- UPCON actuation capabilities are limited to the alleviation measures provided by the PCRF. As such, UPCON is not able to use all the actuation means modern 3GPP networks provide.
- It is always reactive, as only provides information of what is currently happening. Therefore, the application of alleviation measures is performed always before the congestion starts, even when it has become severe.

7   Conclusions

Big Data Analytics is a promising area with a large number of potential applications. Real-time analytics pose additional challenges that need to be afforded so that useful insights are obtained and applied in a faster pace than the traditional approaches. Beyond the intrinsic challenges, valid application frameworks are also needed. Adaptive Quality of Experience aims to provide one of said frameworks, applied to a mobile CSP scenario, going beyond the traditional, off-line approaches, addressing some of the most compelling needs of their users.

242   A. Bascuñana et al.

References

1. Ericsson Mobility Report MWC edition, February 2015. http://www.ericsson.com/res/docs/2015/ericsson-mobility-report-feb-2015-interim.pdf
2. Dillon, E., Power, G., Ramos, M.O., Callejo, M.A., Rodríguez, J.R., Argente, M.F., Tonesi, D.S.: PERIMETER: a quality of experience framework (2009)
3. DSL Forum Architecture & Transport Working Group: DSL Forum Technical Report TR-126: Triple-play Services Quality of Experience (QoE) Requirements, December 2006. https://www.broadband-forum.org/technical/download/TR-126.pdf
4. Vollbrecht, J.R., Holdrege, M., de Laat, C.T., Calhoun, P.R., Gommans, L., Farrell, S., de Bruijn, B., Gross, G.M.: AAA authorization framework (2000)
5. Ericsson SAPC Commercial Presentation. http://archive.ericsson.net/service/internet/picov/get?DocNo=4/22109-FGB101428&Lang=EN&HighestFree=Y
6. Ericsson Expert Analytics. http://www.ericsson.com/ourportfolio/products/expert-analytics
7. GPP TS 23.203 V12.8.0 (2015–03): Policy and Charging Control Architecture (Release 12)
8. Balbas, J.-J.P., Rommer, S., Stenfelt, J.: Policy and charging control in the evolved packet system. Communications Magazine, IEEE **47**(2), 68–74 (2009)
9. Open Dayligh. Service Function Chaining. https://wiki.opendaylight.org/view/Service_Function_Chaining:Main
10. GPP TS 23.402 V12.4.0 (2014–03): Architecture enhancements for non-3GPP accesses (Release 12). GPP TR 23.705 V0.10.0 (2014–04): 3rd Generation Partnership Project; Technical Specification Group Services and System Aspects; Study on system enhancements for user plane congestion management (Release 13)

A Review of Scalable Approaches for Frequent Itemset Mining

Daniele Apiletti, Paolo Garza, and Fabio Pulvirenti[(✉)]

Dipartimento di Automatica e Informatica, Politecnico di Torino, Torino, Italy
{daniele.apiletti,paolo.garza,fabio.pulvirenti}@polito.it

Abstract. Frequent Itemset Mining is a popular data mining task with the aim of discovering frequently co-occurring items and, hence, correlations, hidden in data. Many attempts to apply this family of techniques to Big Data have been presented. Unfortunately, few implementations proved to efficiently scale to huge collections of information. This review presents a comparison of a carefully selected subset of the most efficient and scalable approaches. Focusing on Hadoop and Spark platforms, we consider not only the analysis dimensions typical of the data mining domain, but also criteria to be valued in the Big Data environment.

Keywords: Frequent Itemset Mining · MapReduce · Spark · Data mining

1 Introduction

The increasing capabilities of recent applications to produce huge amounts of information has drastically changed the importance of Data Mining. In both academic and industrial domains, the interest towards data mining techniques, which focus on extracting effective and usable knowledge from large collections of data, has risen. In this paper we focus on Frequent Itemset Mining, which is a data mining technique that discovers frequently co-occurring items. Existing itemset mining algorithms revealed to be very efficient on medium-scale datasets but very resource intensive in Big Data contexts. In general, applying data mining techniques to Big Data has often entailed to cope with computational costs that are likely to become bottlenecks when memory-based algorithms are used. For this reason, parallel and distributed approaches based on the MapReduce paradigm [1] have been proposed. Designed to cope with Big Data, the main idea of the MapReduce paradigm consists in splitting the processing of the data into independent tasks, each one working on a chunk of data. Hadoop is the most scalable open-source MapReduce platform. However, in recent years Apache Spark has grown to become a valid alternative platform, that can run on top of Yarn, the Hadoop resource manager. In this survey, we compare a carefully selected subset of Frequent Itemset Mining algorithms that exploit Hadoop and Spark platforms. The paper presents an overview of the challenges and the algorithms, then compares advantages and drawbacks of each approach.

© Springer International Publishing Switzerland 2015
T. Morzy et al. (Eds): ADBIS 2015, CCIS 539, pp. 243–247, 2015.
DOI: 10.1007/978-3-319-23201-0_27

2 Frequent Itemset Mining

The Frequent Itemset Mining [2] process extracts patterns of items from a transactional dataset $\mathcal{D}$, where items correspond to boolean attributes. The support of an itemset I in $\mathcal{D}$ is defined as the ratio between the number of transactions in $\mathcal{D}$ that contains I and the total number of transactions in $\mathcal{D}$. An itemset I is considered frequent if its support is greater than a minimum support threshold.

3 Hadoop and Spark

This survey compares the most scalable itemset mining algorithms exploiting the Hadoop [3] and Spark [4] platforms. Both stems from a distributed programming model introduced by Google, the MapReduce paradigm [1]. MapReduce applications are divided into two major phases that are known as Map and Reduce, divided by a shuffle phase in which data are sorted and aggregated. The Spark framework, with its cached Resilient Distributed Datasets (RDD), usually outperforms Hadoop MapReduce, in particular when iterative processing is required, as in Frequent Itemset Mining. Furthermore, both platforms offer algorithm libraries such as Mahout [5] for Hadoop and MLLib [6] for Spark.

4 Algorithms

This paper compares five selected Hadoop and Spark implementations based on the most scalable Frequent Itemset Mining algorithms.

[7] is a MapReduce implementations of FP-Growth [8] and it has represented for years the only concrete and effective distributed FIM algorithm based on Hadoop. FP-Growth is based on an FP-tree transposition of the transaction dataset and a recursive divide-and-conquer approach. The parallel version initially builds a set of independent FP-trees that are distributed to the cluster nodes. Then, by applying one instance of the (traditional) FP-growth algorithm on each FP-tree, the complete set of frequent itemsets is generated. Since the generated FP-trees are independent, the mining phase can be performed in parallel, i.e., one independent task for each FP-tree is executed to mine a part of the frequent itemsets.

BigFIM and Dist-Eclat [9], instead, are based on the Apriori and Eclat algorithms, respectively. Both of them consists of two phases. The first has the target to find the k-sized prefixes on which, in the second phase, the algorithms build independent subtrees. While the second phase is the same for both the algorithms, in the first phase they extracts the prefixes exploiting two different strategies. BigFIM exploits the Apriori algorithm [10], which uses a bottom up approach: itemsets are extended one item at a time and their frequency is tested against the dataset. Dist-Eclat, instead, uses an Eclat approach to generate the first phase prefixes: Eclat algorithm [11] is based on equivalence classes (groups of itemsets sharing a common prefix), which are smartly merged to obtain all the candidates. In the second phase, both the algorithms proceed with an Eclat-like independent subtree mining. Dist-Eclat is very fast but, with some prefixes configuration, it assumes that the whole initial dataset (transposed in vertical format) can be stored in nodes main memory. BigFIM proved to be slower than Dist-Eclat but able to process larger datasets, even when Dist-Eclat runs out of memory.

YAFIM in [12] represents, instead, an Apriori distributed implementation developed in Apache Spark. The framework solves the challenges related to the iterative nature of the Apriori algorithm, exploiting Spark RDDs to speed up counting operations.

Finally, Spark PFP [6] represents a pure transposition of FP-Growth to Apache Spark; it is included in MLLib, the Spark Machine Learning library. The algorithm implementation in Spark is very close to the Hadoop sibling, i.e, it first builds independent FP-trees and then invokes the mining step on each tree (one independent task for each FP-tree).

Table 1. Algorithm analysis

Name	Framework	Underlying algorithm	Data distribution	Search Strategy	Communication cost handling	Load balance handling
PFP	Hadoop	FP-Growth	dense	Depth First	Yes	No
Spark PFP	Spark	FP-Growth	dense	Depth First	Yes	No
Dist-Eclat	Hadoop	Eclat	dense	Depth First	Yes (best effort with load balancing)	Yes
BigFIM	Hadoop Hadoop	Apriori and Eclat	dense and sparse	Breadth First and Depth First	Yes (best effort with load balancing)	Yes
YAFIM	Spark	Apriori	sparse	Breadth First	Yes	No

5   Analysis Criteria

In this section we introduce the criteria adopted to evaluate the algorithms. The first set of features are related to the algorithm implementation (e.g., the adopted framework and the underlying algorithm). While Hadoop is an established platform, Spark popularity is growing fast. Hence, Spark implementations are considered more promising and future proof.

Secondly, we consider communication costs and load balancing features. These are two of the most important features in distributed processing but they are often undervalued. Communication cost is a crucial part of the behavior of a parallelized algorithm. It does often overwhelm computation costs and it can become a bottleneck for the overall performance. Load balancing, as well, influences performance and limits the parallelization. Developing an algorithm with a heavy-tailed main reducer that keeps working for a long after all the other nodes have stopped their computation is a common issue.

Finally, we have evaluated the datasets used in the experimental sections of the surveyed papers. Table 1 reports the classification of the five algorithms, based on the criteria described in this section.

6   Frequent Itemset Mining Algorithms Evaluation

From an analytical point of view, BigFIM and Dist-Eclat are the algorithms devoting the most attention to communication costs and load balancing (the other algorithms do not address load balancing at all). For instance, the motivations behind the choice of the length of the prefixes generated during the first step of both algorithms are very interesting. In fact, that choice significantly affects both communication cost and load balancing. The former would benefit of shorter prefixes while the latter would improve with a deeper level of the mining phase before the redistribution of the seeds. Hence, depending on the data distribution and the characteristics of the Hadoop cluster, BigFIM and Dist-Eclat can be tuned to optimize communication cost or load balancing, obviously impacting on the overall execution time.

Parallel FP-Growth (PFP) is based on the generation of independent FP-trees that allow achieving work independence among the nodes. However, the independent FP-trees can have different characteristics (e.g., some are more dense than others) and this factor impacts significantly on the execution time of the mining tasks that are executed independently on each FP-tree. When the FP-trees are significantly different, the tasks are unbalanced, and hence the whole mining process is unbalanced. This problem could be potentially solved by splitting complex trees in sub-trees: however, defining a metric to split a tree is not easy.

YAFIM exploits the Spark architecture and APIs to handle communication costs. Its assumption that all transactions must fit into the RDD may limit its potential. The Spark PFP implementation is integrated in the MLLib collection. It is characterized by dynamic and smooth handling of the different stages of the algorithm, without a strict division in phases. Its main advantage over the Hadoop sibling is the low I/O cost, potentially leading to a single read of the dataset, by loading the transactions in an RDD and processing the data in main memory, whereas the Hadoop-based implementation of PFP performs much more I/O operations.

All surveyed papers show interesting results on very large datasets. Only YAFIM presents results on relatively small datasets, and focuses mainly on the comparison against the Hadoop Apriori implementation. Finally, the number of input parameters, that is another important characteristics of data mining algorithms, is limited for almost all the considered implementations. BigFIM and Dist-Eclat, with their customizable length of first-phase prefixes, could require some experiments to find the proper set of parameters (depending on the dataset distribution and the cluster configuration). However, this parameter allows Big-FIM and Dist-Eclat to handle both communication costs and load balancing. Among Hadoop algorithms, relying on experimental evaluations presented in the papers, BigFIM can be considered as the current baseline in this survey. For future perspective and developments, we consider Spark implementations more promising than Hadoop ones, even if the formers currently appear less mature. Spark algorithms have just started to appear in literature and we expect to find more complete implementations very soon.

7 Conclusions

We presented a critical review of scalable Frequent Itemset Mining implementations based on Hadoop and Spark platforms, extending the analysis to critical dimensions such as load balancing and communication costs. Our future plan is to enrich this analysis with real benchmarking to understand the real scalability of each approach.

Acknowledgments. The research leading to these results has received funding from the European Union under the FP7 Grant Agreement n. 619633 (Integrated Project ONTIC).

References

1. Dean, J., Ghemawat, S.: Mapreduce: simplified data processing on large clusters. In: OSDI 2004, p. 10 (2004)
2. Pang-Ning, T., Steinbach, M., Kumar, V.: Introduction to Data Mining. Addison-Wesley (2006)
3. Borthakur, D.: The hadoop distributed file system: Architecture and design. Hadoop Project **11**, 21 (2007)
4. Zaharia, M., Chowdhury, M., Das, T., Dave, A., Ma, J., McCauley, M., Franklin, M.J., Shenker, S., Stoica, I.: Resilient distributed datasets: a fault-tolerant abstraction for in-memory cluster computing. In: NSDI 2012, p. 2 (2012)
5. The Apache Mahout machine learning library (2013). http://mahout.apache.org/
6. The Apache Spark scalable machine learning library (2013). https://spark.apache.org/mllib/
7. Li, H., Wang, Y., Zhang, D., Zhang, M., Chang, E.Y.: PFP: parallel fp-growth for query recommendation. In: RecSys 2008, pp. 107–114 (2008)
8. Han, J., Pei, J., Yin, Y.: Mining frequent patterns without candidate generation. In: SIGMOD 2000, pp. 1–12 (2000)
9. Moens, S., Aksehirli, E., Goethals, B.: Frequent itemset mining for big data. In: SML: BigData 2013 Workshop on Scalable Machine Learning. IEEE (2013)
10. Agrawal, R., Srikant, R.: Fast algorithms for mining association rules in large databases. In: VLDB 1994, pp. 487–499 (1994)
11. Zaki, M.J., Parthasarathy, S., Ogihara, M., Li, W.: New algorithms for fast discovery of association rules. In: KDD 1997, pp. 283–286. AAAI Press (1997)
12. Qiu, H., Gu, R., Yuan, C., Huang, Y.: YAFIM: a parallel frequent itemset mining algorithm with spark. In: IPDPSW 2014, pp. 1664–1671, May 2014

First International Workshop on Data Centered Smart Applications (DCSA 2015)

A Mutual Resource Exchanging Model and Its Applications to Data Analysis in Mobile Environment

Naofumi Yoshida

Faculty of Global Media Studies, Komazawa University, Setagaya, Japan
naofumi@komazawa-u.ac.jp

Abstract. In this paper, a mutual resource exchanging model in mobile computing environment and its application to data analysis are introduced. Resource exchanging is a key issue in data analysis because the only way for efficient use of limited resources in mobile environment is exchanging them each other. Also this paper presents the applicability of this model by showing applications of (1) a universal battery, (2) bandwidth sharing, (3) a 3D movie production by collective intelligence, and, (4) mobile data analysis for motion sensors. By showing results of data analysis, the applicability of this model will be clarified.

Keywords: Resource exchanging · Data analysis · Mobile environment

1   Introduction

Resource exchanging is a key issue in data analysis because the only way for efficient use of limited resources in mobile environment is exchanging them each other. If there is no chance to generate resources, the only way for efficient use of limited resources is exchanging them each other inside a closed world. The motivative example application is a universal battery [1]. We are now bringing many electronic devices with their batteries. If we collect these electricity [11] from these batteries, we can still survive with some electronic devices even if these batteries are almost empty. If we exchange information resources rather than physical resources in mobile environment, a novel way to collaboration is realized by collective intelligence. The automatic 3D movie generation [2] can be implemented. Also, the immediate data analysis in mobile environment can be done by exchanging resources in a limited environment. In this paper the result of immediate data analysis is shown using by the mutual resource exchanging model.

2   Basic Model of Mutual Resource Exchanging Model

Figure 1 shows the overview of a mutual resource exchanging model [1][2] in mobile computing. In the conceptual layer, all target devices are summarized as a group of devices inside a certain closed world in mobile computing environment. All resources are connected to the group of devices. In the physical layer, all devices are recognized and resources are owned by each device. If one device requires much resource than

© Springer International Publishing Switzerland 2015
T. Morzy et al. (Eds): ADBIS 2015, CCIS 539, pp. 251–258, 2015.
DOI: 10.1007/978-3-319-23201-0_28

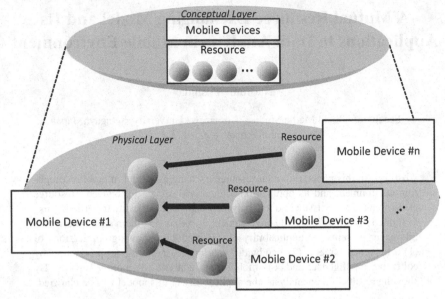

Fig. 1. Overview of a Mutual Resource Exchanging Model in Mobile Computing [1][2]

the amount of its own, resources owned by other devices are exchanged to the target device. Though these exchanging is performed for every kinds of resources, in the conceptual layer there is no exchanging and we have just one group of target devices.

3   Application for Universal Battery

Over the past decade, it was difficult to solve electricity supply problem for mobile and mobile devices. Battery life is still one of the major problems for recent mobile devices. If we are able to move electricity or battery life flexibly among devices, we have a chance to utilize their functionality more efficiently for mobile applications.

In this section, Universal Battery [1] method is introduced as an application of mutual resource exchanging model in mobile computing. This method provides a Universal Battery Function via USB (Universal Serial Bus) among mobile devices by moving flexibly the electricity. In this method, a universal battery provides flexibility of batteries among mobile devices as shown in Figure 2. In this method, we have two types of Universal Battery: Type A (mode oriented, Figure 2) and Type B (many-in and single out oriented). In TYPE A (mode oriented), Universal Battery has three modes. These modes are corresponding three functions. They are (1) collection of electricity, and (2) provision of electricity, charging the electricity from power sources.

(1)   Collection of electricity: It gathers electricity from ubiquitous devices via USB and USB cable by connecting the cable. It must be collection mode.

(2)   Provision of electricity: When it has enough electricity, it is able to charge electricity for the battery of ubiquitous devices. It must be provision mode.

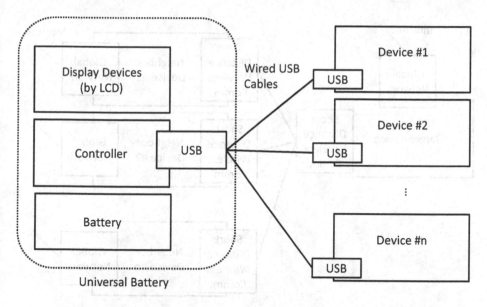

Fig. 2. Overview of a Universal Battery Function among Mobile Devices [1]

(3)   Charging of electricity: When it is plugged in the socket of power sources (at home or office), electricity is charged for the universal battery itself. The mode (1) and (2) are independent from mode (3).

4   Application for Bandwidth Sharing

It was difficult to provide enough network bandwidth for mobile devices. Network bandwidth is the one of the major problem for recent product as mobile or mobile devices. The bandwidth of global network such as 3G (the 3rd Generation of mobile telecommunications technology), LTE (Long Term Evolution), or 4G [10] is getting more faster, but their bandwidths are still not enough in comparison with local networks such as Wi-Fi. In this section, bandwidth sharing method [1] for global network by sharing many short distance local networks among mobile devices is introduced as an application of mutual resource exchanging model. This method is technically innovative because it is easy to implement of bandwidth sharing in a development group. Probability of adoption is likely because it is the combination of promising technology. This will be realized in near future because it is easy to implement. In this method provides flexibility of network bandwidth among mobile devices as shown in Figure 3. All mobile devices are expected to have more than two communication lines, such as 3G cellular networks, Wi-Fi network, Bluetooth etc. Figure 3 shows an example of sharing global network (e.g. 3G/4G network) by short distance wireless communication line [12] (e.g. Wi-Fi), by exchanging limited global network resources among mobile devices through short distance network lines. A target device

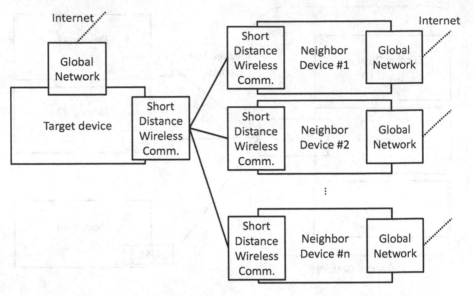

Fig. 3. A Band Sharing among Mobile Devices [1]

has its own 3G network, and it is able to use other many 3G/4G networks simultaneously, if 3G networks of neighbor devices are allowed to use through Wi-Fi network lines. As a result, a target device is able to use n+1 times capacity of 3G/4G network, where n is the number of neighbor devices.

5   Application for 3D Movie Production by Collective Intelligence

Current 3D images or movies are produced by using a pair of fixed cameras. That is also called as stereographic or stereogram images. To provide people with 3D in a highly realistic manner, those two cameras need to be placed so as to align cameras' distance with the distance of people's eyes. It is practically difficult to produce 3D by using multiple different mobile devices like ordinary mobile phones or cameras as the position of mobile devices varies from time to time. The parallax value between the fixed cameras needs to be known. Furthermore, this technology requires special engines or stereo-cameras for producing 3D images or movies. Therefore it is difficult to popularize the conventional 3D and 3D streaming technology in terms of the cost.

In this section 3D movie production by collective intelligence [2] is introduced (Figure 4). This application provides a 3D image and movie producing method from still images or movie taken by several camera-equipped mobile phones as collective intelligence without specific fixed devices. By synthesizing two pictures/movies taken by different mobile phones located within a certain area (e.g. sport stadium, event site), this method enables to produce pseudo 3D pictures or movies and provide them to a third party remotely via the Internet. This application can be implemented on existing mobile phones. The mobile phones equipped with a camera, GPS receiver and 6-axis sensor take a picture of a certain target object, such as sport game, fireworks, auroras, and disasters like fires. These pictures are sent to a server with GPS data and 6-axis data. The server that received these pictures produces 3D images based on these pictures. These pictures have location data (e.g. GPS data) and direction data to the target object (6-axis data) so that the server picks up two images of which differences are the same as parallax. These images are taken by different cameras so that the mobile phone adjusts color, resolution, aspect ratio, frequency in images when 3D images are played. The 3D images can be produced based on adjusted two images. The user can see the produced 3D movie data by a mobile phone, by using real time collective intelligence in mobile and ubiquitous environment without any specific devices for user.

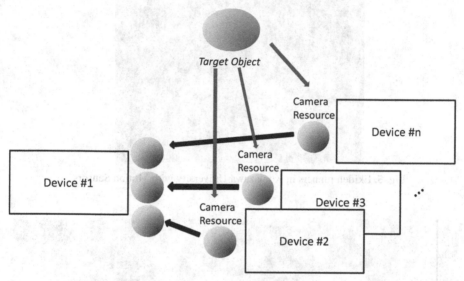

Fig. 4. Overview of this application: 3D Movie Production by Collective Intelligence[2]

6   Application for Mobile Data Analysis for Motion Sensors

Motion sensors or acceleration sensors are widely used in these days. The behavior of human being can be captured as digital data. For example I have a project of data analysis for athlete (Figure 5). We have the Ekiden (a long-distance relay running race) team in Komazawa University, data analysis are effective for the performance by experiments of motion sensors [13], set on to runners' wrists or ankles. The most of all analysis activities are performed after their games or trainings, it is not immediate or on-site analysis.

If we analyze data from motion sensors by exchanging resources among several sensors, immediate data analysis can be done and effectiveness of analysis or performance of analysis improve. Figure 6 shows the example of frequency analysis for athlete number #33 and #39. Horizontal line shows time line, and vertical line represents frequency. They are wearing motion sensors on their wrists. The left side of the figure shows the difference of behavior of arm moving. Figure 7 shows the first half of timeline on Figure 6. In the #33 graph energy is focused on specific frequencies, so the moving of athlete #33's arm is like a sine curve. In the #39 energy is distributed for many frequencies, so the moving of athlete #39's arm is like rectangular. The Figure 8 shows the latter half part of timeline on Figure 6. The right side of the figure shows tiredness of athlete #39 because of comparison with #33 and #39 and distribution of energy for many frequency levels. This kind of analysis can be done immediately by resource exchanging in mobile environment.

Fig. 5. Ekiden runners in Komazawa University with Motion Sensors

Fig. 6. Example of Data Analysis of Motion Sensors

These analysis can be performed on-site immediately after the activities by mutual resource exchanging between devices of #33 and #39.

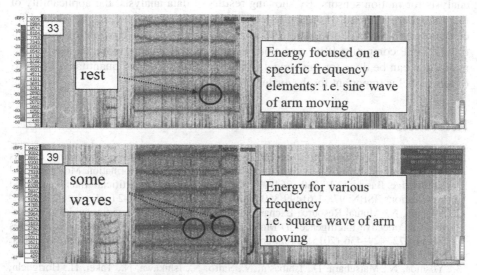

Fig. 7. The Analysis of the First Half of Timeline

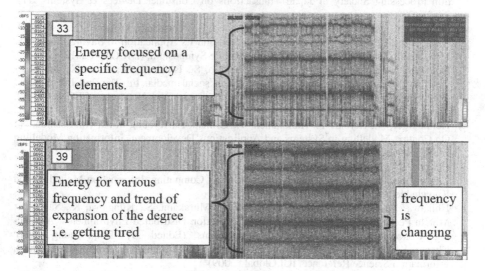

Fig. 8. The Analysis of the Latter Half of Timeline

7   Conclusion

In this paper, a mutual resource exchanging model in mobile computing environment and its application to data analysis are shown. Resource exchanging is a key issue in data analysis because the only way for efficient use of limited resources in mobile environment is exchanging them each other. This paper also shows the applicability of this model by showing applications of (1) a universal battery, (2) bandwidth sharing, (3) a 3D movie production by collective intelligence, and, (4) mobile data analysis for motion sensors. By showing results of data analysis, the applicability of this model are clarified. As a future work, other applications such as healthcare [3], electricity [4], Social Networking [5], context computing [6], cross-cultural computing [7], P2P computing [8], and cloud computing [9]. Furthermore, an algebra formalization can be done for this model. Also, the qualitative and quantitative analysis of this model is important for the evaluation.

References

1. Yoshida, N.: A Mutual Resource Exchanging Model in Mobile Computing and its Applications to Universal Battery and Bandwidth Sharing. Information Modelling and Knowledge Bases XXV **260**, 264–271 (2014). Frontiers in Artificial Intelligence and Applications, ISBN: 978-1-61499-360-5 (print), 978-1-61499-361-2 (online)
2. Yoshida, N.: Mutual Resource Exchanging Model in Mobile Computing and its Application to Collective Intelligence 3D Movies. Information Modelling and Knowledge Bases XXVI **272**, 429–436 (2015). Frontiers in Artificial Intelligence and Applications, ISBN: 978-1-61499-471-8 (print), 978-1-61499-472-5 (online)
3. Yoshida, N., Matsubara, D., Ishibashi, N., Saito, N., Ishikawa, N., Takei, H., Horiguchi, S.: A Health Management Sercice by Cell Phones and its Usability Evaluation. Information Processing Society of Japan Transactions on Consumer Devices & Systems **2**(1), 28–37 (2012)
4. Yoshida, N., Ishibashi, N., Minami, M., Washio, S., Matsubara, D., Saito, N., Ishikawa, N.: A Hybrid Device Profile Detection Method and Its Application to Saving Electricity, Multimedia, Distributed, Cooperative, and Mobile Symposium, (DICOMO 2012) (2012)
5. Ishibashi, N., Yoshida, N., Minami, M., Washio, S., Ishikawa, N., Saito, N.: Machine-machine communications using relationships in social media. In: The 2nd International Conference on Consumer Electronics, Communications and Networks (CECNet 2012), vol. 4, pp. 2688–2691 (2012)
6. Heimburger, A., Kiyoki, Y., Karkkainen, T., Gilman, E., Kyoung-Sook, K., Yoshida, N.: On Context Modelling in Systems and Applications Development. Information Modelling and Knowledge Bases, XXII, 17, May 2011
7. Heimburger, A., Sasaki, S., Yoshida, N., Venalainen, T., Linna, P., Welzer, T.: Cross-Cultural Collaborative Systems: Towards Cultural Computing. Information Modelling and Knowledge Bases **XXI**, 403–417 (2010)
8. Ishikawa, N., Sumino, H., Kato, T., Hjelm, J., Murakami, S., Kitagawa, K., Saito, N.: Mobile Peer-to-Peer Computing for Next Generation Distributed Environments: Advancing Conceptual and Algorithmic Applications (Edited By Boon-Chong Seet): Chapter XVII (Peer-to-Peer Networking Platform and Its Applications for Mobile Phones), Information Science Reference, IGI Global (2009)
9. Mastelic, T., Oleksiak, A., Claussen, H., Brandic, I., Pierson, J.-M., Vasilakos, A.V.: Cloud Computing: Survey on Energy Efficiency. Journal ACM Computing Surveys (CSUR) **47**(2), January 2015
10. Everts, T.: Rules for mobile performance optimization. Communications of the ACM **56**(8), 52–59 (2013)
11. Kugler, L.: The Potential to Provide More Power for Portables. Communication of the ACM, June 4, 2015
12. Pottie, G.J., Kaiser, W.J.: Wireless integrated network sensors. Communications of the ACM **43**(5), 51–58 (2000)
13. Yamashita, S., et al.: A 15 × 15 mm, 1 μA, Reliable sensor-net module: enabling application-specific nodes. In: 5th Int'l Conf. on IPSN (2006)

Detection of Trends and Opinions in Geo-Tagged Social Text Streams

Jevgenij Jakunschin[1]([✉]), Andreas Heuer[2], and Antje Raab-Düsterhöft[1]

[1] Natural Language Processing Laboratory, Department of Electronics
and Information Technology, University of Wismar, Wismar, Germany
j.jakunschin@stud.hs-wismar.de
[2] Department for Databases and Information-Systems, University of Rostock,
Rostock, Germany

Abstract. This paper, describes an application of social media, database and data-mining techniques for the analysis of conflicting trends and opinions in a spatial area. This setup was used to demonstrate the distribution of interests during a global event and can be used for several social media datamining tasks, such as trend prediction, sentiment analysis and social-psychological feedback tracing. To this end the application clusters trends in social text media streams, such as Twitter and detects the different opinion differences within a single trend based on the temporal, spatial and semantic-pragmatic dimensions. The data is stored in a multidimensional space to detect correlations and combine similar trends into clusters, as it is expanded over time. The results of this work provide a system, with the focus to trace several clusters of conflicts within the same trend, as opposed to the common approach of tag-based filtering and sorting by occurrence count.

Keywords: Events · Trends · Clustering · Social text media · Spatial · Temporal · Semantic · Visualization · Social decay

1 Introduction

Online social text media have become an important component of modern society. These resources are used by people from all groups of society and are popular methods to express feelings, opinions and share information. Therefore, in most cases, a fracture of the public social text streams can be viewed as a representation of the subjective opinion on real events of a single person and may contain valuable information. This provides an important resource in social sciences, data mining and other research fields. There are already several works about the analysis on social text streams, such as the analysis of geo-tagged social data or the detection of events in email conversations.

Several of these projects can detect the current trending topic using different social text streams [6]. There are multiple projects that propose ways and methods to analyze this data based on either spatial [6] or temporal [8] or both attributes and extract data from it. Most of these concepts focus on finding

© Springer International Publishing Switzerland 2015
T. Morzy et al. (Eds): ADBIS 2015, CCIS 539, pp. 259–267, 2015.
DOI: 10.1007/978-3-319-23201-0_29

the statistically trending text. However, we suggest that this approach may be insufficient, as it may filter the different threads of thoughts within a single topic - opinions. An example is the current crisis in the Ukraine. People there are reporting about the same topic, using similar tagging (hashtags), words and references, yet they may express very different viewpoints and opinions, depending on both their experience and geographic location. In order to analyse such topics of interest, we suggest an alternative clustering algorithm that allows tracing developments and differences in opinions and enables the classification, clustering and visualization of such "opinion descriptors".

Social text data can be defined as a set of non-structured text. Social text stream data can be therefore described as a collection of social text arriving over time as suggested in [8]. Several modern social text media streams also allow associating additional attributes and properties such as author, recipient and geographic location. Different from the majority of social media analysis tools, our trend detection allows gaining information on trends with the following goals: (1) What is the predominating trend in the area (2) How does the interest of social actors change over time (3) How does the opinion on the trend change over time (4) How does the geographical location affect the opinion on the trend. (5) How do opinion groups behave during their life-cycle. This allows exploring relations between all dimensions of the event and visualizing these on map.

The benefits of such as an algorithm over the regular approach are high customization, tracing and merging of spatial and temporal developments, such as changes over the day-night cycle, while preserving the opinion groups. It also provides additional tools for social analysis and predictive functions. The use cases for such an application that combine the multi-dimensional tracing of opinions, spatial and temporal data, include several social, economical and security problems, such as trend prediction, sentiment analysis, social studies, targeted or focused advertising and security distribution concerns, based on social stability levels. Furthermore this paper presents a method to track the change of the opinion over time, by introducing the "interest decay" simulation - a normalization and tracing concept, based upon the adapted concept of "valid time" from temporal databases [5].

2   Related Work

A detailed description of the analysis and event detection within short social text streams and a definition of social text streams that this projects often relates to can be found in [8]. The related project also uses a multi-dimensional analysis, but focuses on the relationship between social actors, temporal data and text data. Instead this project works with a different set of dimensions. The term "opinion descriptor" and the concept of detecting and using different opinions as a dimension largely relates and uses the background provided by [1][7]. Several temporal concepts and algorithms were adapted from ideas used by temporal databases, such as the concept of valid time [5]. In addition the storage and database methods used by this algorithm largely favor the use of temporal and databases, if the setup is configured towards real time analysis. Several algorithms for the filtering, sorting and detecting the similarity in linguistic problems and tasks relate to the research done in [4].

Modern state of the art social text media stream datamining tools usually rely on spatial data [6] without further exploring the semantic dimension, outside of the main tag. This provides a valid representation of the current trends, by filtering, clustering and sorting the entries. Frequently these algorithms display their results in the form of word cloud graphs hovering over maps. Other setups explore or work with a non-expanding set of data [8] or perform opinion detection within a set of (usually) non-geo-tagged text data [7]. This paper attempts to combine these approaches into a multidimensional text stream clustering algorithm. The application has several uses in the area of sentiment analysis, by providing a pre-clustered dataset and can be used in conjunction with modern sentiment analysis projects [2][3].

3   Definitions

First we split social text stream data into its inherent properties. Social text media data may contain: (1) pieces of text, which can be further split into several data types (2) Meta-data, such as links and "hashtags" (3) Social actor data such as sender, receiver and address. Furthermore social text media data may contain one or multiple of the following attributes: (1) Temporal data, representing the temporal dimension of the text (2) Geo-tagged data, representing the spatial dimension of the text. A set of text data can be represented by the following data-set:

$$(con_n, geo_n, time_n, meta_n) . \tag{1}$$

con_n is the text content converted to string, with removed hyperlinks and meta-data, geo_n is the spatial information, typically represented by longitude and latitude (common with geo-tweets and other social networks), $time_n$ contains the temporal data and $meta_n$ represents the social data and meta-data provided by the messages (such as the username, hash-tags). We understand that the majority of messages won't contain all of the attributes at once, however given the massive data streams, that modern social text media streams generate, it is possible to filter only messages containing this exact content.

Definition (1): (Events and Level-0-list): A set of social text messages can be denoted as

$$L0 = ((con_1, geo_1, time_1, meta_1), (con_2, geo_2, time_2, meta_2), ..., \\ (con_n, geo_n, time_n, meta_n)) . \tag{2}$$

We define each of these messages as an event. This total set $L0$ is the constantly expanded, by adding new messages that are gathered from the social text stream. We define this list as the $level - 0 - list$, it contains pre-filtered text data.

Definition (2): (Potential Trends): Potential trends within events are detected by processing the $level - 0 - list$. We split the con_n component into sub-strings and filter them using one of several available, linguistic algorithms [6] [8]. We then define the core of the message and the several surrounding it as potential trends. Each item in $L0$ may contain none, one or several potential trends.

Definition (3): (Level-1-list and opinion descriptors): Every $t1$ time units new events in the $level - 0 - list$ are analysed and the potential trends and opinion descriptors are extracted. We use this data to generate data sets containing the following information: $(topic_n, opinion_n, geo_n)$. In this data construct $topic_n$ is a subset of con_n - a set of words that defines the detection of the trend, usually represented by the most descriptive or important phrase within con_n; $opinion_n$ is a set of words surrounding $topic_n$ that describes the current opinion and is obtained through an linguistic analysis of the message (derived base forms of directly connected nouns, verbs etc.) and geo_n is the geo-tagged data provided by the social message. It should be noted that extracting messages from the same language (possible using twitter, facebook and other public crawlers) improves the precision of the algorithm. Any single message $L1$ may or may not contain several potential trends $(topic_n, opinion_n, geo_n)$. The majority of the potential trends are false positives that will be filtered out during the next steps. We can define the list of denoted potential trends as the $level - 1 - list$ that is represented by the following structure:

$$L1 = ((topic_1, opinion_1, geo_1), (topic_2, opinion_2, geo_2), ..., \\ (topic_n, opinion_n, geo_n)) . \tag{3}$$

Definition (4): (Confirmed trends and Level-2-list): Repeating potential trends with the same $topic_n$ and similar $opinion_n$, that and have a certain minimum statistical representation within the $level - 1 - list$ are confirmed trends. We define the sum of all confirmed trends, stored within a list as the $Level - 2 - list$. The list can be presented using the following formula:

$$L2 = ((topic_1, \overrightarrow{opinion_1}, \overrightarrow{geo_1}, interest_1), \\ (topic_2, opinion_2, \overrightarrow{geo_2}, interest_2), ..., \\ (topic_n, \overrightarrow{opinion_n}, \overrightarrow{geo_n}, interest_n)) . \tag{4}$$

In this representation $topic_n$ is a set of words representing the trend and equals the to the filtered $\overrightarrow{topic_n}$ from the level-1-list; $opinion_n$ is a weighted set of words, generated from the opinion descriptors from the level-1-list:

$$< (\overrightarrow{w}_1, \overrightarrow{w}_2, \overrightarrow{w}_3, ..., \overrightarrow{w}_n) > . \tag{5}$$

The weighted set represents the local opinion in current instance of $topic_n$; $\overrightarrow{geo_n}$ is the geographical spread of the trend and is represented by an weighted array of coordinates, with the weight describing the spread of messages in a geographic direction.; and finally $interest_n$ is the interest value of the trend, that is decreased over time with the mechanic of interest decay and is increased if the trend/opinion combination will be detected again within a given time. The level-2-list is generated by combining similar trend-opinion combinations from the level-1-list. Synonym detection and a tolerance value are beneficial to improve this process.

Definition (5): (Interest Decay): Human interest is a time-influenced process. Every human is in a connection with a high variety of different problems and topics every day. However, only the things that are above a certain priority/interest threshold receive immediate attention. Other topics are usually either ignored or become the topic of attention for a very brief period of time. Prolonged interest requires the topic to be important or pressuring enough. Assuming that a social media stream is representing a fraction of the current human interests, the number of different topics touched in a massive social stream media is also very overwhelming. In order to recognize the actual and important trends, we adapt this model and use a threshold value that we define as "interest decay". The interest value $interest_n$ of each confirmed trend decreases over time. In this project the "interest decay" is represented as D_i and is representing a value that decreases the interest $interest_n$ in all confirmed trends in the $level - 2 - list$. The value D_i needs to be dynamically adjusted and should contain an absolute value and a percentage value and is dependent upon the setup, goal and number of analyzed messages. The result is that heavily trending messages accumulate more interest and area (during graphical representation) value, while fading trends lose both over time. The result is a normalized data set, adjusted for tracing and evaluating the topic popularity over time. The background and actual implementation for this idea is largely similar to the idea of valid time used in temporal databases.

These definitions will be used to help solve the following problem/process: (1) Gather the basic event, (2) Extract possible trends and corresponding opinion (3) Periodically generate the level-2-list from the level-1-list and apply the interest decay value. (4) Cluster and visualize results from the level-2-list.

4   Workflow

Using the definitions above we can now describe the workflow and the setup. An advantage of the described setup is the possibility to limit the results to an certain geographical area, certain topics or limit, which allows specifying the results and performing targeted queries.

The first step is collecting text messages from the selected social text stream using a social stream crawler. The following workflow was performed and optimized towards the use of twitter.com feeds, but can be adjusted to a different social text stream, assuming it contains geo-tagged data. Currently the prototype implementation of the algorithm is written in $C\#$ and is using the LINQ2Twitter

API. The setup is being currently implemented on a python base, combined with a node.js server for better visualization. This would allow for asynchronous access and control of the data flow, while reducing the system dependency and implementing a modern state-of-the-art GUI. The graphical representation and clustering of different trends and opinions is the primary focus right now. We used this setup to collect messages centered around the Dublin area between the 27.4.2015-28.4.2015 with #nepal being the dominating topic-tag in most cities, after the Nepal earthquake on the 26.4.2015. For the purpose of the experiment 5000 geo-tagged tweets were collected. It should be noted that the application can collect over 100000 geo-tagged tweets per hour.

The messages are collected asynchronously and stored in a database or a list, depending on the goal and query type. Every message $(con_n, geo_n, time_n, meta_n)$ is analyzed and if the contents of every component fit the desired query, added into the $level - 0 - list$ as described above. This step allows inserting additional filtering such as filtering by language, location and tags, which can be frequently performed server-side during the extraction process.

In the example above, this created a simple list with 5000 entries such as:

Table 1. Level 0 list example

ID	time	geo	meta	text
957	27.14 22:58	<....>	#Dublin #YesEquality	Today in Dublin at<....>
958	27.14 22:58	<....>	#Nepal #Terrible	Quake-damage structures mostly<....>
959	27.14 22:59	<....>		Looking fine today at<....>

Then the level-0-list $L0$ is processed. This is done by fragmenting each $content_n$ in the $level - 0 - list$ into strings or stirng groups. Several approaches are viable here such as (1) the text messages is analyzed using a linguistic corpus to detect the primary substantive - verb group (2) every word is treated as a potential trend, resulting in m potential trends per message (3) each word is weighted by it's potential significance or previous occurrence rate. (4) using internal tags from $meta_n$ as the potential trend (in Twitter commonly known as hashtags). Evaluating each method further is still a work in progress, but so far the first app-roach led to the best results at the cost of high calculation costs. Any approach is significantly improved by pre-filtering the level-1-list and removing words such as stop-words and prepositions. The results are stored in the $level - 1 - list$.

Using the example above, the algorithm generated 27400 entries. The level-1-entries generated from the dataset 958 as seen in table 1, can be seen in the table below. The number of entries and their complexity depends on the setup of the linguistic modules. In this case the setup was limiting the opinion to a 4 word-list, while filtering and splitting words. These functions are often provided in modern databases with text support, which can significantly simplify the whole setup.

The next step is the processing of the $level - 1 - list$. The goal now is to combine similar entries in the level-1-list $L1$ using one of several similarity detection algorithms [4]. This is done by iteratively extracting a subset of $L1$ with

Table 2. Level 1 list example

ID	topic	opinion	rest
4475	Quake	damage, structure, hit, Nepal	<....>
4476	Nepal	hit, recently, quake, death	<....>
4477	Death	hundred, hit, death, Nepal	<....>

the common $content_n$. Using a linguistic corpus and weighting, each $opinion_n$ is first compared and classified into several subgroups and then clustered with a $kmeans$ algorithm. The goal is to detect synonyms and common ideas amongst different con_n. When filtering and searching for highly controversial social topics on large scale maps, a second $kmeans$ clustering of the geo_n data is beneficial to detect different groups of interests. The results are then stored in the level-2-list, containing the following entries:

$$topic_n, opinion_n, \vec{geo}_n, interest_n \tag{6}$$

The generated level 2 list can be used in order to visualize the data. $\vec{geo}_n$ representing either the list of geo-coordinates or a vector list to display the gradient distribution on the map, depending on the setup and $interest_n$ represents the number of results in the same cluster.

Using the level-1-list from table 2 we can now process our example generating a level-2-list as seen in the table below. The table had 33 different trending entries with an interest value of at least 10 (messages) after 60 minutes.

Table 3. Level 2 list example

ID	topic	opinions	rest	interest
6	nepal	earthquake/10000/death/hit by	<....>	32
7	nepal	missing/people/english	<....>	17
<....>	<....>	<....>	<....>	<....>

4.1   Interest Decay Cycles

The list conversion is performed every D_i seconds as described in the "Interest decay" definition. During the list conversion the level-0 and level-1 lists are emptied, once they're processed, while the level 2 lists are merged. Once the merging and filtering of the level 2 list is completed and new clusters are generated. Every entry in the level 2 list has it's interest reduced by a flat and a relative value (depending on the geographic scale of the analysis). Entries with a value under a certain threshold are no longer treated trending and are archived. The archive can be extended with an (version) history to observe the temporal development of a single entry over time. Depending on the configuration and the minimum/maximum requirements for cluster density, this setup can be used to

Table 4. Confirmed trends subtable

Group	topic	opinions	rest	interest
Group2.2	nepal	earthquake/10000/death/news	<....>	94
Group2.3	nepal	missing/people/english	<....>	57
Group2.4	nepal	financial/help/support	<....>	120
Group3.1	YesEquality	dublin	<....>	63
Group3.2	YesEquality	mrs/brown/yes	<....>	171

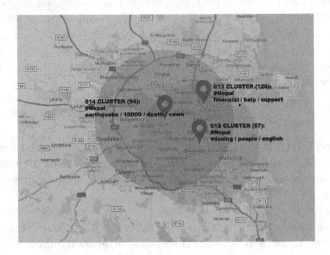

Fig. 1. Clustering centers.

detect the shifts, merges and life-cycle of opinion groups using the interest decay mechanic.

The example above was exposed to the interest decay mechanic every hour, normalizing the results and slowly filtering out the weakened opinion groups.

The following picture shows the rough clustering visualization of the first 3 lines of the table above.

5   Conclusion

In this paper, we proposed a different method of processing geo-tagged social text streams and using the concept of social interest decay to cluster and filter relevant opinions and information. It provides promising results and can be adapted towards different goals and focuses. The presented concept is useful for the analysis of geographically controversial topics, such as elections, politics, crises in comparison to generic trend analysis solutions and therefore provides a powerful social sciences tool. Additionally, depending on the configuration this setup can follow the geographic shift and merge of opinions. This is still a work-in-progress and will enable the support for other functions such as predictive analysis. It should be noted that this concept is a part of a design of a larger system.

References

1. Eisenstein, J., Connor, B., Smith, N.A., Xing, E.P.: A latent variable model for geographic lexical variation. In: Proceedings of the 2010 Conference on Empirical Methods in Natural Language Processing, EMNLP 2010, pp. 1277–1287. ACL, Stroudsburg (2010)
2. Pak, A., Paroubek, P.: Twitter as a corpus for sentiment analysis and opinion mining. In: Proceedings of the Seventh Conference on International Language Resources and Evaluation (2010)
3. Pang, B., Lee, L.: Opinion Mining and Sentiment Analysis. Foundations and Trends in Information Retrieval 2(1–2), 1–135 (2008)
4. Sahami, M., Heilman, T.D.: A web-based kernel function for measuring the similarity of short text snippets. In: Proceedings of the 15th International Conference on World Wide Web, pp. 377–386. ACM Press, New York (2006)
5. Snodgrass, R.T.: The TSQL2 Temporal Query Language, vol. 1. Springer, US (1995)
6. Walther, M., Kaisser, M.: Geo-spatial event detection in the twitter stream. In: Serdyukov, P., Braslavski, P., Kuznetsov, S.O., Kamps, J., Rüger, S., Agichtein, E., Segalovich, I., Yilmaz, E. (eds.) ECIR 2013. LNCS, vol. 7814, pp. 356–367. Springer, Heidelberg (2013)
7. Wu, Y., Liu, S., Yan, K., Liu, M., Wu, F.: Opinionflow: Visual analysis of opinion diffusion on social media. IEEE Transactions on Visualization and Computer Graphics 20(12), 1763–1772 (2014)
8. Zhao, Q., Mitra, P.: Event detection and visualization for social text streams. In: International Conference on Weblogs and Social Media, Colorado, USA (2007)

Software Architecture for Collaborative Crowd-Storming Applications

Nouf Jaafar[✉] and Ajantha Dahanayake

Prince Sultan University – College for Women, King Abdullah Road, Riyadh 11586,
Saudi Arabia
nouf.abdulaziz.j@gmail.com, adahanayake@pscw.psu.edu.sa

Abstract. Diversity in crowdsourcing systems in terms of processes, participants, workflow, and technologies is quite problematic; as there is no standard
guidance to inform the designing process of such crowdsourcing ideation system. To build a well-engineered crowdsourcing system with different ideation
and collaboration components, a software architecture model is needed to guide
the design of the collaborative ideation process. Within this general context, this
paper is focused to create a vision for the architectural design for crowdsourcing collaborative idea generation, with an attempt to provide the required features for coordinating and aggregating ideation of individual participants into
collective solutions.

Keywords: Crowdsourcing · Idea-generation · Collaboration · Crowdstorming ·
Software architecture

1   Introduction

Crowdsourcing is an emerging area that can appear under many names including:
user-generated content, social systems, collective intelligence, crowd wisdom, mass
collaboration, and much more [1]. The term was coined by Jeff Howe in 2006 [2]. In
short, it indicates a paradigm to collect intelligence from the "crowd" to complete a
specific task [3]. Hirth et al. (2012) [4] defines crowdsourcing application as the application that: "Offers an interface for the employer to submit his tasks and an interface for the crowd workers to submit the completed tasks".

1.1   Crowdsourcing Taxonomy

Crowdsourcing application has been classified in various categories according to
different dimensions into taxonomy. This taxonomy classifies a body of knowledge or
collection of objects in a hierarchical manner in order to increase one's ability to understand complex concepts and ideas [5]. Howe (2009) [6] defines four basic categories of crowdsourcing applications which are defined according to the types of activity of the crowd, namely using its wisdom, creativity, opinion, or money [6] as follows:

© Springer International Publishing Switzerland 2015
T. Morzy et al. (Eds): ADBIS 2015, CCIS 539, pp. 268–278, 2015.
DOI: 10.1007/978-3-319-23201-0_30

1- *Crowd wisdom*: this category refers to "wisdom of the crowds" theory which was coined by Surowiecki (2004) in his book *The Wisdom of Crowds* [7]. Hence, this category leverages the crowd to accomplish some activity, such as: solve problems, generate ideas, or make decisions through tapping into what people already know, Howe (2011) [8]. This category is divided into two subtypes as the followings [6]:

    -   *Crowdcasting:* involves broadcasting a problem to the crowd where individuals can upload their submissions and compete for a reward [6].
    -   *Crowdstorming:* involves online brainstorming session allowing the crowd to discuss whatever topics they are interested in, rather than focusing in particular problem to be solved [6]. The difference between crowdcasting and crowdstorming is that in crowdstorming the call for submissions is much more "open ended", and it needs interactions between participants [6].

2- *Crowd creation:* according to Howe (2008) [6] crowd creation refers to "crowdsourcing the activities that use the creative energy of participants". It is usually interchanged with user-generated content [6]. Howe (2011) [8] indicates that this category relies on what the crowd create or what people can make.

3- *Crowd voting:* crowd voting or rating is widely used as an evaluation tool promoting the crowds to express their opinion about something to aid the process of decision making [6]. Howe (2011) [8] demonstrates this category with "what the crowd thinks".

4- *Crowdfunding:* Howe (2008) [9] describes the crowd in crowdfunding as "a funding source for innovative and creative business ideas". Here the crowdsourcing is leveraging crowd's spare money, instead of their spare time or talents [6]. Howe (2011) [8] denotes that "What the Crowd Funds" fits into this category.

1.2   Crowd-Storming

Crowd-storming is one of the crowdsourcing categories that falls under collective intelligence umbrella for harnessing innovation and knowledge of the crowd [6]. It involves online brainstorming sessions that gathering the crowd in one virtual environment; which is typically the crowdsourcing platform [6]. This phenomenon can be discovered with reference to collective innovation [10] that emerges within the intersection between open innovation and collective intelligence [10]. These two terms contribute to collective innovation because the greater the number and diversity of individuals included, the greater the likelihood of generating innovative ideas [11]. Collaboration is a joint problem solving process which is a requirement for collective innovation [12]. Specifically, the collective innovation process describes the entire evolution of an idea through the following four stages [11]: idea generation, idea development, idea prioritization, and idea capitalization. Therefore, crowdstroming can be seen as an application that manages contributors and handle their contributions which involve generating or creating ideas.

1.3   Idea Generation

Idea generation is defined as "The process of generating or conceiving new ideas and concepts that may be useful when addressing a problem or opportunity" [13]. The best-known idea generation methods have evolved from "brainstorming" that developed by Osborn (1957) [14]. Brainstorming is "a group activity with different stages, in which the first stage involves generating a set of number of ideas, and the next stage involves evaluating and pruning the set" [15].

2   Related Works

The crowdsourcing platform is defined as "a mediator that needed by every employer to access the worker crowd" [16]. This platform acts as an interface for the employer to submit his tasks and an interface for the crowd workers to submit their contributions [4]. There are two types of platforms [17]: offline platform [5] [18], and online platform or Internet-based crowdsourcing [17] [19], in which it typically involves: crowdsourcees (workers), crowdsourcers (requesters), and the crowdsourcing platform (intermediary) that connects these two groups [20]. HOWE (2008) & PAPSDORF (2009) [20] indicate that the crowdsourcing platforms are classified into two types: problem solving platforms and idea-generation platforms. The problem solving platforms focus on specific questions and tasks, while idea-generation platforms focus on the creativeness of the crowd, which encompasses two categories including: open competition platforms and co-creation platforms [14]. To reveal how such a platform should be designed, the basic operations need to be set. Vukovim (2009) [21] identifies the set of features for a general-purpose crowdsourcing platform including: registration and specification, initialize crowdsourcing request, carry out crowdsourcing request, and complete crowdsourcing request.

On the other hand, the following section describes different software architectures of existing crowdsourcing applications which are taking into considerations to guide the process of establishing the architectural design for crowdsource collaborative crowdstorming system and aligning it to the requirements of the system. The list of functional and non-functional requirements is illustrated in section 3.

2.1   Cloud-Based Crowdsourcing Architecture

The cloud-based crowdsourcing architecture [22] combines between cloud infrastructure and crowdsourcing requirements. It is widely used to implement the application of crowdsourcing software development. The cloud computing [23] supports the variety of involved software process to accomplish full software development by providing different tools such as: collaboration tools, software development tools, programming language tools, cloud payment tools, board where user can register and upload their profiles information, and repository of software development assets. A reference model involving the required component of cloud-based software crowdsourcing is illustrated in Figure 1, in which HIT is stand for human intelligent tasks.

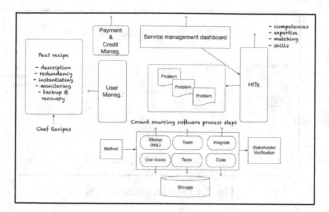

Fig. 1. Reference architecture of cloud-based software crowdsourcing [22].

2.2   Distributed Loosely Coupled Crowdsourcing Architecture

In distributed loosely coupled crowdsourcing architecture [22] the components are deployed in a distributed manner and each one has its own API. These components are combined together into a sole environment to represent the overall crowdsourcing application. Therefore, there will be different APIs, in which first one is dedicated to the crowdsourcing tasks in order to retrieve, update, and create tasks. While, the second one is dedicated to the crowd in order to retrieve, update, and create new crowd member accounts. Additionally, an API is also required to manage the contributions of the crowd member regardless of the crowdsourcing task. The drawback of this architecture approach is the difficulty to control the security measures and scaling options through different APIs.

2.3   Crowdsourcing Data Analytics System (CDAS) Architecture

This architecture is defined as "a framework designed to support the deployment of various crowdsourcing applications" [24]. The overall architecture of this system is presented in Figure 2. As shown in this figure there are four main components which are [25]:

▸   **Job manager:** submits the analytics jobs and transforms them to plan generator, which describes how the other two components (crowdsourcing engine and program executor) should collaborate for the job.

▸   **Crowdsourcing engine:** it works through two different phases to process the human jobs.

▸   **The program executor:** interacts with crowdsourcing engine during the job processing, and then summarizes the results of crowdsourcing engine.

▸   **Crowdsourcing platform:** is the platform of the crowdsourcing application in which the crowd interact with and it is built on the CDAS infrastructure.

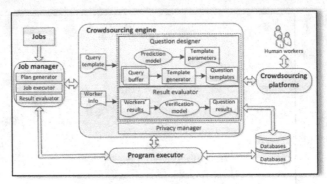

Fig. 2. CDAS architecture [25].

From software deployment perspective, the system is deployed into three different layers as follows [25]: (1) platform layer: that allows the requesters to publish crowd-sourcing tasks on the platform (2) service layer: concerns of the quality aspects of crowdsourcing task's result (3) application layer: concerns of providing effective collaboration between Crowd and Machine.

3   Software Architecture for Collaborative Crowd-Storming System

From technical perspective, using cloud computing to design and implement a crowd-sourcing application is recently directed mainly to software development crowdsourc-ing. Therefore, the effectiveness of using cloud computing in crowdsourcing ideation process is not yet proved. On the other hand, the architecture of CDAS focuses on crowdsourcing complex tasks, moreover the CDAS software is not accessible, and thus the validation cannot be established. While, using the distributed loosely coupled crowdsourcing architecture is a good option in terms of: ease of the implementations, as well as the security in case of crowdsourcing ideation system is not a necessity.

3.1   Structural Elements of the System

The following points illustrate the structural elements of the system (see Figure 3):
- ➤ **Subsystem:** The system is divided into 3 subsystems based on its functional-ity into the followings:
- • *User portal:* this element acts as a portal for the users to access system's re-sources and using its services.
- • *Recruitment:* this element is responsible of handling all related tasks of re-cruiting users and managing user's interaction and crowd coordination.
- • *Ideas jam:* this element handles all ideation tasks.
- ➤ **Components:** Table 1 illustrates the functional requirements of the system and the mapped components for each subsystem:

Table 1. The requirements and the associated subsystems and components.

Requirement	Subsystem	Component
Registration	User portal	*Member management:* holds data related to profiles and contact details.
Log in/out	User portal	*Authentication & authorization:* checks whether the user is eligible to access the system and utilize its facilities either being an administrator, a crowdsourcer, or a member in the crowd.
-Broadcast crowdsourcing task - View crowdsourcing task - Enrolment -Supervise crowd - Coordinate crowd - Opt-out	Recruitment	*Task management:* maintains all crowd-sourcing related tasks involving: task initialization, task broadcasting, task specifications, as well as, crowd coordination.
Interaction	Recruitment	*Interaction management:* maintains all interactions that established between the crowdsourcer and the crowd, as well as between the crowd members, when the crowdsourcing task instance is initiated.
-Supervise crowd's contributions -Add idea -View idea -Aggregate idea -Modify idea	Ideas jam	*Contribution management:* manages the ideation contributions provided by the crowd.

➢ **Connectors:** Table 2 illustrates the connections between the components for each subsystem.

Table 2. The connectors between the components of each subsystem.

Subsystem	Component 1	Component 2	Connection Type
User portal	*Member management*	*Authentication & authorization*	*Blocking connector:* the member management component send a request in a method call or message form to the authentication & authorization component and wait for a response in a form of method return values or message. This request takes place whenever the member wants to utilize system's services. So that the invoked function in member management component will be blocked from further execution until it receives a response from authentication & authorization component. Through this connector the member management component will set restrictions on its functions.
Recruitment	*Task management*	*Interaction management*	*Two-initiator connector:* an initiator is "an incident element of a connector that can make a request to its partner" [26]. Through one initiator connector only the task management component is allowed to make a request to the interaction management, but not vice versa. Once a task instance is initiated, the interaction management component will be activated.
Ideas jam	*Contribution management*	-	Since there is only one component in this subsystem, there is no need for a connector.

> **Configuration (control):** Table 3 illustrates the configurations between the three subsystems that need to be established.

Table 3. The configurations between the subsystems.

Subsystem 1	Subsystem 2	Configurations
User portal	Recruitment	Configurations between user portal and recruitment subsystems to control the members those are able to initiate new crowd-sourcing task, where the recruitment subsystem control the crowd member who recruited in the task through checking their skills suitability in their profiles.
User portal	Ideas jam	Configurations between user portal and ideas jam subsystems to control the members those are already a member in the crowd and assigned to work on the published crowdsourcing task.
Recruitment	Ideas jam	Configurations between recruitment and ideas jam subsystems to control when the idea jam components must be started and ended, and to provide feedback loop through these subsystems.

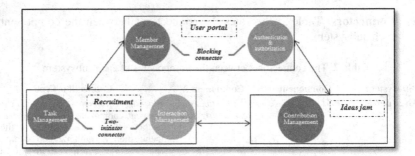

Fig. 3. The Structural elements of the system.

3.2   Architecture Styles of the System

Different architecture styles are discovered for each style's category based on the functional requirements (illustrated in Table 1) and the quality attributes (describe in section 3.3). This is done by making trade-offs between different styles in terms of their applicability, advantages, and disadvantages. These styles are presented in Table 4.

Table 4. Architectural styles of the system.

Category	Style	Description [26] [27]
Data-centered	*Repository*	Based on the functional requirements of the system, this style suits the needs for storing members' data, task information, and logs the interaction activities between the members. Here, we have the same business environment, and users are active while database responds to user's input.
Interaction ori-ented	*Model-View-Controller (MVC)*	Because the system has several members and each member uses a different view of the system based on the same data. Hence, MVC is a more appropriate architectural style. With this style, we can provide the following features: MVC allows separation of presentation and business logic and hence flexibility in extension and modification. Further, this style ensures consistency between the GUI and the model.
Deployment ori-ented	*three-tier Client/ server*	This style can be used to model the system into three main layers: front end, business model, and database. Each service is independent of their application. Three-tier architecture distributes and separates data and processing duties over different tiers so that each tier has its own responsibilities. It supports the centralized data access (Repository style). Besides, It suits web based applications where the system is used by multiple users at a time, and Supports different client types and different devices.
Communication oriented	*Component-Based Architecture (CBA)*	This style is candidate as the subsystems are viewed as exchangeable software units or components with defined interfaces those together build up the overall system and need to interact with each other to perform the required functionalities.

3.3   Quality Attributes of the System

The architecture of the system focuses mainly on the implementation and run-time attributes [28]. For the implementation attributes, the most required properties are:

> *Extensibility:* can be achieved by breaking the system into tiers so that changes to each tier can be performed without affecting the overall system.
> *Availability:* can be achieved through replication of components.
> *Usability:* can be achieved through providing user friendly interfaces to the system components.
> *Performance:* can be achieved by breaking the system into tiers so that the system can serve multiple clients simultaneously in efficient way.

3.4   Views of the System

The "4+1" view model which is developed by P. B. Kruchten provides five fundamental views of the software architecture. They show several views of a software
system, from the perspective of different stakeholders [26]. In this research papers
one view is presented. Figure 4 presents the implementation view. The system generally consists of five components: task management, member management, authentication & authorization, interaction management, and contribution management. Each
component may be deployed on one or more computers, and the overall system is a
distributed system. There are two actors that interact with the system: crowdsourcer
and crowd member. Additionally, *there are* three tiers to model the system: (1) front-
end, (2) business model, and (3) database, with MVC interaction model.

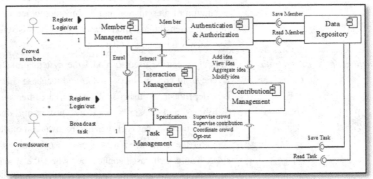

Fig. 4. The implementation view of the system.

4   Conclusion

This paper attempts to highlight the architectural design to develop collaborative
crowdstorming system through describing: architecture styles that have been chosen
according to the functional requirements and quality attributes. In addition, the architectures of existing crowdsourcing applications are discovered to get insights on the
different elements and technologies those make up these architectures. After comparing these architectures to each other and discussing how they could be aligned towards fulfilment of the requirements of the system, the architecture is built accordingly. Furthermore, the future work of this study will emphasise on designing experimental cases in order to validate the proposed architectural design. The experiments
involve the utilization of different applications to represent the collaborative
crowdstorming software architecture, including: social media application to recruit
and broadcast the crowdsourcing task and the ideation application where the crowdsourcing task takes place.

References

1. Doan, A., Ramakrishnan, R., Halevy, A.Y.: Crowdsourcing systems on the World-Wide Web. Communications of the ACM, 86–96 (April 2011)
2. Howe, J.: The rise of crowdsourcing. Wired Magazine **14**(6), 1–4 (2006)
3. Quinn, A.J., Bederson, B.B.: Human Computation: A Survey and Taxonomy of a Growing Field. In: CHI 2011 Proceedings of the SIGCHI Conference on Human Factors in Computing Systems, pp. 1403–1412. ACM (2011)
4. Hirth, M., Hoßfeld, T., Tran-Gia, P.: Analyzing costs and accuracy of validation mechanisms for crowdsourcing platforms. Mathematical and Computer Modelling (2012)
5. Hosseini, M., Phalp, K., Taylor, J., Ali, R.: The four pillars of crowdsourcing: A reference model. In: 2014 IEEE Eighth International Conference Research Challenges in Information Science (RCIS), pp. 1–12. IEEE, Marrakech
6. Geerts, S.A.: Discovering Crowdsourcing:Theory, Classification and Directions for use. Eindhoven University of Technology (February 2009). http://alexandria.tue.nl/extra2/afstversl/tm/Geerts%202009.pdf
7. Westhoff, A.: Using the crowd: an exploration of conditions for crowdsourcing in the idea generation process. University of Twente Student Theses (August 19, 2009). http://essay.utwente.nl/59908/
8. Howe, J.: "Towards A New Taxonomy" By Jeff Howe (July 20, 2011). Retrieved January 29, 2015, from Crowdsourcing.org: The Industry Website: http://www.crowdsourcing.org/editorial/towards-a-new-taxonomy-by-jeff-howe/5458
9. Tripathi, A., Tahmasbi, N., Khazanchi, D., Najjar, L.: Crowdsourcing typology: a review of is research and organizations. In: MWAIS 2014 Proceedings, p. 4. AIS Electronic Library (2014)
10. Wang, K.: Collective innovation: a literature review. technology management in the IT-Driven services (PICMET). In: 2013 Proceedings of PICMET 2013, pp. 608–615. IEEE, San Jose (2013)
11. Strategy & Innovation (March 2008). Retrieved February 5, 2015, from Innosight: http://www.innosight.com/innovation-resources/loader.cfm?csModule=security/getfile&pageid=2518
12. Paulini, M.: Collective Intelligence in Online Innovation Communities (2012). Retrieved November 20, 2014, from academia.edu: https://www.academia.edu/3053107/Collective_Intelligence_in_Online_Innovation_Communities_-_Thesis_Introduction
13. Reinig, B.A., Briggs, R.O.: Measuring the quality of ideation technology and techniques system sciences. In: HICSS 2006 Proceedings of the 39th Annual Hawaii International Conference, p. 20. IEEE (2006)
14. Toubia, O.: Idea Generation, Creativity, and Incentives (September 1, 2006). Retrieved February 8, 2015, from The Institute for Operations Research and the Management Sciences (INFORMS): http://pubsonline.informs.org/doi/abs/10.1287/mksc.1050.0166
15. Nickerson, J.V., Sakamoto, Y.: Crowdsourcing creativity: combining ideas in networks. In: Workshop on Information in Networks. SSRN (2010)
16. Hetmank, L.: Components and Functions of Crowdsourcing Systems – A Systematic Literature Review. Wirtschaftsinformatik Proceedings 2013, p. 4 (2013)
17. Hossain, M.: Crowdsourcing: Activities, incentives and users' motivations to participate. In: 2012 International Conference on Innovation Management and Technology Research (ICIMTR), pp. 501–506. IEEE (May 2012)
18. Brabham, D.C.: Crowdsourcing as a model for problem solving. The International Journal of Research into New Media, 75–90 (2008)

19. Schlagwein, D., Daneshgar, F.: User requirements of a crowdsourcing platform: findings from a serises of focus group. In: Pacific Asia Conference on Information Systems (PACIS), p. 195. AIS Electronic Library (AISeL) (2014)
20. Šundić, M., Leitner, K.: Crowdsourcing as an Innovation Strategy: A Study on Innovation Platforms in Austria and Switzerland. Social Science Research Network (SSRN) (March 31, 2013). http://ssrn.com/abstract=2378250
21. Vukovic, M.: Crowdsourcing for enterprises. In: 2009 World Conference on Services-I, pp. 686–692. IEEE (2009)
22. Huhns, M.N., Li, W., Tsai, W.T.: Cloud-based Software Crowdsourcing. Dagstuhl Seminar Technical Report 13362 (2013)
23. Armbrust, M., Fox, A., Griffith, R., Joseph, A.D., Katz, R., Konwinski, A., … Zaharia, M.: A view of cloud computing. Communications of the ACM 53(4), 50–58 (2010)
24. Liu, X., Lu, M., Ooi, B.C., Shen, Y., Wu, S., Zhang, M.: Cdas: a crowdsourcing data analytics system. Proceedings of the VLDB Endowment 5(10), 1040–1051 (2012)
25. School of Computing, National University of Singapore. (n.d.). Crowdsourcing Data Analytics System. Retrieved April 10, 2015, from Crowdsourcing Data Analytics System: http://www.comp.nus.edu.sg/~cdas/index.html
26. Qian, K.: Software architecture and design illuminated. Jones & Bartlett Learning (2009)
27. Microsoft MSDN Library. (n.d.). Architectural Patterns and Styles. Retrieved April 3, 2015, from Microsoft MSDN Library: https://msdn.microsoft.com/en-us/library/ee658117.aspx

Gamification in Saudi Society: A Framework to Develop Human Values for Early Generations

Alia AlBalawi[⊠], Bariah AlSaawi[⊠], Ghada AlTassan, and Zaynab Fakeerah

Prince Sultan University, Riyadh, Saudi Arabia
Alia.Albluwi@hotmail.com,
{Bariah.Saawi,Ighadah.Altassan,Zaynab.Fakeerh}@gmail.com

Abstract. In a technology era, where the evolution is faster than we can imagine, it is without no doubts that our daily life is mostly driven by the technology, in learning, arranging to do lists, in our cars, mobile phones and even in games. In fact if we want to talk about gaming nowadays, it is also used not just for entertainment and there is a new revolution on a young field called gamification, where it combines learning with entertainment. It is used in many fields such as social media, schools, sports, marketing and even in NASA, so you can imagine how gamification is spreading like fire in the past few years that even oxford has added the word "gamification" to its dictionary. We are aiming to use gamification in teaching values to the young generation. Research title will be "Using Gamification to Develop Human Values for Children in Saudi Arabia".

Keywords: Gamification · Children development · Capability maturity model · Values Adoption · Saudi arabia · Learning · Society

1 Introduction

A heterogeneous area of research is covered in this paper which is Gamification and human development. Gamification is a new field which includes applications and techniques such as electronic games to motivate people. Human Development science studies how a person can adopt changes in a good manner throughout the life span. The main contribution our research will add is teaching the children in the early ages some values that they would not learn at school in an interesting and attractive way. The research aims to help parents, teachers and game developers in the childcare process using games. Electronic learning is a growing subject at this age. Ministries of education are leaning in that direction in schools. Currently, in growing countries, they are implanting the new schools with coverage of networking to support the e-learning methods for students. [1]

Our research is targeting the children category and their parents as well, so they benefit and get value in the human development which will improve the society by producing a well-raised generation.

The main problem that will be addressed in this research is the time wasted in useless games and how to convert that into a positive point and use technology in a right

© Springer International Publishing Switzerland 2015
T. Morzy et al. (Eds): ADBIS 2015, CCIS 539, pp. 279–287, 2015.
DOI: 10.1007/978-3-319-23201-0_31

way as the technology nowadays is in every house and every family. Also the research will aim to help the teachers in teaching the human values.

2   Literature Review

2.1   Gamification

Sebastian Deterding, Dan Dixon, Rilla Khaled and Lennart Nacke investigate Gamification and the historical origins of the term in relation to precursors and similar concepts. The goal is to distinguish between gamification gameful and playful concepts. From this, the research aims to propose a proper industry standard definition of gamification, which is the use of game design elements in none-game contexts. The research also examines gamification industry origins, precursors, parallels, and digs deeper into its terminology and factors such as game, element, design, non-game contexts, and finally use all of this info to set the record straight on what gamification is and how it would be best used. [1]

Hee Jung Park and Jae Hwan Bae have studied the concept of gamification and its actual applications cases. Going through gamification techniques, history, design elements and models then analyzing those models and comparing them to traditional game methods all with the goal of improving gamification and its application into various fields including finance, entertainment, shopping, production and education. The research follows a methodology of case study examination and evaluation leading up to proposing design elements and utilization area of gamification models to improve its uses and applications in the future. [2]

Additionally, Alaa AlMarshedi, Gary B. Wills, Vanissa Wanick and Ashok Ranchhod define gamification as the use of game elements in a non-gaming context. Their research is oriented around designing a framework that includes all necessary components aiming to increase the sustainability of the desired impact of Gamified applications. These components are; flow, relatedness, purpose, autonomy and mastery. The research examines and investigates other gamification research to identify and apply motivational and engagement factors of human behavior in the newly proposed framework and demonstrated how these components work together to balance and design the best sustainable Gamified system experience for the user. [3]

Finally, Barryl Herbert, Darryl Charles, Adrian Moore and Therese Charles present a new Gamified learning system called Reflex which takes into account learning variations in relation to learner motivation all with the ultimate goal to improve the design of future Gamified learning systems. The research starts by examining other research sources to identify gamification typologies, user types, temperaments and motivational drivers then proceeds to linking them together. The research utilizes a 3D browser-based virtual world as a learning environment embedded with both learning content and feedback. The data from the system, in addition to other data collected from specifically designed questionnaire, are extracted and analyzed using data mining and cluster analysis. The results demonstrates an approach to distinguish between learners in terms of a gamification typology, which can potentially be used to provide tailored motivational features for enhancing engagement of learners in Gamified learning systems and thus enhancing those systems themselves. [4]

2.2   Human Development

In a human development research conducted in India on children, Aditya Ponnada, K V Ketan and Pradeep Yammiyavar, present a final persuasive game concept, which is intended to bring about positive changes in children diffidence and shyness behaviors. Core of the protocol process adopted involved semi-structured interviews with nineteen local families, where parents have reported certain behaviors which they wished to change in their children in a positive way. [5] In our research, we can learn and get from the above experience.

In the second research, they find a relation between having fun in gamification and the human nature and development side. The human society has been playing games to socialize and to gain satisfaction leading to a change in the interaction. This research shows the human behavior from cultural and social standpoint with regards to the beliefs. Also it shows that life is associated or resembles a game where people play it in search for fun, to be happy and succeed. In fact in the track of a game, the player will have to lead, make choices, sense and know in order to win in a game. [6]

The last research of Preeti Srivastava focuses on the technology and how to use it nowadays to benefit the students as well as the teachers. It talks about applying ICT in education, whether it is a management education, learning, administration, teaching, education outcome evaluation, counseling or distance education. [7]

2.3   Framework

Frameworks are used to set procedures based on best practices in order to increase quality of any output, save resources and ensure covering best techniques and steps. As for gamification, there are multiple defined frameworks for different areas. For instance, framework to implement gamification in healthcare, schools and at home. [8]

Michael Zyda defined serious games as "games that was built with sense of entertainment that aims to support trainings, education, healthcare, policies, using technology". [8]

However, many researches focused on adopting gamification through different frameworks in different areas to meet certain needs such as education, staff motivation, awareness, etc. For instance, Both Aida Azadegan and Riedel Johann collected 256 cases studies from GaLA Network Colleagues, Recent IFIP SIG Proceedings and other online studies. [8]

Among these studies, 39% of serious games were used for business and management whereas, the other distributed on other industry such as healthcare and education. However, Aida and Johann classify the type of serious games into four groups which are: Training, Intervention, Viral Diffusion and Gamification. [8]

Oscar Wongso, Yusep Rosmansyah and Yoanes Bandung present a framework to adopt Gamification in education. The framework consists of five stages which are: identifying problem and motivation, defining objectives, designing the game, implementing the design, evaluating the game, and publishing the game and evaluation result. [9]

Moreover, Fabio Conceicao, Alan Silva, Ananias Filho and Reinaldo Cabral propose a best practice to adopted gamification in IT environment. The procedure starts first by defining game elements and characteristics, then design the game, define the scope and system operation, deploy any changes based on cultural needs and then evaluate the quality of the game. [9]

3   Gamification Framework in Human Development

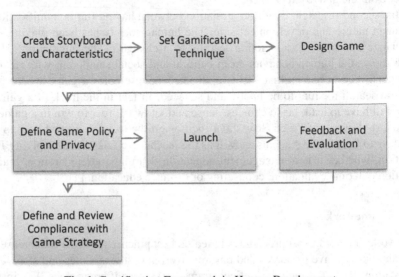

Fig. 1. Gamification Framework in Human Development

This research aims to define framework which combines recommended rules and procedures that can be used to build human values using games for age 4 to 12. The framework consists of eight steps as following:

1. Create Storyboard and Characteristics
The most important part in designing a framework is planning and creating a clear structure to follow and to choose a method that can be executed with an ROI that will benefit the user of this framework. Such frameworks in our research should be utilized to give a rich user experience that will make the user feels comfortable and somehow attached to come back and use the framework many times.

The key element is to set the main goals that should be reached behind creating this framework. As researchers that are doing their research in such a domain, we are thinking to reach an environment that is fun and educational in improving skills and manners. Our aim here is to take advantage of the entertainment aspect that is controlling this century especially the young generation, and make use of that time spent in gaming, into time spent on learning.

In order to create a fun learning environment, we have to deduct some of our research time to study and understand the psychology of our target users, to design a framework that suits their mentality, to make them reach the highest level of enthusiasm that will make their brains understand the values and lessons given in that application or framework designed, and by that we would have achieved the goal of this framework.

2. Setting Gamification Technique

There is a large number of creative techniques that could be used to motivate users and to meet valuable determined goal.

In the game, parents or game mentors can participate with the kids and create a group play. So by that it creates the spirit of challenge and the kid will like to play that game. Another way, the player could play and challenge with the system itself. The player could play and challenge with another player in particular. You can make levels that will not be earned without team group (leader/group member). The group can choose one of the players to be their leader. As the player achieves more, he or she can have access to leaderboards and see how they rank amongst other players. Top players can share their knowledge and experience with the low ones and can take the winners title. [7]

Achievement badges and recognition boards are techniques to encourage and appreciate users who reach higher levels or higher scores. Achievement badge should be attractive, show the real achievement and be linked to the player either in person or in user profile. Moreover, the game designer can implement one recognition board where it contains all players with high scores and who has something that needs to be recognized. For instance, players can nominate a player to be recognized in the beard based on determined aspects such as leadership, helpfulness, respectfulness, creativeness, etc. To give the board more attractive style, recognized players can have unusual labels such as Hulk Player, Geek Player, Player of the Week, etc.

Progress bar is an attractive way to show the gamer the progress of playing, add some excitement to the game and what the actual effort was. Progress bar shall be shown while user is playing and in a visible location. Moreover, it is important to have progress bar that runs on time.

A good technique to encourage user participation and collaboration is to launch virtual gifts that can be shared among users. Gifts can be varied in each level. A good initiative is to create virtual gifts based on occasions for example birthdays, national days, or any other special day.

3. Designing the Game

Now, when we want to start designing a game, we have to first create the characters that we want to use in our environment, let's say, we are designing a framework for teaching children the value of sharing, we will be having a main character, which will transfer him/ her into an animated character, that he can react to the game, and go on level by level with him or her as a character in the game.

Levels will vary to be suitable to the age of the user, manners or lessons that a 12 year old should learn is different from what a 4 year old should learn, such as the language used, and the games that should be designed. So, basically, it will adapt to the age and gender of the user. [10]

4. Defining Game Policies and Privacy
In every game, there are regulations to protect the player and the owner of the game as well. Also regulations are set to make the game more useful and achieve its goals.

The Gamification application is accessible to all ages except for the ones below 12 must be supervised. In this case, the account has to be created by a parent

The ads used in the game should not be distracting and suitable for the targeted users.

The user should not spend excessive time on the game. A reasonable amount of time can be enjoyed. Anything excessive might affect the sight and energy of the player.

The game should comply with COPPA, which is the US regulator of children's on-line privacy content. It is governed by the Federal Trade Commission. We will use it in order to know what to include in the privacy policy and what not.

5. Launching
After designing and working on the prototype, we have been working on a way to launch the game, working on campaigns and advertising the work, and making beta versions to be tested for a sample of users, to know how and what to say to the market when doing the official launching to the market.

6. Feedback and Evaluation
After any launching of a new product, people's impression is very important, as we have to see how they are reacting towards the game and what the outputs measured are. In our case, how much they learned in a period of time and what manners they have gained and applied after using this game. [10]

7. Defining and Reviewing Compliance with Game Strategy
Use the Game Maturity Model as a platform to identify the position of human values in regard to Gamification and help the organization identify steps to proceed to the next maturity level.

After launching and collecting feedback, we recommend to determine which level of maturity the games are in the following:

1. **No-exist**: the game has no value and though, no Gamification is implemented.
2. **Pleasure**: Game is used for pleasure but there is no learning outcomes
3. **Passion**.
4. **Purpose**: the game seeks a determined human value. Actual human development is done through this game.
5. **Optimized**: Continues improvement of the game is in present. Quality of the game is under consideration

The results of maturity model must be oriented towards continues improvement.

4   Analysis and Review

To help us on our research, a survey was done utilizing social network channels as a distribution medium to identify the most absent set of behavioral values that were missing from Saudi children between the ages of 4 to 12 all with the aim to create our framework to best focus on those missing values and maximize the potential for the research. The values that we focused our survey on were satisfaction, leadership, independence, obedience, organization, respect and honesty.

The survey was constructed in such a way to link between different behavioral activities that were observed in children in Saudi Arabia with corresponding values that drive these behaviors. The survey questions attempts to describe a behavior and ask the respondent about the degree of observation he/she believes this behavior is present in his child or not. Strongly Agree= 4, Agree= 3, Disagree = 2, Strongly Disagree = 1, Not Sure = 0.The survey was distributed among 143 participants and it showed that value of Organization is one of the most absent value in children in Saudi Arabia. Furthermore, Obedience was found to come in 2nd place in terms of absent values as well. A percentage of 41.90% of participants are mothers and 62.99% hold undergraduate degrees and 31.50% hold graduate degrees.

As for the children that were studied in this survey, 93.33% live in Saudi Arabia and 82.33% of those were found to attend schools.

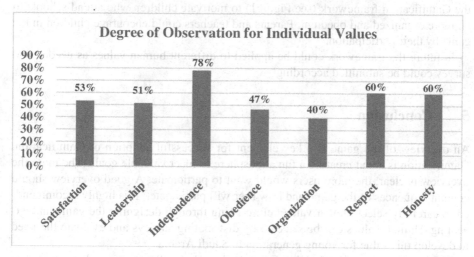

Fig. 2. Degree of Observation for Individual Values

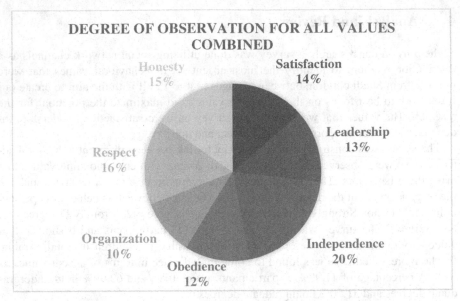

Fig. 3. Degree of Observation for All Values Combined

Based on these statistics (see Figure 2, Figure 3), it is highly recommended to use the Gamification framework (see Figure1) to motivate children who attend schools to be more organized and obedient. Parents and teachers could encourage children in the game by their participation.

In future the framework could be applied to different human values as needed and survey could be submitted accordingly.

5   Conclusion

An overview of the game is a key element for successful adoption of Gamification. Gamification is about creating a fun atmosphere to meet valuable goals. The more the overview is clear, the more users would want to participate. A good overview should contain sequences of the game and how user will participate. It is highly recommended to carefully select human value before going through designing the game and executing. Human values can be selected by distributing surveys and evaluate the need to develop this value for young generation in Saudi Arabia.

This research focuses on giving best practices that will encourage distributing the concept of Gamification in Saudi Arabia. A proposed framework was an output of reviewing multiple Gamification frameworks in different field such as education, healthcare, and enterprises. Proposed framework consists of seven main steps starting with identifying the game and ending by continuously enhcancing the game by adopting Capability Maturity Model.

As part of future work, there will be a plan to adopt proposed framework in Arabic language to support more children, teachers and parents.

In conclusion, the analysis shows that Gamification can be used for children who live in Saudi Arabia to enhance values of organization and obedience. Games can be sponsored by schools, ministry of education or even private enterprises that want to participate in the social responsibility.

References

1. Deterding, S., Dixon, D., Khaled, R., Nacke, L.: From game design elements to gameful-ness: defining gamification. In: Proceedings of the 15th International Academic MindTrek Conference: Envisioning Future Media Environments, pp. 9–15. ACM (September 2011)
2. Park, H.J., Bae, J.H.: Study and Research of Gamification Design. International Journal of Software Engineering & Its Applications **8**(8) (2014)
3. AlMarshedi, A., Wills, G.B., Wanick, V., Ranchhod, A.: Towards a sustainable gamification impact. In: 2014 International Conference on Information Society (i-Society), pp. 195–200. IEEE (November 2014)
4. Herbert, B., Charles, D., Moore, A., Charles, T.: An investigation of gamification typologies for enhancing learner motivation. In: 2014 International Conference on Interactive Technologies and Games (iTAG), pp. 71–78. IEEE (October 2014)
5. Ponnada, A., Ketan, K.V., Yammiyavar, P.: A persuasive game for social development of children in Indian cultural context—a human computer interaction design approach. In: 2012 4th International Conference on Intelligent Human Computer Interaction (IHCI), pp. 1–6. IEEE (December 2012)
6. Normal, M.J., MdNor, K., Ishak, B.I.: Fun beliefs in digital games from the perspective of human nature: a systematic review. In: 2014 International Symposium on Technology Management and Emerging Technologies (ISTMET), pp. 359–364. IEEE (May 2014)
7. Srivastava, P.: Educational informatics: An era in education. In: 2012 IEEE International Conference on Technology Enhanced Education (ICTEE), pp. 1–10. IEEE (January 2012)
8. Azadegan, A., Riedel, J.C.K.H.: Serious games integration in companies: a research and application framework. In: 2012 IEEE 12th International Conference on Advanced Learning Technologies (ICALT), pp. 485–487. IEEE (July 2012)
9. da Conceicao, F.S., da Silva, A.P., de Oliveira Filho, A.Q., Silva Filho, R.C.: Toward a gamification model to improve IT service management quality on service desk. In: 2014 9th International Conference on the Quality of Information and Communications Technology (QUATIC), pp. 255–260. IEEE (September 2014)
10. Wongso, O., Rosmansyah, Y., Bandung, Y.: Gamification framework model, based on social engagement in e-learning 2.0. In: 2014 2nd International Conference on Technology, Informatics, Management, Engineering, and Environment (TIME-E), pp. 10–14. IEEE (August 2014)
11. Ponnada, A., Ketan, K.V., Yammiyavar, P.: A persuasive game for social development of children in Indian cultural context—a human computer interaction design approach. In: 2012 4th International Conference on Intelligent Human Computer Interaction (IHCI), pp. 1–6. IEEE (December 2012)
12. da Conceicao, F.S., da Silva, A.P., de Oliveira Filho, A.Q., Silva Filho, R.C.: Toward a gamification model to improve IT service management quality on service desk. In: 2014 9th International Conference on the Quality of Information and Communications Technology (QUATIC), pp. 255–260. IEEE (September 2014)
13. Simões, J., Redondo, R.D., Vilas, A.F.: A social gamification framework for a K-6 learning platform. Computers in Human Behavior **29**(2), 345–353 (2013)
14. Kapp, K.M.: The gamification of learning and instruction: game-based methods and strategies for training and education. John Wiley & Sons (2012)

Fourth International Workshop on GPUs in Databases (GID 2015)

Big Data Conditional Business Rule Calculations in Multidimensional In-GPU-Memory OLAP Databases

Alexander Haberstroh[✉] and Peter Strohm

Jedox AG, Bismarckallee 7a, 79106 Freiburg, Germany
{alexander.haberstroh,peter.strohm}@jedox.com

Abstract. The ability to handle Big Data is one of the key requirements of today's database systems. Calculating conditional business rules in OLAP scenarios means creating virtual cube cells out of previously stored database entries and precalculated aggregates based on a given condition. It requires passing several steps such as source data filtering, aggregation and conditional analysis, each involving storing intermediate results which can easily get very large. Therefore, algorithms allowing to stream data instead of calculating the results in one step are essential to process big sets of data without exceeding the hardware limitations. This paper shows how the evaluation of conditional business rules can be accelerated using GPUs and massively data-parallel streaming-algorithms written in CUDA.

Keywords: MOLAP · Business rules · Business Intelligence · CUDA · GPU · Database · Streaming · Big Data

1 Introduction

In Business Intelligence (BI) systems, applying business rules allows users to adapt the stored data dynamically to their needs without changing the actual data model or Extract, Transform and Load (ETL) processes. This is an important feature because external business rules e.g. coming from legal regulations as well as internal business rules aiming to insure competitive advantages in the market may require frequent updates [1]. Changing the existing data model or ETL processes would interrupt daily business being at least extremely time consuming if not completely impossible e.g. due to restrictions of the preceding data warehouse systems.

In our solution, the commercially available Jedox OLAP Server [2], business rules can be applied to dimension elements E of a multidimensional OLAP (MOLAP) cube in the form of binary arithmetical ($*$) or ternary conditional ($\bullet$) operators, i.e.

$$* : T \times T \to E \tag{1}$$

© Springer International Publishing Switzerland 2015
T. Morzy et al. (Eds): ADBIS 2015, CCIS 539, pp. 291–304, 2015.
DOI: 10.1007/978-3-319-23201-0_32

in the case of arithmetical operations and

$$\bullet : T \times T \times \delta \to E$$
$$(t_1, t_2, \delta) \to \delta \cdot t_1 + (1 - \delta) \cdot t_2 \qquad (2)$$
$$\text{for } t_1, t_2 \in T \text{ and}$$
$$\delta = \begin{cases} 1, & \text{if } c \\ 0, & \text{else} \end{cases}$$

in the case of conditional operations, with T being a set of arithmetical terms and c being a Boolean expression, each involving constants and/or cube dimension elements different from E as primitives (Fig. 1). The dimensional elements used in a business rule don't necessarily have to originate from dimensions of only one single cube: Combining dimensional elements of different cubes in one rule enables users to connect different data sets like it is done e.g. in SQL table joins. However, this type of cube-interconnected rules is only applicable if the dimensions of the involved cubes other than the ones used in the rule coincide to be able to find a unique solution.

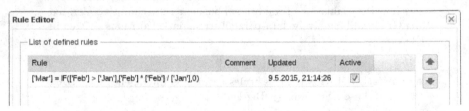

Fig. 1. Defining a rule in the Jedox OLAP Server: Rules can be used in planning scenarios e.g. to implement linear regression: If the sales figures of a product have increased in February, the same growth factor is assumed for March. Otherwise the sale of the product is discontinued (expected earnings = 0):
(Mar = Feb $\cdot \frac{\text{Feb}}{\text{Jan}}$, if Feb > Jan, 0 otherwise)

Applying a business rule to a dimension element means that the value of each cube cell containing this element in its identifier (cell path) is henceforth calculated by evaluating the according business rule (Fig. 2). Due to the dynamic nature of business rules mentioned above, the affected target cells aren't stored in the database. Hence, one could refer to them as virtual cells being created and calculated on the fly when required and immediately deleted afterwards. Evaluating the rule involves a) looking up all the references defined in the rule, b) aggregations, if consolidated elements are referenced, and finally c) applying the rule's operation which in most cases includes matching the operands' cell paths. Business rules can also be concatenated by using several operands in one rule, or nested, if elements are referenced which are again afflicted by business rules, leading to recursive evaluation. Each of these steps requires storing intermediate results which can easily get very large.

8		Jan	Feb	Mar
9	Desktop L	287,091.87	397,990.97	551,728.67
10	Desktop Pro	203,993.01	387,346.45	735,502.02
11	Desktop Pro XL	152,777.60	265,375.13	460,957.37
12	Desktop High XL	105,610.69	123,923.26	145,411.17
13	Desktop High XQ	85,957.38	99,462.42	115,089.30
14	Server Power XC	115,414.60	157,204.32	214,125.41
15	Server Power TT	116,275.53	186,877.95	300,350.12
16	Server Dual C	122,450.43	135,473.48	149,881.58
17	Server Dual XC	94,974.53	109,716.31	126,746.30
18	Server Lion RX	130,199.12	110,743.11	0.00
19	Notebook SX	388,312.30	371,739.65	0.00
20	Notebook GT	368,948.51	516,943.49	724,303.16
21	Notebook LXC	304,958.23	395,818.19	513,749.19

Fig. 2. The effect of the rule defined in Fig. 1: The sales figures of the products in March differ from the according sales figures in February by the same factor the sales figures in February differ from the sales figures in January, if the growth factor is greater than 1. Else, the sale of the product is discontinued and the expected earnings in March are 0 (lines 18 and 19).

The amount of data generated by applying a business rule is highly dependent on the business rule itself: Factors like the depth of recursion, the number of references i.e. dimension elements used in the rule, the rule's operation type and the type of its operands (meaning dimension elements or constants) have a great impact on the evaluation costs. But also the sparsity of the cube and the number of virtual target cells, which have to be created temporarily as a result, have an influence on the data size being processed while evaluating the business rule. It is not unusual that hundreds of thousands or even millions of cells have to be created several times as intermediate results in the evaluation process. Instead of calculating the virtual target cells in one step, a stepwise streaming approach prevents intermediate results of business rule evaluations and their required preprocessing steps from exceeding the hardware constraints of GPUs, both, memory limitations and computational overload of the GPU's stream multiprocessors (SM).

This paper shows how big amounts of data can be processed during OLAP calculations without exceeding the GPU's memory limitations at the example of business rule evaluations.

- Preprocessing steps like source data filtering and path matching are introduced in order to prepare the operands for the rule evaluation in section 4.1 and 4.2.
- The calculation of the rule's operations is described in section 4.3.
- A streaming approach is introduced in section 4.4, dividing the data into small portions and streaming them through the various stages of the rule evaluation in order to avoid memory overflows on the GPU.
- Finally, the tremendous performance gain of our prototypical rule evaluation implementation on GPU in comparison to our sequential CPU-based solution

is shown in section 5, using business rules on a cube with 500 million filled cube cells and the results are being discussed.

W.l.o.g., the paper focuses on the evaluation of conditional business rules, i.e. business rules with a conditional operator. Arithmetical business rules can be evaluated analogously by calculating the respective operation with the basic concepts of preprocessing, streaming and evaluation being unaffected.

2   Motivation: Computing Business Rules Using GPUs

In times of Big Data, data sizes are getting so large that traditional data processing applications are inadequate. They require advanced and unique data storage, management, analysis and visualization technologies to efficiently process data within tolerable elapsed times [3,4]. It has been shown in previous work [5–8] that using GPUs for the calculation of OLAP operations like aggregations and data filters can lead to enormous performance boosts up to 110x in comparison to a sequential CPU- based solution. Speedups of this magnitudes are only reached in suitable operations, which are able to exploit the massively parallel SIMD-architecture of modern GPGPU cards. This is in particular the case if the same simple instruction like an arithmetical or conditional calculation is applied to a large number of data points. This is exactly what happens in the evaluation of business rules and its preprocessing steps: At first, the relevant source data has to be filtered out of the entire set of persistent data stored in the database, which can be done using one GPU-thread per source cell with simple comparison operations. Then, depending on the output size, the source-driven or target-driven aggregation – as introduced in [8] – takes care of generating possibly needed aggregates, which are then used together with the non-aggregated source cells as operands to calculate the rule's target cells. This is done by bringing all the operands' cell paths into accordance in a preprocessing step and then by applying simple arithmetical or boolean operations pairwise on the operands – with again only one GPU-thread used for one single pair of cells.

3   Related Work

In [7], simple binary, arithmetic business rule calculations in the GPU-accelerated Jedox OLAP server were introduced by S. Wittmer et al. The work described in this paper extends these first implementations by adding more advanced rule operators, like ternary conditional (IF) and logical (AND, OR, etc.) operators. Also, an arbitrary number of rule operations can now be used in one single rule with this paper's solution. Another difference to the previous work is that big amounts of data are now handled with a streaming approach instead of falling back to the CPU-based solution in the case of GPU-memory overflows.

In [8], the authors of this paper and S. Wittmer introduce two different aggregation algorithms and a massively data-parallel approach to filter OLAP cube dimensions by a given condition. Parts of the insights that have been gained during the implementation of these dimension filters have been reused in the implementation of the conditional rule calculations introduced in this paper.

J. He et al. present in [9] an alternative to our in-GPU-memory approach by storing the data of an OLAP database in the shared memory of coupled CPU-GPU architectures. This strategy reduces the risk of memory overflows, since the size of the shared memory in those coupled architectures is usually bigger than the memory size of current GPGPU-cards. The big problem of this approach is the low performance of GPUs in such coupled architectures and their low memory bandwidth, leading to frequent memory stalls. The authors propose an in-cache query co-processing paradigm to reduce the likelihood of memory stalls.

Also J. Power et al. show in [10] that integrated GPUs sharing their memory with the CPU can reduce the overheads of CPU-to-GPU memory transfers in a database environment. Performance gains of 3x in comparison to discrete GPUs are shown in their scan-aggregate query implementation, although the computational capabilities of discrete GPUs are 4x better than those of integrated GPUs.

N. Govindaraju et al. present in [11] GPU-algorithms for performing fast computation of several common database operations, like predicates, boolean combinations and aggregations in comparison to a CPU-based implementation.

4   The Conditional Business Rule Evaluation Process

Evaluating conditional business rules starts with calculating the required sets of cells serving as operands in the rule operation. Depending on the type of the operands, this means filtering the storage and searching for persistent cells contributing to the rule calculation, aggregations, if consolidated elements are included in the rule, or calculating the results of sub-rules, if multiple operators are used in the rule. Since the aggregation has been introduced in detail in [8], this section focuses on source data filtering and the actual rule operation and how the different sources are combined with a streaming approach. As an example, we use the following rule:

$$
\text{if}([B] > [C]) :
$$
$$
[A] := [D] \tag{3}
$$
$$
\text{else} :
$$
$$
[A] := [3]
$$

This rule assigns the value of each cell in [D] to the value of the corresponding (that is, all other dimension elements coincide) target cell in [A] if the condition [B] > [C] is satisfied. This is the case if the corresponding cell in [B] has a value greater than the corresponding cell in [C]. Else, the value of the target cell is set to 3 (Fig. 3). As mentioned above, all the cells used as operands in a rule (in the example: [B], [C], [D] and [3]) can either be the result of source data filtering, aggregation or sub-rule calculation, including the creation of cells if a constant is involved.

4.1 Source Data Filtering

Source filtering in the business rule context primarily means searching for cells that contain the dimension element in their path, which are used as a rule operand. This is done on the GPU by checking with one CUDA-thread for each cell in the storage whether it is relevant for the rule or not. Each CUDA-thread hereby loops over the dimension elements in the cell path to test for each element if it corresponds to one of the dimension elements used in the rule.

To find the cells in the storage that contribute to the rule, we hand over an already high-level prefiltered copy of the cells in storage in the GPU's global memory and the source area of the rule to the filter kernel. The high-level prefilter isn't elaborated in detail in this paper. Since the rule's source area is accessed several times during the filter process (namely #dimensions · #number of source cells being checked) we avoid slow global memory accesses by storing the source area information in the GPU's faster constant memory as sets of sorted dimension element ranges – one set for each cube dimension.

In the following, our filter kernel is listed in a simplified version:

```
1: // get cell index of current cell
2: // (= basically the CUDA thread index)
3: uint32_t cellIdx = getCellIdx();
4:
5: // store if the cell is relevant for the rule or not
6: // (initially: relevant)
7: uint32_t isCellRelevant = 1;
8:
9: // read current cell's path information from global memory;
10: // cell paths are compressed in our solution by storing
11: // individual dimension elements together in a uint64 variable
12: uint64_t cellPath = readPathOfCell(cellIdx);
13:
14: // iterate over dimension elements in the cell path
15: for (uint32_t dimIdx = 0; dimIdx < dimCount; ++dimIdx) {
16:
17:     // a bitmask is used to extract
18:     // the dimension element out of the path
19:     uint64_t bitmask = getDimensionBitmask(dimIdx);
20:
21:     // extract the current dimension
22:     // element out of the entire cell path
23:     uint32_t dimValue = extractDimension(cellPath, bitmask);
24:
25:     // search the dimension element in the current
26:     // dimension's ranges in constant memory via binary search
27:     uint32_t rangeIdx = findInSourceAreaRanges(dimValue);
28:
29:     // if dimension element couldn't be found in the
30:     // range: set "isCellRelevant" to 0
31:     if (rangeIdx == -1)
32:         isCellRelevant = 0;
33: }
34:
35: // remember if the cell is relevant in a global flag array
36: uint32_t* globalIsRelevantFlagArray = getGlobalFlagArray();
37: globalIsRelevantFlagArray[cellIdx] = isCellRelevant;
```

Listing 1: The (simplified) filter kernel: each CUDA-thread checks for a cell if it is relevant for the rule or not. The results are stored in a flag array with the same size as the array of input cells, which is later used to compact the cells to close the gaps between relevant and irrelevant cells.

All cells that are relevant for the calculation of the rule are marked in a separately stored flag-array with 1, the irrelevant cells are marked with 0. The irrelevant cells are removed in a post-processing step by at first calculating the prefix-sums of the flag- array and therefore the new positions of the relevant cells. Then, a compaction step copies the relevant cells to these positions, removing the irrelevant cells. As a result, we get an array of source cells which are relevant for the rule evaluation.

4.2 Matching the Paths of the Operands and Creating Constants

As another preprocessing step before calculating the actual rule operation, the sets of cells serving as rule operands have to be aligned in order to find the matching pairs of cells to apply the rule's operation. This is done in our approach by translating the cell paths of the operands to the according cell path of the target cell. Though, it is not guaranteed that every possible cell in the source areas of cells exists, meaning that there can be zero-valued cells in the operands' sets of cells. Those aren't stored in our solution. This means that during the rule calculation, the matching pairs of cells have to be searched in the operands before the rule operation can be applied.

We require the operand sets to be sorted to be able to use a binary search in the rule calculation step. Usually, our preprocessing steps like source filters or aggregations already provide sorted results. But in some rare cases, e.g. when there has been a writeback process lately and the idle job wasn't able to resort the database yet, it may occur that the source filter provides unsorted results which have to be sorted before applying the rule's operation.

If constants are involved in the rule, we temporarily have to create cells with cell paths matching the other operand and the used constant as values. In our example this is the case in the part, where values of 3 are assigned to [A] if [B] < [C]. This doesn't apply if these constants are used in the conditional part of the rule because the operands in a condition aren't used in later calculations or in an assignment anymore. In this case, the constant is used in the condition without having to create cells in advance.

4.3 Applying the Rule's Operation

As a result of the preprocessing steps introduced above, we now have sets of cells serving as operands for the rule calculation. In conditional rule calculations, the first step is to evaluate the conditional part of the rule. In our example, this would be calculating the condition [B] > [C]. A CUDA-kernel searches via binary search with one CUDA-thread per cell in [B] if there's a matching cell in [C]. If such a cell is found, the condition is evaluated and the result is stored separately in a flag-array. If a matching cell isn't found in [C], the algorithm treats the missing cell in [C] as an existing but empty cell. The condition is then evaluated as if that cell was existing with value 0. In the next step, the resulting flag-array is

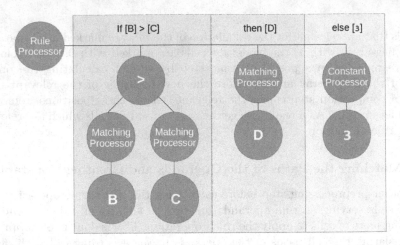

Fig. 3. The rule evaluation process of our example in a tree graph. Based on the comparison of B and C, cells from D or the cells with value = 3 provided by the constant processor are used as a result of the rule. The matching processor makes sure that the operands can be matched during the rule evaluation by translating the operand's cell paths to the cell path of the rule's result. The matching processor and the constant processor are introduced in section 4.2.

then used to decide, if the value of the according cell in [D] is assigned to the according cell in [A] or if the according cell in the set of constant cells in [3] is chosen instead.

4.4   A Streaming Approach

All the previous steps of the rule evaluation process introduced above create sets of temporary cells as intermediate results. Potentially, these sets can get very large, i.e. hundreds of thousands or even millions of cells. To avoid a memory overflow, we use a floating frame approach streaming only parts of the entire data through the various stages in the rule evaluation process instead of calculating all results at once. There are processors for every preprocessing step as well as for the rule calculation: A source processor to filter persistent data for later calculations, an aggregation processor taking care of generating consolidated cells, a constant processor creating temporary sets of cells with a constant value and finally a rule processor calculating the final result of a business rule (Fig. 4). Each of these processors guarantees that a predefined size of its resulting data isn't exceeded (in our implementation: 250MB) and that its result set is available in a compact representation by eliminating gaps between relevant cells, as it is done in the source filter step. As mentioned above, we also require the processor results to be sorted in order to ease calculations in later steps and enable them e.g. to use binary search.

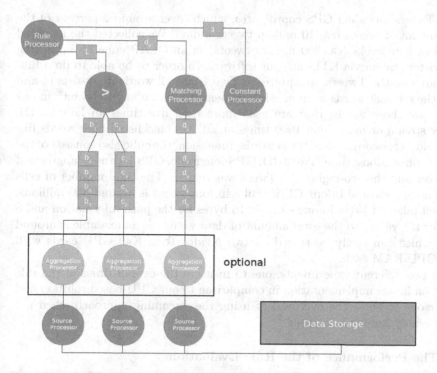

Fig. 4. Only parts of the data are streamed through the processors and combined in the rule processor to the final result. Source Processors extract data out of the data storage. If consolidated elements are involved in the rule, Aggregation Processors take care of aggregations and hand portions of the data over to the Matching Processors or the Rule Processor. Matching Processors align operands by translating their cell's paths to the respective target paths of the rule. The Rule Processor decides then, which cell is taken as a result cell depending on the rule's condition: The cell in D or the constant (3) generated by the Constant Processor.

5 Performance of Our Implementation

5.1 The Test System

We used a test system with the following hardware specifications:

CPU	Intel Xeon E5-2643
GPU	2x Nvidia Tesla K40
RAM	256GB
GPU memory	24GB

5.2 The Data Used for the Experiments

As a Big Data example, we chose to use a time-limited subset of the Twitter stream, since Twitter provides a large amount of data (around 500 million Tweets a day) which is easily accessible and also easy to relate to. To be able to show Tweets on a map in our software, we selected only the localized Tweets, that means Tweets including GPS coordinates, which cover around 2 percent of the total amount of tweets (ca. 10 million tweets a day). We collected the localized Tweets of four weeks (ca. 300 million tweets) in an OLAP cube by importing the Twitter stream via ETL into our software. In order to be able to do a full-text search on the Tweets, we split each Tweet into all words it consists of and added the relevant words as dimension elements to our cube. "Relevant" in our context are those words, that are used more than five times an hour in the Twitter stream or more than 1000 times at all. We blacklisted filler words like "and" and swearwords. Besides the words-dimension, the cube also consists of the following dimensions: time, Tweet-ID, GPS-longitude, GPS-latitude, language of the Tweet and the geo-region the Tweet was sent in. The total number of cells that is hereby created in our OLAP cube in four weeks is around 500 millions. Each cell takes 24 bytes in our storage: 16 bytes for the path information and 8 bytes for the value. So the total amount of data we use as an example is around 11GB, which can easily be stored on two Nvidia Tesla K40 GPU cards with 12GB GPU RAM each.

We did two different experiments, one to measure the overall times of the rule evaluation in our implementation in comparison to our CPU-based solution and another one to show the overhead of using the streaming approach when the GPU memory limits are reached.

5.3   The Performance of the Rule Evaluation

The conditional rule defined in this experiment calculates a simplified version of the tonality of certain words within Tweets. The tonality in our context means that a word that was used in combination with words like "love" or "adore" is marked as a positive reference, otherwise it is treated as a neutral appearance and is not being considered further. We first search for existing base cells in the cube for the words to be tonality-interpreted with the condition "Tweet-ID > 0". In the "then"-part of the rule, we assign 1 to the result, if the words "love" or "adore" appeared in combination with the respective word in a Tweet. In the "else"-part, we assign 0 to the result. Finally, we sum up all those appearances for every word to be interpreted to get an impression, how often a certain word was used in a positive context within the Tweets in a preselected interval of time.

For the experiment, we selected 120 random words to be tonality-interpreted and increased the number of input cells for the rule stepwise by increasing the time interval the Tweets were sent in. We measured the response times of queries with different numbers of input cells for the conditional part of the rule together with the aggregation part of the tonality calculations:

	CPU	GPU	speedup	#input cells (rule)	#input cells (aggregation)
Query 1	628.89	1.01	622.66	3,000,000	200,000
Query 2	821.66	1.31	627.22	2,800,000	198,888
Query 3	953.87	1.51	631.7	3,800,000	388,720
Query 4	1042.73	1.64	635.81	4,500,000	480,000

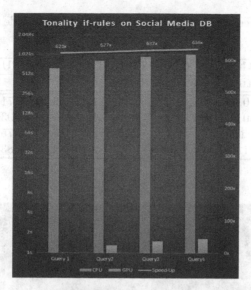

Fig. 5. The performance of our GPU rule evaluation implementation in comparison to our sequential CPU-based solution at the example of tonality calculations in a social-media context: The performance of both approaches is highly depending on the number of input cells of the query (#input cells for the rule and #input cells for the aggregation). The speedup factor scales linearly with increasing query complexity.

5.4   The Performance of the Streaming Approach

The streaming approach is a trade-off between performance and memory consumption. If the data being processed doesn't fit into memory anymore, we accept a certain performance drop in order to be able to still calculate the results on the GPU. Of course, the performance of the GPU-calculations with streaming enabled still has to be better than the performance of the CPU-based solution. Otherwise, it would be better to just fall back to the CPU-based solution in these cases. To test the performance of the streaming approach, we artificially decreased stepwise the threshold of the amount of data (frame size) that has to be reached in a calculation in order to activate streaming in our implementation, leading to different numbers of data frames used for streaming. We used four simple aggregations with different input sizes for our tests instead of going through the whole rule evaluation to exclude influences other than streaming from our results. The input data in the smallest case was slightly greater than 500 MB, so that streaming even occured with the biggest of the chosen frame sizes (500 MB). A small performance drop is noticable in the first three performance measurements with frame sizes of 500MB (1 frame used), 100MB (5 frames used) and 10MB (50 frames used). If we go down to a frame size of 1MB (500 frames used), the overhead of the streaming approach is so high that the performance drops to a level that is even worse than the sequential CPU-based solution in most of the cases. In the other cases, the performance with streaming is still much better than the performance of the CPU based solution and provides speedup factors of up to 36x.

frame size	GPU no streaming	GPU 500MB	GPU 100MB	GPU 10MB	GPU 1MB	CPU -	#input cells
Query 1	0.68	0.72	0.84	1.23	53.19	15.71	12,700,000
Query 2	0.84	0.9	1.01	1.57	51.04	20.93	21,100,000
Query 3	1.2	1.3	1.45	2.17	46.99	39.75	48,500,000
Query 4	1.51	1.63	1.67	2.49	44.67	58.83	64,200,000

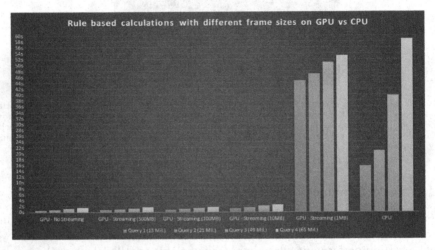

Fig. 6. This graph shows the performance of our streaming approach in comparison to our CPU-based solution and in comparison to the GPU-based solution without streaming. Several frame sizes were used in the streaming approach: 500 MB (1 frame), 100MB (5 frames), 10MB (50 frames), 1MB (500 frames). The performance drops noticeably when using the streaming approach in comparison to the results without streaming. In extreme cases (frame size = 1MB; 500 frames used), the performance is even worse than the CPU's performance. In all other cases, the performance of the GPU-based solution is still much better than the performance of the CPU-based solution.

5.5   Discussion

We showed that using GPUs to evaluate business rules in our solution can provide tremendous speedups of up to 635x in comparison to our sequential CPU-based solution. Both, the GPU-based and the CPU-based solution scale linear in dependency of the size of the data used as input for our algorithms. The massively data-parallel architecture of GPUs is better suited for large amounts of data than the architecture of the CPU. This is why the speedup factor even gets bigger, the more input cells were used in our experiments.

The streaming approach we introduced in this paper avoids memory-overflows, so that GPUs with even small memory sizes can be used to process huge amounts of data without having to fall back on the CPU-based solution.

The performance drops noticeably when using the streaming approach in comparison to calculating all results at once and also with increasing number of frames used in streaming. In extreme cases, when the overhead of using streaming with very small frame sizes gets too high, the performance is even worse than the CPU's performance.

6   Conclusion

In this paper we have introduced a massively parallel GPU-approach implemented in CUDA to evaluate conditional business rules including the preprocessing steps which are needed to create the rule's operands. We have shown how memory overflows can be avoided using a streaming approach with a tolerable performance drop and we pointed out the tremendous performance gain of our GPU-based solution in comparison to our sequential CPU algorithms.

Future work comprises developing a highly parallel way of writing back large amounts of cells to our OLAP server in bulks, like it is e.g. frequently done in ETL processes. The advantages of GPUs are mainy used in these cases to merge the new data with the already existing cells in the database.

Acknowledgment. Thanks to the other members of the Jedox GPU-R&D team: Steffen Wittmer and Leo Mehlig and to our former member Tobias Lauer (University of Offenburg, Germany).

References

1. Mircea, M., Andreescu, A.: Using Business Rules in Business Intelligence. Journal of Applied Quantitative Methods (2009)
2. Jedox Olap. www.jedox.com/en/product
3. Wikipedia, Big Data. http://en.wikipedia.org/wiki/Big_data
4. Chen, H., Chiang, R., Storey, V.: Business Intelligence And Analytics: From Big Data to Big Impact (2012)
5. Govindaraju, N.K., Lloyd, B., Wang, W., Lin, M., Manocha, D.: Fast computation of database operations using graphics processors. In: Proceedings of SIGMOD, Paris, France, June 2004, pp. 206–217. ACM (2004)
6. Lauer, T., Datta, A., Khadikov, Z., Anselm, C.: Exploring graphics processing units as parallel coprocessors for online aggregation. In: Proceedings of DOLAP 2010, Toronto, Canada, October 2010
7. Wittmer, S., Lauer, T., Datta, A.: Real-time computation of advanced rules in OLAP databases. In: Eder, J., Bielikova, M., Tjoa, A.M. (eds.) ADBIS 2011. LNCS, vol. 6909, pp. 139–152. Springer, Heidelberg (2011)
8. Strohm, P.T., Wittmer, S., Haberstroh, A., Lauer, T.: GPU-accelerated quantification filters for analytical queries in multidimensional databases. In: Bassiliades, N., Ivanovic, M., Kon-Popovska, M., Manolopoulos, Y., Palpanas, T., Trajcevski, G., Vakali, A. (eds.) New Trends in Database and Information Systems II. AISC, vol. 312, pp. 229–242. Springer, Heidelberg (2015)
9. He, J., Zhang, S., He, B.: In-cache query co-processing on coupled CPU-GPU architectures. Proc. VLDB Endow. **8**(4), 329–340 (2014)

304 A. Haberstroh and P. Strohm

10. Power, J., Li, Y., Hill, M., Patel, J., Wood, D.: Toward GPUs being mainstream in analytic processing. an initial argument using simple scan-aggregate queries. In: Proceedings of the Eleventh International Workshop on Data Management on New Hardware, DaMoN 2015, June 2015
11. Govindaraju, N., Lloyd, B., Wang, W., Lin, M., Manocha, D.: Fast computation of database operations using graphics processors. In: Proceedings of the 2004 ACM SIGMOD International Conference on Management of Data, SIGMOD 2004, pp. 215–226. ACM (2004)

Optimizing Sorting and Top-k Selection Steps in Permutation Based Indexing on GPUs

Martin Kruliš[1]([⊠]), Hasmik Osipyan[2], and Stéphane Marchand-Maillet[3]

[1] Charles University in Prague, Prague, Czech Republic
krulis@ksi.mff.cuni.cz
[2] State Engineering University of Armenia, Yerevan, Armenia
hasmik.osipyan.external@worldline.com
[3] University of Geneva, Geneva, Switzerland
stephane.marchand-maillet@unige.ch

Abstract. Permutation-based indexing is one of the most popular techniques for the approximate nearest-neighbor search problem in high-dimensional spaces. Due to the exponential increase of multimedia data, the time required to index this data has become a serious constraint of current techniques. One of the possible steps towards faster index construction is the utilization of massively parallel platforms such as the GPGPU architectures. In this paper, we have focused on two particular steps of permutation index construction – the selection of top-k nearest pivot points and sorting these pivots according to their respective distances. Even though these steps are integrated into a more complex algorithm, we address them selectively since they may be employed individually for different indexing techniques or query processing algorithms in multimedia databases. We also provide a discussion of alternative approaches that we have tested but which have proved less efficient on present hardware.

Keywords: Indexing · Permutation · GPU · Top-k · Sorting · Bitonic sort

1 Introduction

Searching for images inside large-scale multimedia databases is very problematic due to the rapid increase of digital features. Similarity search [4], which is an important paradigm employed across many fields of science, is an extraction process that selects the most similar objects to a given query. The most simple way of finding similar objects within large-scale data set is to compare features of the query object and of the database elements. For high dimensional data, these algorithms suffer from the unfavorable time complexity of the problem [11]. Hence, there is a real demand for approximate similarity search solutions [9].

One of the most popular techniques of approximate search is the permutation-based indexing [3]. In such index, each object is represented by a permutation

© Springer International Publishing Switzerland 2015
T. Morzy et al. (Eds): ADBIS 2015, CCIS 539, pp. 305–317, 2015.
DOI: 10.1007/978-3-319-23201-0_33

of reference object indices. The similarity between two objects is subsequently defined as a comparison of the permutation lists of corresponding objects.

The main limitation of indexing large-scale datasets containing millions of data objects is the index construction time. This type of problems may be solved by either developing faster indexing techniques (i.e., more efficient algorithms) or by porting existing ones on massively parallel architectures, such as GPUs.

In this paper, we have focused on the second part (postprocessing) of the index construction, which comprises top-k selection of nearest reference points and sorting. Unlike the first part which computes Euclidean distances between database objects and reference objects, the postprocessing step is much less suitable for Single Instruction Multiple Threads (SIMT) execution model employed by the GPUs. Originally, this step was performed by the CPU whilst the distances were computed by the GPU, which presents a bottleneck for future scaling. Therefore, we propose an updated version of the indexing algorithm that shifts some parts of the postprocessing step to the GPU as well.

The paper is organized as follows. In Section 2, we review the related work and the Bitonic sort algorithm. Section 3 overviews the GPU fundamentals, which are essential for our implementation. The index construction algorithm and previous results are outlined in Section 4. We propose our solution and discuss the alternatives in Section 5. Section 6 presents our empirical results and Section 7 concludes the paper.

2 Related Work

In this section, we are going to present recent techniques of similarity search indexing algorithms for GPU architectures. Krulis et al. [6] proposed a GPU implementation of feature extraction process which can be used as an indexing technique that reflects similarity of color and texture of images. Due to the limitation of GPU shared memory and limited means of communication/synchronization between thread groups the best performance is obtained by computing each signature in one thread group. Experiments showed that GPU implementation achieved in two orders of magnitude speed up compared to a CPU implementation. The main difference of this work from the previous ones is that it addresses the issue of data indexing, rather than acceleration of the search process.

Mohammed et al. [8] demonstrated several approaches for permutation-based indexing algorithm on multi-core CPU and GPU architectures. In the paper, authors suggested three different approaches for distance calculation and sorting algorithm on GPUs. To overcome the problem of non-suitability of sorting algorithm on multi-core architectures, authors suggested "Parallel Indexing on Fly" approach which reduced the memory usage as only the $\tilde{n}$ references were stored and sorted for the rest of work. But even with these approaches, due to increase of latency for getting the distances from the global memory the multi-disk access approach of multi-core CPU (implemented via OpenMP) gave much better performance compared to all GPU implementations with CUDA technology.

The proposed permutation-based indexing algorithm extends our previous work [7]. It has been established that the CPU implementation of the postprocessing step may create a bottleneck when modern GPUs are used. In our newly proposed method, we have utilized a GPU implementation of modified bitonic sort to work as a top-k selection algorithm Presented solution outperforms the previous work several times.

2.1   Bitonic Sort

The most popular sorting algorithms (like Quicksort [5]) are quite unsuitable for GPUs. In fact, it is not easy to implement a parallel version of these algorithms that would scale optimally on multi-core CPUs. On the other hand, Bitonic sort [2] algorithm was designed from the start to be perfectly parallelizable. It was originally developed for specialized sorting networks that would use comparators, but the technique can be easily adopted for SIMD and SIMT environment – i.e., for GPUs [10] as well. The main drawback of Bitonic sort is that the algorithm has $O(n \log(n)^2)$ time complexity.

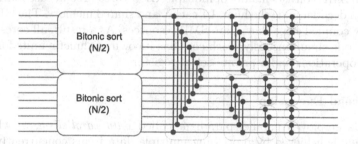

Fig. 1. Bitonic sorting network

Let us review the fundamentals of this algorithm, since we have used it and its modified version in our solution. The sorting network is depicted in Figure 1. We used traditional visualization of the algorithm by the means of *comparators* attached to wires. The horizontal lines symbolize items of the sorted array and the comparators (vertical lines) symbolize compare operations. A comparator ensures the correct order of elements – i.e., if its upper element is greater than its lower element, the elements are swapped.

The algorithm works similarly to *mergesort*, but unlike mergesort, it operates in place. First, it recursively applies itself to sort first half and second half of the input. Then, both halves are combined using the *upward bitonic merge*, so that the smallest item from the first half is compared with the largest of the second half, second item is compared with second last item and so on. Finally, item subgroups of sizes $N/2$, $N/4$, $N/8$, etc. are merged using regular *bitonic merge*.

3   GPU Fundamentals

In this section, we review the GPU architecture fundamentals with particular emphasis on aspects, which have great importance in the light of the studied problem. We will focus mainly on the NVIDIA Maxwell [1], since this architectures was used in our experiments.

3.1   GPU Device

A GPU card is a peripherial device connected to the host system via the PCI-Express (PCIe) bus. The GPU is quite independent, since it has its own memory and processing unit. Therefore, the host system may offload computations to the GPU while the CPU performs other tasks. On the other hand, the GPU processor cannot access the host memory directly, so all the data must be transferred to/from the GPU internal memory.

The GPU processor consists of several *streaming multiprocessors* (SMPs), which can be roughly related to the CPU cores. The SMPs share the main memory bus and the L2 cache of the GPU, but otherwise they are almost independent. Each SMP consists mainly of multiple *GPU cores* (128 in case of Maxwell), instruction decoder and scheduler, L1 cache, and shared memory. The GPU cores are tightly coupled since they share SMP resources. As a result, all cores execute the same code simultaneously. Each core has its own arithmetic units for integer and float operations and a private set of registers.

3.2   Thread Execution

The GPUs employ a parallel paradigm called *data parallelism*, in which the concurrency is achieved by processing multiple data items concurrently by the same routine (called *kernel* in the case of GPUs). When a kernel is invoked, the caller specifies, how many threads are spawned for this kernel. Each thread executes the kernel code, but has an unique thread ID, which is used to identify its portion of the workload.

The threads are grouped together into blocks of the same size. The blocks are assigned non-preemptively to available SMPs. Threads from different blocks are not allowed to communicate with each other directly, since it is not even guaranteed that any two blocks will be executed concurrently. On the other hand, threads within one block may synchronize their work using barriers and closely cooperate using internal resources of the SMP.

Furthermore, threads in a block are divided into subgroups called *warps*. The number of threads in a warp is fixed for each architecture (current architectures use 32 threads per warp). Threads in a warp are executed in a *lockstep*, which means they are all issued the same instruction at a time. When a warp is forced to wait (e.g., when transferring data from the memory), SMP simply schedules instructions of another warp, so the GPU cores are kept occupied.

The lockstep execution suffers from branching problems. When threads in a warp choose different code branches – for instance when executing 'if' or 'while' statement – all branches must be executed by all threads. Thread masks instruction execution according to their local conditions to ensure correct results, but heavily branched code does not perform well on GPUs.

3.3   Memory Organization

Another important issue is the GPU memory organization which is depicted in Figure 2.

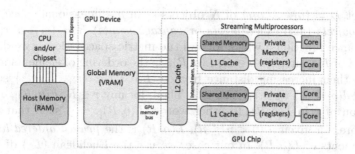

Fig. 2. Host and GPU memory organization scheme

The *host memory* is the operational memory of the computer. It is directly accessible by the CPU, but it cannot be accessed by any peripheral devices such as the GPU. Input data needs to be transferred from the host memory (RAM) to the graphic device global memory (VRAM), and the results need to be transferred back when the kernel execution finishes. For the transfer the PCI-Express bus is used, which is rather slow (8 GB/s) when compared to the internal memory buses.

The *global memory* can be accessed from the GPU cores, so the input data and the results computed by a kernel are stored here. The global memory bus has both high latency and high bandwidth. It is capable of transferring wide blocks of aligned consecutive data in each transaction. In order to access the global memory optimally, threads in a warp are encouraged to address data which are close together, so they can be transferred by as few transactions as possible.

The *shared memory* is shared among threads within one group. It is rather small (tens of kB) but almost as fast as the GPU registers. The shared memory can play the role of a program-managed cache for the global memory, or it can be used to exchange intermediate results by the threads in the block. The memory is divided into several (usually 16 or 32) banks, so that subsequent 4-byte words are stored in subsequent banks (modulo the number of banks). When multiple threads access the same bank (except if they read the same address), the memory operations are serialized which creates undesirable delay for all threads in the warp due to the SIMT execution model.

The *private memory* belongs exclusively to a single thread and corresponds to the GPU core registers. Private memory size is very limited (tens to hundreds of words), therefore it is suitable just for a few local variables.

Finally, there are two levels of cache present. The L2 cache is shared by all SMPs and it transparently caches all access to global memory. The L1 cache is private to each SMP and it caches data from global memory selectively.

4   Index Construction

Given a set $D = \{o_1, \ldots, o_N\}$ of N objects o_i in m-dimensional space, a set of *reference objects* (also called *pivots* or *vantage points*) $R = \{r_1, \ldots, r_n\} \subseteq D$, and a distance function d which follows the metric space axioms, we define the *ordered list* of R relative to $o \in D$, $L(o, R)$, as the ordering of elements in R with respect to their increasing distance from o: $L(o, R) = \{r_{i_1}, \ldots, r_{i_n}\}$ such that $d(o, r_{i_j}) \leq d(o, r_{i_{j+1}}) \; \forall j = 1, \ldots, n-1$. Then, for any $r \in R$, $L(o, R)_{|r}$ indicates the position of r in $L(o, R)$. In other words, $L(o, R)_{|r} = j$ such that $r_{i_j} = r$. Furthermore, for given $\tilde{n}$ $(0 < \tilde{n} \leq n)$, $\tilde{L}(o, R)$ is the *pruned ordered list* of the first $\tilde{n}$ elements of $L(o, R)$. In our case, we use the Euclidean (L_2) distance to determine the nearest pivots.

The index construction performs the following step for each object in the database. First, the distances between the object and all pivots are computed. Then, the distances are sorted and nearest $\tilde{n}$ pivot indices are taken. The second step may be also performed in reverse order, so that $\tilde{n}$ nearest pivots are selected first (by the means of heap data structure, for instance) and the sorting is performed afterwards. The latter approach may be significantly better, if the size of the selection is much smaller than the number of pivots.

The indexing operations may be performed independently for each database object, which makes it a trivial data parallel problem. On the other hand, the indexing operations are quite memory-intensive, so they need to be implemented in a cache-aware manner to achieve optimal performance, especially if the objects are represented in high-dimensional space.

In our previous work [7], we have focused on optimizing the distance computations on GPUs. Since the time complexity of a distance computation is linearly proportional to the dimension of the feature space, these computations are the most computationally demanding operation of the index construction, especially if the dimension exceeds one hundred. We have designed a hybrid solution that computes distances on the GPU while the CPU performs the postprocessing step (i.e., selecting nearest $\tilde{n}$ distances for each point and sorting them). Thus the algorithm has four independent steps:

1. Copy a batch of database objects to the GPU.
2. Compute pivot-object distances on the GPU.
3. Copy the distances back to the host.
4. Postprocess the distances and create the index.

As mentioned in Section 3, modern GPUs are capable of performing duplex data transfers whilst running a kernel computation. Therefore, all four steps may be executed as concurrent stages of a linear pipeline. Such a pipeline would perform optimally if all steps take approximately the same amount of time (for a fixed-size batches).

4.1 Profiling Results

In order to better understand the performance of the pipeline, we have profiled the individual operations from our previous implementation. Figure 3 presents results measured for 128k objects and 2,048 pivots on a hardware configuration described in Section 6. The *dist_208* results indicate times required for distance computations (in 208-dimensional feature space) and *post_ñ* are times required for selecting $\tilde{n}$ nearest distances using regular heap and subsequent sorting.

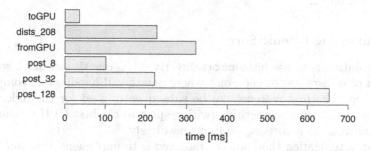

Fig. 3. Time complexity summary of individual steps

Our optimized implementation of distance computations performed on Maxwell architecture become so efficient, that it is no longer the most time consuming operation of the index construction (especially, when $\tilde{n}$ is larger). Furthermore, the computations become even faster than the transfer of the distances back to the host memory. In case that multiple GPUs are employed, the workload ratio would shift even more and the CPU would become a serious bottleneck. In the following section, we present several techniques, how to offload some more work to the GPU and restore balance of the pipeline.

5 Postprocessing Distances on the GPU

Basically, we can choose three different approaches to accelerate the postprocessing step:

- Create pre-sorted runs of distances, so the postprocessing step could use simply merge them.
- Select only nearest $\tilde{n}$ (or at least nearest k, where $\tilde{n} < k << n$) distances for each point.
- Implement the whole postprocessing step on the GPU.

The first option is also the least favorable, since it does not reduce the time required to transfer the distances to the host system. In fact, it inevitably increases this time, since the distances would have to be transferred along with the indices of the pivots (or with some other additional information from which the indices can be reconstructed). Therefore, we consider only the latter two approaches.

We have implemented and tested multiple methods of selecting smallest $\tilde{n}$ distances and for sorting the distances on the GPU. For our configuration (i.e., when one or two thousand distances per each point are processed), the best performance was achieved with modified bitonic sorting. Hence, we present our solution first and the alternative methods are discussed at the end of this section.

Let us emphasize that all presented methods (sorting or top-k selections) have to operate not only with distances, but with pairs distance-index. The distance value is used for ordering and the pivot index is then stored in the constructed permutation index.

5.1   Employing Bitonic Sort

A naïve solution is to use bitonic sort directly – i.e., one thread block sorts the whole set of n distances of one point where one thread handles one comparator in each step. The first $\tilde{n}$ items can be subsequently used for the index. Even though this solution is competitive (with respect to the best CPU version), it is far from optimal as it sorts $n - \tilde{n}$ items needlessly.

A first optimization that can be employed is to implement a *partial Bitonic sort*. It initially starts as a regular sort, but the final bitonic merge phases are performed only on items which can fall within the first $\tilde{n}$ items of the result. In order to minimize the warp divergence, we have implemented this approach as follows. The distances of one point are divided into $\tilde{n}$ blocks and each block is sorted using bitonic sort. Then, an upward bitonic merge is performed on each two adjacent blocks. After that, only odd blocks contain items which are candidates for top-$\tilde{n}$, so we discard even blocks and compact the array. The procedure is repeated until only one block remains. The algorithm is visualized in Figure 4.

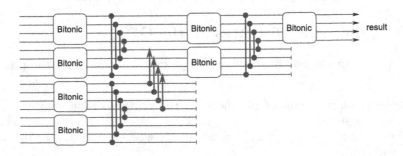

Fig. 4. Partial Bitonic sort that selects top $\tilde{n} = 4$ items

The distances and their corresponding pivot indices are kept in the shared memory. The distances are loaded from the global memory before the sorting and the final $\tilde{n}$ pivot indices are copied back when it is completed. We can make an observation, that most of the memory is not utilized for the whole time as the amount of processed data is halved in every phase of the algorithm.

Since shared memory is a precious resource, we have designed a more memory efficient version of this algorithm. The algorithm maintains only $2\tilde{n}$ items in the shared memory. Once they are sorted and the upward bitonic merge is performed, the second block is replaced with another block of distances from the global memory.

The memory optimized version may seem less effective as it provides less opportunities for parallelism. On the other hand, the lowered amount of allocated shared memory allows multiple points to be processed by one SMP (either automatically as multiple blocks can be assigned to one SMP, or by extending the implementation). If the occupancy of the cores is maintained, the optimized version may perform better as it requires less synchronization and it better interleaves computations with global memory transactions.

5.2   Alternative Algorithms

We have tested multiple different approaches to our problem. A complete survey is beyond the scope of this paper, so we have selected two of them which are the most intriguing. Both approaches are designed to reduce the number of distances per point, so that we save time transferring the data as well as post-processing them on the CPU.

Parallel Top-k Algorithm. is based on an idea to parallelize traditional top-k filtering that maintains a regular heap or sorted list of k smallest items while it process the items one by one. The problem can be parallelized in two ways either, the items are processed concurrently while each thread maintains its own top-k list, or the top-k structure is updated concurrently. The first approach is not feasible, as there is not enough shared memory to fit a top-k structure for reasonable number of threads, so we have tried the second approach.

The algorithm can be implemented so that one warp is processing one point – i.e., operate one top-k structure. Implementing parallel updates in regular heaps is quite impossible and we are expected the size of the top-k structure to be small, so we have chosen a sorted list. The list is initially sorted by Bitonic sort algorithm and maintained by parallel insert sort. The threads in the warp compare newly inserted item cooperatively and exchange their results using `ballot` instruction. Then they cooperatively move the items in the sorted list and the first thread inserts the new item.

Unfortunately, this approach was much slower than the Bitonic sort approach. We have discovered that it can slightly outperform the Bitonic sort, but only for very small top-k structures (when $\tilde{n} = 8$) and when multiple structures are handled by one warp.

Approximate Filtering. is a technique that is based on an idea that we can prefilter the distances with a carefully selected value. Ideally, if we knew the distance of $\tilde{n}$-th item, we could perform a very quick filtering sweep to select only smallest $\tilde{n}$ items. This sweep can be performed in parallel using the prefix-sum algorithm.

Finding exactly the $\tilde{n}$-th value is virtually as difficult as finding all first $\tilde{n}$ values, so the approximate filtering finds a k-th element, where $\tilde{n} \leq k << n$, so that we can still prune the distance set significantly while ensuring that it contains at least $\tilde{n}$ items. In order to do so, we have implemented modified median-selection algorithm that works as follows.

The distances are divided into blocks of 32 and each block is sorted by a Bitonic algorithm. Then s_i-th item is selected from each group (so $n/32$ items are taken) and the selected items are sorted again. Then s_j-th value from the selected items is taken as the pruning value. The difficult part is to select s_i and s_j correctly, so that we prune as many distances as possible while maintaining at least $\tilde{n}$ items in the result. To do that, we select s_i and s_j as the smallest values that fulfill $s_i \cdot s_j \geq \tilde{n}$ inequality and where $|s_i - s_j|$ value is also the smallest possible.

Even though this technique is slightly faster than Bitonic sorting, we have not used it as the final solution. The main reason is that it produces unpredictable amounts of distances for each point. This is quite tedious since it complicates memory allocation and creates workload imbalance. Furthermore, it is significantly more complicated for implementation, which makes it less attractive for possible applications.

6   Experimental Results

Due to a limited scope, we present results using fixed set of parameters. The results were measured for 128k objects represented in 208-dimensional feature space and 2,048 pivots. The index size ($\tilde{n}$) parameter was tested with values 8, 32, and 128 which represents a rather typical values for most indices. Only last two are included in the results as $\tilde{n} = 8$ benefits from our optimizations the least.

6.1   Hardware and Methodology

Our experiments were conducted on a desktop PC with an Intel Core i7 870 CPU (4 physical cores with HyperThreading) clocked at 2.93 GHz and 16 GB of RAM. The PC was equipped with two NVIDIA GeForce GTX 980 GPUs (Maxwell architecture), which have 2,048 CUDA cores (16 SMPs) and 4 GB of global memory. The host used Red Hat Enterprise Linux 7.1 as operating system and CUDA 6.5 framework for the GPGPU computations.

The experiments were timed using the real-time clock of the operating system. Each experiment was conducted 5× and the arithmetic average of the measured values is presented as the result. All measured values were within 1% deviation from their respective averages.

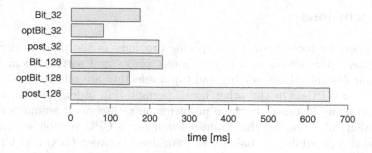

Fig. 5. Comparing Bitonic sort postprocessing on GPUs with baseline CPU version

6.2   Bitonic Sort Results

Figure 5 presents the comparison of our solution to the optimal postprocessing algorithm on CPU. The *Bit* denotes partial Bitonic sorting algorithm and *optBit* is its memory optimized version. The *post* denotes CPU postprocessing algorithm that employs heaps for top-k selections and standard Quicksort algorithm (C++ std::sort). The number in the suffix is the value of $\tilde{n}$ parameter.

The comparison of individual steps that employ GPU postprocessing is summarized in Figure 6. The *dists* is the time taken solely by the distance computations. The *Bit* and *optBit* denote the time taken by GPU to compute the distances and to postprocess them using partial Bitonic sort algorithms. The *fromGPU* is the time required to copy all $n \cdot N$ distances back to host memory, while *fromGPU_32* and *fromGPU_128* are the times of transferring the index of 32 and 128 pivots respectively.

The presented solution shifts all the workload to the GPU, so that we can test the limits of this approach. On the other hand, we can easily process some parts of that workload on the CPUs to achieve a better workload balance in the pipeline. For instance, we could terminate the Bitonic sort steps prematurely and finalize the postprocessing on the host.

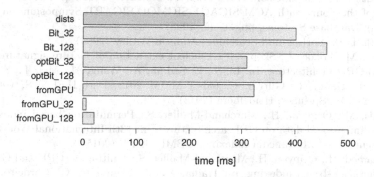

Fig. 6. Measured times of individual steps with GPU postprocessing

7   Conclusions

In this paper, we have studied two specific problems of the permutation index construction – the selection of top-k nearest pivots and sorting them according to their distance. Both sorting and top-k selection are relatively difficult to parallelize on GPUs. On the other hand, permutation index creation requires to perform many instances of these problems on rather small amounts of data. Our solution outperformed the optimized multi-core CPU version several times and provide a possibility to balance the workload between GPU and CPU. We have also discussed alternative approaches to summarize our experience gathered when solving this problem.

In our future work, we would like to finalize our prototype implementation and integrate it into a database framework that also handles persistent memory data transfers. Furthermore, we would like to extend our implementation to utilize multiple GPUs for permutation based index and compare the performance of our state-of-the-art GPUs with Xeon Phi devices.

Acknowledgments. This paper was supported by Czech Science Foundation (GAČR) project no. P103-14-14292P and by the Swiss State Secretariat for Education and Research (SER) under grant C11.0043, in relation to the European COST Action on Multilingual and Multifaceted Interactive Information Access (MUMIA).

References

1. NVIDIA: Maxwell GPU Architecture. http://developer.nvidia.com/maxwell-compute-architecture
2. Batcher, K.E.: Sorting networks and their applications. In: Proceedings of the Spring Joint Computer Conference, April 30-May 2, 1968, pp. 307–314. ACM (1968)
3. Gonzalez, E.C., Figueroa, K., Navarro, G.: Effective proximity retrieval by ordering permutations. IEEE Trans. Pattern Anal. Mach. Intell. **30**(9), 1647–1658 (2008)
4. Jagadish, H.V., Mendelzon, A.O., Milo, T.: Similarity-based queries. In: Proceedings of the Fourteenth ACM SIGACT-SIGMOD-SIGART Symposium on Principles of Database Systems, pp. 36–45 (1995)
5. Knuth, D.E.: Sorting and Searching. Addison-Wesley (2003)
6. Kruliš, M., Lokoč, J., Skopal, T.: Efficient extraction of feature signatures using Multi-GPU architecture. In: Li, S., El Saddik, A., Wang, M., Mei, T., Sebe, N., Yan, S., Hong, R., Gurrin, C. (eds.) MMM 2013, Part II. LNCS, vol. 7733, pp. 446–456. Springer, Heidelberg (2013)
7. Krulis, M., Osipyan, H., Marchand-Maillet, S.: Permutation based indexing for high dimensional data on GPU architectures. In: 13th International Workshop on Content-Based Multimedia Indexing (CBMI). IEEE (2015)
8. Mohamed, H., Osipyan, H., Marchand-Maillet, S.: Multi-core (CPU and GPU) for permutation-based indexing. In: Traina, A.J.M., Traina Jr., C., Cordeiro, R.L.F. (eds.) SISAP 2014. LNCS, vol. 8821, pp. 277–288. Springer, Heidelberg (2014)

9. Patella, M., Ciaccia, P.: Approximate similarity search: A multi-faceted problem. J. of Discrete Algorithms **7**(1), 36–48 (2009). http://dx.doi.org/10.1016/j.jda.2008.09.014

10. Peters, H., Schulz-Hildebrandt, O., Luttenberger, N.: Fast in-place sorting with CUDA based on bitonic sort. In: Wyrzykowski, R., Dongarra, J., Karczewski, K., Wasniewski, J. (eds.) PPAM 2009, Part I. LNCS, vol. 6067, pp. 403–410. Springer, Heidelberg (2010)

11. Samet, H.: Foundations of Multidimensional and Metric Data Structures (The Morgan Kaufmann Series in Computer Graphics and Geometric Modeling). Morgan Kaufmann Publishers Inc., San Francisco (2005)

First International Workshop on Managing Evolving Business Intelligence Systems (MEBIS 2015)

E-ETL Framework: ETL Process Reparation Algorithms Using Case-Based Reasoning

Artur Wojciechowski[✉]

Institute of Computing Science, Poznan University of Technology, Poznań, Poland
artur.wojciechowski@cs.put.poznan.pl,
http://calypso.cs.put.poznan.pl/projects/e-etl/

Abstract. External data sources (EDSs) being integrated in a data warehouse (DW) frequently change their structures/schemas. As a consequence, in many cases, an already deployed ETL workflow stops its execution, yielding errors. Since structural changes of EDSs are frequent, an automatic reparation of an ETL workflow after such a change is of high importance. In this paper we present a framework, called *E-ETL*, for handling the evolution of an ETL layer. In the framework, an ETL workflow is semi-automatically or automatically (depending on a case) repaired as the result of structural changes in data sources, so that it works with the changed data sources. *E-ETL* supports three different reparation methods, but in this paper we discuss the one that is based on case-based reasoning. The proposed framework is being developed as a module external to an ETL engine, so that it can work with any engine that supports API for manipulating ETL workloads.

Keywords: Data source evolution · ETL evolution · Case-based reasoning

1 Introduction

A data warehouse (DW) system has been developed in order to provide a framework for the integration of heterogeneous, distributed, and autonomous data storage systems (typically databases) with a company and multiple external data feeds like Wikipedia, Twitter, Facebook, and LinkedIn. They will further be called data sources (DSs). The integrated data are subjects of advanced data analysis, e.g., trends analysis, prediction, and data mining. Typically, a DW system is composed of four layers: (1) a data sources layer, (2) an Extraction-Translation-Loading (ETL) layer (or its ELT variant) responsible among others for extracting data from DSs, transforming data into a common data model, cleaning data, removing missing, inconsistent, and redundant values, integrating data, and loading them into a DW, (3) a repository layer (a data warehouse) that stores the integrated and summarized data, and (4) a data analytics layer. The ETL or ELT layer, is implemented as a workflow of tasks managed by a dedicated software. The ETL workflow is interchangeably called the ETL process.

© Springer International Publishing Switzerland 2015
T. Morzy et al. (Eds): ADBIS 2015, CCIS 539, pp. 321–333, 2015.
DOI: 10.1007/978-3-319-23201-0_34

An inherent feature of DSs is their evolution in time with respect not only to their contents (data) but also to their structures (schemas) [19,21]. According to [8,16], structures of data sources change frequently. For example, the Wikipedia schema has changed on average every 9-10 days in years 2003-2008 [2]. As a consequence of DSs changes, in many cases, an already deployed ETL workflow stops its execution with errors. As a consequence, an ETL process must be repaired, i.e., redesigned and redeployed.

In practice, in large companies the number of ETL workflows may exceed 20,000. Thus, manual modification of ETL workflows is complex, time-consuming, and prone-to-fail. Since structural changes in DSs are frequent, an automatic or semi-automatic reparation of an ETL workflow after such changes is of high importance.

Handling and incorporating structural changes to the ETL layer received so far little attention from the research community [7,10,14]. Moreover, none of the commercial or open-source ETL tools existing on the market supports this functionality.

In this paper, we propose a reparation algorithm for an ETL workflow after structural changes in DSs. The algorithm is based on the case-based reasoning method. The algorithm is applied in the Evolving ETL (*E-ETL*) framework that we are developing [18].

The paper is organized as follows. Section 2 discusses how case-based reasoning can be applied to an evolving ETL layer. Section 3 describes the *E-ETL* library of cases. Section 4 presents a procedure of building a case. Section 5 introduces a method for comparing cases and shows how to apply modifications to an ETL process. Section 6 outlines research related to the topic of this paper. Section 7 summarizes the paper and outlines issues for future development.

2   Case-Based Reasoning for ETL Reparation

Case-based reasoning (CBR) solves a new problem, say P, by adapting and applying to P previously successful solutions that were similar to P [15]. This approach is inspired by human reasoning and the fact that people commonly solve problems by remembering how they solved similar problems in the past. Further in this paper, a problem and its solution will be called a *case*.

In the area of an evolving ETL layer, the problem arises with structural changes in DSs. A desired solution is a recipe for the reparation of an ETL workflow. According to the case-based reasoning method in an evolving ETL, in order to find a solution, i.e., a reparation of an ETL workflow, similar case should be found and adapted to the current ETL problem.

In the ETL process reparation, a case can be described as a pair of: (1) changes in data sources and (2) a sequence of modifications of the ETL process that adjust the ETL process to the new state, so that it works without errors on the modified data sources. Cases can be found in two dimensions of an ETL project: (1) fragments of the ETL process and (2) time. The first dimension means that a case represents not only the same change in the same data source but also a similar change in different part of the ETL process that can use a similar data source. The time dimension allows to look for cases in the whole evolution history of the ETL process. Therefore, if there was a change in the past and now a different part of the ETL process changes in the same way, then the first change can be a case for the ETL process reparation algorithm.

The similarity of data sources can be considered on four levels, namely: (1) a group of tables, (2) a table, (3) a group of columns and (4) a column.

An example of the *group of tables* similarity level can be found in an ETL process that loads data from several instances of an ERP system. Every instance can use a different version of an ERP system, hence each instance (treated as a data source) is similar to the other instances but, not necessarily the same. For each instance, there can be a dedicated ETL subprocess. When there are updates to some of the ERP systems, the structures of data sources that are associated with them may change. That causes the need of repairing the relevant ETL subprocesses. Changes in one data source and a manual reparation of an associated ETL subprocess may be used as a case for a reparation algorithm to fix other ETL subprocesses associated with the updated data sources.

The *table* similarity level occurs when a data source consists of multiple tables that are logical partitions of the same data set. For example, a set of clients data can be spread across three tables: (1) individual clients, (2) enterprise clients, and (3) institutional clients. Each of them may have some specific columns important only for one client type but, all of them have a common set of the same columns so, they are similar. If there is a change in data sources that modifies the common set of columns (e.g., adding a contact person to all three types of clients) then the change of one of these tables and the reparation of the ETL process associated with this table can be used as a case for a reparation algorithm to fix other ETL parts associated with the other two tables.

The *group of columns* similarity level can be found in data sources that consist of multiple tables containing the same set of columns, e.g., multiple tables storing columns State, City, and Street. Replacement of those three columns in one of these tables for a reference to table Address and modifying an ETL fragment associated with the columns can be used as a case for a reparation algorithm to fix other ETL fragments starting from tables where the same change has been applied. The difference between the *group of columns* and *table* similarity levels is in the number of common columns. In the former level, most of the columns are the same in the modified tables. In the latter level, few columns are the same in the modified tables. This distinction is subtle, but we decided to distinguish them to ease the explanation of our ETL reparation algorithm.

The *column* level is similar to the previous level but considers changes to only one column in a table. For example, a change of a column that contains one string describing an address to an integer column which is a reference to an entry in a table with addresses.

3   Library of Reparation Cases in E-ETL

The case-based reasoning method for repairing ETL workflows is based on a library of reparation cases. The library is constantly augmented with new cases encountered during the execution of the *E-ETL* framework [18].

3.1   E-ETL Framework

E-ETL allows to detect structural changes in DSs and handle the changes at an ETL layer. Changes are detected either by means of Event-Condition-Action (triggers) mechanism or by means of comparing two consecutive DS metadata snapshots. Detection of a DS schema change causes a reparation of the ETL workflow that interact with the changed DS. The reparation of the ETL workflow is guided by several customizable reparation algorithms.

The framework is customizable and it allows to:

- work with different ETL engines that provide API communication,
- define the set of detected structural changes,
- modify and extend the set of algorithms for managing the changes,
- define rules for the evolution of ETL processes,
- present to the user the impact analysis of the ETL workflow,
- store versions of the ETL process and history of DS changes.

One of the key features of the *E-ETL* framework is that it stores the whole history of all changes in the ETL process. This history is the source for building the library of reparation cases. After every modification of the ETL process a new case is added to the library. Thus, whenever a data source change (DSC) is detected and the procedure of evolving the ETL process repairs it by using one of the supported reparation methods new cases are generated and stored in the library. The reparation methods include: the *User Defined Rules*, the *Replacer algorithm*, and the *Case-based reasoning*.

The *E-ETL* framework is not supposed to be a fully functional ETL designing tool but an external extension to complex commercial tools available on the market. The proposed framework was developed as a module external to an ETL engine. Communication between *E-ETL* and the ETL engine is realized by means of the ETL engine API. Therefore, it is sometimes impossible to apply all the necessary modification of the ETL process within the *E-ETL* framework (i.e., new business logic must be implemented). For that reason the *E-ETL* framework has a functionality of updating the ETL process. When a user introduces manually some modification in the ETL process using the external ETL tool, then the *E-ETL* framework loads the new definition of the ETL process and compares it with the previous version. The result of this comparison is a set of all modifications made manually by the user. Those modifications together with detected DSCs also provide reparation cases.

E-ETL provides a graphical user interface for visualizing ETL workflows and impact analysis.

3.2 Reverse Cases

For each DSC A it is possible to define reverse DSC A' that restores the data source to the state before DSC A was applied. For example, for a column addition - it would be a column deletion, for a table renaming *Persons* to *Employees* - it would be a table renaming *Employees* to *Persons*. Analogically, for each modification B of the ETL process it is possible to define a reverse modification B'. Therefore, for each reparation case C that consists of DSC A and modification B it is possible to define its reverse case C' that consists of DSC A' and modification B'.

The reverse cases are useful when a user has to handle some undone DSCs and associated ETL process modifications but he/she cannot use an old version of the ETL process. The restriction that forbids the usage of the old version may be caused by some other important changes that have been made after the change which the user wants to undo. In such a situation the *E-ETL* framework tries to repair the ETL process as if it was a regular DSC and not an undo change. As a result, the *case-based reasoning* algorithm tries to find the best case. For example, in order to undo data source change A, DSC A' must be applied and to undo ETL process modification B, ETL process modification B' must be applied. Therefore, if the *E-ETL* framework detects DSC A' then the *case-based reasoning* algorithm will use case C' and modification B' will be applied.

3.3 Library Scope

For each ETL process, there are three scopes of the library of reparation cases, namely: (1) a process scope, (2) a project scope, and (3) a global scope. Since the *case-based reasoning* algorithm works in the context of the ETL process, the first scope is the *process scope* and it covers cases that originate from the same ETL process.

Huge companies maintain ETL projects that consist of hundreds or even several thousands of ETL processes. Within a project there are usually some rules that define how to handle DSCs. Therefore, DSCs from one ETL process probably should be handled in a similar way as in other ETL processes from the same ETL project. The *project scope* covers cases that originate from the same ETL project.

The purpose of the *global scope* is to support the *case-based reasoning* algorithm with cases for the most common changes. The global scope consists of cases build into the *E-ETL* framework.

4   Case Detection in ETL Process

As mentioned in Section 2, the case is a pair of: (1) a set of changes in data sources and (2) a set of modifications that adjust an ETL process to the new state of the modified data sources (repair an ETL process). The modifications are defined on ETL activities, i.e., tasks of an ETL workflow. Both sets that create the case should be complete and minimal.

Completeness means that the definition of the case should contain all DSCs and all modifications that are connected. In other words, the set of DSCs should contain all changes that are handled by the set of modifications. Moreover, the set of modifications should contain all modifications that repair the ETL process affected by the set of DSCs. An incomplete set of DSCs causes that the set of modifications cannot be applied since it is based on a different state of the DS (not fully changed).

Minimality means that the definition of the case should be as small as possible without violating the completeness constraint. In other words, the set of DSCs should contain only these changes that are handled by the set of modifications. Additionally, the set of modifications should contain only these modifications that repair the ETL process affected by the set of DSCs. Moreover, minimality means that it is impossible to separate from one case another smaller case that remain complete. If the set of DSCs is not minimal, then:

- the case-based reasoning algorithm may incorrectly skip the case, since it will not match a new problem because of redundant DSCs,
- the reparation of the ETL process might be excessive and modifications that do not repair the ETL process might be applied.

Figure 1 shows an example of DSCs and associated ETL process modifications. The exemplary ETL process reads data from the *OnLineSales* table and the *InStoreSales* table. Next, the data are merged by the union activity. The result is joined with data read from the *Customers* table. Finally, the result is stored in the *TotalSales* table.

Let us assume that as a result of a data source evolution three DSCs, namely: *dsc1*, *dsc2*, and *dsc3* were introduced. *dsc1* is the addition of column *Discount* to table *OnLineSales*. *dsc2* is the addition of column *Discount* to table *InStoreSales*. *dsc3* is the renaming of table *Customers* to *Clients*.

In order to handle the three DCSs, three ETL process modifications, namely: *mod1*, *mod2*, and *mod3* were introduced. *mod1* is the addition of new attribute *Discount* to activity *Union*. *mod2* is the addition of new activity that replaces old *TotalValue* with *TotalValue - Discount*. It was added between activities *Union* and *Join*. *mod3* is the update of the component that reads the *Customers* table - it replaces the old table name with the new one, which is *Clients*.

For this evolution scenario, two new cases can be defined. The first one (*case1*) consist of two DSCs, namely: *dsc1*, *dsc2* and two ETL process modifications, namely: *mod1* and *mod2*. The second case (*case2*) consist of DSC *dsc3* and modification *mod3*.

These cases are both, minimal and complete. An addition of *dsc3* or *mod3* to *case1* would violate the minimality constraint. Moreover, a removal of any item from *case1* would violate the completeness constraint. We can remove neither *dsc1* nor *dsc2* from *case1* because *mod1* is connected with them. Moreover, we can remove neither *mod1* nor *mod2* because they are the results of *dsc1* and *dsc2*.

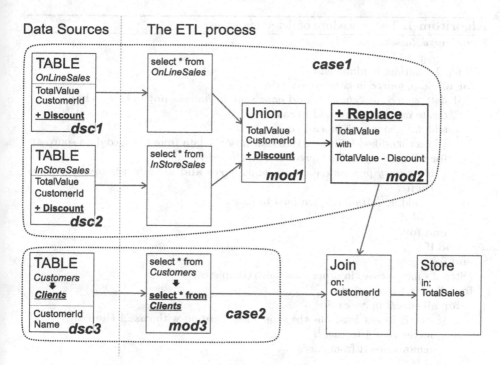

Fig. 1. An example of DSCs and ETL process modifications

In the *E-ETL* framework there is an algorithm that detects and builds new minimal and complete cases for the ETL reparation algorithm, based on case-based reasoning. The algorithm consists of three main steps presented in the Algorithm 1. In first step, for each data source that has any changes, a new case is initiated. A set of modifications for the new case consists of all modifications from all ETL activities that are preceded by an activity reading the changed data source. At this point, cases are minimal but they are not complete. In the second step, each case is compared with other cases and if any two cases have at least one common modification then they are merged into one case. The purpose of the second step is to achieve the completeness. The third step, i.e., the last *for all* loop in Algorithm 1 is clarified in section 5.

5   Choosing the Right Case

One of the key points in the case-based reasoning is to choose the most appropriate case for solving a new problem. In an ETL environment, a new problem is represented by changes in data sources. Therefore, to solve the problem it is crucial to find the case whose set of DSCs is similar to the problem. Moreover, the set of DSCs in the case must be applicable to the ETL process where the problem is meant to be solved.

Algorithm 1. The procedure of detecting cases

Require: *data_sources*
  cases ← []
  {Step 1 - initiate minimal cases}
  for all *data_source* in *data_sources* **do**
    if *data_source* has *changes* **and** *data_source.changes* **not** in cases **then**
      create new *case* **and** add to *cases*
      add *data_source.changes* to *case*
      {detect modified ETL activities that process data from changed data sources}
      for all *activity* in *ETL_process_activities* **do**
        if *activity* process data from *data_source* **and** *activity* has *modifications*
        then
          add *activity.modifications* to *case*
        end if
      end for
    end if
  end for
  {Step 2 - merge cases in order to achieve completes}
  for all *caseA* in *cases* **do**
    for all *caseB* in *cases* **do**
      if *caseB* has at least one the same *modification* with *caseA* **then**
        merge *caseA* to *caseB*
        remove *caseA* from *cases*
      end if
    end for
  end for
  {Step 3 - compute maximum FSAs}
  for all *case* in *cases* **do**
    compute maximum *FSAs* for the *case*
  end for
  return *cases*

In this section, we contribute the algorithm for searching and selecting the best case for a given problem. The algorithm is illustrated with an informal description. The formal notation is not elaborated here due to space limits.

The similarity of sets of DSCs means that although types of DSCs have to be the same, they can modify non-identical data source elements. For example, if a new DSC is an addition of column *Contact* to table *EnterpriseClients*, then a similar change can be an addition of column *ContactPerson* to table *InstitutionClients*. Both changes represent additions of columns but the names of the columns and the names of the tables are different. Therefore, a procedure that compares changes not only has to consider element names but also their types and subelements. If an element is an ETL activity, then subelements are its attributes. An ETL activity attribute does not have subelements. Added columns should have the same types and if they are foreign keys they should refer to the same table.

It is possible to decompose a given problem into smaller problems. Therefore, it is sufficient that the set of DSCs in the case is a subset of DSCs in a new problem. The opposite situation (the set of DSCs in the case is a superset of the set of DSCs in a new problem) is forbidden because it is not possible to apply to the ETL process modifications that base on missing DSCs.

In order to repair an ETL process it has to be modified. Not all modifications can be applied to all ETL processes. Let us assume that there is an ETL process modification that adds an ETL activity that sorts data by column *Age* and the ETL process does not process such a column. It is impossible to add an ETL activity that functionality is based on a non-existing column. Such a modification of the ETL process is not applicable. Thus, for each candidate case it is necessary to check if all modifications of the ETL process can be applied to the ETL process. In order to validate a single modification, the element (i.e., an ETL activity or its attribute) of the ETL process that would be modified, must exist in the ETL process. An exception for that is the modification that adds a new element. What's more, the modified element should process a set of columns similar to the one processed by the case. In other words, the ETL activity should have a similar set of inputs and outputs to this defined in the case. Therefore, each modification contains an information about the surrounding elements and an information about activities between the modified element and a data source.

It is possible that more than one case will be similar to a new problem. Thus, there is a need for defining how similar is the reparation case to the new problem. The more similar is the reparation case to the new problem the more appropriate is to solve the problem. Therefore, the case that is the most similar to the new problem should be chosen. To this end, for each case, a Factor of Similarity and Applicability (FSA) is calculated. The higher the value of FSA the more appropriate is the case for the new problem. If DSCs does not have the same types or modifications are not applicable, then FSA automatically equals 0.

The more common features of the data sources (column names, column types, column type lengths, table names, table columns) the higher FSA is. Also the more common elements in the surroundings of the modified ETL process elements the higher FSA is. The scope of the case also influences FSA. Cases form process scope have the increased value of FSA and cases form global scope have the decreased value of FSA. If there are two cases with the same value of FSA, the case that has been already used more times is probably the more appropriate one. For that reason, the next element that is part of FSA is the number of accepted (accepted and verified by the user) usages of the case.

Finding the case for a new problem requires a calculation of FSA for every case from the library of cases. Defining the similarity of sets of DSCs requires the comparison of every DSC in one set of DSCs with every DSC in another set of DSCs in order to map equivalent DSCs. Analogically, sets of modifications have to be mapped. Those two operations cause that the calculation of FSA is computationally expensive. For this reason, it is crucial to limit the set of potential cases for the new problem out of the whole library. The set of cases can be limited to only those that have less or equal number of DSCs, as compared to DSCs in a new problem. For the rest of the cases the value of FSA will equal 0.

The final step of Algorithm 1 computes the maximum value of FSAs for each library scope. Maximum can be achieved if the case perfectly matches a new problem. Having such maximum values, the calculation of FSA can be omitted (FSA=0) if its maximum value is lower than the FSA value of the best case that has already been checked. In order to get the best results from the usage of the maximum FSAs, the library of reparation cases should be sorted by the maximum value of the case FSA. Searching for the most appropriate case should start from the case with the highest maximum FSAs. This allows to stop the computation as soon as the first case is rejected because of its maximum FSAs.

After the most appropriate case has been chosen, the next step is to apply its modifications to the ETL process. During the procedure of checking the applicability of the modifications, all modifications have been mapped to activities existing in the ETL process. As a result, the procedure of applying the ETL process modification is reduced to executing the modification. After that, the repaired ETL process is being presented to the user for the acceptance.

As mentioned before, a problem can be decomposed into smaller problems and the ETL process can be repaired using more than one case. In such scenario each case usage is processed separately. This means that the user can independently decide about acceptance of each application of the case modifications.

The last step is the increase of case usage counters. If the used cases did not match perfectly (e.g., data source tables had different names) then the procedure of detecting cases is executed.

Currently we are developing this solution as one of the modules in the *E-ETL* framework. We are also preparing experiments that will examine the applicability of the proposed algorithms in a real world environment.

6 Related Work

The work presented in [15] is considered d to be the origin of the Case-based reasoning. The idea was developed, extended, and reviewed in [1,4–6,17]. Since the algorithm presented in this paper uses only a key concept of this methodology, the mentioned publications will not be discussed here. In this section we focus on related approaches to handling the evolution of the ETL layer.

Detecting structural changes in data sources and propagating them into the ETL layer have not received much attention from the research community. One of the first solution of this problem is Evolvable View Environment (EVE) [14]. EVE is an environment that allows to detect an ETL workflow that is implemented by means of views. For every view, it is possible to specify which elements of the view may change. It is possible to determine whether a particular attribute both, in the *select* and *where* clauses, can be omitted, or replaced by another attribute. Another possibility is that for every table, which is referred by a given view, a user can define whether this table can be omitted or replaced by another table.

Recent development in the field of evolving ETL workflows includes a framework called *Hecataeus* [9,10]. In Hecataeus, all ETL activities and DSs are modeled as a graph whose nodes are relations, attributes, queries, conditions, views, functions, and ETL steps. Nodes are connected by edges that represent relationships between different nodes. The graph is annotated with rules that define the behavior of an ETL workflow in response to a certain DS change event. In a response to an event, Hecataeus can either propagate the event, i.e. modify the graph according to a predefined policy, prompt an administrator, or block the event propagation.

A drawback of the above two approaches is that they handle ETL workflows that are developed as sequences of SQL queries. As a consequence, the application of these approaches is limited, as commercially deployed ETL workflows not only use SQL queries but also tasks defined in multiple procedural languages. Another disadvantage is that both approaches require from a user that he/she will define evolution rules "by hand" in advance. Whereas, E-ETL was designed from scratch as a module external to the existing ETL development tools and manages workflows designed in these tools. If an external tool supports ETL tasks defined in languages other than SQL, E-ETL will manage such tasks as well. Moreover, the ETL reparation algorithm that we proposed is based on case-based reasoning and, as such, does not require the existence of user-defined rules.

In [7], the authors proposed a method for the adaptation of evolving data-intensive ecosystems. An ETL layer is example of such ecosystem. The key idea of the method is to maintain alternative variants (old versions) of data sources and data flows. Data flow operations can be annotated with policies that instruct whether they should be adapted to an evolved data source or should use the old version of the data source. A week point of this approach is that in practice, data sources are independent and usually it is impossible to maintain its alternative variants for the purpose of using them in an ETL layer.

In [11], the authors proposed several metrics based on the graph model (the same as in *Hecataeus*) for measuring and evaluating the design quality of an ETL layer with respect to its ability to sustain DSCs. In [12], the set of metrics has been extended and tested in a real world ETL project. The authors identified three factors that can be used in order to predict a system vulnerability to DSCs. Those factors are: (1) *schema sizes*, (2) *functionality of an ETL activity*, and (3) *module-level design*. The first factor indicates that if the DS consist of tables with many columns then the ETL process is probably more vulnerable to changes. The second factor indicates that activities that depend on many columns and outputs many columns are more vulnerable to changes. According to the last factor, activities that reduce a number of processed attributes should be placed early in the ETL process.

The factors define only the probability that an ETL process will need the reparation but the estimation of reparation costs was not considered. Some changes (e.g. renaming a column) cause a simple reparation of an ETL process and some others (e.g. column deletion) may cause changes of the business logic in the ETL process. The second weak point is the size of the experiments. The authors analyzed only 7 ETL processes that consist of 58 activities in total. In comparison with enterprise ETL projects it seems to be too small.

In [20] the authors proposed a prototype system that automatically detect changes in DSs and propagate them into a DW. The prototype allows to define changes that are to be detected and associates actions with the changes to be executed in a DW. The main limitation of the prototype is that it does not allow ETL workflows to evolve. Instead, it focuses on propagating DSs changes into a DW. Moreover, the presented solution is restricted to only relational databases as DSs. The next drawback is a detection of changes which depends on triggers, and as such in practice it will not be allowed to be installed in a production database.

7   Summary

This paper presents the *E-ETL* framework for detecting structural changes in DSs and repairing an ETL workflow accordingly to the detected changes. The framework repairs automatically an ETL workflow using reparation algorithms based on the case-based reasoning methodology.

Currently, we are implementing the presented framework and we are developing the described algorithms. We are also preparing tests in an environment including structural changes that appeared in a real production DW systems. *E-ETL* API is currently under development for communicating with Microsoft SQL Server Integration Services.

The approaches outlined in Section 6 handle structural changes in DSs. However, as stressed in [3,13] even ordinary content (data) changes of a DS may cause structural changes at a DW or changes to the structure of dimension data in a DW. Neither Hecataeus, nor EVE, nor [7], nor *E-ETL* provide support for handling appropriately such content changes. In the future, we will focus on handling such kinds of content changes at the ETL layer and on correctly propagating them into a DW.

References

1. Aamodt, A., Plaza, E.: Case-based reasoning: Foundational issues, methodological variations, and system approaches. AI Comunications **7**(1), 39–59 (1994)
2. Curino, C., Moon, H.J., Tanca, L., Zaniolo, C.: Schema evolution in wikipedia - toward a web information system benchmark. In: Proc. of Int. Conf. on Enterprise Information Systems (ICEIS), pp. 323–332 (2008)
3. Eder, J., Koncilia, C., Morzy, T.: The COMET metamodel for temporal data warehouses. In: Pidduck, A.B., Mylopoulos, J., Woo, C.C., Ozsu, M.T. (eds.) CAiSE 2002. LNCS, vol. 2348, pp. 83–99. Springer, Heidelberg (2002)
4. Hammond, K.J.: Case-based planning: A framework for planning from experience. Cognitive Science **14**(3), 385–443 (1990)
5. Kolodner, J.L.: An introduction to case-based reasoning. Artificial Intelligence Review **6**(1), 3–34 (1992)

6. Kolodner, J.L.: Case-Based Reasoning. Morgan-Kaufmann Publishers, Inc. (1993)
7. Manousis, P., Vassiliadis, P., Papastefanatos, G.: Automating the adaptation of evolving data-intensive ecosystems. In: Ng, W., Storey, V.C., Trujillo, J.C. (eds.) ER 2013. LNCS, vol. 8217, pp. 182–196. Springer, Heidelberg (2013)
8. Moon, H.J., Curino, C.A., Deutsch, A., Hou, C., Zaniolo, C.: Managing and querying transaction-time databases under schema evolution. In: Proc. of Int. Conf. on Very Large Data Bases (VLDB), vol. 1, pp. 882–895 (2008)
9. Papastefanatos, G., Vassiliadis, P., Simitsis, A., Sellis, T., Vassiliou, Y.: Rule-based management of schema changes at ETL sources. In: Grundspenkis, J., Kirikova, M., Manolopoulos, Y., Novickis, L. (eds.) ADBIS 2009. LNCS, vol. 5968, pp. 55–62. Springer, Heidelberg (2010)
10. Papastefanatos, G., Vassiliadis, P., Simitsis, A., Vassiliou, Y.: Policy-regulated management of ETL evolution. In: Spaccapietra, S., Zimányi, E., Song, I.-Y. (eds.) Journal on Data Semantics XIII. LNCS, vol. 5530, pp. 147–177. Springer, Heidelberg (2009)
11. Papastefanatos, G., Vassiliadis, P., Simitsis, A., Vassiliou, Y.: Design metrics for data warehouse evolution. In: Li, Q., Spaccapietra, S., Yu, E., Olivé, A. (eds.) ER 2008. LNCS, vol. 5231, pp. 440–454. Springer, Heidelberg (2008)
12. Papastefanatos, G., Vassiliadis, P., Simitsis, A., Vassiliou, Y.: Metrics for the prediction of evolution impact in etl ecosystems: A case study. J. Data Semantics 1(2), 75–97 (2012)
13. Rundensteiner, E.A., Koeller, A., Zhang, X.: Maintaining data warehouses over changing information sources. Communications of the ACM 43(6), 57–62 (2000)
14. Rundensteiner, E.A., Koeller, A., Zhang, X., Lee, A.J., Nica, A., Van Wyk, A., Lee, Y.: Evolvable view environment (EVE): non-equivalent view maintenance under schema changes. In: Proc. of ACM Int. Conf. on Management of Data (SIGMOD), pp. 553–555 (1999)
15. Schank, R.C.: Dynamic Memory: A Theory of Reminding and Learning in Computers and People. Cambridge University Press (1983)
16. Sjøberg, D.: Quantifying schema evolution. Information and Software Technology 35(1), 35–54 (1993)
17. Watson, I., Marir, F.: Case-based reasoning: A review. The Knowledge Engineering Review 9(4), 327–354 (1994)
18. Wojciechowski, A.: E-ETL: framework for managing evolving ETL processes. In: Proc. of ACM Information and Knowledge Management Workshop (PIKM), pp. 59–66 (2011)
19. Wrembel, R.: A survey on managing the evolution of data warehouses. International Journal of Data Warehousing & Mining 5(2), 24–56 (2009)
20. Wrembel, R., Bębel, B.: The Framework for Detecting and Propagating Changes from Data Sources Structure into a Data Warehouse. Foundations of Computing & Decision Sciences 30(4), 361–372 (2005)
21. Wrembel, R., Bębel, B.: Metadata management in a multiversion data warehouse. In: Spaccapietra, S., et al. (eds.) Journal on Data Semantics VIII. LNCS, vol. 4380, pp. 118–157. Springer, Heidelberg (2007)

Handling Evolving Data Warehouse Requirements

Darja Solodovnikova[✉], Laila Niedrite, and Natalija Kozmina

Faculty of Computing, University of Latvia, Riga, Latvia
{darja.solodovnikova,laila.niedrite,natalija.kozmina}@lu.lv

Abstract. A data warehouse is a dynamic environment and its business requirements tend to evolve over time, therefore, it is necessary not only to handle changes in data warehouse data, but also to adjust a data warehouse schema in accordance with changes in requirements. In this paper, we propose an approach to propagate modified data warehouse requirements in data warehouse schemata. The approach supports versions of data warehouse schemata and employs the requirements formalization metamodel and multiversion data warehouse metamodel to identify necessary changes in a data warehouse.

1   Introduction

Effective organization of business processes guarantees the achievement of institution's goals. Organizations use performance measures to align daily activities to strategic objectives [1] and a data warehouse could be used for storing them.

Data warehouses are databases designed for querying and analysing data. Developing a data warehouse that fits all requirements of potential users is a challenging task. In the case of a data warehouse we speak about information requirements [2]. These requirements also express the information needs to define performance indicators of an organization; they are gathered by interviewing users and defined on various levels of formality depending on particular method used. Some proposals on how to specify formally performance indicators exist and are described in [3], [4]. Formal notation helps capture all details that are important in later development of a data warehouse.

A data warehouse should provide accurate and historically correct information to users to support the decision making. This means that the data warehouse must reflect all changes that occur in the analysed business process. Besides, information systems that serve as data sources for the data warehouse can evolve during the time. Changes in data sources or business requirements can invalidate existing schemata of a data warehouse, data extraction, transformation and loading processes and pre-defined reports. To avoid the loss of history and keep the data warehouse usable, data warehouse schema versions can be used. According to [5], 'schema version is a schema that reflects the business requirements during a given time interval, called its validity, that starts upon schema creation and extends until the next version is created.'

The main contribution of this paper is the approach for semi-automatic generation of data warehouse evolution changes that create new data warehouse schema versions based on the changes in business requirements.

© Springer International Publishing Switzerland 2015
T. Morzy et al. (Eds): ADBIS 2015, CCIS 539, pp. 334–345, 2015.
DOI: 10.1007/978-3-319-23201-0_35

The rest of this paper is organized as follows. In Section 2 the related work is presented. In Section 3 the proposed data warehouse framework is outlined. The requirements formalization metamodel and multiversion data warehouse metamodel together with schema changes are described in Sections 4 and 5. The main contribution of this paper is presented in Section 6, where the algorithm for propagation of changed requirements in data warehouse schema versions is described. We conclude with directions for future work in Section 7.

2   Related Work

The researchers have proposed different solutions for the problem, how to adjust the data warehouse schema to the changes in source systems and business requirements. The two most popular approaches are schema evolution and schema versioning. The schema evolution approaches, (e.g., [6]), update the data warehouse schema, to reflect the changes that have occurred. These approaches provide a set of evolution operations (e.g. add dimension etc.). The second direction is schema versioning and temporal data warehouses. According to these approaches, not only the schema of a data warehouse is updated, but also the history of previous schemas or validity of schema elements is kept (e.g., [5],[7],[8]). But these works do not describe formally how the information requirements affect the evolution changes. Some research is done to get such methods [9], [10], [11].

In [9] the authors present a semi-automatic method that incrementally integrates the end-user information requirements into existing schema design. The requirements are expressed as SQL queries over the relational data sources and after processing a set of multidimensional logical schemas is gained. The proposed method produces semi-automatically in each iteration a set of integration operations, e.g. "insertFact" and others that perform certain tasks to implement the new information requirement into the current multidimensional schema.

In the paper [10] the authors present an approach to automate the evolution of a data warehouse schema. The ontology is used to represent the data sources, requirements and data warehouse schema. According to the approach proposed in paper [10], changes to requirements are propagated in the data warehouse schema ontology and no history of schema versions is accumulated. The requirements ontology is based on the i* framework and is highly conceptual. Besides, the data warehouse ontology describes the data warehouse schema at the physical level and does not include information about such elements as hierarchies and levels.

In [11] a requirement driven framework for data warehouse evolution is proposed. The framework is divided into 4 levels. The requirement level integrates requirements gathered from users into a checklist. The change management level controls the changes. The design level implements changes in the data warehouse schema or data in order to incorporate the changed requirements. The view level enables the users to get customized reports. This framework exploits concepts of the framework described in [12]. Our paper extends the original framework described in [12] with the requirements management component.

3   Data Warehouse Evolution Framework

Information requirement definition, processing and subsequent data warehouse metadata changes are integrated into the data warehouse framework depicted in Figure 1, which is an extended version of the framework given in [12]. This framework handles not only information requirement changes, but also data warehouse evolution problems such as maintenance and adaptation. The framework is composed of the development environment, where the metadata repository is located and ETL processes and change processing is conducted, and the user environment, where reports on one or several data warehouse versions are defined and executed by users.

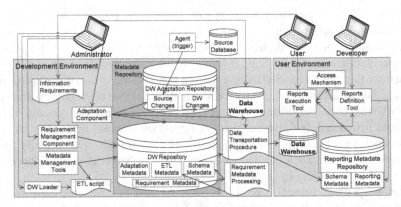

Fig. 1. Data warehouse evolution framework

All operation of the data warehouse framework is based on the metadata in data warehouse repository. Schema metadata describe data warehouse schema versions, their storage in the relational database, and semantics of data stored in the data warehouse. ETL metadata define the logics of ETL processes. Adaptation metadata are used to adapt a data warehouse after changes in data sources. Requirement metadata represent information requirements with formal terms.

Information requirements are gathered during the interviews with a client. The requirement management component processes the obtained set of requirements in natural language according to the requirement formalization metamodel (see Section 4) and formalization principles [13]. The result of the requirement formalization is requirement metadata. The requirement metadata processing procedure keeps track of the requirement metadata evolution and by means of exploiting the Algorithm (see Section 6) and ETL metadata makes up-to-date changes in data warehouse schema (e.g. deletion of an attribute, creation of a new measure).

4   Requirements Formalization Metamodel

The risk of interpreting information requirement erroneously is threefold: a client might be imprecise in formulating the needs, an interviewer might capture them incorrectly, and a developer might construct a conceptual model that does not fully comply with stated information requirements. The requirement formalization metamodel (Fig. 2) serves to minimize the risk at all three stages. We employ a revised version of the requirement formalization metamodel, which is an outcome of the case study described in [14], with some classes omitted (e.g. Priority) as they are out of scope.

A *Requirement* is classified either as *Simple* or *Complex* and is related to a certain *Theme* (e.g. Finance, Education, Customer Focus [1]). A complex requirement is composed of two or more requirements joined with either an *Arithmetical Operator* or a *Comparison*. A simple requirement may consist either of an *Expression* or an *Operation* that denotes a command applied to an *Object*, which is either an instance of *Quantifying Data* (measurements) or *Qualifying Data* (properties of measurements) depending on the requirement, and an optional *Typified Condition*.

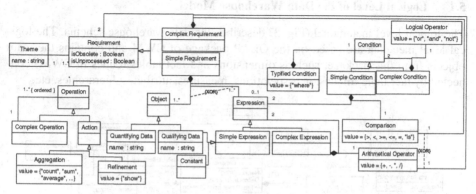

Fig. 2. Requirement formalization metamodel

The algorithm (see Section 6) can be deployed, when changes in requirement metadata take place. All requirements (both new and existing) are analyzed, and *isUnprocessed* changes from "false" to "true" once the requirement is processed. Also, each requirement can be either up-to-date (*isObsolete*="false"), or no longer valid (*isObsolete*="true"). In terms of the algorithm in Section 6, all schema elements that correspond to obsolete requirements are removed from the current data warehouse schema version.

A *Complex Operation* consists of two or more *Actions* divided into *Aggregation* ("roll-up"; for calculation and grouping) and *Refinement* ("drill-down"; for information selection). Just like requirements, conditions and expressions are either *Simple* or *Complex*. A *Complex Condition* joins two or more conditions with a *Logical Operator*. A *Simple Condition* (e.g. "year > 2014", "grade is NULL") consists of a *Comparison* of two expressions. A *Complex Expression* contains two or more expressions with an arithmetical operator in between, whereas a *Simple Expression* belongs either to *Qualifying Data* (e.g. "year") or to *Constants* (e.g. "2014").

5   Multiversion Data Warehouse Metamodel and Schema Changes

The metadata of the data warehouse repository of the framework correspond to the formal model for multiversion data warehouse. Common Warehouse Metamodel (CWM) [15] was used as a basis of the formal model. CWM consists of packages, which describe different aspects of a data warehouse. The formal model describes a data warehouse at the logical, physical and semantic levels. The brief overview of the metamodel of the logical and physical levels follows, however the detailed description of the full metamodel is available in the paper [16].

5.1   Logical Level of the Data Warehouse Model

The logical level metamodel (Fig. 3) describes the data warehouse schema. The logical level metadata are based on the OLAP package of CWM and contains the main objects from this package, such as dimensions and fact tables (cubes in CWM) connected by fact-table-dimension associations, measures, attributes, hierarchies, etc.

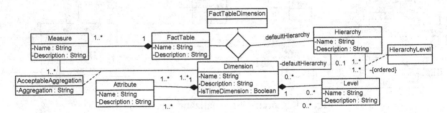

Fig. 3. Logical level model

To reflect multiple versions of a data warehouse schema, two objects were introduced: *SchemaVersion* and *VersionTransformation* (Fig. 4), which are not included in CWM. An object *SchemaVersion* corresponds to a data warehouse schema version, which is created as a result of some change in a data warehouse schema. Each schema version has a validity period defined by attributes *ValidFrom* and *ValidTill*. Each version, except for the first one, has a link to a previous version.

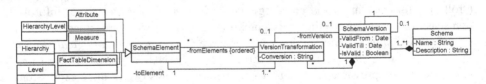

Fig. 4. Schema versions at the logical level

Elements of a version are connected to the schema version by the *VersionTransformations*. The association *toElement* connects an element of the current version. A *Schema Element* can be any element of the logical metamodel, for example,

Measure, FactTable, etc. If an element remains unchanged it is connected to several versions through the version transformations. The attribute *Conversion* stores a function that obtains a changed element from elements of other version. Elements of other version, which are used to calculate the changed element of a new version, are connected to version transformation by the association *fromElements* and the corresponding version is connected by the association *fromVersion*.

5.2   Physical Level of the Data Warehouse Model

The physical level of the data warehouse model describes relational database schema of a data warehouse and mapping of a multidimensional schema to relational database objects from the logical level. The model of physical level is shown in Figure 4. Physical metadata do not include versioning information because in the database there is only one schema version and versioning is implemented at the logical level. The physical level metadata are based on the Relational package of CWM. The objects of physical and logical levels are connected by objects defined in the Transformation package of CWM. This means that attributes of dimensions and measures of fact tables are defined by *Mappings*, which specify formulas that obtain attributes and measures from one or several columns of physical tables and views.

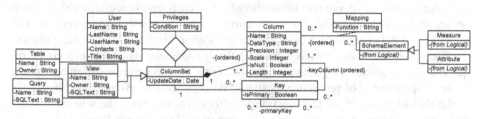

Fig. 5. Physical level model

5.3   Data Warehouse Schema Changes

The data warehouse evolution framework supports three types of data warehouse schema changes, which are distinguished according to their operation: physical, logical and semantic changes. Physical changes (i.e. addition of a new attribute to a dimension) operate with database objects and an instance of the data warehouse physical level model. Logical changes (i.e. connection of a dimension to a fact table) modify mainly an instance of the data warehouse logical level model. Semantic changes (i.e. change of a meaning of an attribute) can adjust an instance of the data warehouse semantic level model, as well as an instance of the logical level model.

In the evolution framework, each schema change is implemented as a special procedure or function that makes all necessary changes in the metadata and the database. The operation of each change procedure and its impact on the formal model is formally described in the paper [16].

6   Algorithm for Handling Requirements

We propose an algorithm for semi-automatic generation of data warehouse evolution changes based on the changes in requirements. The algorithm utilizes the requirements formalization metamodel to obtain new data warehouse requirements and find out about requirements that became obsolete.

The main idea of the algorithm is to check whether there is a mapping of objects used to describe a requirement in the existing data warehouse schema, and to create a new data warehouse schema version that corresponds to the evolved requirements, if objects used in a requirement are not mapped. To propagate changes in requirements in the data warehouse schema, the algorithm takes advantage of the data warehouse logical and physical level metamodel and the requirement formalization metamodel, as well as the information about available source database elements.

6.1   Processing of New Requirements

At first, the set of unprocessed new (not obsolete) simple requirements of a certain theme is processed by the procedure *ProcessNewRequirements* (Fig. 6). Complex requirements are not reviewed, because objects, operations, expressions, and conditions are defined in simple requirements only. For each simple requirement, the procedure searches for mapping of it in the current data warehouse schema version. The mapping is a schema element described in the logical level metamodel that contains data necessary to satisfy the requirement. We assume that names of objects of requirements are unique and in case of similar terms used to describe different objects, some pre-processing of requirements should be conducted to avoid erroneous results of the algorithm operation. The function *MapRequirement* returns the schema element that maps the given simple requirement (or part of it) if anything is found.

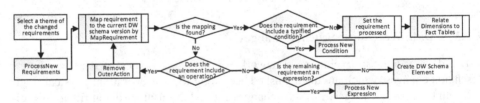

Fig. 6. Procedure Process New Requirements

If the mapping is found, i.e. the object with the operation used in the requirement is already present in the schema, nothing should be done. If the corresponding mapping is not found, the procedure *RemoveOuterAction* removes "outer" operation (for example: "sum") and the check is repeated. This is done with the purpose to construct a data warehouse schema with the lowest possible granularity sufficient for the requirement. Finally, if no operation that is applied to an object or expression remains in the simple requirement and the mapping of it is still not found in the current schema version, then it is checked whether the simple requirement is expression or just an object.

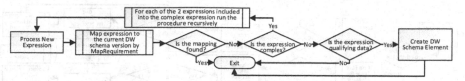

Fig. 7. Procedure Process New Expression

If the processed requirement is an expression then the procedure *ProcessNewExpression* (Fig. 7) searches for mappings of the whole expression or parts of it (if the expression is complex) in the logical level model. If the mapping is not found and the expression is complex, i.e. it is calculated from 2 expressions by arithmetic operator, the procedure *ProcessNewExpression* is executed recursively for both expressions-operands. If the expression is simple, i.e. it does not consist of other expressions that may be mapped, the procedure checks weather the expression is qualifying data and in such a case, the procedure *CreateDWSchemaElement* (Fig. 8.) is executed. This procedure is universal and is used in other parts of the algorithm, therefore, it identifies the type of the input object, which can be qualifying data or quantifying data.

If the object is qualifying data, it corresponds to the attribute and, therefore, the physical change 'Addition of a new attribute to a dimension' should be applied. The function *AddNewAttributeToDimension*, which implements the change, requires the dimension, to which the new attribute is added. Therefore, for the qualifying data, the function *FindDimensionForAttribute* searches for the existing dimension that can contain the processed qualifying data. To do this, the function can use two approaches. First, the dimension with the same name as the processed object can be used if such exists. Second, the function can take advantage of the information about the source tables in the ETL metadata and search for the dimension that contains attributes that are mapped to the columns of the data warehouse table at the physical level model, which corresponds to the source table, which contains the column with the same name as the processed object. If no dimension is found using both approaches, the new dimension with the same name as the attribute is created by the function *CreateNewDimension*.

After adding a new attribute to a dimension, a new data warehouse schema version must be created that contains the new attribute. To make it possible to run reports that use the new attribute and span several schema versions, including ones that do not include the added attribute, the version transformation, which defines how the new attribute is calculated from other attributes, is obtained by the function *GetTransformation*. This algorithm step is not automatic and requires developer's participation. If it is impossible to calculate the new attributes from other attributes, the function returns null and no transformation is constructed.

After that the schema change function *AddNewAttributeToDimension* can be executed with the existing or the new dimension and the version transformation as input parameters. If the dimension is the existing one and contains hierarchies, it is possible that the new attribute should be added to the new or existing level of a hierarchy. Therefore, the procedure *UpdateHierarchies* appends the new attribute to the dimension hierarchies. Here the information about the relationships between the source tables can be used and some data-driven algorithm (e.g. [17]) can be applied.

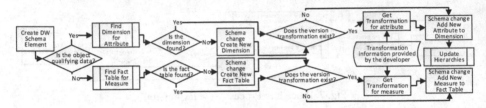

Fig. 8. Procedure Create DW Schema Element

If the object is quantifying data, it corresponds to the measure, and, therefore, the physical change 'Addition of a new measure to a fact table' should be applied. Similar to the addition of a new attribute, the function *AddNewMeasureToFactTable*, which implements the change, requires the fact table, to which the new measure is added. Therefore, for the quantifying data, the function *FindFactTableForMeasure* searches for the existing fact table that can contain the processed quantifying data. The same approaches used by the function *FindDimensionForAttribute* can also be followed to find the corresponding fact table. If no existing fact table is found for the measure, the function *CreateNewFactTable* is executed to create a new fact table with the same name as the name of the measure. The function *GetTransformation* is run to obtain the possible version transformation for the new measure and finally the function *AddNewMeasureToFactTable* creates a new measure with or without version transformation and adds it to the new or existing fact table.

The procedure *CreateDWSchemaElement* is also used to create a new data warehouse schema element in case when after removal of all operations from the simple requirement processed by the procedure *ProcessNewRequirements* (Fig. 6), only the object remains.

After processing the object or expression of the requirement, the procedure *ProcessNewRequirements* deals with conditions, if the requirement includes those. For that purpose, the procedure *ProcessNewCondition* (Fig. 9) is executed. If the condition is complex, i.e. it relates 2 other conditions by the logical operator, the procedure is launched recursively for each of these parts of the complex condition. Otherwise, if the processed condition is simple, i.e. it compares 2 expressions, the procedure *ProcessNewExpression* is executed for each of these expressions.

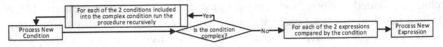

Fig. 9. Procedure Process New Condition

After processing all parts of the simple requirement, new dimensions and fact table may be created that are not mutually connected by the *FactTableDimension* association in the logical level model. Therefore, the procedure *RelateDimensionsFactTables* (Fig. 6) is run to execute the logical schema change 'Connection of a dimension to a fact table' for every new dimension and fact table created when processing the simple requirement. Finally the requirement status is set to "Processed".

6.2   Processing of Obsolete Requirements

When all schema elements necessary for the new requirements are added to the data warehouse schema, the next step is to process obsolete requirements. For this purpose the set of unprocessed obsolete simple requirements is treated by the procedure *ProcessObsoleteRequirements* (Fig. 10). At first, the procedure searches for the mapping of the requirement with or without operations in the current data warehouse schema version the same way as the procedure *ProcessNewRequirements* does it. If the mapped schema element is found, it is added to the array of schema elements used in the requirement.

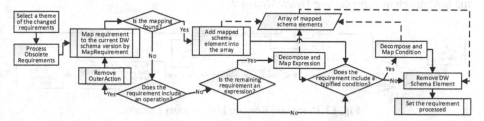

Fig. 10. Procedure Process Obsolete Requirements

If the mapping is not found and the requirement consists of an expression, the function *DecomposeAndMapExpression* (Fig. 11) is executed. This function returns the array of data warehouse schema elements used to calculate the expression. The function searches for the mapping of the input expression and if the mapping is not found and the expression is complex, the function is executed recursively for both expressions that are used in the calculation of the input complex expression.

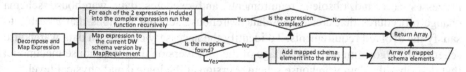

Fig. 11. Function Decompose and Map Expression

After the processing of the object or expression of the requirement, the procedure *ProcessObsoleteRequirements* launches the function *DecomposeAndMapCondition* (Fig. 12), which returns the array of schema elements used to build a condition if the requirement includes one. This function checks weather the condition is complex and if it is, the function is executed recursively for both conditions that are united into the input condition. If the condition is not complex, then it compares 2 expressions and schema elements used in both of these expressions are obtained by the function *DecomposeAndMapExpression*.

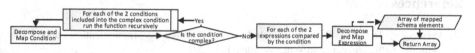

Fig. 12. Function Decompose and Map Condition

When all schema elements used to obtain the object or expression of the requirement as well as the condition are gathered into the array in the previous steps, every
schema element in the array is processed by the procedure *RemoveDWSchemaEle-
ment* (Fig. 13). At first, this procedure verifies that there are no actual (not obsolete)
requirements where the processed schema element is used and it is possible to remove
the schema element from the schema version. In such case, the further processing
depends on the type of the element. If it is an attribute, it is disconnected from all
related hierarchy levels by the schema change procedure *DisconnectAttributeFrom-
Level* and is deleted from a dimension by the schema change procedure *DeleteDimen-
sionAttribute*. If the schema element is a measure, it is deleted from the fact table by
the schema change procedure *DeleteMeasure*.

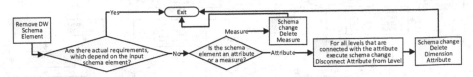

Fig. 13. Procedure Remove DW Schema Element

7   Conclusions

In this paper we have presented an approach to handle changes in data warehouse
requirements that are formalized according to the requirements formalization metamodel constructed based on the investigation of different key performance indicators.

The contribution of this paper is the semi-automatic algorithm for adjustment of
data warehouse in accordance with evolving business requirements. The algorithm
processes new and obsolete requirements and executes data warehouse schema
change procedures that create new data warehouse schema versions that are able to
satisfy the evolved requirements. To identify the necessary changes in data warehouse
schema, the algorithm takes advantage of the multiversion data warehouse metamodel
that describes the data warehouse schema version at the logical and physical level.

The algorithm, the requirements formalization metamodel and the multiversion data
warehouse metamodel are elements of the data warehouse evolution framework that is
able to handle not only changes in business requirements and in data sources, but also it
allows users to run reports on one or several data warehouse schema versions.

Several directions for future research related to the presented issue are automatic or
semi-automatic generation of reports based on data in the requirements formalization
metamodel and incorporation of non-functional requirements of a data warehouse into
our proposed approach.

References

1. Parmenter, D.: Key Performance Indicators: Developing, Implementing, and Using
   Winning KPIs, 2nd edn. Jon Wiley & Sons, Inc. (2010)

2. Winter, R., Strauch, B.: A method for demand-driven information requirements analysis in data warehousing projects. In: Proceedings of the 36th Annual Hawaii International Conference on System Sciences, p. 9. IEEE (2003)

3. Frank, U., Heise, D., Kattenstroth, H., Schauer, H.: Designing and utilizing business indicator systems within enterprise models - outline of a method. In: Loos, P., Nüttgens, M., Turowski, K., Werth, D. (eds.) Modellierung Betrieblicher Informationssysteme (MobIS 2008) - Modellierung zwischen SOA und Compliance Management, LNI, Saarbrücken, Germany, vol. 141, pp. 89–105 (2008)

4. Popova, V., Treur, J.: A Specification Language for Organizational Performance Indicators. Applied Intelligence Journal **27**(3), 291–301 (2007)

5. Golfarelli, M., Lechtenbörger, J., Rizzi, S., Vossen, G.: Schema versioning in data warehouses: Enabling cross-version querying via schema augmentation. Data Knowl. Eng. **59**(2), 435–459 (2006)

6. Vaisman, A.A., Mendelzon, A.O., Ruaro, W., Cymerman, S.G.: Supporting dimension updates in an OLAP server. Inf. Syst. **29**(2), 165–185 (2004)

7. Bebel, B., Eder, J., Koncilia, C., Morzy, T., Wrembel, R.: Creation and management of versions in multiversion data warehouse. In: SAC. pp. 717–723 (2004)

8. Malinowski, E., Zimányi, E.: A Conceptual Model for Temporal Data Warehouses and Its Transformation to the ER and the Object-Relational Models. Data Knowl. Eng. **64**(1), 101–133 (2008)

9. Jovanovic, P., Romero, O., Simitsis, A., Abelló, A.: ORE: an iterative approach to the design and evolution of multi-dimensional schemas. In: Proceedings of the 15th International Workshop on Data Warehousing and OLAP, pp. 1–8. ACM (2002)

10. Thenmozhi, M., Vivekanandan, K.: An Ontological Approach to Handle Multidimensional Schema Evolution for Data Warehouse. International Journal of Database Management Systems **6**(3), 33–52 (2014)

11. Thakur, G., Gosain, A.: DWEVOLVE: a Requirement Based Framework for Data Warehouse Evolution. ACM SIGSOFT Software Engineering Notes **36**(6), 1–8 (2011)

12. Solodovnikova, D.: Data warehouse evolution framework. In: Proceedings of Spring Young Researcher's Colloquium on Database and Information Systems, Moscow, Russia (2007)

13. Niedritis, A., Niedrite, L., Kozmina, N.: Performance measurement framework with formal indicator definitions. In: Grabis, J., Kirikova, M. (eds.) BIR 2011. LNBIP, vol. 90, pp. 44–58. Springer, Heidelberg (2011)

14. Kozmina, N., Niedrite, L.: Extending a metamodel for formalization of data warehouse requirements. In: Johansson, B., Andersson, B., Holmberg, N. (eds.) BIR 2014. LNBIP, vol. 194, pp. 362–374. Springer, Heidelberg (2014)

15. Object Management Group. Common Warehouse Metamodel Specification, v1.1. http://www.omg.org/cgi-bin/doc?formal/03-03-02

16. Solodovnikova, D.: The Formal Model for Multiversion Data Warehouse Evolution. Selected Papers from the 8th International Baltic Conference DB&IS, pp. 91–102. IOS Press (2008)

17. Golfarelli, M., Maio, D., Rizzi, S.: Conceptual design of data warehouses from E/R schemes. In: Proc. of 31st Hawaii Int. Conf. on System Sciences, pp. 334–343. IEEE, USA (1998)

Querying Multiversion Data Warehouses

Waqas Ahmed[1,2](✉) and Esteban Zimányi[1]

[1] Department of Computer & Decision Engineering (CoDE),
Université libre de Bruxelles, Brussels, Belgium
{waqas.ahmed,ezimanyi}@ulb.ac.be
[2] Institute of Computing Science, Poznan University of Technology, Poznan, Poland

Abstract. Data warehouses (DWs) change in their content and structure due to changes in the feeding sources, business requirements, the modeled reality, and legislation, to name a few. Keeping the history of changes in the content and structure of a DW enables the user to analyze the state of the business world retrospectively or prospectively. Multiversion data warehouses (MVDWs) keep the history of content and structure changes by creating multiple data warehouse versions. Querying such DWs is complex as data is stored in multiple schema versions. In this paper, we discuss various schema changes in a multidimensional model, and elaborate their impact on the queries. Further, we also propose a system to support querying MVDWs.

1 Introduction

Data warehouses (DWs) store historical, subject-oriented, and often heterogeneous data that is fed by external data sources (EDSs). An inherent characteristic of these data sources is that they change in their content and structure independently of the DW that integrates data from them.

Data warehouses are modeled as multidimensional (MD) cubes and analytical applications query these cubes to produce reports. A MD cube consists of facts, measures and dimensions. A *fact* is a focus of analysis and a *measure* is an associated numerical value that quantifies the fact. For example, the analysis of sales in a store is a fact and the quantity of items sold can be a measure to analyze the sales. *Dimensions* provide various perspectives to analyze a fact. For example, a dimension customer can be used to analyze the sales made by a store to a particular age group of customers. Dimensions consist of discrete alphanumeric attributes which are organized in *hierarchies*. Each set of distinct attributes of a dimension hierarchy is called a *level*. Hierarchies allow decision-makers to analyze measures at various levels of detail. An example of a hierarchy that belongs to dimension geography is store→city→region where store, city, and region are the levels of hierarchy and each level stores specific characteristics about the dimension. Instances of a level are called *members*.

W. Ahmed—This research is partially funded by the Erasmus Mundus Joint Doctorate IT4BI-DC

The MD model was based on the assumption that the content and structure of a DW remain fixed, but the practice has proved this assumption wrong. The content and structure of a MD cube may change due to changes in feeding sources, the business requirements, the modeled reality, the analysis requirements, and legislation, to name a few. Maintaining the history of changes in the content and structure of a DW enables users to analyze the state of the business world in the past or future, and in some cases, it may be required for audit and accountability purposes. A change in the value of dimension attribute is an example of content change, whereas a change in the structure of a hierarchy is an example of schema change. Three alternative approaches, namely slowly changing dimensions (SCDs) [10], temporal data warehouses (TDWs) [2,11], and multiversion data warehouses (MVDWs) [17,18] have been proposed to address the issue of content and schema changes in DWs. SCDs only handle changes in dimension members. TDWs provide support for storing and querying time-varying dimension members and facts. MVDWs maintain the history of content and schema changes by creating multiple DW versions.

Answering queries that require data from multiple schema versions (called cross-version queries) is not trivial, in particular because of the data in multiple versions may have different structure or the data may be present in one version but missing in another. To address the issue of querying MVDWs, in this paper (1) we provide a detailed discussion about schema changes in the MD model and their impact on user queries, and (2) based on our discussion we propose a system to support querying MVDWs.

The rest of this paper is organized as follows. Section 2 reviews related work. In Sect. 3, we introduce a running example that will be used throughout the rest of the paper. In Sect. 4, we discuss schema changes in the MD model and their impact on user queries. In Sect. 5, we provide three possible approaches to manage schema changes in relational databases, highlight the differences among queries for each approach, and propose a system to support querying MVDWs. Finally, Sect. 6 concludes the paper and provides considerations for future research.

2 Related Work

The challenge of managing schema changes in database systems and querying data across multiple schema versions is not new and the issues related to the problems are discussed in [14]. In the literature, the solutions to managing schema changes in databases are classified into three broad categories [6,14]. *Schema modification* allows changes in the schema but as a result, the existing data may become unavailable. *Schema evolution* supports schema changes while preserving existing data. Finally, *schema versioning* enables to store data in multiple schema versions.

In [3] is presented a system to automate database schema and integrity constraint evolution. The presented PRISM and PRISM++ systems automatically rewrite queries to support legacy applications and data migration. The

PRIMA [13] system also provides a mechanism to archive and query historical data. In [8], the authors examined data and schema evolution in a branched environment where data can evolve simultaneously under various schema versions. The schema definitions are stored as XML-based documents and the data records are stored in relational columns. In all of the these approaches, the challenge of querying data from multiple schema versions is not addressed.

Golfarelli et al. [5] presented an approach that uses a graph-based metaschema to create and query multiple schema versions in a DW. They introduced the concept of an augmented schema to handle the issue of missing data between versions. When the user creates a new schema version from an existing one, an augmented schema is also associated with the old version: It is the most generic schema containing all the elements from both the new and the old versions. In [19], the authors also presented a metadata-based version management system for MVDWs. In both of the above mentioned approaches, to answer a cross-version query, firstly the user query is converted into individual queries against each version, and then the results of these individual queries are combined and presented to the user.

The model presented in [4] supports changes in the structure of dimension members and also provides a list of integrity constraints to maintain the consistency of data across multiple versions. Although the authors mentioned about querying multiple schema versions, the question of how to answer cross-version queries remained unanswered.

3   Running Example

Multiversion DWs manage the evolution of their content and structure by creating multiple DW versions. Each DW version consists of a schema version and an instance version. The *schema version* defines the structure of data during a specific period, whereas the *instance version* consists of the data stored using a particular schema version. At a given instant, only one DW version is used to store data and it is called the *current version*. Although it is possible to derive multiple schema versions from the current version, for the sake of simplicity, we only consider the sequential versioning approach [19], in which a new version can be derived by applying changes to the current version only.

Each version has an associated begin application time (BAT) and end application time (EAT) that represent a close-open interval during which a version is used to store data. The interval [BAT, EAT) is called the *validity period* of the version. The EAT of the current version is set to a special value "*UC*" (until changed). A more detailed and formal definition of a MVDW can be found in [1]. Figure 1 shows the multiple versions in an example MVDW (for the moment, ignore the query types in the figure).

We will use the following motivating example throughout the rest of the paper. Our example MVDW is used to analyze the sales of a company. The cube can be modeled as either a star or snowflake schema. In our example, we consider a snowflake schema as it is more difficult to manage its evolution as compared

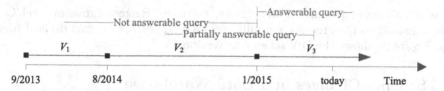

Fig. 1. Three versions of a DW and possible cases for a query computing the value of a schema element present in the current version only

to a star schema [9]. The initial version $V_1[9/2013, UC)$ of the DW was created in September 2013 and consisted of fact Sales and dimensions Time, Geography, and Product. Dimension Geography consisted of levels Store, City, and Region. Dimension Product consisted of levels Product and Category. Figure 2a shows the schema of the DW in version V_1.

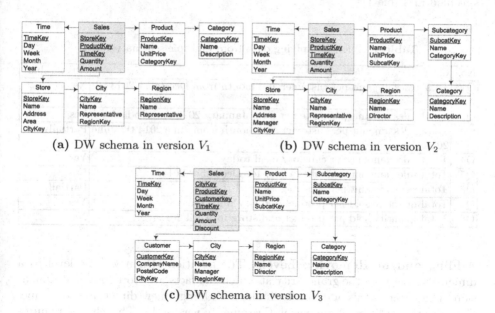

(a) DW schema in version V_1 (b) DW schema in version V_2

(c) DW schema in version V_3

Fig. 2. Schema versions in the example MVDW

In August 2014, the user extended the Product dimension by inserting a new level Subcategory between Product and Category. Also, attribute Area was removed and attribute Manager was added to level Store. These changes resulted version $V_2[8/2014, UC)$. As a consequence, the validity period of V_1 was changed to $[9/2013, 8/2014)$. Figure 2b shows the schema of the DW in version V_2. In January 2015, the user deleted level Store from Geography and made City the base level of the dimension. She also added a new dimension Customer and a measure Discount to Sales. Furthermore, attribute Representative of level City was renamed to Director. As a result, version $V_3[1/2015, UC)$ was derived from V_2. Figure 2c shows the DW schema in version V_3.

4   Schema Changes in a Data Warehouse

The possible changes to a dimension schema include: (1) adding a new level to a dimension, (2) deleting a level from a dimension, (3) adding an attribute to a level, (4) removing an attribute from a level, (5) renaming an attribute of a level, and (6) changing the domain of an attribute of a level. Similarly, the possible changes to a fact schema include: (1) adding a new dimension to the fact, (2) deleting a dimension from the fact, (3) adding a new measure to the fact, (4) deleting a measure from the fact, (5) renaming a measure in the fact and, (6) changing the domain of a measure in the fact. We discuss next the impact of these changes on querying a MVDW. For the discussion, we will use the queries specified in Table 1.

Table 1. Queries implying data from multiple schema versions

No.	Query	Answerable
Q1	Total sales amount per customer and month from December 2013 till August 2014.	No
Q2	Total sales amount per customer from January 2015 until today.	Yes
Q3	Total sales amount per customer and month from June 2014 till June 2015.	Partially
Q4	Total sales amount per category until today.	Yes
Q5	Total sales amount per subcategory for December 2014.	No
Q6	Total sales amount per subcategory until today.	Partially
Q7	Total quantity sold per product and city until today	Yes
Q8	Total quantity sold per product and store in April 2015	No

Adding and/or Removing Levels. The addition of a new base level to a dimension increases the granularity at which the facts are stored for that dimension. The effect of this addition is similar to adding a new dimension to a cube. Consider the addition of dimension Customer in version V_3. This change requires not only adding a new dimension but also a new version of the fact so that all newly added facts may have a customer dimension member associated with them. It is worth noticing that the customer information related to sales will not be available for the facts stored before January 2015 because the earlier versions did not include the customer dimension. Query Q1 cannot be answered because dimension Customer did not exist between December 2013 and August 2014 and the sales facts stored for this period do not have any customer members associated with them. Query Q2 can be answered without any problem because sales facts from January 2013 until today have associated customer members. However, query Q3 can be answered partially, only from January 2015 onwards.

The addition of a new level to a dimension other than at the base level does not require the creation of a new fact version. Consider the schema change where level Subcategory was added between Product and Category. We assume that the user does not assign the existing products to the subcategories and only newly added products are assigned to the subcategories. Now consider query Q4, which computes the total sales amount per category until today. Since the product to category relation exists in all the versions of DW, the query can be answered and the result set can be computed by traversing through two different lattices, i.e., from September 2013 till July 2014 this information is directly available, whereas from August 2014 until today, the products' categories can be obtained from the subcategories.

The effect of deleting a base level is almost similar to adding a new dimension because a new base level of the dimension has to be defined. In our example, in version V_3, level Store was deleted from dimension Geography and City was made the new base level. Consider query Q7, which requires to compute the total quantity of a given product sold per city. The query can be answered completely because in earlier versions V_1 and V_2, measure Quantity in fact Sales can be related to the cities through level Store, whereas in the current version this information is directly available. In case of query Q8, level Store ceased to exist after January 2015, hence it cannot be answered.

Adding and/or Removing Attributes. The schema of dimensions change by adding or removing attributes in a level. For example, in version V_2, attribute Area is deleted from level Store and Manager is added to the same level. The value of attribute Area will not be available for store members that were stored after August 2014. Similarly, the information about the managers of stores, which existed in V_1, will not be available.

Other changes in the attributes such as, changes in the domain of an attribute, or renaming an attribute are trivial and can be handled either by modifying the existing attribute or adding a new attribute for each change. However, in case of renaming an attribute, it is important to keep track of attribute names in each version so that cross-version queries could be supported. In V_2, attribute Representative of level City was renamed to Director and it is important to keep the record of this change to support the queries asking for Representative and/or Director during the validity period of any of these two versions.

Adding or Removing Dimensions. The addition or removal of a dimension from a fact changes the schema of the fact. It is similar to adding or removing a base level of a dimension. The effect of changes in the measures is similar to changes in the attributes of dimensions and such changes can be managed in the same manner as in case of changes in the attributes.

To sum up our discussion in this section, consider a query Q that computes the value of a schema element E over a period P, such that E is present in the current version only. There are three possibilities for Q based on P: (1) if P is contained within the validity period of the current version then Q is answerable, (2) if P overlaps with the validity period of current version, Q is

Store Key	Product Key	Time Key	Quantity	Amount
s1	p1	t1	5	20
s2	p1	t1	3	15
s3	p2	t2	2	18
s4	p2	t3	3	9

Store Key	City Key	Product Key	Time Key	Cust. Key	Quantity	Amount	Discount
s1	null	p1	t1	null	5	20	null
s2	null	p1	t1	null	3	15	null
s3	null	p2	t2	null	2	18	null
s4	null	p2	t3	null	3	9	null
null	c1	p3	t4	cu1	6	8	0.10

(a) Sales in V_2 using the STV approach   (b) Sales in V_3 using the STV approach

City Key	Product Key	Time Key	Cust. Key	Quantity	Amount	Discount
c1	p3	t4	cu1	6	8	0.10

(c) Sales in V_3 using the MTV approach

Fig. 3. State of the Sales table in various versions using the STV and MTV approaches

partially answerable, and (3) otherwise, it is not possible to answer Q because the value of E is unavailable. Figure 1 shows these three cases.

5   Manipulating Multiversion Data Warehouses

Three approaches [1,16] have been proposed for storing multiple versions of a data warehouse, namely, (1) single-table version (STV), (2) multiple-table version (MTV), and (3) hybrid-table version (HTV). In this section, we briefly discuss each of these approaches and their impact on multiversion queries.

In the STV approach, each table has only one version throughout the lifetime of the DW. Each table contains all attributes that have ever been defined for it. This means that it is an append-only approach and the new attributes are added to the table. In this approach, the schema of the DW is always growing. For every new record, a default or null value is stored for the deleted attributes. Figures 3a and 3b show fact Sales in versions V_2 and V_3, respectively. Notice that new attributes CityKey and CustomerKey are added to link new levels City and Customer. Similarly, a new measure Discount is added. The sales records stored using earlier versions were not linked to the customers and cities, therefore for those records null values are stored in these attributes. Although level Store is deleted in the new version, the corresponding attribute StoreKey in Sales is not deleted and all new records store a null value for it.

In the MTV approach, each change in the schema of a table produces a new version of the table and new data is stored using this new version. Contrary to the STV approach, attributes can be added or removed in the new version. Figure 3c shows the Sales in V_3 using the MTV approach. Notice that the deleted attribute StoreKey does not exist in the new version. The MTV approach avoids the space overhead which is incurred in case of the STV approach but in some scenarios, it may require creating of fact versions because of new dimension versions. For example, if a new version of a table corresponding to the base level of a dimension is created, this requires to create a new fact version as well. Furthermore, all valid dimension members from the previous dimension version must be inserted into the new version so that they could be linked to the incoming facts, if needed.

The HTV approach combines the advantages of the STV and the MTV approaches. It creates a new version only in case of the schema changes in facts but maintains a single version for dimension tables. In this way, the problem of space overhead and complexity of managing multiple dimension versions and members loading can be avoided. In our example, in version V_3, fact Sales consists of two tables shown in Fig. 3c and 3a, respectively. However, the Product and Store tables will consist of a single version as in the STV approach.

Analytical queries are complex in nature since they often involve aggregation of data and joins across multiple dimensions. Querying a MVDW is even more complex since the data may be stored across multiple versions with different structure. Furthermore, as discussed above, queries to a MVDW also depend upon the approach used for storing the various versions. For example, if the user of our example MVDW is interested in average yearly sales for each city, then for the STV approach, it can be computed by query shown in Fig. 4a. The same query is different for the MTV approach because as a result of the first schema change, a new version of Store was created. To link this new version with the fact, a new version of the fact was also created. The second schema change also produced another version of Sales. The query for the MTV approach should consider all these three versions of the fact to aggregate the yearly sales amount per city. Figure 4c shows the SQL query for the MTV approach. In case of HTV, only the new fact versions are created therefore, the query shown in Fig. 4b is simple as compared to the one for the MTV approach.

To make the user queries independent of the version management approach, we propose that each DW version is defined as set of views [12]. Figures 5a and 5b show the views defining the schema of level Store in version V_1 and V_2 in the STV and the HTV approaches, respectively. Since in case of the MTV approach a new table is created, the view definition will be different. Figure 5c shows the view defining the schema of level Store in version V_2. Obviously, the view definitions depend upon the approach used for storing the various versions (i.e., STV, MTV, or HTV), but once defined, such views may serve as virtual tables and thus user queries can be rewritten in term of such views [7,15] without considering the underlying the version management approach.

In addition to the versions defined as views, the system also maintains the metadata to support the coross-version queries. The metadata consists of the mappings between views, columns in the views, columns data types, and column names from one version to the other. In Fig. 6 we shows the partial metadata of our example MVDW. The mapping V_2 : Sales_V1 $\rightarrow$ Sales_V2 denotes that in version V_2, view Sales_V2 represents the same level that was represented by view Sales_V1. Similarly, the mapping V_3 : City_V1.Representative $\rightarrow$ City_V2.Manager denotes that in version V_3 column Representative of view City_V1 is renamed to Manager. Notice that this renaming resulted in the creation of a new version of level City and consequently a view City_V2. In our adopted numbering convention,

```
SELECT      AVG(Amount) as "Avg. Sales Amount", C.Name , T.Year
FROM        Sales S, Store E, City C, Time T
WHERE       S.TimeKey = T.TimeKey and S. StoreKey = E. StoreKey and E. City = C. CityKey or
            S. CityKey=C. CityKey
GROUP BY    C.Name, T.Year
```

```
WITH        SalesByCity AS (
SELECT      S.Amount, C.Name , T.Year FROM Sales_V1 S, Store E, City C, Time T
WHERE       S.TimeKey = T.TimeKey and S. StoreKey = E. StoreKey and E. City = C. CityKey
UNION ALL
SELECT      S2.Amount, C.Name , T.Year FROM Sales_V2 S2, City C, Time T
WHERE       S2.TimeKey = T.TimeKey and S2. CityKey = C.CityKey )
SELECT      AVG(Amount) as "Avg. Sales Amount", C.Name , T.Year FROM SalesByCity
GROUP BY    C.Name, T.Year
```

```
WITH        SalesByCity AS (
SELECT      S.Amount, C.Name , T.Year FROM Sales_V1 S, Store_V1 E, City C, Time T
WHERE       S.TimeKey = T.TimeKey and S. StoreKey = E. StoreKey and E. City = C. CityKey
UNION ALL
SELECT      S2.Amount, C.Name , T.Year FROM Sales_V2 S2, Store_V2 E2, Time T
WHERE       S2.TimeKey = T.TimeKey and S2.StoreKey = E2.StoreKey and E2.CityKey = C.CityKey
UNION ALL
SELECT      S3.Amount, C.Name , T.Year FROM Sales_V3 S3, City C, Time T
WHERE       S3.TimeKey = T.TimeKey and S3. CityKey = C.CityKey )
SELECT      AVG(Amount) as "Avg. Sales Amount", C.Name , T.Year FROM SalesByCity
GROUP BY    C.Name, T.Year
```

Fig. 4. Average sales amount per city per year: SQL queries for the STV, MTV, HTV approaches, respectively

the version number in the view is independent of the schema version number. Consider the query Q7 from Table 1, which computes total quantity of each product until today. Suppose that the user writes this query corresponding to current schema version but actually it requires data which was stored using all three schema versions of the DW. The schema mappings in the metadata provides the information that views Sales_V1, Sales_V2, and Sales_V3 represent the same fact Sales in all the schema versions, thus Q7 can be rewritten in term of these views.

Now consider query Q3 which requires total sales amount per customer and month from June 2014 till June 2015. The customer information exists only for the facts starting from January 2015. Ideally, the system should inform the user that though he/she asked the total sales for the customers from June 2014 , the answer is partial and it excludes the facts stored before January 2015 because dimension Customer did not exist in versions V_1 and V_2.

To support querying MVDWs we propose the architecture of a system shown in Fig. 7. As a first step to answer a cross-version query, the system determines the schema versions to be used. Either a user can explicitly specify a version(s) or they can be obtained from the time interval mentioned in the query. Depending upon the schema elements in the query and versions involved, the system determines whether the query is answerable, partially answerable, or not answerable at all. The metadata will be used to determine if there are any renamed

```
CREATE VIEW Store_V1 AS (
SELECT  StoreKey, Name, Area, CityKey
FROM    Store )
```

```
CREATE VIEW Store_V2 AS (
SELECT  StoreKey, Name, Manager, CityKey
FROM    Store )
```

(a) Store in version V_1 using the STV, MTV, and HTV approaches

(b) Store in version V_2 using the STV and HTV approaches

```
CREATE VIEW Store_V2 AS (
SELECT  StoreKey, Name, Manager, CityKey
FROM    Store_2 )
```

(c) Store in version V_2 using the MTV approach

Fig. 5. Views defining the schema of level Store in two schema versions

View Mappings	Column Mappings
V_2 : Sales_V1 → Sales_V2	V_2 : Sales_V1.Amount → Sales_V2.Amount
V_2 : Store_V1 → Store_V2	V_2 : ...
V_2 : Time_V1 → Time_V1	V_3 : City_V1.Representative → City_V2.Manager
...	...

Fig. 6. Metadata of the example MVDW

User query   Result set

Query Rewriter: Answers user query using defined views and annotates a result set with metadata	
Versions defines as Views	Metadata
Versions stored in RDBMS using STV, MTV or HTV	

Fig. 7. Proposed system to answer cross-version queries

attributes in the involved versions. Since each DW version is defined as a set of views, in the next step the user query is rewritten in term of the views belonging to the schema versions, which were determined in the first step. Finally, the query is executed and the result set is annotated with information such as missing or partial data.

6 Conclusions

The capability to maintain the history of changes in content and structure of a DW enables the user to recreate the state of the business world in the past or simulate the effect of a prospective change. Multiversion data warehouses (MVDWs) provide such capability by creating multiple schema and data versions. However, querying data from multiple schema versions is not a straightforward task. Further, the creation of each version may change the queries written for the already existing versions. In this paper, we discussed how changes in a multidimensional model impact on queries. Further, we showed that queries in a MVDW are dependent upon the version management approach. Since it would be convenient from the end user viewpoint to query a DW independently of the version management approach, we proposed a system to query data from the MVDWs.

One disadvantage of MVDWs is that they create multiple data versions to manage the content changes. This can further complicate the tasks of maintaining and querying a MVDWs, and may negatively impact performance of the system. One natural solution to the challenge of managing content and schema changes in a DW is to combine both, the temporal and multiversion data warehouses. As future work, we plan to combine the two approaches as a single solution which will be able to query multiple data versions stored in multiple schema versions. Further we want to present our solution as a data warehouse with temporal and multiversion functionality.

References

1. Ahmed, W., Zimányi, E., Wrembel, R.: A logical model for multiversion data warehouses. In: Bellatreche, L., Mohania, M.K. (eds.) DaWaK 2014. LNCS, vol. 8646, pp. 23–34. Springer, Heidelberg (2014)
2. Ahmed, W., Zimányi, E., Wrembel, R.: Temporal data warehouses: Logical models and querying. In: Proc. of EDA, pp. 33–47 (2015)
3. Curino, C., Moon, H.J., Deutsch, A., Zaniolo, C.: Automating the database schema evolution process. VLDB Journal 22(1), 73–98 (2013)
4. Eder, J., Koncilia, C., Morzy, T.: The COMET metamodel for temporal data warehouses. In: Pidduck, A.B., Mylopoulos, J., Woo, C.C., Ozsu, M.T. (eds.) CAiSE 2002. LNCS, vol. 2348, p. 83. Springer, Heidelberg (2002)
5. Golfarelli, M., Lechtenbörger, J., Rizzi, S., Vossen, G.: Schema versioning in data warehouses: Enabling cross-version querying via schema augmentation. Data & Knowledge Engineering 59(2), 435–459 (2006)
6. Golfarelli, M., Rizzi, S.: A survey on temporal data warehousing. International Journal of Data Warehousing and Mining 5(1), 1–17 (2009)
7. Halevy, A.Y.: Answering queries using views: A survey. VLDB Journal 10(4), 270–294 (2001)
8. Huo, W., Tsotras, V.J.: Querying transaction–time databases under branched schema evolution. In: Liddle, S.W., Schewe, K.-D., Tjoa, A.M., Zhou, X. (eds.) DEXA 2012, Part I. LNCS, vol. 7446, pp. 265–280. Springer, Heidelberg (2012)
9. Kaas, C., Pedersen, T.B., Rasmussen, B.: Schema evolution for stars and snowflakes. In: Proc. of ICEIS, pp. 425–433 (2004)
10. Kimball, R., Ross, M.: The Data Warehouse Toolkit: The Definitive Guide to Dimensional Modeling. John Wiley & Sons (2013)
11. Malinowski, E., Zimányi, E.: A conceptual model for temporal data warehouses and its transformation to the ER and the object-relational models. Data & Knowledge Engineering 64(1), 101–133 (2008)
12. Medeiros, C.B., Bellosta, M., Jomier, G.: Multiversion views: Constructing views in a multiversion database. Data & Knowledge Engineering 33(3), 277–306 (2000)
13. Moon, H.J., Curino, C., Ham, M., Zaniolo, C.: PRIMA: archiving and querying historical data with evolving schemas. In: Proc. of SIGMOD, pp. 1019–1022 (2009)

14. Roddick, J.F.: A survey of schema versioning issues for database systems. Information & Software Technology **37**(7), 383–393 (1995)
15. Srivastava, D., Dar, S., Jagadish, H.V., Levy, A.Y.: Answering queries with aggregation using views. In: Proc. of VLDB, pp. 318–329 (1996)
16. Wei, H.-C., Elmasri, R.: Schema versioning and database conversion techniques for bi-temporal databases. Annals of Mathematics and Artificial Intelligence **30**(1–4), 23–52 (2000)
17. Wrembel, R.: A survey on managing the evolution of data warehouses. International Journal of Data Warehousing & Mining **5**(2), 24–56 (2009)
18. Wrembel, R.: On handling the evolution of external data sources in a data warehouse architecture. In: Taniar, D., Chen, L. (eds.) Data Mining and Database Technologies: Innovative Approaches. IGI Group (2011)
19. Wrembel, R., Bębel, B.: Metadata management in a multiversion data warehouse. In: Meersman, R. (ed.) OTM 2005. LNCS, vol. 3761, pp. 1347–1364. Springer, Heidelberg (2005)

CUDA-Powered CTBE Algorithm
for Zero-Latency Data Warehouse

Marcin Gorawski[✉], Damian Lis, and Anna Gorawska

Institute of Computer Science, Silesian University of Technology,
Akademicka 16, 44-100 Gliwice, Poland
{Marcin.Gorawski,Damian.Lis,Anna.Gorawska}@polsl.pl

Abstract. The systems dedicated for Zero-Latency Data Warehouses
must meet the growing requirements for the most up-to-date data. The
currently used sequential algorithms are not suited to deal with the pressure on receiving the freshest data. The one-module architecture implemented in current solutions, limits the development opportunities and
increases the risk of critical system failure. In this paper we propose a
new, innovative, multi-modular system that is based on parallel Choose
Transaction by Election (CTBE) algorithm. Additionally we utilize the
CUDA architecture to boost system efficiency, using computing power
of multi-core graphic processors. The aim of this paper is to highlight
pros and cons of such a solution. Performed tests and results show the
potential and capabilities of the multi-modular system, using CUDA
architecture.

Keywords: ETL process · Cuda · Ctbe · Zero-Latency Data Warehouse · Distributed system · Workload balancing unit

1 Introduction

The information extracted from different devices or additional sources are not
always useful for every data warehouse user at the same time. We need to extract
adequate data from the data flow, which requires determination of relations
between queries and updates awaiting to be processed. Based on these relations
it is possible to establish a map of correlations between transactions; however,
to use the map we need to outline which transactions should be fed to the data
warehouse. The order of queries and updates processing and their prioritization
were presented in [4] and [7]. The election algorithms presented in these works
prioritize and set the transactions order based on information such as:

- time of transaction arrival to the system,
- data warehouse partition to which transaction is assigned,
- maps of correlations between transactions,
- user preferences concerning the response time,
- user preferences concerning the relevance of returned data.

© Springer International Publishing Switzerland 2015
T. Morzy et al. (Eds): ADBIS 2015, CCIS 539, pp. 358–367, 2015.
DOI: 10.1007/978-3-319-23201-0_37

The initial concept of the Choose Transaction By Election (CTBE) algorithm was connected with research on a Workload Balancing Unit (WBU) and corresponding ETL process [2] [3] [6] [8] [12] that would meet requirements of Zero-Latency Data Warehouses (ZLDW) and real-time processing constraints [13]. Works carried out in these area resulted in a WBU-based ZLDW system where cloud computing with Windows Azure and Google App Engine were used. Mentioned system was presented in [4], while in [7] we have presented a different approach to the same problem. In [7] we have abandoned the cloud computing concept and proposed parallel processing with additional cache memory usage. Results have shown that the cache memory decreases resource utilization and enables effortless communication between system modules [7]. In the same time we were able to limit vulnerability to intrusions and attacks, while cloud computing did not ensure privacy nor data protection.

In this paper the next generation of the CTBE algorithm and the WBU is presented. We have focused our attention on enhancing computing abilities. Moreover, to increase privacy protection we propose additional data encoding between source devices and Hostimp servers, Hostimp servers identification and verification by web service, and new methods of protection against unauthorized access to the communication schema. In the WBU calculations are performed on the matrices while we have introduced processing on the CUDA architecture [5].

Although, in this paper we have focused on the next generation of the CTBE algorithm and the WBU, the presented system may be adjusted to any structure through proper Hostimp servers implementation.

2   Parallel CTBE Algorithm

The CTBE (Choose Transaction By Election) algorithm [4] [7] is dedicated for the Zero-Latency Data Warehouse systems [1] [9] [10]. It originated from combining the main assumptions of the WINE algorithm [7] with FIFO algorithms. We have modified some assumptions to negative algorithms features. The CTBE algorithm is implemented in a Workload Balancing Unit (WBU), where all transactions are analysed before they are fed to the data warehouse. Differently than in the WINE algorithms, the CTBE algorithm uses additional cache memory, through which all information is transmitted, both going out and going in the WBU.

The CTBE algorithm is based on processing of two types of transactions, i.e. queries and updates, stored in three queues. First queue stores read-only transactions, i.e. queries. While write-only transactions, i.e. updates, are localized in the second queue. The third queue is filled with both types of transactions; hence, it contains all updates and queries stored in first two queues. Queries and updates queues are processed in one thread. The transaction queue is processed independently of two remaining ones in a separate thread. Threads are independent; hence, it is possible to launch only one of them, but to fully utilize capabilities of the CTBE algorithm it is necessary to run both threads at the same time.

Queries processed in the WBU are stored in two separate queues, query Q and transaction T, that is $q_i : q_i \in Q \vee q_i \in T$. Whereas, updates are stored in the T (transaction) queue and in the update queue U, which gives in total $u_j : u_j \in U \vee u_j \in T$. That means that each and every transaction is stored in two queues at the same time, i.e. the transaction queue and update or query queue depending on the transaction type. On figure 1, a division of queries and updates in the queues is presented.

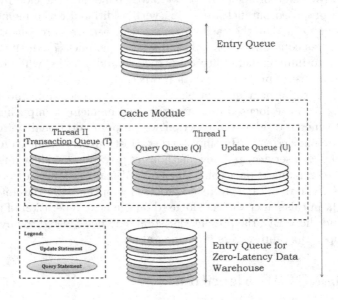

Fig. 1. Schedule of queues used in the system

While placing new updates in the queues, transactions placed in queues T and U are denoted with a timestamp T_u, which determines the time of update arrival to the cache memory. Queries fed to the query queue are additionally denoted with values representing quality of service ($QoS(q_i)$) and quality of data ($QoD(q_i)$). The $QoS(q_i)$ value represents q_i query execution speed expected by the user (expressed as a priority value), while $QoD(q_i)$ denotes expected degree of results 'freshness'. It is assumed that $QoS(q_i) + QoD(q_i) \leq 1$.

In addition, each update in the update queue is analysed in scope of to which data warehouse partition they are assigned [4] [11] (assuming that one update is assigned to a single partition). Queries from the query queue are also analysed in scope of to which data warehouse partition transaction is assigned; however, due to their relation with updates, they may be assigned to more than one partition at the same time. Another assumption is that updates may be referring to only one partition.

2.1 Operations Based on the CUDA Architecture

After the initial analysis stage, transaction's scheduling begins. After determining the number of transactions in queues T, Q and U, fulfilment of the condition $|T| = |Q| + |U|$ is verified. Next stage can be divided into three independent threads; however, during one iteration only one thread can be processed. Threads are selected with respect to the situation:

1. the update queue U or the query queue Q is empty,
2. the update queue U and the queue of queries Q are not empty and $\sum QoS(q_i) \geq \sum QoD(q_i)$,
3. the update queue U and the queue of queries Q are not empty and $\sum QoS(q_i) \leq \sum QoD(q_i)$.

When at least one queue is empty (criteria number 1), transactions in the T queue are scheduled according to the timestamp (T_u if there are only updates to process or T_q if the query queue is the non-empty one). Timestamp denotes the time of arrival of transaction to the cache memory. Then, transaction with the smallest timestamp is transferred to the WBU's output queue and removed from the T queue and from the Q or U queue - depending on type of the transaction, i.e. query or update.

When criteria from 2 are met, the first operation is checking if the sum of values of QoS or QoD is larger for the number of queries in the Q queue. If $\sum QoS(q_i) \geq \sum QoD(q_i)$ is met, queries are granted higher priority then updates. Higher $\sum QoS(q_i)$ value informs that, in a given time, system users are oriented on retrieving information with less emphasis on receiving the most up-to-date data. Therefore, number of queries is increased and number of performed updates is reduced.

When criteria from 3 are met, the problem is the most computationally challenging. When $\sum QoS(q_i) \leq \sum QoD(q_i)$, system users are more oriented on receiving the most up-to-date results. Therefore, updates are granted higher priority and ought to be performed before queries that are associated with the same partition. Due to updates to queries relations creating a map of correlation is essential. The key value in updates scheduling is $w(u_j)$, which depends on 'freshness of data' ($QoD(q_i)$ value) as well as on position in the query queue (pos_{qi}) of queries that are associated with the same partition as u_j update. With the increase in $w(u_j)$ value the update will be inserted to the U queue closer to its head; hence, will be executed faster. The $w(u_j)$ value is calculated according to the following formula:

$$w(u_j) = \sum_{\forall q_i, |Par_{qi} \cap Par_{uj}|=1} \frac{QoD(q_i)}{1 + pos_{qi}} \tag{1}$$

where:

- Par_{qi} : partitions associated with a q_i query,
- Par_{uj} : partition associated with a u_j update,
- pos_{qi} : position of a q_i query in the query queue, i.e. Q.

When value of the intersection $|Par_{qi} \cap Par_{uj}|$ equals 1, it denotes that transactions q_i and u_j are associated with the same partition and are dependant of each other.

Calculation of the $w(u)$ value is the most time-consuming part of the process. The comparison of queries and updates correlations maps is resource demanding as well. It resulted in adjusting the outline of the algorithm to calculations on arrays, which is allowed by parallel calculations using the CUDA architecture. Two main arrays, A and B, are used to calculate the $w(u_j)$ value. Array A, with one row and $|Q|$ columns, is filled with the $QoD(q_i)$ values for each subsequent query from the query queue Q:

$$a_{1i} = QoD(q_i) \tag{2}$$

Array B consists of $|Q|$ rows and $|U|$ columns, and is filled with the values according to formula:

$$b_{ij} = \begin{cases} 0 & \text{,if } Par_{qi}! = Par_{uj} \\ \frac{1}{1+pos_{qi}} & \text{,if } Par_{qi} = Par_{uj} \end{cases} \tag{3}$$

Result array C consists of $|U|$ columns and one row; therefore, for each query we receive separately the $w(u_j)$ value, which is calculated as follows:

$$w(u_j) = \sum_{k=1}^{p} a_{ik}b_{kj} \text{where} \quad i = 1; \quad j = 1, 2, 3, ..., n \tag{4}$$

After calculating the $w(u)$ value, updates are scheduled - descending in relation to the $w(u)$ value and ascending in relation to the timestamp T_u. Figure 1 presents the process of update choice and how the data is transmitted to the data warehouse.

The array calculations enabled utilization of the graphic card processor and reduction of the processing time.

3   Structure of Multi-modular System

The system architecture bases on a modular approach and is presented on Figure 2. Thanks to the multi-modular structure, it is possible to extend or replace any module on the fly, without resetting the whole system. It ensures continuity of work, as well as increases the capacity of specific modules, for example, by increasing its computing power.

Properly programmed system should function without problems, as long as there are no external factors, which may cause its failure. The multi-modular approach allows quick pinpointing of eventual problems and introducing solutions. We have identified following risks that may possibly occur in the presented system:

– module failure,
– modules connection failure,
– an attempt of data theft through unauthorized module access.

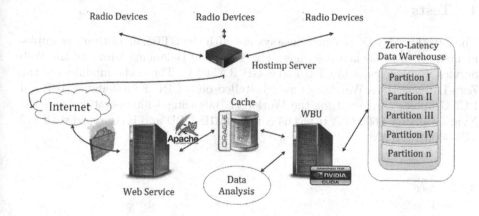

Fig. 2. Outline of the multi-modular system structure

The first module, that starts the data operations sequence, is the Hostimp server. Each Hostimp module should download data stream from external devices, decode it and prepare adequate update. Use of the Hostimp modules allows integration of many types of devices, encoded in a different manner depending on the producer or company responsible for data reading.

Such coding of devices, where only a data stream is sent, makes it impossible to extract data without knowing a table stored on the server. The data can be decoded and read only through direct access to the server. Moreover, Hostimp servers construction and configuration enables data extraction from a wide range of sources (databases, Excel files etc), while preserving privacy.

Second module with the Apache Web Service is responsible for communication within the system. The Web Service allows to construct distributed systems, where modules communicate through computer network, with the use of appropriate remote access protocols. In the presented system, we use the Simple Object Access Protocol (SOAP) protocol approved by the World Wide Web Consortium (W3C). As the additional protection mechanism, system authorizes incoming calls. Each Hostimp server has a login and a password for the remote access to the service. When data authorization fails, i.e. login and password are incorrect, announcement is rejected. However, a source of the announcement always receives a random, false response.

Cache memory is used as in buffer for all transactions (updates and queries) incoming from external systems via the Web Service. All data is stored in tables in a database created on SSD discs. Each stored transaction receives a unique identifier. Therefore, after receiving a response from the data warehouse, it is possible to assign responses to appropriate queries and then transmit them to users. Using the additional memory, cache module allows to reduce load of the main WBU module and consequently enables its better capacity.

4   Tests

The Zero-Latency Data Warehouse system with the CTBE algorithm was implemented on two PC-class machines. The Workload Balancing Unit and the Web Service were run on a i5 CPU and 8 GB RAM PC. The cache module and the Zero-Latency Data Warehouse were installed on a CPU E8400 and 3GB RAM PC. Graphics card used for the Workload Balancing Unit operations was the Nvidia GeForce 9600 GT. It has 64 core 600MHz CPU and is equipped with 512 MB 900MHz RAM.

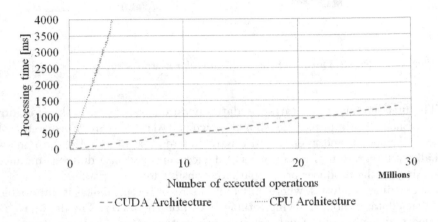

Fig. 3. Comparison of time execution on CUDA and CPU

Over 600 tests were performed during research phase, using diverse data input. Initial tests have shown the capacity of the whole system and time required to perform an operation. Data from the graph in figure 3 allows to observe increase of the CUDA based system capacity. Despite the use of the CPUs multithreading capabilities, result is not satisfactory for the second system. Parallel computational CUDA architecture ensures radical increase of the calculations efficiency, through the GPU power (multi-thread processing).

System based on a classic multi-thread processing CPU, with over 5 million of operations to perform, looks very poor in relation to the new system, and time of processing is increasing over 8 times. The new system, with over 30 million of operations, requires shorter calculation time, with about 2,5 millions of operations to perform.

Another experiment was conducted in order to analyse behaviour of two different types of transactions and their processing time. Figure 4 shows results for the CTBE algorithm, which include data concerning update queue and query queue. For the query queue, number of operations to perform was substantially smaller than the number of operations performed during analysis of the update queue.

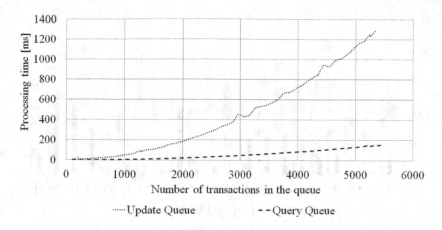

Fig. 4. Comparison of processing time used for Queries and for Updates

On the basis of performed tests and presented results, we have noticed that for operations on queries, it is sufficient to use only the CPU. In addition, for a small number of operations, we should not use graphic cards, due to too much time needed for transfer of data to memory of the card, what was already discussed in the previous publications [4] [7].

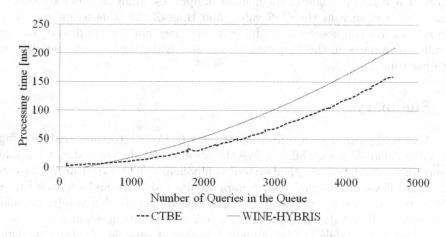

Fig. 5. Comparison of processing times for the WINE-HYBRIS and the CTBE algorithm

The last conducted experiment (figure 5) was a comparison of the previous solution, based on the WINE-HYBRIS algorithm, with a system based on the CTBE algorithm. For this test, we have used an additional set of test data, with no more than 5-6 new queries and updates in total in every iteration. The same data was used for both systems.

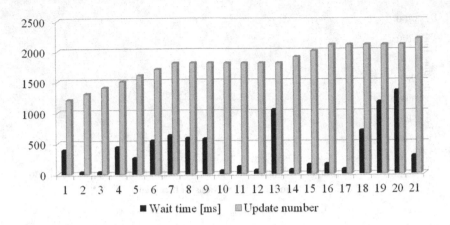

Fig. 6. An impact of number of update on waiting time of queries in the system

Test proved, that the CTBE algorithm is quite close to the WINE-HYBRIS algorithm in terms of processing time. However, 50 ms difference, during in long perspective may save hours and even days . Using the CTBE algorithm leads to minimization of processing time and limits demands for the computing power.

Moreover, wait time for queries in the system was measured (figure 6) and it shown that it is dependant of the number of updates. Many factors have impact time efficiency, such as the QoS value, and time of the transaction arrival to the system. We can observe that the wait time does not depend directly on the number of updates in the U queue. The reason is the data warehouse division into partitions.

5   Summary

The CTBE algorithm with utilization of the CUDA architecture, enables using the GPU calculation capabilities, free the resources of the computers CPU and divert these resources to other system components. The use of arrays in calculations allowed additional reduction of results wait time and decreased time of transaction processing. Using the cache memory enabled free intermodular communication and streamlined constant inflow of new transactions.

In case of a module failure presented system is capable of proper reacting to this failure. If a module was permanently damaged, it can be replaced very quickly without loss of sensitive data, which are the updates flowing to the system, as well as queries coming from the users.

Additional advantage of the multi-modular system is its compatibility with different types of data sources. The Workload Balancing Unit does not work directly on the data contained in transactions but on the parameters describing a given type of transaction. Such approach allows to adjust them to any data warehouse in a very short time, without the need to modify its structure.

References

1. Bruckner, R.M., Tjoa, A.M.: Capturing Delays and Valid Times in Data Warehouses - Towards Timely Consistent Analyses. J. Intell. Inf. Syst. **19**(2), 169–190 (2002)
2. Gorawski, M., Gorawska, A.: Research on the stream ETL process. In: Kozielski, S., Mrozek, D., Kasprowski, P., Małysiak-Mrozek, B. (eds.) BDAS 2014. CCIS, vol. 424, pp. 61–71. Springer, Heidelberg (2014)
3. Gorawski, M., Gorawski, M., Dyduch, S.: Use of grammars and machine learning in ETL systems that control load balancing process. In: HPCC 2013 Fourth International Workshop on Frontiers of Heterogeneous Computing, FHC 2013, vol. 370, pp. 1709–1714, Institute of Electrical and Electronics Engineers, Piscataway (2013)
4. Gorawski, M., Lis, D., Gorawski, M.: The use of a cloud computing and the CUDA architecture in zero-latency data warehouses. In: Kwiecień, A., Gaj, P., Stera, P. (eds.) CN 2013. CCIS, vol. 370, pp. 312–322. Springer, Heidelberg (2013)
5. Gorawski, M., Lorek, M., Gorawska, A.: CUDA powered user-defined types and aggregates. In: 27th International Conference on Advanced Information Networking and Applications Workshops, WAINA 2013, pp. 1423–1428. IEEE Computer Society (2013)
6. Gorawski, M., Wardas, R.: The Workload Balancing ETL System Basing on a Learning Machine. Studia Informatica **31**((2A (89))), 517–530 (2010)
7. Gorawski, M., Lis, D., Gorawska, A.: Zero–latency data warehouse system based on parallel processing and cache module. In: Corchado, E., Lozano, J.A., Quintián, H., Yin, H. (eds.) IDEAL 2014. LNCS, vol. 8669, pp. 465–474. Springer, Heidelberg (2014)
8. Karagiannis, A., Vassiliadis, P., Simitsis, A.: Scheduling Strategies for Efficient ETL Execution. Inf. Syst. **38**(6), 927–945 (2013)
9. Nguyen, T.M., Brezany, P., Tjoa, A.M., Weippl, E.: Toward a Grid-Based Zero-Latency Data Warehousing Implementation for Continuous Data Streams Processing. IJDWM **1**(4), 22–55 (2005)
10. Nguyen, T.M., Tjoa, A.M.: Zero-latency data warehousing (ZLDWH): the state-of-the-art and experimental implementation approaches. In: 4th International Conference on Computer Sciences: Research, Innovation and Vision for the Future RIFV, pp. 167–176. IEEE (2006)
11. Thiele, M., Fischer, U., Lehner, W.: Partition-based Workload Scheduling in Living Data Warehouse Environments. Inf. Syst. **34**(4–5), 382–399 (2009)
12. Waas, F., Wrembel, R., Freudenreich, T., Thiele, M., Koncilia, C., Furtado, P.: On-Demand ELT Architecture for Right-Time BI: Extending the Vision. IJDWM **9**(2), 21–38 (2013)
13. Gorawski, M., Marks, P., Gorawski, M.: Collecting data streams from a distributed radio-based measurement system. In: Haritsa, J.R., Kotagiri, R., Pudi, V. (eds.) DASFAA 2008. LNCS, vol. 4947, pp. 702–705. Springer, Heidelberg (2008)

Fourth International Workshop on Ontologies Meet Advanced Information Systems (OAIS 2015)

Mobile Co-Authoring of Linked Data in the Cloud

Moulay Driss Mechaoui[1]([⊠]), Nadir Guetmi[2]([⊠]), and Abdessamad Imine[3]([⊠])

[1] University of Sciences and Technology Oran 'Mohamed Boudiaf' USTO-MB,
Oran, Algeria
moulaydriss.mechaoui@univ-usto.dz
[2] LIAS/ISAE-ENSMA, Poitiers University, Chasseneuil, France
nadir.guetmi@ensma.fr
[3] Université de Lorraine and INRIA-LORIA Grand Est, Nancy, France
abdessamad.imine@loria.fr

Abstract. The powerful evolution of hardware, software and data connectivity of mobile devices (such as smartphones and tablets) stimulates people to publish and share their personal data (like social network information or sensor readings) independently of spatial and temporal constraints. To do this, the development of an efficient semantic web collaborative editor for mobile devices is needed. However, collaboratively editing a shared semantic web document in real-time through ad-hoc peer-to-peer mobile networks requires increasing amounts of computation, data storage and network communication. In this paper, we propose a new cloud service-based model that allows a real-time co-authoring of Linked-Data (LD) as a RDF (Resource Description Framework) graph using mobile devices. Our model is built upon two layers: (i) cloning engine that enables users to clone their mobile devices in the cloud to delegate the overload of collaborative tasks and provides peer-to-peer networks where users can create ad-hoc groups; (ii) Collaborative engine that allows updating freely and concurrently a shared RDF graph in peer-to-peer fashion without requiring a central server. This work represents a step forward toward a practical and flexible co-authoring environment for LD.

Keywords: Linked data · RDF · Mobile collaboration · Mobile cloud computing

1 Introduction

The massive development of structured data on semantic web and their growing utilization require high availability. Thus, the replication of semantic web data enables services and applications to run independently of network connection quality and central server availability. Recently, and due to the rapid development of mobile devices (e.g. smartphones and tablets), users can download semantic data from central servers and process/query it locally on their mobile

© Springer International Publishing Switzerland 2015
T. Morzy et al. (Eds): ADBIS 2015, CCIS 539, pp. 371–381, 2015.
DOI: 10.1007/978-3-319-23201-0_38

devices. This replication leads to the distribution of the computation among a large number of mobile devices, and accordingly, a great level of scalability can be achieved [13]. In addition, users can process semantic data (e.g. their personal data like social network information or sensor readings) even in off-line. Several mobile applications such as DBpedia Mobile [2], HDTourist [9] and RDF On the Go [13] that allow using RDF (Resource Description Framework) documents in mobile devices. However, the mobile user is *passive* in such applications since she/he stores RDF graphs and interrogates them using locally SPARQL queries. But, if the mobile user goes *active* (i.e. she/he modifies the local semantic data): what about updating and synchronizing copies of RDF graphs?

Updating and synchronizing semantic web data is a serious problem that has been raised by Berners-Lee in [17]. Linked Data (LD) enables semantic web data to be created in online mode and to be accessible to a large public; it allows replacing collections of offline RDF data [17]. The goal of LD is to enable people to share structured data on the web as easily as they can share documents today. It uses RDF technology that (i) relies on HTTP URIs to denote things; (ii) provides useful information about a thing at that things URI; and (iii) includes in that information other URIs of LD. Tabulator [3] is a LD browser, designed to provide the ability to navigate the web of linked things. In [17], Berners-Lee et al. raise some interesting challenges when adding collaborative co-authoring mode in Tabulator. This mode consists in collaboratively editing the LD which is represented by a RDF graph.

Using mobile environments for supporting collaborative editing of LD is not without problems. Indeed, mobile devices are resource-poor despite their continuous development (less secure, unstable connectivity, and constrained energy). Consequently, computation on mobile devices will always involve a compromise [16]. For example, peer-to-peer (P2P) collaborative editing using mobile devices is often costly since it requires important energy consumption in order to synchronize multiple copies of shared data to preserve consistency and handle the group scalability (i.e. with join/leave events). Furthermore, ensuring a continuous collaboration with frequent disconnections is impossible.

To attenuate the resource limitation of mobile devices, one simple solution is to harness cloud computing, that is an emerged model based on virtualization for efficient and flexible use of hardware assets and software services over a network. It extends the mobile device resources by offloading execution from the mobile to the cloud where a *clone* (or *virtual machine*) of the mobile is running. Cloud computing allows building Virtual Private Networks (VPN) such as peer-to-peer where a mobile device can be continuously connected to other mobiles to achieve a common task.

Contribution. In this paper, we present a model aimed at the combination of mobile and cloud environments where the management of real-time collaborative editing service plays central role. Our objective is to provide computer support for modifying concurrently shared RDF documents by dispersed mobile users. To improve availability of data, each user has a local copy of the shared RDF documents. The collaboration is performed as follows: the updates of each user are locally executed in non-blocking manner and then are propagated to other users in order to be executed on other copies. Our model is a two levels system. The first level provides self-protocol to create clones of mobiles, manage users groups and recover failed clones in the cloud. The second level provides the group collaboration mechanisms in real-time, without any role assigned to the server. To illustrate our proposed model, we give the following use case:

Assisting Conference Attendees. Suppose after attending the ADBIS 2015 conference, a group of participants want to make a tour in Poitiers city. They are already provided with mobile application based on our proposed model helping them to visit the city using a map. One of the participants creates a collaborative group in the cloud and downloads the RDF document with relevant information about Poitiers city from DBPedia[1]. The other members join the created group and share the Poitiers RDF document. During the journey one of them realizes that, instead of the restaurant appearing on his map, there is a bookshop. Hence, she/he corrects/updates the description of this localization in his map, and then he sends it to his clone in the cloud in order to be broadcast to other group members. At the meanwhile, one participant was disconnected when sending the update information. In this case, her/his clone will notify the update information after her/his re-connection to the group as if she/he did not quit it. Moreover, when visiting museums and restaurants, participants can share in real-time their thoughts by writing their reviews in the local RDF document.

Outline. The remainder of this paper is organized as follows: our model description is given in Section 2. Section 3 presents the cloning engine. In Section 4, we describe our collaborative editing protocols for consistency maintenance of shared RDF graph. We discuss the related work in Section 5 and conclude in Section 6.

2 Model Description

We propose a new collaboration model for manipulating semantic web data, regardless of spatial and temporal constraints, where mobile users can edit collaboratively shared RDF documents in peer-to-peer mode. The advantages of our model are (i) the availability of semantic web data anytime and anywhere, and (ii) the optimal use of mobile devices resources. In fact, the collaboration and communication tasks are seamlessly turned on the cloud.

As illustrated in Figure 1, our model consists of two levels. The first level provides self-protocol to create clones of mobiles, manages users groups and recover failed clones in the cloud. It is based on a cloning engine protocol that (i) instantiates clones for mobile devices, (ii) builds virtual peer-to-peer networks across collaborative groups, (iii) manages seamlessly the join and the leave of clones inside the groups, and (iv) creates a new instance of a clone when a failure appears.

---

[1] http://fr.dbpedia.org/page/Poitiers

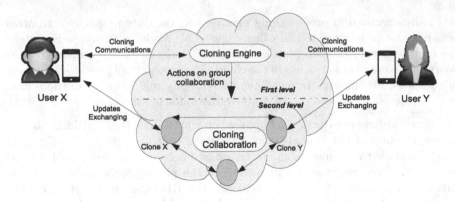

Fig. 1. Our model presentation.

The second level provides the group collaboration mechanisms in real-time, without any role assigned to the server. It provides a pure *peer-to-peer virtual private network* platform where users can form ad-hoc groups based on their clones to achieve a common objective. This platform is equipped with mechanisms to transparently manage the user departure, the arrival of new users joining the group and the collaboration spots. Our mobile co-authoring level allows users to cooperate as follows: each user has a local copy of the shared RDF document in her/his mobile and another copy of the same document in the clone; the user's updates are locally executed in mobile device and then are sent to its clone in order to be executed on other mobile devices (via their clones). It should be noted that concurrent execution of updates may lead to document inconsistency. Therefore, to maintain consistency, two synchronization protocols are proposed for clone-to-clone and mobile-to-clone interactions, respectively.

In the sequel, both levels of our model (see Figure 1) will described in Sections 3 and 4, respectively.

3   Cloning Engine

Our cloning engine involves a set of mobile users and a set of clones (or virtual machines) in such a way each mobile user owns her/his clone in the cloud. The clone has its own machine features (e.g. CPU frequency, memory size) and a virtual image to be set up (e.g. softwares like the operating system and synchronization protocols).

A set of web services is deployed to manage user groups and create clones for mobiles in the cloud. To deal with server failure and/or migration to a new cloud provider, clones are equipped with a pre-configuration to get a fast redeployment. In this case, the web services play a crucial role to ensure the continuity of clone-mobile connection (and *a fortiori* the continuity of collaboration between mobile users). The cloning engine is therefore responsible of the smooth running of clones; its main role is to:

Instantiate Clones for Mobile Devices: A new clone is built for a mobile user when she/he registers for the first time. The cloning process consists of creating a new virtual machine (e.g. x86 Android) as well as its configuration and autostart. This process is presented as follows:

- *Saving clone parameters*: It saves information related to the user and its clone (e.g. username, password, email, new clone identifier and IP address). During this step, the user can create a new group or join an existing one.
- *Clone startup and initialization*: A new virtual machine (e.g. Android x86[2]) is created for each mobile device. When the clone starts (i.e. the clone boot), it gets its IP address from a host server and calls the web service to receive the required data for any collaboration, such as clone identifier, group identifier, multicast address of the group and the IP address of the mobile. This data is important for the communication and collaboration of mobile-to-clone and clone-to-clone.

Build P2P VPNs Across Collaborative Groups: it consists of (a) assigning a new identifier and a multicast address to a new group, (b) building a new VPN for clones to communicate between them in the group, and (c) managing seamlessly the join and leave of clones inside the groups.

Recover Clone State After Failure: It supports many failure situations (clone, host server and network failure) and allows any failed clone to restore its consistent state and re-join its collaborative group. For instance, in the case of the host server failure, the clone calls automatically the web service using its initial settings to provide the new IP address of the new host server after migrating.

4   Collaborative Engine

Our collaborative engines manages all concurrent access for editing collaboratively a shared RDF graph and preserving the consistency of all copies. In this section, we describe how to map an RDF graph into a sequence data structure and we present our concurrency control procedures.

4.1   Mapping RDF to a Sequence

On the web, any information published about resources of LD is represented using the RDF. Each expression in RDF is a collection of triples, a triple consists of a *subject*, a *predicate* (also called property) and an *object*. The subject of a triple is the URI describing resource. The object can either be a simple literal value (e.g.,a string, a number) or the URI of another resource. The predicate indicates what kind of relation exists between subject and object. The predicate is a URI too. A set of such triples is called an RDF graph. This can be illustrated

---

[2] http://www.android-x86.org/

by a node and directed-arc diagram, in which each triple is represented as a node-arc-node link.

Since a set is implemented by means of a list, then we can use operations such as insert and delete to edit a shared list. Thus, we can reuse the state-of-the-art of collaborative editing systems [10]. For instance, the following group of statements "there is a Person" identified by http://www.w3.org/People/EM/contact#me (this example is taken from [8]):

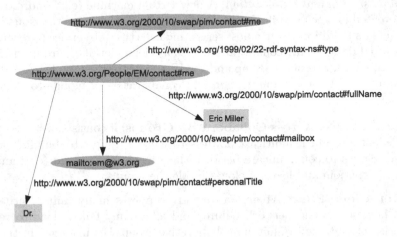

Fig. 2. An RDF Graph

- http://www.w3.org/People/EM/contact#me has a name whose value is Eric Miller
- http://www.w3.org/People/EM/contact#me has a title whose value is Dr
- http://www.w3.org/People/EM/contact#me has a email address whose value is em@w3.org

An RDF graph can be serialized into a sequence of triples and considered as a text where each line corresponds to a simple triple of subject, predicate and object. For example, the first statement shown in Figure 2 would be written as a text line (subject, predicate, object):

```
http://www.w3.org/People/EM/contact\#me
    http://www.w3.org/2000/10/swap/pim/contact\#fullName "Eric Miller"
```

Since we consider an RDF graph as a sequence, then each triple is addressed simply by a position within the sequence. Therefore, we assume that the sequence of triples can be modified by the following primitive operations: $Ins(p, T)$ which adds triple T at position p and $Del(p, T)$ which deletes triple T at position p. Updating a triple (e.g., by modifying the predicate URI) can be expressed by a sequence of delete (by removing the old triple) and insert (by adding the new one) operations.

4.2 Synchronization Protocols

We inspired from Optic [10] that allows managing collaborative editing works in P2P environments. We have designed a new protocol for coordinating the same works in the cloud. Our protocol supports an unconstrained editing work. Using optimistic replication scheme (shared RDF documents are replicated at mobiles and their clones), it provides simultaneous access to shared RDF documents. Using operational transformation approach for maintaining consistency, reconciliation of divergent copies is done automatically in decentralized fashion (i.e., without the necessity of central synchronization). Moreover, it supports dynamic groups where users can leave and join at any time. There are two layers in our synchronization protocol: the first layer ensures synchronization between clones and the second one consists in synchronizing the mobile with its clone. Figure 3 presents a general view of our proposal model.

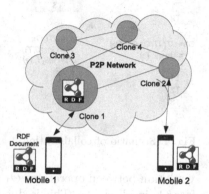

Fig. 3. Scenario of collaboration.

It runs in split mode: the light computing task is executed on mobile device (as front-end) and the heavy computing task is performed on the clone (as back-end) in order to delegate the maximum consuming-resources tasks (such as operational transformation and communication) to the cloud.

Clone-Clone Synchronization. Each clone has a local copy of RDF graph and a unique identity. It generates operations sequentially and stores these operations in a buffer called a *log*. When a clone receives a remote operation O, the integration of this operation proceeds by the following steps:

1. from the local log it determines the sequence *seq* of operations that are concurrent to O;
2. it calls the transformation component in order to get operation O' that is the transformation of O according to *seq*;
3. it executes O' on the current state;
4. it adds O' to the local log.

Illustrative Example. Given two clones, Clone 1 and Clone 2 editing a shared RDF graph as described in Figure 4. Initially, each Clone has an empty copy. Two local insertion operations O_1 and O_2 have been executed by Clone 1. Concurrently, Clone 2 has executed another insertion operation O_3. The added triples T_1, T_2 and T_3 are respectively as follows (where UR1 is http://www.w3.org/People/ EM/contact#me and UR2 is mailto:em@w3.org):

```
<UR1> <http://www.w3.org/2000/10/swap/pim/contact#fullName> "Eric Miller".
<UR1> <http://www.w3.org/2000/10/swap/pim/contact#mailbox> <UR2>.
<UR1>  <http://www.w3.org/2000/10/swap/pim/contact#personalTitle> "Dr".
```

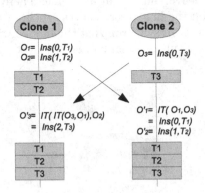

Fig. 4. Scenario of collaboration.

There is a dependency relation between operations O_1 and O_2 in such a way O_1 must be executed before O_2 in all clones. This is due to the fact that their added triples are adjacent (positions 0 and 1) and created by the same clone (from more details see [10]). This dependency relation is minimal in the sense that when O_2 is broadcast to all clones it holds only the identity of O_1 as it depends on directly. As illustrated in Figure 4, the execution order is as follows.

At Clone 1, O_3 is considered as concurrent. It is then transformed against O_1 and O_2. The sequence $[\ O_1 = \text{Ins}(0,T_1),\ O_2 = \text{Ins}(1,T_2),\ O_3' = \text{Ins}(2,T_3)]$ is executed and logged in Clone 1, where O_3' results from transforming O_3 to include the effect of operations O_1 and O_2 (i.e. $O_3' = \text{IT}(\text{IT}(O_3,O_1),O_2)=\text{Ins}(2,T_3)$ using transformation function IT given in [10]).

At Clone 2, O_1 and O_2 are concurrent with respect to O_3. They must be transformed before to be executed after O_3 according to their dependency relation. Thus, the following sequence is executed and logged in Clone 2: $O_3 = \text{Ins}(0,T_3)$ and $O_1' = \text{IT}(O_1,O_3) = O_1$ and $O_2' = O_2$.

Clone-Mobile Synchronization. Mobile and its clone are two entities that are physically separated. Therefore, a delay of applying the same operations on both sides with the same state is possible. This may lead to document inconsistency.

To overcome this problem, we propose a solution based on distributed mutual exclusion between the mobile and its clone. The shared RDF document will be considered as a critical section (CS) when the mobile tries to commit/synchronize w.r.t its clone. Only one of them will have the exclusive right to access in synchronizing mode to its document copy. This distributed mutual exclusion protocol is achieved by the exchange of messages (i.e., token). Initially, the mobile device has the right to be the first to commit/synchronize with its clone. Whatever where the exclusive access right is, the mobile device and its clone can edit independently their local copies. The clone continues to receive remote operations from other clones to integrate them later on the local state. At the meanwhile, the mobile user can work on his copy in unconstrained way. But, once she/he decides to synchronize with its clone, all local editing operations are sent to its clone and the exclusive access right is released to enable the clone to start the synchronization with the mobile device.

5   Related work

In this section, we review the most important works done in the context of our proposal based on the main facets: (i) P2P Semantic Web replication, (ii) Mobile Semantic Data Replication and (iii) Cloud Collaborative Editing.

P2P Semantic Web Replication. Except C-Set [1] that is a CRDT (Commutative Replicated Data Types) designed to be integrated within a semantic store in order to provide P2P synchronisation of autonomous semantic store, all previous work on replication in semantic P2P systems such Delta [17], RDFGrowth [18] and RDFPeers [4] are focused on sharing RDF resources. However, C-Set is not suitable for mobile devices since it incurs some overhead and does not consider directly a set as a list (or a sequence).

Mobile Semantic Data Replication. The replication of semantic data (as RDF) in mobile devices allows services and applications to operate independently of the network connection quality. Currently, several mobile applications make use of semantic stored in mobile devices [2,5,13,15]. However, mobile devices cannot use natively semantic datasets due to their limited resources. In [19] authors propose a framework based on contextual information and user description to selectively replicate data from external datasets in order to reduce the amounts of replicated data in the mobile device. This idea is interesting for our model since it allows updating just a subgraph of the shared RDF graph.

Cloud Collaborative Editing. To overcome resource constraints on mobile devices, several approaches have been proposed to offload parts of resource-intensive tasks to the cloud. Since execution in the cloud is considerably faster than the one on mobile devices and mobile-cloud offloading mechanisms delegate heavy mobile computation to the cloud. Therefore, it extends the battery life and speeds up the application [7,8,14]. CloneDoc [12] enables collaboration for mobile devices that are cloned in the cloud in order to alleviate the burden of collaborative editing works on mobile devices. CloneDoc is implemented upon C2C platform [11].

It is based on Operational Transformation (OT) approach and a single server to enforce a continuous and global order to avoid the divergence of client's document view from the server. Unfortunately, a server failure could stop the collaboration between mobile devices. Moreover, CloneDoc needs additional treatment of OT on the mobile side to ensure convergence between the mobile and its clone. However, this will cause supplementary energy consumption.

6   Conclusion

In this paper, we have presented a step forward toward a practical linked data co-authoring system combining mobile and cloud environments. Our cloud-based service enables mobile users to benefit of a huge computing power and data storage of the cloud and manages an efficient and scalable real-time editing tasks. Our service is a two levels system: (i) a cloning engine which manages the complete life cycle of clones (creation, group management and failure recovery); (ii) a collaborative engine for co-authoring shared RDF graphs in fully decentralized way.

As future work, we plan to propose a distributed access control to protect and secure the shared linked data according to the optimistic access control model given in [6].

References

1. Aslan, K., Molli, P., Skaf-Molli, H., Weiss, S.: C-set: a commutative replicated data type for semantic stores. In: Fourth International Workshop on REsource Discovery RED (2011)
2. Becker, C., Bizer, C.: Dbpedia mobile: a location-enabled linked data browser. In: LDOW (2008)
3. Berners-Lee, T., Hollenbach, J., Lu, K., Presbery, J., Pru d'ommeaux, E., Schraefel, M.: Tabulator redux: writing into the semantic web. In: LDOW (2008)
4. Cai, M., Frank, M.: Rdfpeers: a scalable distributed rdf repository based on a structured peer-to-peer network. In: Proceedings of the 13th International Conference on World Wide Web, pp. 182–197 (2004)
5. Cano, A., Dadzie, A., Hartmann, M.: Whos whoa linked data visualisation tool for mobile environments. In ISWC, pp. 451–455 (2011)
6. Cherif, A., Imine, A., Rusinowitch, M.: Practical access control management for distributed collaborative editors. Pervasive and Mobile Computing **15**, 62–86 (2014)
7. Chun, B.-G., Ihm, S., Maniatis, P., Naik, M., Patti, A.: Clonecloud: elastic execution between mobile device and cloud. In: EuroSys, pp. 301–314 (2011)
8. Cuervo, E., Balasubramanian, A., Cho, D.k., Wolman, A., Saroiu, S., Chandra, R., Bahl, P.: Maui: making smartphones last longer with code offload. In: MobiSys, pp. 49–62 (2010)
9. Hervalejo, E., Martínez-Prieto, M.A., Fernández, J.D., Corcho, Ó: Hdtourist: exploring urban data on android. In: ISWC (2014)
10. Imine, A.: Coordination model for real-time collaborative editors. In: Field, J., Vasconcelos, V.T. (eds.) COORDINATION 2009. LNCS, vol. 5521, pp. 225–246. Springer, Heidelberg (2009)

11. Kosta, S., Perta, V.C., Stefa, J., Hui, P., Mei, A.: Clone2clone (c2c): peer-to-peer networking of smartphones on the cloud. In: HotCloud (2013)

12. Kosta, S., Perta, V.C., Stefa, J., Hui, P., Mei, A.: Clonedoc: exploiting the cloud to leverage secure group collaboration mechanisms for smartphones. In: IEEE INFO-COM, vol. 13 (2013)

13. Le-Phuoc, D., Parreira, J.X., Reynolds, V., Hauswirth, M.: Rdf on the go: an rdf storage and query processor for mobile devices. In: ISWC (2010)

14. Liu, F., Shu, P., Jin, H., Ding, L., Yu, J., Niu, D., Li, B.: Gearing resource-poor mobile devices with powerful clouds: architectures, challenges, and applications. IEEE Wireless Commun. **20**(3), 1–10 (2013)

15. Ostuni, V.C., Gentile, G., Di Noia, T., Mirizzi, R., Romito, D., Di Sciascio, E.: Mobile movie recommendations with linked data. In: Cuzzocrea, A., Kittl, C., Simos, D.E., Weippl, E., Xu, L. (eds.) CD-ARES 2013. LNCS, vol. 8127, pp. 400–415. Springer, Heidelberg (2013)

16. Satyanarayanan, M., Bahl, P., Caceres, R., Davies, N.: The case for vm-based cloudlets in mobile computing. IEEE Pervasive Computing **8**(4), 14–23 (2009)

17. Tim, B.-L., Dan, C.: Delta: an ontology for the distribution of differences between rdf graphs (2004)

18. Tummarello, G., Morbidoni, C., Petersson, J., Puliti, P., Piazza, F.: Rdfgrowth, a p2p annotation exchange algorithm for scalable semantic web applications. In: P2PKM (2004)

19. Zander, S., Schandl, B.: A framework for context-driven rdf data replication on mobile devices. Semantic Web **3**, 131–155 (2012)

Ontology Based Linkage Between Enterprise Architecture, Processes, and Time

Marite Kirikova[✉], Ludmila Penicina, and Andrejs Gaidukovs

Institute of Applied Computer Systems, Riga Technical University, Riga, Latvia
{marite.kirikova,ludmila.penicina,andrejs.gaidukovs}@rtu.lv

Abstract. In an highly dynamic social and business environment time becomes one of the most treasured recourses of companies and individuals. However, there are no many research works devoted to explicit analysis of time issues. Therefore, despite of a large number of enterprise and business process representation and analysis tools, it is still impossible to address time to full extent in systems modeling and analysis. This paper proposes linkage between enterprise architecture, process, and time models to promote development of methods for managing time issues in systems development, maintenance and change management. The linkage roots in Bunge's systems ontology and concerns time ontology edited by Hobbs and Pan.

Keywords: Upper ontology · Enterprise architecture · Business process · Time ontology

1   Introduction

Time is one of the most treasured resources in highly dynamic business environment. Time also becomes an important issue in development of different information technology solutions, because their impact on users' time shall be considered. For instance, by development of information systems for administrative purposes the time of administration is saved, however the time of other employees might be misused to such an extent that the new administrative information system causes losses instead of benefits in the company in terms of financial resources, job satisfaction, and ability to serve the customers of the company. On the other hand, often the business is run on the global scale and companies from different time zones have to cooperate in providing services and products to their customers. This means that their business process management systems must be supported by information technology solutions that can easily calculate process parameters taking into consideration time zone differences with respect to the current geographical location of the companies and temporal locations of their employees. We assume that, to ensure good possibility to analyze time issues, it is necessary to develop the following two approaches:

- The approach how to model time as one of the sub-models of enterprise architecture.
- The approach how to relate time model to other models of the enterprise architecture, especially to the business process model.

© Springer International Publishing Switzerland 2015
T. Morzy et al. (Eds): ADBIS 2015, CCIS 539, pp. 382–391, 2015.
DOI: 10.1007/978-3-319-23201-0_39

In this paper we discuss the time model proposed in [1] and the time ontology [2] with respect to possibilities to model time, and we propose to relate time, enterprise architecture, and business process models via Bunge's Wand and Weber information systems ontology (BWW model) [3], which is based on Bunge's systems ontology [4].

The paper is structured as follows. The related work is discussed in Section 2. The proposed approach of linkage between enterprise architecture, business process and time models is presented in Section 3. The proposed approach is illustrated with an example in Section 4. Brief conclusions are provided in Section 5.

2 Related Work

In this section we describe related work in modeling of time, representation of time in enterprise architecture and business process models, and linkage between the enterprise architecture and business process models to set the ground for the approach for linking time, enterprise architecture, and business process models.

2.1 Time Model and Time Ontology

Taking into consideration different usages of time in information logistics and databases, the time model was proposed in [1]. On the basis of the code of general time ontology [5] this model (transferred into the form of ontology) was manually compared to the publically available W3C Time Ontology [2] (see Fig. 1).

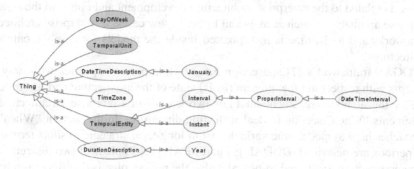

Fig. 1. Representation of W3C Time Ontology [2] which was used for comparison (simplified).

In this paper, (1) on the basis of above-mentioned comparison and (2) taking into consideration other related works discussed in Section 2.3, new time model is proposed, which amalgamates the issues reflected in the previous version of the time model [1] and W3C Time Ontology [2]. It also includes sequencing elements *Next* and *Last* (Fig. 2). The model covers all time issues referred to in Sections 2.2 and 2.3 of this paper; however, some duration variations, dependencies, and constraints discussed in [6] are not included in the model.

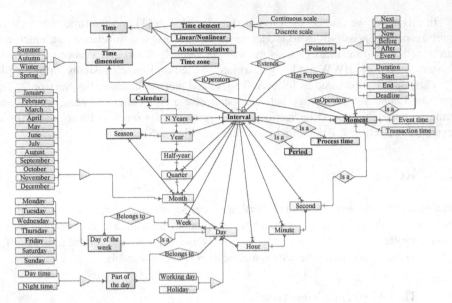

Fig. 2. New version of the time model presented in [1] (simplified).

2.2 Time in Enterprise Architecture Frameworks and Tools

In the field of Enterprise Architecture, time is considered in two ways. In the most of cases it is related to the enterprise architecture development and points to the states of enterprise architecture, such as as-is and to-be. In few cases, in enterprise architecture frameworks and tools, time is incorporated inside the models forming the enterprise architecture.

TOGAF framework's [7] subsection 5.5.3 "Time period" operates with only two concepts with respect to time: present (as-is) state of the architecture and future(to-be) state of the architecture. Zachman Framework (ZF) of Enterprise Architecture [8] implements "When" section to deal with time dimension. In the section "When" the approaches how to model time variables from long-term strategic to short-term every day periods are described. DODAF [9] and FEAF [10] implement two discrete states of the architecture (as-is and to-be), and also the project plan is described as a transition from as-is to to-be states. Project plan usually is implemented as Gantt diagram. Enterprise architecture modeling language ArchiMate [11] does not support time attributes.

Enterprise architecture tools differ one from another regarding time features available in them. For instance, ADOit supports Gantt diagram. ARIS supports BPMN and EPC standards, where BPMN ensures particular temporal features [12]. There is the research work reported that aims to update ARIS EPC meta-model with time concepts using Unified Foundational Ontology [13]. Cameo Enterprise Architect [14] supports several enterprise architecture frameworks but only ZF has expanded modeling capabilities for time constraints. Cameo Enterprise Architect supports ZF "When" section using UML activity and sequence diagrams.

2.3   Time in Business Process Models

BPMN implements only Time Event entity for temporal constraints modeling. In many cases this is sufficient. However, BPMN lacks such important temporal constraints as minimum and maximum time [15], and it is not possible to define time when tasks cannot be done [16]. There are difficulties to map time constraints and other resources in BPMN [16]. There are two possible solutions in this situation. One approach is to define time constraints using existing BPM tools. For instance, delay and wait constraints can be incorporated in the tools [17]. Another approach is to add new tools for BPMN, like many temporal constraints are offered in [16].

There are few offers [15], [16], [18] that extend BPMN standard with additional graphical elements visualizing temporal constraints. Before verifying, the models with the extensions must be transformed into formal representations (Petri nets or timed automata). Additionally, there is a possibility to model temporal constraints directly using Petri nets [19], [20] or automata [21], [22]. In this case there is no need for additional transformation, but such models have specific visualization not so common as BPMN.

2.4   BWW based Linkage Between Enterprise Architecture and Business Processes

The previous research [23] has shown the gap between business process models and state spaces of structural elements in business process models. In this paper mainly BPMN is considered as a notation for business processes modeling and ArchiMate - as a language for enterprise architecture modeling due to reasons described in [23].

Existing modelling languages BPMN and ArchiMate lack elements for capturing states of structural elements [23]. This gap hinders ensuring compliance between states of objects in business process models and other enterprise architecture elements. The gap could be closed by using BWW model [3] and linking structural elements of BPMN and ArchiMate using the following BWW concepts [24]:

- States - the states are defined via properties of elements.
- Conceivable State Space that is a set of all states the element can assume.
- Lawful State Space that is a set of states that are lawful for the element.
- History that is the chronologically-ordered states of the element.
- Events - an event is a change in the state of the element.
- Conceivable Event Space that is a set of all events that can occur to the element.
- Lawful Event Space that is a set of all events that are lawful to the element.

BWW model [3] that roots in Bunge's systems ontology [4] allows straightforwardly to address the lawful and conceivable state spaces of elements in business process models. BWW linkage with BPMN and ArchiMate potentially has a capacity to support compliance of business process models with other enterprise architecture elements or external regulations that define legal states of business process elements.

However, BWW, BPMN, and ArchiMate lack temporal concepts that are required to analyse state transition and history of business process objects for compliance purposes. Temporal concepts are required to describe temporal content in business process models, such as data objects and changes of their states, and temporal properties of objects. Defining temporal concepts in business process models will provide a capacity for defining history of states and events for further analysis. Ability to capture temporal concepts of business processes will support monitoring business process compliance regarding temporal restrictions during the run-time and analysing executed business process models for bottlenecks and other deficiencies, e.g., recognizing business objects in not allowed states for not allowed period of time, events that happen at not allowed time, objects that are idle, etc. The model for linking time concepts to enterprise architecture and business process concepts, which themselves are linked via BWW model based on Bunge's ontology, is proposed in the next section.

3   Linking Time to Business Process Models and Enterprise Architecture

Business processes always contain temporal information – time of order placement, time of error, time of received message, intended delay of the process, unintended delay, specific date to start a business process, etc. However, BPMN, ArchiMate, and BWW lack a vocabulary for expressing facts about time. W3 Time Ontology [2] provides a set of temporal concepts for expressing facts about topological relations among instants and intervals, together with information about durations and time points. Basically, W3 Time Ontology defines two subclasses of temporal concept: Instant and Interval. Intervals are things with extent and instants are point-like in that they have no interior points [2].

We propose to think that Events correspond to instants and States correspond to intervals. Event is a happening that instantly triggers a state transformation, however, objects are in a particular state for a particular interval of time, e.g., message is received on Monday at 6 am (Event), and Message is in a State *Received* for an interval of time since receiving of the receipt until it is read. Fig. 3 shows the proposed linkage between BWW states and events, ArchiMate and BPMN structural elements, and W3 Time Ontology. This linkage applies also to time model reflected in Fig. 2. In case of time model *Moment* stays instead of *Instant*.

The following issues are considered in Fig. 3:

- BPMN and ArchiMate elements representing the structure elements having BWW States.
- BWW States are changed in the case of a happening of an Event. The happening of any Event has a particular point in time, but since BWW model does not include the concept of time, we propose to use W3C Time Ontology Instant concept (or in time model - Moment), which will record the time of Event occurrence.
- BWW States have BWW History – chronologically ordered States of structural elements, but since BWW model does not include the concept of time, we propose to use W3C Time Ontology Interval concept, which will record the period of being in a particular State and History of States.

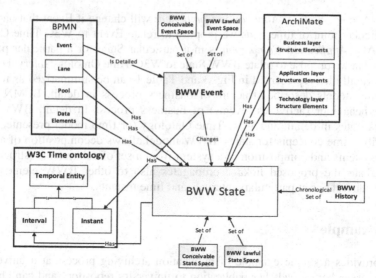

Fig. 3. The model for linking time, enterprise architecture and business processes.

Table 1 describes in detail the relationships between BWW, BPMN, ArchiMate, and W3C Time ontology.

Table 1. Detailed relationships between BWW, BPMN, ArchiMate, and W3C Time Ontology

BWW Element	BPMN Element	ArchiMate Element	W3C Time Ontology Element
State, Conceivable State Space, Lawful State Space	Lane, Participant (Pool), Data Object, Data Store	Business actor, Business role, Business interface, Location, Business object, Contract, Product Application component, Application interface, Data object, Node, Device, Infrastructure interface, System software, Artifact	Interval (interval of time BPMN and ArchiMate elements are in particular State)
History	Lane, Participant (Pool), Data Object, Data Store	Business actor, Business role, Business interface Location, Business object, Contract, Product, Application component, Application interface, Data object, Node, Device, Infrastructure interface, System software, Artifact	Set of chronologically ordered Intervals
Event, Conceivable Event Space, Lawful Event Space	Start event, Intermediate event, End event	Business Event	Instants (a particular point of time BPMN and ArchiMate Event happens)

Table 1 shows for which BPMN and ArchiMate elements we propose to define W3C Instants or Intervals. For example, ArchiMate *Application Component* can be related to BPMN Pool, which shows in detail the process *Application Component* performs, and *Application Component* is capable of having *States*: BWW State spaces – Conceivable and Lawful, e.g., *Conceivable state* can be "Not responding (Timeout)", and Lawful

State - "Active". State of *Application Component* will change if Event that happens is in a particular point of time, thus we propose to relate Event to W3C Time Ontology Instant. Also *Application Component* is in a particular State for a particular period of time, for that we propose to relate BWW State to W3C Time Ontology Interval element.

The overall idea represented in Fig. 3 and Table 1 can be summarized as follows – ArchiMate models describe structure of business process in detail, BPMN models describe behaviour (activities, events) of business process in detail, BWW model describes states in detail, and W3C Time Ontology (or time model presented in Fig. 2) describes time concepts for states. BWW model allows decomposition of a system into subsystems and composition of a system from a set of subsystems (structure elements). Thus the proposed linkage propagates also to other BWW elements, e.g., states of subsystems at particulars intervals and time instants.

4   Example

Fig. 4 provides a simple example of publication archiving process at a university in which a researcher uploads her publication to university repository and can choose an option to publish her work as Open Access item. If a Researcher agrees to archive her publication as Open Access item, she must choose (1) a licence under which she wishes to publish her publication, (2) a version of the full text, which the publisher permits to archive in the institutional repository. The possible versions of the publication's full texts are: pre-print, post-print, or published version. Next, Librarian must check the Publisher's policy for Open Access archiving in repositories. Three outcomes are possible: (1) Publisher does not allow to archive the publication as Open Access item, (2) Publisher allows to archive the Publication as Open Access item after a specific period of time (embargo period, e.g., 6 months), or (3) the Publisher permits Open Access archiving. Uploaded publication can assume several states: (1) Publication [Registered] – publication is registered in the system; (2) Publication [Internal] – publication is available only for internal university users; (3) Publication [Confirmed Open Access] – publication is confirmed by Librarian to be archived as Open Access item; (4) Publication [EMBARGO Open Access] – publication has an embargo period before it can be published as Open Access item; (5) Publication [Cancelled Open Access] – publication has been cancelled for Open Access archiving by Librarian.

Lawful event is allowing to start to show a full text of the publication publicly if its state has been changed to [Confirmed Open Access]. Unlawful event is to start to show a full text of the publication publicly if the state has been changed to [Cancelled Open Access]. Also, an unlawful event is to start to show a full text of the publication publicly when embargo period is still active. Here the time concepts Instant and Interval are essential to provide a mechanism for embargo period control and to change the state of publication from [EMBARGO Open Access] to [Confirmed Open Access]. Table 2 describes in detail the relationships between BWW, BPMN, and W3C Time Ontology concepts for the above-described process (the linkage with ArchiMate elements is the same as in Table 1). Fig. 5 shows how the business process model can be transformed to new time-oriented form of representation using relationships depicted in Table 1 and Table 2. In Fig. 5 the instance of the archiving process has the following state history:{Registered 15/04/2015 6:00 AM - 15/04/2015 10:00 AM;

EMBARGO Open Access 15/04/2015 10:00 AM-15/07/2015 10:00 AM; Confirmed Open Access 15/07/2015 10:00 AM-15/04/2020 10:00 AM}.

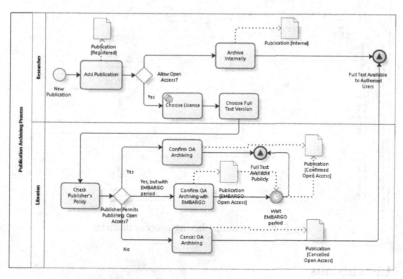

Fig. 4. Publication archiving process example.

Table 2. Relationships between BWW, BPMN, and W3C Time Ontology in publication archiving process.

BWW element	BPMN element	W3C Time Ontology
State space for Data Object {Registered, Internal, Confirmed Open Access, EMBARGO Open Access, Cancelled Open Access}	Data Object {Publication}	Instant - a particular moment of time to initiate a particular State of a Publication, e.g., 01/04/2015 6:00 AM [Registered] Interval - a period of time in which Publication is in a particular State, e.g., 15/04/2015 6:00 AM – 15/07/2015 10:00 AM [EMBARGO Open Access]
Event	Timer Event {Wait EMBARGO period}	Instant - a particular moment of time to start a countdown of EMBARGO period, e.g., 15/04/2015 6:00 AM However, an Event {Wait EMBARGO period} must stop the process for a period of time (embargo period), and only after that to initiate a State transition from [EMBARGO Open Access] to [Confirmed Open Access].
History	Data Object {Publication}	Chronologically ordered States for a particular instance of the process, e.g., {Registered, EMBARGO Open Access, Confirmed Open Access} A set of chronologically ordered Intervals of Publication States: {15/04/2015 6:00 AM - 15/04/2015 10:00 AM; 15/04/2015 10:00 AM-15/07/2015 10:00 AM; 15/07/2015 10:00 AM-15/04/2020 10:00 AM}

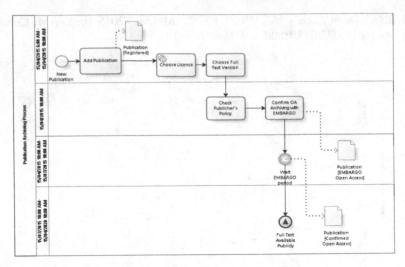

Fig. 5. Publication archiving process instance.

5   Conclusions

The paper concerns the linkage between time, enterprise architecture, and business processes models. This linkage is necessary to address more extensively time issues in different enterprise modeling and analysis tasks as the time becomes one of the most treasured resources of companies in highly dynamic and global business environment. The paper contributes the new version of the time model and, in the Bunge's systems ontology rooted, model for linking time concepts to enterprise architecture and business process models via entities of W3C Time Ontology. The elements of W3C Time Ontology map to the corresponding elements in the proposed time model. Thus both (W3C Time Ontology and the time model) can be related to the enterprise architecture and business process models.

For further experiments the inclusion of time model and time ontology in enterprise modeling tools would be helpful to ensure efficiency of modeling efforts. The development of extensions for existing tools or designing new tools and experiments with a number of different time-sensitive scenarios is the matter of further research.

References

1. Gaidukovs, A., Kirikova, M.: The time dimension in information logistics. In: Proceedings of ILOG@ BIR., pp. 35–43. CEUR Press (2013)
2. Hobbs, J.R., Pan, F.: Time Ontology in OWL: W3C Working Draft (2006). http://www.w3.org/TR/owl-time/
3. Wand, Y., Weber, R.: On the ontological expressiveness of information systems analysis and design grammars. Inf. Syst. J. **3**, 217–237 (1993)
4. Bunge, M.: Treatise on Basic Philosophy: Vol. 4: Ontology II: A World of Systems (1979)

5. Knowledge Craver: Guidelines for Using the W3C Time Ontology (2014). http://knowledgecraver.blogspot.com/2014/07/guidelines-for-using-w3c-time-ontology.html
6. Trudel, A.: The Temporal Perspective: Expressing Temporal Constraints and Dependencies in Process Models, pp. 1–14 (2008)
7. TOGAF® Version 9.1 (2011). http://www.opengroup.org/togaf/
8. Zachman International® - The Official Home of The Zachman FrameworkTM (2014). http://www.zachman.com/
9. DODAF - DOD Architecture Framework Version 2.02 - DOD Deputy Chief Information Officer (2012). http://dodcio.defense.gov/Portals/0/Documents/DODAF
10. Federal Enterprise Architecture (FEA), The White House (2013). https://www.whitehouse.gov/omb/e-GOV/FEA
11. Archimate (2012). http://www.opengroup.org/subjectareas/enterprise/archimate
12. Betke, H., Kittel, K., Sackmann, S.: Modeling controls for compliance - an analysis of business process modeling languages. In: 27th International Conference on Advanced Information Networking and Applications Workshops, pp. 866–871. IEEE (2013)
13. Santos Jr., P.S., Almeida, J.P.A., Guizzardi, G.: An ontology-based semantic foundation for ARIS EPCs. In: Proceedings of the 2010 ACM Symposium on Applied Computing, pp. 124–130. ACM (2010)
14. CAMEO Enterprise Architecture (2015). http://www.nomagic.com/products/cameo-enterprise-architecture.html
15. Watahiki, K., Ishikawa, F., Hiraishi, K.: Formal verification of business processes with temporal and resource constraints. In: 2011 IEEE International Conference on Systems, Man, and Cybernetics, pp. 1173–1180. IEEE (2011)
16. Cheikhrouhou, S., Kallel, S., Guermouche, N., Jmaiel, M.: Toward a time-centric modeling of business processes in BPMN 2.0. In: Proceedings of International Conference on Information Integration and Web-based Applications & Services, pp. 154–163. ACM (2013)
17. Stiehl, V.: Process-Driven Applications with BPMN. Springer (2014)
18. Gagne, D., Trudel, A.: Time-BPMN. In: 2009 IEEE Conference on Commerce and Enterprise Computing, pp. 361–367. IEEE (2009)
19. Huai, W., Liu, X., Sun, H.: Towards trustworthy composite service through business process model verification. In: 7th International Conference on Ubiquitous Intelligence & Computing and 7th International Conference on Autonomic & Trusted Computing, pp. 422–427. IEEE (2010)
20. Makni, M., Tata, S., Yeddes, M., Ben Hadj-Alouane, N.: Satisfaction and coherence of deadline constraints in inter-organizational workflows. In: Meersman, R., Dillon, T.S., Herrero, P. (eds.) OTM 2010. LNCS, vol. 6426, pp. 523–539. Springer, Heidelberg (2010)
21. Kallel, S., Charfi, A., Dinkelaker, T., Mezini, M., Jmaiel, M.: Specifying and monitoring temporal properties in web services compositions. In: Seventh IEEE European Conference on Web Services. pp. 148–157. IEEE (2009)
22. Guermouche, N., Zilio, S.D.: Towards timed requirement verification for service choreographies. In: Proceedings of the 8th IEEE International Conference on Collaborative Computing: Networking, Applications and Worksharing, pp. 117–126. IEEE (2012)
23. Peņicina, L., Kirikova, M.: Towards controlling lawful states and events in business process models. In: IEEE 16th Conference on Business Informatics (CBI 2014), pp. 15–23. IEEE (2014)
24. Rosemann, M., Vessey, I., Weber, R., Wyssusek, B.: On the Applicability of the Bunge – Wand – Weber Ontology to Enterprise Systems Requirements (2004)

Fuzzy Inference-Based Ontology Matching Using Upper Ontology

S. Hashem Davarpanah[1], Alsayed Algergawy[2,3]([✉]), and Samira Babalou[1]

[1] Department of Computer Engineering, University of Science and Culture,
Tehran, Iran
[2] Institute for Computer Science, Friedrich Schiller University of Jena,
Jena, Germany
alsayed.algergawy@uni-jena.de
[3] Department of Computer Engineering, Tanta University, Tanta, Egypt

Abstract. Bio-ontologies are characterized by large sizes, and there is a large number of smaller ontologies derived from them. Determining semantic correspondences across these smaller ones can be based on this "upper" ontology. To this end, we introduce a new fuzzy inference-based ontology matching approach exploiting upper ontologies as semantic bridges in the matching process. The approach comprises two main steps: first, a fuzzy inference-based matching method is used to determine the confidence values in the ontology matching process. To learn the fuzzy system parameters and to enhance the adaptability of fuzzy membership function parameters, we exploit a gradient discriminate learning technique. Second, the achieved results are then composed and combined to derive the final match result. The experimental results show that the performance of the proposed approach compared to one of the famous benchmark research is acceptable.

1 Introduction

Biomedical ontologies such as the Gene Ontology (GO)[1] [24], ChEBI[2] [17], and other OBO Foundry ontologies [21] are characterised by their large sizes. A large number of smaller ontologies have been constructed using sub-sets of these larger ontologies. However, in some applications, there is a growing need to identify and determine corresponding concepts across these ontologies. The process that identifies similar concepts across a set of ontologies called *ontology matching* [5], which plays a central role to deal with semantic heterogeneity. For its importance, myriad of ontology matching approaches have been proposed and developed [1,10–12,16].

It is obvious that the ontology matching process is a tedious, time-consuming, and error-prone process due to the inherent heterogeneities of the involved ontologies [19,20]. Therefore, almost all of existing matching systems attempt to

---

[1] http://geneontology.org/
[2] http://www.ontobee.org/browser/index.php?o=CHEBI

© Springer International Publishing Switzerland 2015
T. Morzy et al. (Eds): ADBIS 2015, CCIS 539, pp. 392–402, 2015.
DOI: 10.1007/978-3-319-23201-0_40

automate the matching process. This can be achieved by involving user/expert in the process [8,9], exploiting external sources and/or dictionaries [7]. However, most of these existing approaches focus on direct matching between ontologies, ignoring the fact that some of these ontologies are derived from the same upper ontology. Surprisingly, a few attempts to exploit "upper ontologies" in a systematic way have been made so far, such as [16] and BLOOM+ [10].

The approach in [16] proposes a set of algorithms that exploit upper ontologies as semantic bridges in the matching process. It implements three matchers: un-structure, structural, and mixed methods to compute the similarity between concepts. However, this approach does not consider the uncertainty in the similarity computation. The BLOOM+ approach presents a solution for automatically finding schema-level links between two Linked Open Data (LOD) ontologies [10]. It first uses a set of measures to calculate the similarity between ontology classes and it then uses an external information source, which is organized as a class hierarchy, to further support an alignment. However, BLOOM+ employs a similarity measure, which is a combination of the class overlap and the contextual similarity.

To cope with these challenges, we introduce a fuzzy inference-based matching approach, which also exploit upper ontologies as semantic bridges in the matching process. The approach depends on a fuzzy inference system to compute the degree of similarity between ontology entities. It makes use of a gradient discriminator learning method in order to learn and adapt the fuzzy system parameters and to increase the adaptability of the fuzzy membership function parameters. To validate the effectiveness of the proposed approach, we conducted a set of experiments utilizing two upper ontologies, namely Cyc[3] and DOLCE. The preliminary results demonstrate encouraging results compared to the approach proposed in [16].

The rest of the paper is organized as follows: the next section presents basic concepts and background. We then introduce the approach in Section 3. Experimental evaluation and results are reported in Section 4. Finally, conclusion and future directions are summarized in Section 5.

2 Background

2.1 Related Work

Semantic heterogeneity is a key challenge in basically all data-sharing applications [2,18]. Ontology matching plays a central role to discover and reconcile the semantic heterogeneity looking for semantic correspondences between similar entities across ontologies. Therefore, myriad of approaches have been proposed and developed [3,5,19,20]. Most of these systems match ontology entities based only on the internal properties of ontology entities. Some other utilize some external resources, such as external dictionaries and vocabularies and user feedback to improve the matching quality [7,8,12]. However, a few of them adapt upper ontologies as semantic bridges in the matching process [10,16]

---

[3] http://www.cyc.com/

SAMOB is a system for matching and merging biomedical ontologies [12]. It computes similarity values between terms of the source ontologies utilizing several matchers: linguistic-based, structure-based, constraint-based, and instance-based matchers. Each matcher captures information from one or more auxiliary sources, such as instance corpora, general dictionaries, and domain thesauri.

The approach proposed in [16] is a real effort to adopt upper ontologies as semantic bridges in the matching process as well as to fully automate the process. However, it uses also Wordnet[4] as an auxiliary source to resolve term mismatches. The main similarity function is called upper ontology match, *Uo_match*, which mainly depends on three auxiliary similarity functions. 1) *Aggregate (a,a')* produces the alignment obtained by making the union of all the correspondences in a and a', and choosing the correspondence with highest confidence measure. 2) *Parall_match(O,O',res,par)* computes an alignment between O and O' by applying substring, n-gram, SMOA, and language-based methods. 3) *compose(a,a')* computes the alignment a'' given the two alignments a and a', where O and O' are ontologies, a and a' are alignments, *res* are external resources, and *par* are parameters[16].

BLOOM+ is another recent work that aims to adopt the usage of upper ontology in the matching process [10]. It first employs Wikipedia to construct a set of hierarchy trees for each class in the source and target ontologies. It then uses a more sophisticated measure to compute the similarity between source and target classes based on their category hierarchy trees and computes the contextual similarity between these classes to further support (or reject) an alignment. Finally, BLOOMS+ aligns classes with high similarity based on the class and contextual similarities [10].

2.2 Upper Ontology

Upper ontologies have become more and more compelling for sharing and integrating heterogenous knowledge from different sources. Therefore, a set of upper ontologies has been developed and constructed [15]. Examples are Basic Formal Ontology (BFO)[5], General Formal Ontology (GFO)[6], Descriptive Ontology for Linguistic and Cognitive Engineering (DOLCE)[7], and Cyc. An elegant review comparing these ontologies can be found in [15]. In the following, we present a short overview of two of the largest available upper ontologies, which we used later in the experimental evaluation.

- *Descriptive Ontology for Linguistic and Cognitive Engineering (DOLCE).* DOLCE is the first module of a Library of Foundational Ontologies being developed within the WonderWeb project [6]. It aims at capturing the ontological categories underlying natural language and human commonsense.

---

[4] http://wn-similarity.sourceforge.net/
[5] http://www.ifomis.org/bfo.
[6] http://www.onto-med.de/ontologies/gfo.html.
[7] http://www.loa-cnr.it/DOLCE.html.

DOLCE is an ontology of particulars, in the sense that its domain of discourse is restricted to them. The taxonomy of the most basic categories of particulars assumed in DOLCE includes, for example, abstract quality, abstract region, amount of matter, physical quality, temporal region. DOLCE is implemented in First Order Logic and OWL. Its OWL version contains around 250 concepts and 320 properties, divided into 10 sub-ontologies [15].

- *Cyc.* Cyc is an artificial intelligence project that attempts to assemble a comprehensive ontology and knowledge base of everyday common sense knowledge [13]. It is a formalised representation of facts, rules of thumb, and heuristics for reasoning about the objects and events of everyday life. Cyc consists of terms and assertions relating those terms. These assertions include both simple ground assertions and rules. Cyc is a commercial product, but Cycorp also released OpenCyc[8], the open source version of the Cyc technology, and ResearchCyc[9]), namely the Cyc ontology delivered with a research-only license. The Opencyc contains 500,000+ concepts, forming an ontology in the domain of human consensus reality, nearly 5,000,000 assertions (facts and rules), using 26,000+ relations. It also includes an Ontology Exporter that makes it simple to export specified portions of the knowledge base to OWL files.

3   Fuzzy Inference-Based Matching

In this section, we introduce a fuzzy-based ontology matching approach. As shown in Fig. 1, the proposed method comprises two main steps. The first is to apply a fuzzy-based matching method to each input ontology and the upper ontology to produce an intermediate set of correspondences a and a'; and the second is to compose these correspondences to generate the final alignment. The combination is done using the *compose* method in [16].

In order to align each ontology to the input upper ontology, a fuzzy inference engine is introduced, where three string similarity measures are used to compute the similarity between input entities and combination of the results are carried out using a fuzzy logic inference engine. In particular, as shown in Fig. 2, we employ three well-know string measures, namely *substring*, *n-gram*, and *SMOA* [5,23]. The sub-string similarity is a similarity measure such that for each string pairs s_1 and s_2 and l is the longest common substring between s_1 and s_2 is defined as

$$sim_{sub}(s_1, s_2) = \frac{2|l|}{|s_1| + |s_2|} \tag{1}$$

n-grams are typically used in approximate string matching by sliding a window of length n over the characters of a string to create a number of n length grams for finding a match. The *n-gram* similarity between the two strings s_1 and s_2 is defined as

---

[8] http://www.cyc.org/
[9] http://research.cyc.com/

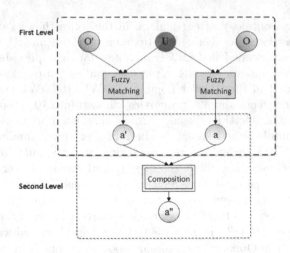

Fig. 1. Fuzzy matching scheme using upper Ontology.

$$sim_{n-gram}(s_1, s_2) = \frac{2 \times |n-gram(s_1) \cap n-gram(s_2)|}{|n-gram(s_1)| + |n-gram(s_2)|} \qquad (2)$$

where $n-gram(s)$ is the set of n-grams for the string s. The string metric for ontology alignment (SMOA) is based on the following principle: the similarity among two entities is related to their commonalities as well as to their dierences. Thus, the similarity should be a function of both these features [23]. Therefore, the similarity between two string s_1 and s_2 can be defined as:

$$sim_{SMOA}(s_1, s_2) = Comm(s_1, s_2) - Diff(s_1, s_2) + winkler(s_1, s_2) \qquad (3)$$

where $Comm(s_1, s_2)$ represent the commonality between s_1 and s_2, $Diff(s_1, s_2)$ is the difference , and $winkler(s_1, s_2)$ is the improvement for the original method introduced by Winkler.

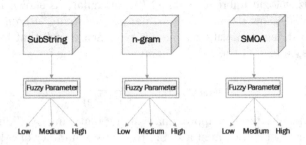

Fig. 2. The 9 fuzzy terms used in fuzzy alignment.

The output of each similarity function is considered as a fuzzy variable. As shown in Fig. 2, for each of these fuzzy variables, three fuzzy terms (*low*, *medium*, and *high*) are designed. Thus, the output of this phase of the algorithm is a set of 9 membership degrees which are calculated by applying the output of the triple estimation distance functions on the 9 fuzzy membership functions. These values construct the input of the fuzzy rules. In the next step, correspondences for each fuzzy rule is obtained and output of the fuzzy engine which is equivalent to the correspondences degree of the two input entities (two classes, objects, or components) according to one of the rules is identified.

3.1 Fuzzy Inference System

A multi-input-output fuzzy system is considered as the fuzzy inference system. Let define a set of input-output pairs [4, 14]:

$$(x(n), y(n)), n = 1, 2, ..., N \tag{4}$$

where $x(n) \in R^m$ and $y(n) \in R$ and N refers to number of input-output pairs. A fuzzy rule in the system can be formulated as follows:

$$if x_1 is A_{i1} ... \ and \ x_m \ is \ A_{im}, \ then \ y \ is \ w_i \tag{5}$$

where $x_1, x_2, .., x_m$ stands for M input variables. A_{ij} is a selected fuzzy set from $\{A_j^1, ..., A_j^{M_j}\}$, where M_j refers to the number of fuzzy sets of input variables x_j, $i = 1, .., K$ stands for the index of the fuzzy rule, and w_i is the numeric consequence.

Fuzzy Variables. As it was mentioned before, there are three fuzzy variables in the condition section of the proposed fuzzy rules which are called Fp_sub_str, Fp_n_gram, and Fp_SMOA. These names refer to their corresponding similarity methods, *sub-string*, *n-gram*, and *SMOA*. Each of these variables has three fuzzy terms which are as follows:

```
Fp_sub_str_low, Fp_sub_str_medium, Fp_sub_str_high
Fp_n_gram_low, Fp_n_gram_medium, Fp_n_gram_high
Fp_SMOA_low, Fp_SMOA_medium, Fp_SMOA_high
```

In the action piece of fuzzy rules, there is a fuzzy variable which shows the alignment degree resulted by that rule. In order to reduce the complexity of the fuzzy system, this variable is designed such a non-fuzzy variable so that the output of each rule is a crisp number between 0 and 1 explaining the matching degree directly. Each of these fuzzy terms has a *specific fuzzy membership function*; three low terms have descending linear membership functions, three medium terms have triangular membership functions, and three high terms have ascending linear membership functions.

Fuzzy Rules. There are 27 rules in the proposed fuzzy inference system. These rules present all possible combinations of three existing fuzzy terms. A number of rules are as follows:

```
Rule1: If Fp_sub_str=Ft_sub_str_low &&
Fp_n_gram=Ft_n_gram_low && Fp_SMOA= Ft_SMOA_low
 Then Ft_matching= w_1
...


...
Rule27: If Fp_sub_str= Ft_sub_str_high &&
 Fp_n_gram= Ft_n_gram_high && Fp_SMOA= Ft_SMOA_high
 Then Ft_matching= w_27
```

In the definition of these rules, all possible states of the terms which are assignable to the fuzzy variables are considered. The membership degree of the condition section of each rule is obtained by multiplying the degrees of the triple fuzzy terms. A database is designed containing only one table with 189 records so that each record presents the last value of a parameter; there are 27 rules with 27*2 parameter values for descending linear membership functions, the same numbers for ascending linear membership functions and triangular membership functions and 27 parameter values for the output parameter of the rules.

Given a set of data input $x(n) = (x_1(n), ...x_m(n))$, the ancestors of the ith fuzzy rule are combined using product operation to determine the membership degree, $\mu_i(n)$, of the output $\mu_i(n) = \prod_j^m \mu_{A_{ij}}(x_j(n))$. The estimated corresponding output is calculated using defuzzification as follows [4, 14]:

$$\hat{y}(n) = \frac{\sum_{i=1}^K \mu_i(n)w_i}{\sum_{i=1}^K \mu_i(n)} \tag{6}$$

where the membership function A_{ij} is determined by the center a_{ij} and the width b_{ij} of the triangle as follows: $\mu_{A_{ij}} = 1 - \frac{2|x_j - a_{ij}|}{b_{ij}}$.

Learning of Fuzzy Inference System. In the current implementation, we use the gradient descent algorithm to determine an appropriate combination for membership functions, and also to create a set of fuzzy rules while the estimated result of the fuzzy system will be truth (more truth) when the inputs are unknown, complex or nonlinear. Through the learning phase of the fuzzy system, a descending gradient function is utilized and all the existent 189 parameters are initialized. The learning process is done based on the descent gradient algorithm introduced in [22] and using *least square error* as the goal function.

We define a membership function (MF) for descending, ascending, and the triangular cases. Each function is based on the central point a_{ij} and the width b_{ij} of the triangle. For each MF, a learning rate is used. We also use a learning rule for the output parameter of the fuzzy system. This learning process is used to update these parameters adaptively. The process is repeated until the *least square error* is either sufficiently low or zero.

3.2   Alignments Mixing Algorithm

Once determining the components alignments of each input ontology with the upper ontology, we deploy them to drive direct alignments between input ontologies. For this purpose, the compose function presented in reference [16] is implemented. $Compose(a, a')$ computes the alignment a'' in such a way that a correspondence $< id, c, c', r, conf >$ belongs to a'' iff $\exists c_u$ such that $< id_1, c, c_u, r, conf_1 > \in a$, $< id_2, c', c_u, r, conf_2 > \in a'$, and $conf = conf_1 * conf_2$. In the compose algorithm, a and a', are alignments between O and the upper ontology U, and O' and U, respectively. Thus, the second concept c_u in the correspondences belongs to U. If both $c \in O$ and $c' \in O'$ correspond to the same concept $c_u \in U$, then c and c' are related with a confidence value of $conf_1 \times conf_2$.

4   Experimental Evaluation

To validate the quality of the proposed approach, we conducted a set of experiments. To this end, we make use of the Alignment API[10]. All the method presented by API including ours accept all ontologies presented in OWL, RDF, and RDFS. The fuzzy-based system has been implemented using Java platform , however, we use MySQL to implement the required database.

We designed ten experiments, which were executed on 16 standard ontologies and two upper ontologies. The results were compared with the average of the results achieved in research [16]. The experiments are conducted on a 3.0 GHz Intel Core Dual processor with 4 GB memory and using the Windows XP SP 3.0 operating system. *Precision, Recall* and *F-measure* are used to measure the quality of the matching result.

We run ten experiments using different input ontologies and two upper ontologies. Results are reported in Figs. 3 a & b. Each figure gives the number of matching tasks, input ontologies involved in the task, the matching results of our proposed fuzzy-based system (abbreviated in the figure as B) and the system developed in [16]. Fig. 3a represent the result using the openCyc ontology as an upper ontology, while Fig. 3b the result using DOLCE as an upper ontology. Figs. 3 a & b show our achived results, where our results are extracted from our experiments directly and the results of reference [16] equal to the average of the values explained in [16]. The averages are calculated separately for each test from the ten different experiments carried out. Results indicate that in general using an upper ontology as a semantic bridge in the matching process improves the matching quality for both systems (except one case using the OpenCyc ontology). Furthermore, the figure shows that our proposed method achieves better average matching quality compared to the system A. This can be explained as using the fuzzy inference-based system is able to cope with uncertainty in the matching process and achieve better result, given that in the current implementation, we use only a set of string-based measures. This means that this matching quality could be further improved when we exploit more similarity measures, such as structural matchers.

---

[10] http://alignapi.gforge.inria.fr

Test No.	Ontology	Matching Method	Precision	Recall	F-measure
1	Ka	A	0.362	0.07	0.117
	Bibtex	B	0.308	0.05	0.086
2	Biosphere	A	0.086	0	0.000
	Top-bio	B	0.133	0	0.000
3	Space	A	0.217	0.07	0.106
	Geofile	B	0.332	0.08	0.129
4	Restaurant	A	0.266	0.06	0.098
	Food	B	0.332	0.05	0.087
5	MPEG7	A	0.216	0.18	0.197
	Subject	B	0.251	0.2	0.223
6	Travel	A	0.190	0.04	0.066
	Vacation	B	0.226	0.05	0.082
7	Resume	A	0.240	0.06	0.096
	Agent	B	0.226	0.05	0.082
8	Resume	A	0.203	0.16	0.179
	HL7_RBAC	B	0.188	0.15	0.167
9	Ecology	A	0.141	0.08	0.102
	Top-bio	B	0.139	0.08	0.102
10	Vertebrate	A	0.163	0.35	0.222
	Top-bio	B	0.175	0.38	0.240
Average		A	0.215	0.107	0.143
		B	0.246	0.109	0.151

(a) Using OpenCyc upper ontology

Test No.	Ontology	Matching Method	Precision	Recall	F-measure
1	Ka	A	0.123	0.07	0.089
	Bibtex	B	0.121	0.08	0.096
2	Biosphere	A	0.142	0.01	0.019
	Top-bio	B	0.165	0.02	0.036
3	Space	A	0.094	0.07	0.080
	Geofile	B	0.086	0.06	0.071
4	Restaurant	A	0.074	0.01	0.018
	Food	B	0.120	0.01	0.018
5	MPEG7	A	0.120	0.16	0.137
	Subject	B	0.158	0.17	0.164
6	Travel	A	0.113	0.04	0.059
	Vacation	B	0.115	0.03	0.048
7	Resume	A	0.148	0.06	0.085
	Agent	B	0.147	0.06	0.085
8	Resume	A	0.118	0.11	0.114
	HL7_RBAC	B	0.135	0.14	0.137
9	Ecology	A	0.110	0.08	0.093
	Top-bio	B	0.114	0.08	0.094
10	Vertebrate	A	0.056	0.09	0.069
	Top-bio	B	0.119	0.11	0.114
Average		A	0.117	0.070	0.087
		B	0.136	0.076	0.097

(b) Using DOLCE upper ontology

Fig. 3. Average matching results: A is the system in [16], B is the propsed system.

5   Conclusion

In this paper, we deal with the problem of matching ontologies using an upper ontology as a semantic bridge in the matching process. To cope with the uncertainty in the process, in this research, we introduced a fuzzy inference-based ontology matching approach. First, the parameters of the proposed fuzzy inference engine are learned using a descent gradient algorithm. Then, the matching process is done in two steps; the correspondences of each component of these ontologies with an upper ontology are determined and in the second step, the achieved results are combined to determine the correspondences between two ontologies. To validate the quality of the proposed approach, we conducted a set of experiments held on any pairs of 16 standard ontologies and using two standard upper ontologies. The preliminary results are encouraging compared to recent work. However, more and different similarity measures should be involved and validated. Furthermore, we need to investigate optimization solution for the learning process.

Acknowledgments. A. Algergawy'work is partly funded by DFG in the INFRA1 project of CRC AquaDiva.

References

1. Algergawy, A., Massmann, S., Rahm, E.: A clustering-based approach for large-scale ontology matching. In: Eder, J., Bielikova, M., Tjoa, A.M. (eds.) ADBIS 2011. LNCS, vol. 6909, pp. 415–428. Springer, Heidelberg (2011)

2. Algergawy, A., Nayak, R., Siegmund, N., Köppen, V., Saake, G.: Combining schema and level-based matching for web service discovery. In: Benatallah, B., Casati, F., Kappel, G., Rossi, G. (eds.) ICWE 2010. LNCS, vol. 6189, pp. 114–128. Springer, Heidelberg (2010)

3. Bellahsene, Z., Bonifati, A., Rahm, E.: Schema Matching and Mapping. Springer Verlag (2011)

4. Castro-Schez, J.J., Murillo, J.M., Miguel, R., Luo, X.: Knowledge acquisition based on learning of maximal structure fuzzy rules. Knowledge-Based Systems **44**, 112–120 (2013)

5. Euzenat, J., Shvaiko, P.: Ontology Matching, 2nd edn. Springer (2013)

6. Gangemi, A., Guarino, N., Masolo, C., Oltramari, A., Schneider, L.: Sweetening ontologies with DOLCE. In: Gómez-Pérez, A., Benjamins, V.R. (eds.) EKAW 2002. LNCS (LNAI), vol. 2473, p. 166. Springer, Heidelberg (2002)

7. Hertling, S., Paulheim, H.: WikiMatch - using wikipedia for ontology matching. In: 7th International Workshop on Ontology Matching (2012)

8. Hung, N.Q.V., Tam, N.T., Mikls, Z., Aberer, K., Gal, A., Weidlich, M.: Pay-as-you-go reconciliation in schema matching networks. In: ICDE (2014)

9. Hung, N.Q.V., Wijaya, T.K., Mikls, Z., Aberer, K., Levy, E., Shafran, V., Gal, A., Weidlich, M.: Minimizing human effort in reconciling match networks. In: ER 2013, pp. 212–226 (2013)

10. Jain, P., Yeh, P.Z., Verma, K., Vasquez, R.G., Damova, M., Hitzler, P., Sheth, A.P.: Contextual ontology alignment of LOD with an upper ontology: a case study with proton. In: Antoniou, G., Grobelnik, M., Simperl, E., Parsia, B., Plexousakis, D., De Leenheer, P., Pan, J. (eds.) ESWC 2011, Part I. LNCS, vol. 6643, pp. 80–92. Springer, Heidelberg (2011)

11. Jimenez-Ruiz, E., Grau, B.C., Zhou, Y., Horrocks, I.: Large-scale interactive ontology matching: algorithms and implementation. In: 20th European Conference on Artificial Intelligence, pp. 444–449 (2012)

12. Lambrix, P., Tan, H.: SAMBOA system for aligning and merging biomedical ontologies. Web Semantics: Science, Services and Agents on the World Wide Web **4**, 196–206 (2006)

13. Lenat, D.B.: Cyc: a large-scale investment in knowledge infrastructure. Communications of the ACM **38**(11), 33–38 (1995)

14. Li-Quan, Z., Cheng, S.: An adaptive learning method for the generation of fuzzy inference system from data. Acta Automatica Sinica **34**(1) (2008)

15. Mascardi, V., Cord, V., Rosso, P.: A comparison of upper ontologies. WOA **2007**, 55–64 (2007)

16. Mascardi, V., Locoro, A., Rosso, P.: Automatic ontology matching via upper ontologies: A systematic evaluation. IEEE Trans. Knowl. Data Eng. **22**(5), 609–623 (2010)

17. Matos, P., Alcntara, R., Dekker, A., Ennis, M., Hastings, J., Haug, K., Spiteri, I., Turner, S., Steinbeck, C.: Chemical entities of biological interest: an update. Nucleic Acids Res. **38**, D249–D254 (2010)

18. Noy, N.F., Klein, M.: Ontology evolution: Not the same as schema evolution. Knowledge and Information Systems **6**, 428–440 (2004)

19. Rahm, E., Bernstein, P.A.: A survey of approaches to automatic schema matching. VLDB Journal **10**(4), 334–350 (2001)

20. Shvaiko, P., Euzenat, J.: Ontology matching: State of the art and future challenges. IEEE Trans. Knowl. Data Eng. **25**(1), 158–176 (2013)

21. Smith, B., Ashburner, M., Rosse, C., Bard, J., Bug, W., et al.: The OBO foundry: coordinated evolution of ontologies to support biomedical data integration. Nat Biotechnol **25**, 1251–1255 (2007)
22. Snyman, J.A.: Practical mathematical optimization: an introduction to basic optimization theory and classical and new gradient-based algorithms. Springer (2005)
23. Stoilos, G., Stamou, G., Kollias, S.D.: A string metric for ontology alignment. In: Gil, Y., Motta, E., Benjamins, V.R., Musen, M.A. (eds.) ISWC 2005. LNCS, vol. 3729, pp. 624–637. Springer, Heidelberg (2005)
24. The Gene Ontology Consortium: Gene ontology: tool for the unification of biology. Nat. Genet. **25** (2000)

An Ontology-Based Approach for Handling Explicit and Implicit Knowledge over Trajectories

Rouaa Wannous[1], Cécile Vincent[2], Jamal Malki[1(✉)], and Alain Bouju[1]

[1] Laboratoire L3i, Pôle Sciences & Technologie, Avenue Michel Crépeau,
17042 La Rochelle, France
Jamal.Malki@univ-lr.fr
[2] CNRS UMR 7372, Centre d'Etudes Biologiques de Chizé,
79360 Villiers-en-Bois, France

Abstract. The current information systems manage several, different and huge databases. The data can be temporal, spatial and other application domains with specific knowledge. For these reasons, new approaches must be designed to fully exploit data expressiveness and heterogeneity taking into account application's needs. As part of ontology-based information system design, this paper proposes an ontology modeling approach for trajectories of moving objects. Consider domain, temporal and spatial knowledge gives a complexity to our system. We propose optimizations to annotate data with these knowledge.

Keywords: Trajectory data model · Spatial data model · Time data model · Ontology inference · Domain spatial temporal knowledge · Ontology rules

1 Introduction

At present, most information systems are facing the problem of analysis, interpretation and querying huge masses of data. In these systems, knowledge plays a central role in such processes. This knowledge can be expressed in form of explicit and implicit semantics. An important part of the explicit semantics comes from the application domain. Also, we need to consider intrinsic data properties. Otherwise, there are others explicit semantics resulting from the nature of the data. Furthermore, users needs or requirements also form another part of explicit semantics that must be taken into account. These different forms of knowledge are involved at different levels of data modeling. In general, any approach considering this issue, should provide answers to three issues: data models; explicit semantics models; connecting explicit semantics and data models.

This article presents an ontological based approach to solve this problem by respecting the previous three requirements. The central idea of our approach is data and semantics models transformation into ontologies, considering two steps:

© Springer International Publishing Switzerland 2015
T. Morzy et al. (Eds): ADBIS 2015, CCIS 539, pp. 403–413, 2015.
DOI: 10.1007/978-3-319-23201-0_41

- declarative mapping: concepts and relationships of the resulting ontologies are mapped to a higher and generic predefined ontologies;
- imperative mapping: user defined rules on generic predefined ontologies are defined to decide semantics of new created data.

In this case, input models can be formalized and transformed independently from the generic ontologies which can be extended to accomplish the declarative mapping. The imperative mapping leads to implicit semantics and then to new knowledge.

Application domain considered in this work is trajectories. So, we present a trajectory modeling framework to enrich a high-level data layer. To meet challenges imposed by a semantic trajectory notion, we present a modeling approach based on a generic trajectory ontology centered on a triple (object, trajectory, activity). Considering this workspace, we show how we can use our modeling approach in a particular domain application: marine mammal trajectories. This paper provides solutions to several scientific problems to be considered in a trajectory's modeling approach based on ontologies taking into account spatial, temporal and domain knowledge.

2   Related Work

Data management approaches including modeling, indexing, inferencing and querying large data have been actively investigated during the last decade [3,11]. Most of these techniques are only interested in representing and querying moving object trajectories [4,14,15]. The problem of using implicit or explicit knowledge associated with these data, or the domain context is not considered. In [12], authors propose a conceptual view on trajectories. In this approach, a trajectory is a set of stops, moves. Each part contains a set of semantic data or knowledge. Based on this conceptual model, several studies have been proposed [3,15]. In [3], authors proposed a trajectory data preprocessing method to integrate trajectories with spatial knowledge. Their application concerned daily trips of employees from home to work and back. However, this work is limited to the formal definition of semantic trajectories with spatial and time knowledge without any implementation and evaluation. In [15], authors proposed a trajectory computing platform which exploits a spatio-semantic trajectory model. One of the layers of this platform is a data preprocessing layer which cleanses the raw GPS feed, in terms of preliminary tasks such as outliers removal and regression-based smoothing. Based on a space-time ontology and events approach, in [5] authors proposed a generic meta-model for trajectories to allow independent applications. They processed trajectories data benefit from a high level of interoperability, information sharing. Their approach is inspired by ontologies, however the proposed resulting system is a pure database approach. This work elaborated a meta-model to represent moving objects using an ontology mapping for locations. In extracting information from the instantiated model, this work seems to rely on a pure SQL-based approach not on semantic queries.

3 Modeling Approach

3.1 Design and Methodology

Our work is based on moving objects trajectories. This requires a trajectory data model and a moving object model. Moreover, to enrich data with knowledge, a semantic model should be taken into consideration. Therefore, we need a generic model to consider the trajectory, moving object and semantic models simultaneously. This is represented by a semantic trajectory model shown in Figure 1. This model can consume captured data of trajectories and other external data

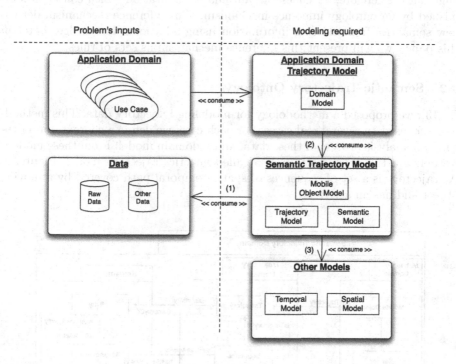

Fig. 1. Problem and its modeling required

as shown in Figure 1 link (1). These data are related to an application domain. This requires an application domain trajectory model which consists of domain model, as shown in Figure 1 link (2). The latter will support semantics related to users' needs. In the domain model, we also find the necessary semantics related to the real moving object, its trajectories, its activities and others. This semantics is often designed by a domain expert. In general, considering various facets of data, the semantic trajectory model must be extended by other models: application domain, temporal and spatial models. Then, the main issue is to build and design the semantic trajectory model with its required components.

The semantic trajectory modeling approach is tightly related to the problem of a semantic gap between this model and raw data. Link (1) in Figure 1 presents this gap. Moreover, our approach involves multiple models and then must establish semantic mappings among them, to ensure interoperability. In Figure 1, links (2) and (3) match the domain, temporal and spatial models with the semantic trajectory model. This matching extends the capabilities of our approach. For more efficient semantic capabilities, we want to annotate the data with domain, temporal and spatial knowledge. These knowledge are defined by experts representing users' needs. Annotating data with these knowledge could be done automatically or manually. We cannot use a manual annotation over huge data. Therefore, we choose an automatic annotation which can be accomplished by an ontology inference mechanism. This inference mechanism derives new semantics from existing information using additional knowledge. Later in this paper, we will present this inference mechanism as sets of rules.

3.2 Semantic Trajectory Ontology

In [13], we proposed a methodology for modeling trajectory data. This methodology focused on several real cases. For each case, we define a context, data capture, an analysis process of these data, and a domain model. From these models, we define a trajectory pattern also called generic trajectory model, Figure. 2. A trajectory is a set of sequences of spatio-temporal path covered by a moving object and has an activity.

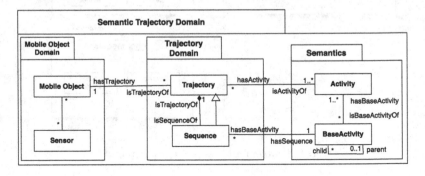

Fig. 2. Semantic trajectory modeling approach

The declarative part of our semantic trajectory ontology is presented in Figure. 3, named `owlSemanticTrajectory`. This ontology contains three models: mobile object, trajectory and semantic. By definition, a trajectory is a set of spatio-temporal concepts. Spatial and temporal models can be reused to enrich description of the concepts in the trajectory ontology to represent their spatial and temporal localization.

3.3 Reusing Time Ontology

The requirements of an ontology of time highlight the temporal concepts: `instant` and `interval`. The identification of temporal relationships leads to consider Allen temporal algebra [2]. In our approach, we consider an ontology of temporal concepts named `OWL-Time` [1] [8] developed by the World Wide Web Consortium (W3C).

`OWL-Time` ontology has a precise specification of temporal concepts and relationships as defined in the theory of Allen, formalized in OWL. An extract of the declarative part of this ontology is shown in Figure. 4 and described in detail in [9].

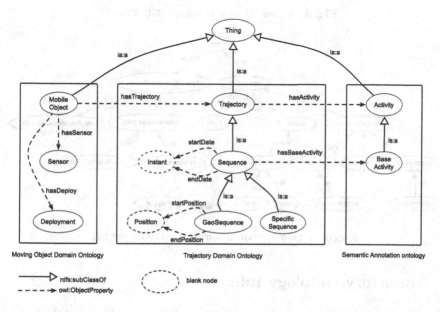

Fig. 3. A view of the semantic trajectory ontology `owlSemanticTrajectory`

3.4 Reusing Spatial Ontology

The requirements of spatial ontology highlight spatial concepts, such as `point`, `line` and `polygon` concepts and others. The identification of spatial relationships leads to consider spatial relationships such as: `Equals, Within, Touches, Disjoint, Intersects, Crosses, Contains` and `Overlaps`.

The standard `OGC OpenGIS` [10] presents spatial objects and functions over these objects. This standard contains a precise definition of spatial classes and reference systems. We transform this model into a formal ontology called `owlOGCSpatial` ontology. Figure 5 presents an extract of this ontology.

---

[1] http://www.w3.org/2006/time

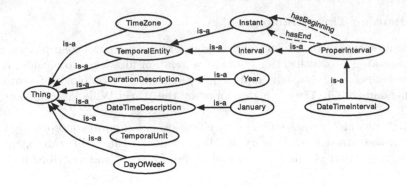

Fig. 4. A view of time ontology OWL-Time

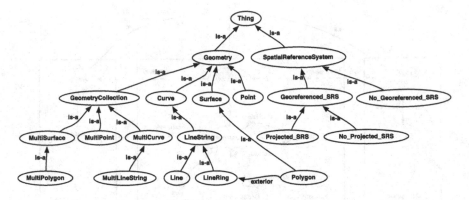

Fig. 5. A view of spatial ontology owlOGCSpatial

4   Trajectory Ontology Inference

Inference is the ability to make logical deductions based on ontologies, and optionally individuals. It derives new knowledge based on rules. A rule's definition, Figure. 6, has an antecedent, filters and a consequent. If knowledge are represented using RDF triples, then the antecedent is a set of triples, filters apply restrictions, and finally consequent is a new derived triple.

Inference Using Standard Rules. Our trajectory ontology owlSemantic Trajectory is based on RDF, RDFS and OWL constructs. Inference mechanism associates with each construct a rule. The results sets are called standard rules. An example of standard rules is OWLPrime in Oracle RDF triple store [1].

Inference Using Temporal Rules. Our trajectory ontology owlSemantic Trajectory uses temporal relationships as defined by Allen's algebra [2]. Each relationship is defined as a rule, intervalAfter, intervalBefore, intervalDuring, etc.

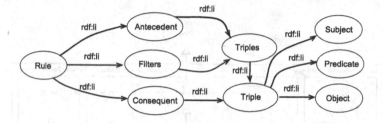

Fig. 6. Rule's definition

Inference using spatial rules . Our ontology `owlSemanticTrajectory` uses spatial relationships as defined by the Dimensionally Extended Nine-Intersection Model (DE-9IM) [6, 7]. Each relation is defined as a rule, `Contains`, `Overlaps`, etc.

5   Trajectory Ontology Inference Using Domain Rules

Our application domain is seals' trajectories, where a seal is considered as a mobile object. The captured data comes from the LIENSs laboratory[2] in collaboration with SMRU[3]. We consider three main states of a seal : `Dive`, `Haulout` and `Cruise`. Every state is related to a seal's activity, like `Resting`, `Traveling` and `Foraging`.

The captured data can also contain some meta-data CTD (Conductivity-Temperature-Depth) about the marine environment such as water conductivity, temperature and pressure. Starting from our trajectory ontology `owlSemantic Trajectory`, we define the seal trajectory ontology, named `owlSealTrajectory`, Figure 7. Formally, each activity is declared in the ontology and associated to a domain rule.

6   Implementation

Our implementation framework uses Oracle RDF triple store [1]. Based on a graph data model, RDF triples are persisted, indexed and queried, like other object-relational data. In this framework, we create the following models and rulebases (a set of rules):

- `owlTrajectory, owlTime, owlOGCSpatial, owlSealTrajectory`: declarative part of the trajectory, time, spatial and seal ontologies;
- `OWLPrime`: rulebase of the standard rules;
- `Time_Rules, Spatial_Rules, Seal_Rules`: rulebase of the temporal, spatial and seal rules, respectively.

---

[2] Lab. CNRS/University of La Rochelle - http://lienss.univ-larochelle.fr
[3] SMRU: Sea Mammal Research Unit - http://www.smru.st-and.ac.uk

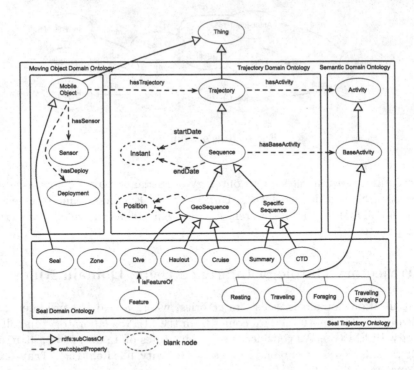

Fig. 7. Overview of the seal trajectory ontology with their activities

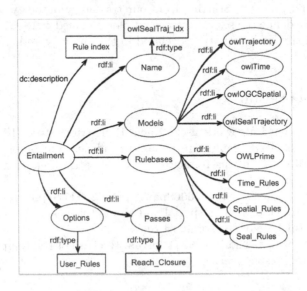

Fig. 8. Seal ontology inference: entailment

In our framework, inference mechanism creates a rule index, Figure 8. A rule index (entailment) is an object containing pre-computed triples from applying a specified set of rulebases to a specified set of models. If a graph query refers to any rulebases, a rule index must exist for each rulebase-model combination in the query. The USER_RULES=T option is required while applying user-defined rules. The default number of rounds that the inference engine should run is SEM_APIS.REACH_CLOSURE.

7   Trajectory Ontology Inference Refinement

The inference mechanism computes data relationships and annotates each one with activities of the moving object. This mechanism is needed for queries on the spatio-temporal trajectory ontologies. Our objective is to enhance the inference mechanism as much as possible, particularly, when using user-defined rules.

The inference engine has many computation cycles. The inference time increases seemingly out of proportion when using user-defined rules. To control this number of cycles, we define an inference refinement, named **passes refinement**. In this refinement, algorithm 1 limits these cycles into one pass. However, we keep the assurance of the results quality. The idea is to persist the inference results into a database from the first pass and use these results for the other passes. The algorithm 1 considers the list of objects (L_SP) and list of relationships (L_R). For each two objects, we check the existence of a relationship between them in the database, otherwise we compute it and persist it in the database.

---

input   : List of the objects: L_SP
input   : List of Relationships: L_R
initialization;
for $S_r, S_a \in L\_SP$ **do**
   if L_R between $(S_r, S_a) \notin database$ **then**
      $Res :=$ calculate L_R between S_r and S_a;
      Save Res in the $database$;
   end
end

Algorithm 1. Passes refinement algorithm

8   Evaluation

Experiments and evaluations are performed over real seal trajectories, around 410 690 raw data as seal's dives and 1 255 raw data as seal's haulout. So, in our system we have at least 5 749 660 triples in the triple store. In this work, we will focus on spatial rules. We evaluate the proposed passes refinement. The relationships, considered in algorithm 1, are the spatial rules. The experimental results are shown in Figure 9. The impact results are shown by the following experiments:

1. Spatial rules: presents the number of executions of the spatial rules;
2. Spatial rules - passes refinement: displays the number of executions of the spatial rules when applying the passes refinement.

In this experiment, the number of spatial rules executions decreases while applying the passes refinement. For example, considering 300 dives, the spatial rules are executed 2 000 000 times. However, in the passes refinement case, the spatial rules are executed 500 000 times. The passes refinement enhances 4 times in terms of executions and time computations.

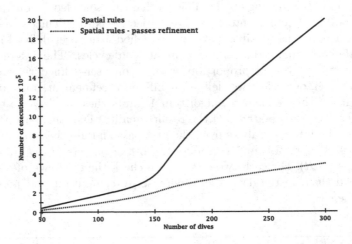

Fig. 9. Evaluation of the spatial ontology inference over the passes refinement

9   Conclusion

Trajectories are usually available as raw data. The data lack semantics which is fundamental for their efficient use. Our work is based on an ontology modeling approach for semantic trajectories. However, the domain part of the trajectory ontology focuses on mobile object's characteristics and its trajectory's activities. Our model considers a trajectory as a spatio-temporal concept, then we map it to temporal and spatial models. We reuse the W3C OWL-Time ontology and the spatial ontology based on the OpenGIS Simple Features Interface Standard (SFS). We apply our modeling approach to an application domain: marine mammal tracking, in particularly seal trajectories. In this implementation, we consider RDF triple store. Technically, we use Oracle Semantic Data Technologies. The objective is to annotate data with the considered knowledge: the domain, temporal and spatial knowledge. Over huge data, an ontology inference mechanism can do this automatic annotation. The inference mechanism therefore becomes an expensive mechanism in terms of time computations. To reduce the inference complexity, we propose a passes refinement. We evaluate it on real-world trajectory data. The experimental results highlight the positive impact of the passes refinement. We approximately reduce half of the time computation of the inference mechanism.

References

1. Oracle Database Semantic Technologies Developers guide 11G Release 2. Technical report (2012)
2. Allen, J.F.: Maintaining knowledge about temporal intervals. Communications of the ACM 832–843 (1983)
3. Alvares, L.O., Bogorny, V., Kuijpers, B., Macedo, A.F., Moelans, B., Vaisman, A.: A model for enriching trajectories with semantic geographical information. In: Proceedings of the 15th annual ACM international symposium on Advances in geographic information systems, pp. 22:1–22:8. ACM (2007)
4. Baglioni, M., Macedo, J., Renso, C., Wachowicz, M.: An ontology-based approach for the semantic modelling and reasoning on trajectories. In: Song, I.-Y., Piattini, M., Chen, Y.-P.P., Hartmann, S., Grandi, F., Trujillo, J., Opdahl, A.L., Ferri, F., Grifoni, P., Caschera, M.C., Rolland, C., Woo, C., Salinesi, C., Zimányi, E., Claramunt, C., Frasincar, F., Houben, G.-J., Thiran, P. (eds.) ER Workshops 2008. LNCS, vol. 5232, pp. 344–353. Springer, Heidelberg (2008)
5. Boulmakoul, A., Karim, L., Lbath, A.: Moving object trajectories meta-model and spatio-temporal queries. International Journal of Database Management Systems (IJDMS) 35–54 (2012)
6. Clementini, E., Di Felice, P., van Oosterom, P.: A small set of formal topological relationships suitable for end-user interaction. In: Abel, D.J., Ooi, B.-C. (eds.) SSD 1993. LNCS, vol. 692, pp. 277–295. Springer, Heidelberg (1993)
7. Clementini, E., Sharma, J., Egenhofer, M.J.: Modelling topological spatial relations: Strategies for query processing. Computers and Graphics 815–822 (1994)
8. Hobbs, J.R., Fang, P.: Time ontology in OWL. W3C recommendation (2006)
9. Jerry, R.H., Feng, P.: An ontology of time for the semantic Web. ACM Transactions on Asian Language Information Processing 66–85 (2004)
10. Herring, J.R.: OpenGIS implementation standard for Geographic information - simple feature access - part 2: SQL option. Open Geospatial Consortium Inc. (2011)
11. Malki, J., Bouju, A., Mefteh, W.: An ontological approach modeling and reasoning on trajectories. Taking into account thematic, temporal and spatial rules. In: TSI. Technique et Science Informatiques, vol. 31/1-2012, pp. 71–96 (2012)
12. Spaccapietra, S., Parent, C., Damiani, M., Demacedo, J., Porto, F., Vangenot, C.: A conceptual view on trajectories. Data and Knowledge Engineering 126–146 (2008)
13. Wannous, R.: Trajectory ontology inference considering domain, temporal and spatial dimensions. Application to marine mammals. Ph.D. thesis, La Rochelle University (2014)
14. Wannous, R., Malki, J., Bouju, A., Vincent, C.: Time integration in semantic trajectories using an ontological modelling approach. In: New Trends in Databases and Information Systems, pp. 187–198. Springer, Heidelberg (2013)
15. Yan, Z., Parent, C., Spaccapietra, S., Chakraborty, D.: A hybrid model and computing platform for spatio-semantic trajectories. In: Aroyo, L., Antoniou, G., Hyvönen, E., ten Teije, A., Stuckenschmidt, H., Cabral, L., Tudorache, T. (eds.) ESWC 2010, Part I. LNCS, vol. 6088, pp. 60–75. Springer, Heidelberg (2010)

Interpretation of DD-LOTOS Specification by C-DATA*

Maarouk Toufik Messaoud[1]([✉]), Saidouni Djamel Eddine[2], Mahdaoui Rafik[1], and Houassi Hichem[1]

[1] Faculty of ST, ICOSI Lab, University Khenchela, BP 1252 EL Houria,
40004 Khenchela, Algeria
tmaarouk@gmail.com, {mehdaoui.rafik,houassi_h}@yahoo.fr
[2] Faculty of NTIC, MISC Lab, Univ Constantine 2, Constantine, Algeria
saidouni@hotmail.com

Abstract. The DD-LOTOS language is defined for the formal specification of distributed real-time systems. The peculiarity of this language compared to existing languages is its taken into account of the distributed aspect of real-time systems. DD-LOTOS has been defined on a semantic model of true concurrency ie the semantics of maximality. Our work focuses on the translation of DD-LOTOS specifications to an adequate semantic model. The destination model is a communicating timed automaton with durations of actions, temporal constraints and supports communication between localities; this model is called C-DATA*.

Keywords: Real time system · Duration of actions · Formal specification · DD-LOTOS language · C-DATA*

1 Introduction

The specification of real-time systems is very important step, these systems are everywhere in our environment. Moreover, they are often part of critical systems in various fields, such as aviation, industrial process control, control of nuclear power plants, etc.

Formal methods play a fundamental role in the various stages of the process engineering of computer systems especially in real-time systems. Formal methods allows to specify a critical system using formalisms, langanges and models defined on a formal semantics. The essential feature in this approach is unambiguously specify critical systems in the different phases of its life cycle, and to validate a number of its requirements. In recent years the proposal of models, languages and formalism defined on a formellle semantics, able to specify critical real-time systems have known a lot of progress. Most research focuses on the extension of existing models and in particular on the process algebra. Among these works focus on the extension of LOTOS[3][6][4][5][10]. The process algebra is a formal framework for the specification and analysis of complex systems in general, and in particular real-time systems.

© Springer International Publishing Switzerland 2015
T. Morzy et al. (Eds): ADBIS 2015, CCIS 539, pp. 414–423, 2015.
DOI: 10.1007/978-3-319-23201-0_42

Our aim is to design an environment for the formal specification and compilation of concurrent systems. The specification is described in formal language DD-LOTOS, this language allows to describe distributed systems with temporal constraints, then the specification is translated into the semantic model C-DATA*[7] to be verified by the formal verification tools.

2   Distributed D-LOTOS Language

2.1   Syntax

The DD-LOTOS[7] language represents an extension of D-LOTOS language to support the distribution and communication between the localities, it was enriched with the following features:

- The explicit distribution,
- Remote communication.

Distribution is ensured by introducing the notion of locality. Localities exchange information by the message exchange paradigm. The syntax of DD-LOTOS is defined as follows:

$$E ::= \textbf{Behaviors}$$
$$stop \mid exit\{d\} \mid \Delta^d E \mid X[L] \mid$$
$$g@t[SP]; E \mid i@t\{d\}; E \mid hide\, L\, in\, E \mid$$
$$E[]E \mid E|[L]|E \mid E \gg E \mid E[> E$$
$$a!v\{d\}; E$$
$$a?xE$$
$$S ::= \textbf{Systems}$$
$$\phi \mid S \mid S \mid l(E)$$

Fig. 1. *Syntax of DD-LOTOS*

Let PN, ranged over by $X, Y...$, be an infinite set of process identifiers, and let $\mathcal{G}$, ranged over by g, set of gates (observable actions). $i \notin \mathcal{G}$ is the internal action and $\delta \notin \mathcal{G}$ is the successful termination action. $Act = \mathcal{G} \cup \{i, \delta\}$, ranged over by α, is the set of actions. L denotes any finite subset of $\mathcal{G}$. The terms of DD-LOTOS are named behavior expressions, $\mathcal{B}$ ranged over by $E, F, ...$ denotes set of behavior expressions.

Let $\mathcal{D}$ be a domain of time. $\tau : Act \rightarrow \mathcal{D}$ is the duration function which associates to each action its duration. We assume $\tau(i) = \tau(\delta) = 0$. Let g be an action, E a behavior expression and $d \in \mathcal{D}$ a value in the temporal domain.

The main syntax concerns the syntax of systems S and behavior expression E. The informal semantics of syntactic items is the following:

- Informally $a\{d\}$ means that action a has to begin its execution in a temporal interval $[0, d]$. $\Delta^d E$ means that no evolution of E is allowed before the end of a delay equal to d. In $g@t[SP]; E$ (resp. $i@t\{d\}; E$) t is a temporal variable recording the time taken after the sensitization of the action g (resp. i) and which will be substituted by zero when this action ends its execution.
- The basic operators of process algebras as: nondeterministic choice $E[]E$, parallel composition $E|[L]|E$, the interiorization $hide\ L\ in\ E$, sequential composition $E \gg E$, and preemption $E\ [> E$.
- The expression $a!v\{d\}; E$, specifies the emission message v via the communication channel a. This emission operation must occur in the temporal interval $[0, d]$.
- On the other side, the behavior expression $a?xE$ specifies the message receiving on channel a. The received message substitutes the variable value x. This variable is used in the behavior expression E.
- A system may be either:
  - Empty, expressed by ϕ,
  - The composition of sub systems $S \mid S$, or
  - A behavior expression E in a locality l expressed by $l(E)$.

Definition 1. *(actions) The actions in global system are:*

- Set of communication actions between localities: are emission or receiving messages through a communication channel $Act_{com} ::= a!m\ |a?x\ |\tau$ (output actions, input actions and the silent action).
- Set $Act = \mathcal{G} \cup \{i, \delta\}$ previously defined.

Definition 2. *(Localities and channels): The set $\mathcal{L}$ ranged over by l, denotes set of localities. ϑ an infinite set of channels defined by users ranged over by $a,b,...$ channels are used for communication message between localities.*

2.2   Structured Operational Semantics

The operational semantics of behaviors are given by the operational semantics of D-LOTOS. This semantic is extended to DD-LOTOS by giving the semantics rules for communicated systems as follows:

Process $a!v\{d\}; E$: Let us consider the configuration $_M[a!v\{d\}; E]$, the emission of the message v begins once the actions indexed by the set M have finished their execution, conditioned by the condition $Wait(M)$ which must be equal to *false* in rule 1. Rules 2 and 3 express the fact that the time attached to the process of sends cannot begin to elapse until all the actions referenced by M are finished. Rule 4 imposes that the occurrence of the action of sends takes place for the period d, otherwise the process is transformed to *Stop*.

1. $\dfrac{\neg Wait(M)}{_M[a!v\{d\};E] \xrightarrow{Ma!v\,x} {}_{\{x:a!v:t\}}[E]}$ $x = get(\mathcal{M})$

2. $\dfrac{Wait(M^{d'})\ or\ (\neg Wait(M^{d'})\ and\ \forall \varepsilon>0.\ Wait(M^{d'-\varepsilon}))\qquad d'>0}{_M[a!v\{d\};E] \xrightarrow{d'} {}_{Md'}[a!v\{d\};E]}$

3. $$\frac{\neg Wait(M)}{_M[a!v\{d'+d\};E] \xrightarrow{d} {}_M[a!v\{d'\};E]}$$

4. $$\frac{\neg Wait(M) \ and \ d'>d}{_M[a!v\{d\};E] \xrightarrow{d'} {}_M[stop]}$$

Process $a?xE$: Let us consider the configuration $_M[a?xE]$, the following rule expresses that the receiving starts once the action indexed by set M have finished their execution.

$$\frac{\neg Wait(M)}{_M[a?xE] \xrightarrow{M^{a?x}y} {}_{\{y:a?x:0\}}[E]}$$

Remote Communication

Distributed activities exchange messages between them, the expression $l(a!v\{d\})$, expresses that the message v is offered for a duration d. By an activity at the locality l, the message v should be sent on channel a. On the other side, $k(a?xE)$ specify that activity E in locality k is ready to receive a message on channel a. The following rule defines the remote communication between two distributed activities via the channel a. In this case communication will be specified by silent (i) evolution as follows: (action silencieuse τ:

$$\frac{-}{_M[l(a!v\{d\};E1)]|_{M'}[k(a?xE2)] \xrightarrow{\tau} {}_M[l(E1)]|_{M'}[k(E2\{v/x\})]}$$

Time Evolution on System

$$\frac{E \xrightarrow{d} E'}{l(E) \xrightarrow{d} l(E')}$$

$$\frac{S_1 \xrightarrow{d} S'_1 \quad S_2 \xrightarrow{d} S'_2}{S_1 \mid S_2 \xrightarrow{d} S'_1 \mid S'_2}$$

3 Distributed Semantic Model for Distributed Realtime Systems

The behavior of a real system can be represented by a transition system under certain assumptions of abstraction. The formalism of timed automata (TA's) was introduced by Rajeev Alur and David Dill in [1]. Its definition provides a simple way to provide transitions systems a set of temporal constraints expressed using real variables called clocks. The model of timed automata is constructed in conformity with the hypothesis of structural and temporal atomicity actions.

The model Durational Action Timed Automata (DATA)[2] is introduced with the aim of take into account the explicit durations of action, an extension of this approach to systems with constrained time in order to take into account the temporal constraints and the urgency of action, this model is called DATA*[9]. The objective of this section is the presentation of a semantic models that can support the distributed aspect of real-time distributed systems.

3.1   Durational Action Timed Automata(DATA*)

In this section, we describe a method for taking into account of non-atomicity temporal and structural of actions in timed automata, through the DATA model. In general, temporal constrained systems can not be completely specified if we do not consider concepts such urgency, deadlines, constraints, etc. To account for these concepts, the DATA* model's was defined in[9].

Let H, ranged over by $x, y...$ be a set of clocks with non-negative (in a time domain T as Q^+ or R^+). The set $\Phi_t(H)$ of temporal constraints γ Over H, is defined by the syntax $\gamma ::= x \sim t$, where x is a clock in H, $\sim \in \{=, <, >, \le, \ge\}$ and $t \in Q^+$. A DATA * A is a quintuplet (S, L_S, s_0, H, T) tel que:

1. S is a finite set of states, and
2. $L_S : S \to 2_{fn}^{\Phi_t(H)}$ is a function which corresponds to each state s the set F of ending conditions(duration conditions) of actions possibly in execution in s,
3. $s_0 \in S$ is the initial state,
4. H is a finite set of clocks, and
5. $T \subseteq S \times 2_{fn}^{\Phi_t(H)} \times 2_{fn}^{\Phi_t(H)} \times Act \times H \times S$ is the set of transitions. A transition (s, G, D, a, x, s) represents switch from state s to state s, by starting execution of action a and resetting clock x. G is the corresponding guard which must be satisfied to fire this transition. D is the corresponding deadline which requires, at the moment of its satisfaction, that action a must occur. (s, G, D, a, x, s) can be written $s \xrightarrow{G, D, a, x} s'$.

The semantics of a DATA* $A = (S, L_S, s_0, H, T)$ is defined by associating to it an infinite transitions system S_A over $Act \bigcup T$. A state of S_A (or configuration) is a pair $< s, v >$ such as s is a state of A and v is a valuation for H. A configuration $< s_0, v_0 >$ is initial if s_0 is the initial state of A and $\forall x \in H$, $v_0(x) = 0$. Two types of transitions between S_A configurations are possible, and which correspond respectively to time passing and launching of a transition from A.

Example of DATA * is given in figure 2.

3.2   Communicating Durational Action Timed Automata (C-DATA)

C-DATA*[7] is a semantic model that allows taking into account of all the aspects present in the DATA* s model, such as non-atomicity temporal and structural of the actions, urgency of the actions, deadlines and temporal constraints. In the C-DATA* each locality is represented by a DATA*, the global system is represented by the set of DATA* s locals, which communicate by exchanging messages through communication channels

Definition 3. *A Communicating DATA (C-DATA) $A(S, L_S, s_0, \vartheta, H, \Pi, T_D)$ represents a subsystem with:*

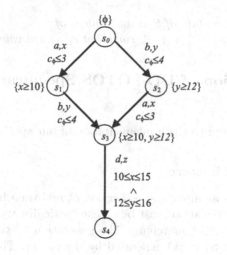

Fig. 2. Example of DATA*

- S is a finite set of states,
- $L_S : S \to 2_{fn}^{\Phi_t(H)}$ is a function which corresponds to each state s the set F of ending conditions(duration conditions) of actions possibly in execution in s,
- $s_0 \in S$ is the initial state,
- ϑ is the alphabet of the channels on which messages flow between the subsystems.
- H is a finite set of clocks,
- $\Pi = Act_{com} \cup Act$, is the set of internal and communication actions of A, and
- $T_D \subseteq S \times 2_{fn}^{\Phi_t(H)} \times 2_{fn}^{\Phi_t(H)} \times \Pi \times H \times S$ is the set of transitions.
  A transition $(s, G, D, \alpha/(a(!/?)v)/i, z, s\prime)$ represents switch from state s to state $s\prime$, by starting execution of action $\alpha \in Act$ or actions (Sending or Receiving) or synchronization for the accomplishment of communication (silent action) and updating clock z.
  G is the corresponding guard which must be satisfied to fire this transition.
  D is the corresponding deadline which requires, at the moment of its satisfaction, that action α must occur.
  $(s, G, D, \alpha/(a(!/?)v)/\tau, z, s\prime)$ can be written
  $$s \xrightarrow{\;\;G\,,\;D\,,\;alpha\;/\;(\;a\;(\;!/\;?\;)v)\;/i\;),\;z\;\;} s'.$$

Definition 4. *1. System:* A system of n C-DATA is a tuple $S = (A_1, \ldots, A_n)$, with $A_i = (S_i, L_{S_i}, s_{0_i}, \vartheta, H_i, \Pi_i, T_{iD})$ a C-DATA.
2. *States:* $GS(S) = (s_1, v_1) \times \ldots \times (s_n, v_n) \times (\vartheta^*)^p$, is The set of states.
3. *Initial State:* The initial state of S is: $q_0 = ((s_{01}, 0), \ldots, (s_{0n}, 0) : \epsilon_1, \ldots, \epsilon_p)$ such as ϵ is the empty word on the alphabet ϑ
4. *System States:* Let $S = (A_1, \ldots, A_n)$ a system of n C-DATA, $A_i = (S_i, L_{S_i}, s_{0_i}, \vartheta, H_i, \Pi_i, T_{iD})$:
   A global state of S is defined by the state of each subsystem and the states

of each channel, a state of S is an element of
$(s_1, v_1) \times \ldots \times (s_n, v_n) \times (\vartheta^*)^p$ *such that $v_i(h)$ are valuations on H.*

4   Interpretation of DD-LOTOS Specifications to C-DATA*

This section is devoted to the implementation of our specification environment.

4.1   System Architecture

The system receives as input a specification of real-time behavior expressed in DD-LOTS. This specification must be checked lexically and syntactically before generating the C-DATA* matching. The generation of the C-DATA* is from AST (Abstract Tree Synrtax) generated by the parser. The architecture of our system is as follows:

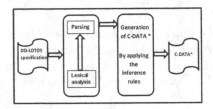

Fig. 3. System architecture

4.2   The Analysis Steps

From the syntax of the DD-LOTOS, we define the grammar below: the terminals are bold, non-terminals in italics. The axiom of the grammar is the non-terminal SPECIFICTION.

DD-LOTOS Specification:
$< specification >::= SYSTEM < entete >:=< code > ENDSYS$
$< entete >::= id[< params >]$
$< params >::= id < last\_params > |id[nombre] < last\_params > |epsilon$
$< last\_params >::=, id < last\_params > |, id[nombre] < last\_params > |epsilon$
$< code >::=< expression >< where\_exp > |epsilon$
$< where\_exp >::= WHERE < decl\_proc >< other\_decl > |epsilon$
Process:
$< processus >::= PROCESS < entete >:=< code > ENDPROC$
$< other\_proc >::=< processus >< other\_proc > |epsilon$
Behaviors:
$< expression >::= STOP|EXIT\{Duree\}|NIL$

$| < expression >< opr >< expression >$
$|HIDE < hide\_params > IN < expression >$
$|(< expression >)$
$| < expression >< para - op >< expression >$
$|DELAY < Duree >< expression >$
$|G[< params >]$
$| < act >; < expression >$
$|G@ < Entier > [SP]; < expression >$
$|i@ < Entier > \{Duree\}; < expression >$
$< hide\_params >::= id < last\_params>$
And grammar of operators, identifiers, and durations of actions.

4.3 Representation of a C-DATA*

The C-DATA* will be represented by a graph whose nodes represent the states and the edges represent transitions. An abstract syntax tree AST nodes and leaves.

4.4 AST Transformation Algorithm to a Graph

This algorithm transforms an AST into a graph.
Input: AST.
Output: Graph that represents the list of configurations representing the initial specification.
Begin
While there are lines to visit in the AST Do

- Apply the corresponding operational rule, calls the implementation procedure of each rule,
- Extract the resulting configuration,
- Skip to the next line.

End While
End

4.5 Example (sender2receivers)

In this section we will try to apply our application on example of sender2receivers[8]. DD-LOTOS specification of this example is as follows:

Specification $sender2receivers[a, b]$
  Behavior
      $l(E)|k(P)|n(Q).$

  Where
      Process $E ::= E1|||E2$
      Where

$$E1 ::= (a!v\{5\}|b!v\{5\}) >> DELAY\,5E1$$
$$E2 ::= (c?xB) >> E2$$

Endproc

Process $P ::= (a?xB\,[]\,DELAY\,5; c!nack\{5\}) >> P$

Endproc

Process $Q ::= (b?xB\,[]\,DELAY\,5; c!nack\{5\}) >> Q$

Endproc

Endspec

Once the compilation completes successfully, we generates the corresponding C-DATA*.

The execution of this example is shown in the figure 4:

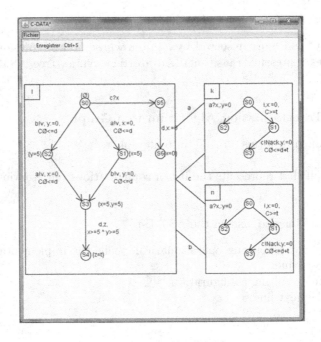

Fig. 4. Final C-DATA*

5   Conclusion

In this paper we presented a contribution to the specification of real-time distributed systems, with explicit duration of actions. In a previous work[7], we introduced the notion of locality required for modeling the distributed aspect of distributed systems.

The main interest of our approach is the proposal of a language defined on true concurrency semantics: semantics of maximality[10] which allows the explicit expression of durations, and it supports temporal constraints including urgency of actions. Concerning communication, we have defined local and remote communication, when two processes want to communicate, so are on the same locality, then the communication is ensured through the gates which are defined locally. If both processes are on two different localities then the message exchange is the way of communication.

References

1. Alur, R., Dill, D.L.: A Theory of Timed Automata. Theoretical Computer Science **126**, 183–235 (1994). Elsevier
2. Belala, N., Saïdouni, D.E.: Non-Atomicity in Timed Models, International Arab Conference on Information Technology, Al-Isra Private University, Jordan, LIRE Laboratory, University of Mentouri, 25000 Constantine, Algeria (December 2005)
3. Bolognesi, T., Lucidi, F.: LOTOS-like process algebras with urgent or timed interactions. In: Parker, K.R., Rose, G.A. (eds.) FORTE. IFIP Transactions, vol. C-2, pp. 249–264. North-Holland (1991)
4. Courtiat, J.P., de Oliveira, R.C.: On RT-LOTOS and its Application to the Formal Design of Multimedia Protocols. Annals of Telecommunications **50**, 11–12 (1995)
5. Courtiat, J.P., Santos, C.A.S., Lohr, C., Outtaj, B.: Experience with RT-LOTOS, a temporal extension of the LOTOS formal description technique. Computer Communications **23**, 1104–1123 (2000). Elsevier
6. Léonard, L., Leduc, G.: A Formal Definition of Time in LOTOS - Extended Abstract. Formal Aspects of Computing **10**, 248–266 (1998). BCS
7. Maarouk, T.M., Saïdouni, D.E., Khergag, M.: DD-LOTOS: a distributed real time language. In: Proceedings 2nd Annual International Conference on Advances in Distributed and Parallel Computing, Special Track: Real Time and Embedded Systems, pp. 45–50. Singapore (2011)
8. Maarouk, T.M.: Modèles formels pour la conception des systèmes temps réel, thèse de Doctorat. Laboratoire MISC, Constantine, Algérie (2012)
9. Saïdouni, D.E., Belala, N.: Actions duration in timed models. In: International Arab Conference on Information Technology. Yarmouk University, Irbid, December 2006
10. Saïdouni, D.E., Courtiat, J.P.: Prise En Compte Des Durées D'action Dans Les Algèbres deProcessus Par L'utilisation de la Sémantique de Maximalité. Ingénierie des Protocoles, Hermes (2003)

First International Workshop on Semantic Web for Cultural Heritage (SW4CH 2015)

Knowledge Representation in EPNet

Alessandro Mosca[1]([✉]), José Remesal[2], Martin Rezk[3], and Guillem Rull[2]

[1] SIRIS Lab, Research Division of SIRIS Academic, Barcelona, Spain
a.mosca@sirisacademic.com
[2] CEIPAC, University of Barcelona, Barcelona, Spain
[3] KRDB Research Centre, Free University of Bozen-Bolzano, Bolzano, Italy

Abstract. Semantic technologies are rapidly changing the historical research. This paper focuses on the knowledge representation and data modelling initiative that has characterised the first year of the EPNet project in the context of the historical research. The so-called EPNet CRM and Ontology are introduced here, and put in connection with existing modelling standards. The formal specification of the domain expert knowledge is the preliminary step toward the design of an innovative 'Virtual Research Environment' for scholars of the Roman Empire. Potential and actual benefits coming from the knowledge representation initiative are also discussed, with the main aim of encouraging experts in the humanities in embracing an innovative paradigm whose proved efficacy can positively affect their current research practices.

1 Semantic Technologies Enter the Humanities

Semantic technologies [1–3] are rapidly changing the historical research and, more in general, the humanities. Over the last decades, an immense amount of new quantifiable data have been accumulated, and made available in interchangeable formats, from social sciences to economics, opening up new possibilities for solving old questions and posing new ones [4]. Historians, especially in Digital Humanities, are starting to use new datasets to aggregate information about history: collections of data, information, and knowledge that are devoted to the preservation of the legacy of tangible and intangible culture inherited from previous generations. Moreover, the recent advances in computing and computational tools make feasible to meaningfully manipulate, manage, and analyse these datasets.

Several public initiatives and projects have been funded to address the issue of building, and make public through the web historical and cultural data. Among other, the following are worth to be mentioned here, since they represent pioneering efforts in the application of semantic technologies toward the development of e-culture portals providing multimedia access to distributed collections of cultural heritage objects: EUROPEANA[1], EAGLE[2], CIDOC CRM[3], GETTY Vocabularies[4],

---

[1] http://www.europeana.eu
[2] http://http://eagle-network.eu
[3] http://www.cidoc-crm.org
[4] http://www.getty.edu/research/tools/vocabularies

© Springer International Publishing Switzerland 2015
T. Morzy et al. (Eds): ADBIS 2015, CCIS 539, pp. 427–437, 2015.
DOI: 10.1007/978-3-319-23201-0_43

CULTURESAMPO[5], STICH CATCH[6], MultimediaN N9C[7], CHIP[8], INCONCLASS[9], EPI-DOC[10], ARIADNE[11]..

The computational outcomes of these and similar projects resulted mostly into web-based virtual museum applications, collecting heterogeneous contents (e.g., paintings, music, movies, books) from various distributed sources. There, the users experience a digital environment where, for instance, they can look for a painting of a historic event, find information on the event along with other artwork depicting it, geo-locate it, and see where nearby events occurred and how they are represented in artwork [5]. Due to the lack of space, we cannot provide a detailed analysis of the above mentioned projects, but a more comprehensive list of projects, tools, ontologies, lexical resources, and online resources in the area of 'Semantic Web and history' can be found in [6].

The emphasis of the EPNet Project is on providing historians with computational tools to compare, aggregate, measure, geo-localise, and search data about latin and greek inscriptions on amphoras for food transportation. The EPNet approach relies on the Ontology-based Data Access (OBDA) paradigm [7,8], a specific semantic technology where different datasets are virtually integrated by a conceptual layer (an ontology). Differently from providing access to virtual museums or digitalised collections, the usage the OBDA implementation, by means of state-of-the-art principles coming from the Knowledge Representation area [9], is meant to support scholars in *experimentally verifying* theoretical hypotheses, and to formulate new ones. The example below intuitively shows the rationale behind choosing an OBDA approach for data access and integration:

Example. *Suppose the user looks for all the amphoras produced in 'La Corregidora' and its geo-coordinates. The EPNet dataset contains information about amphoras and some (potentially incomplete) information about geo-coordinates. On the other hand, the Pleiades dataset (http://pleiades.stoa.org) contains complete geo-coordinates but has no information about amphoras. In EPNet there are hundreds of types of amphoras such as Dressel 1, Dressel 2-4, Leptiminus 1, etc., each of them represented by a alphanumeric code such as "DR1C-BTIR". Thus creating a query for this simple information need is not only extremely complex, but requires the user to know the DB encoding of each type, the schemas in the datasources, and manually merge the information obtained from each of them. Ideally the user should be able to execute a single simple query that does no require any specific knowledge about the underlying data sources, and get all the available information coming from both datasets.*

Historical Research and the Impact of Technology in EPNet. The Roman Empire trade system is generally considered to be the first complex

---

[5] http://www.kulttuurisampo.fi

[6] http://www.cs.vu.nl/STITCH

[7] http://e-culture.multimedian.nl

[8] http://chip.win.tue.nl

[9] http://www.iconclass.nl

[10] http://epidoc.sourceforge.net

[11] http://www.ariadne-infrastructure.eu

European trade network. It formed an integrated system of interactions and interdependences between the Mediterranean basin and northern Europe. Over the last couple of centuries, scholars have developed a variety of theories to explain the organisation of the Roman Empire trade system, the majority of them continue to be speculative and difficult to *falsify* [10,11]. The ERC Advanced Grant EPNet ("Production and distribution of food during the Roman Empire: Economics and Political Dynamics", ERC-2013-ADG 340828), started in March 2014, aims at setting up an innovative framework to investigate the mechanisms and characteristics of the commercial trade system during the Roman Empire. The main objective of EPNet is to create an interdisciplinary experimental laboratory (the project team includes specialists from Social Sciences and Humanities and from Physical, and Computer Sciences) for the exploration, validation and falsification of existing theories, and for the formulation of new ones. This approach is made possible by (i) a large dataset of existing empirical data about Roman amphorae and their associated epigraphy that has been created during the last 2 decades and (ii) the front line theoretical research done by historians on the political and economic aspects of the Roman trade system.

The computational infrastructure of the EPNet Project takes the form of a 'Virtual Research Environment' offering: (i) a conceptual layer (ontology) driving the access to datasets stored into fragmented, heterogeneous and distributed digital repositories; (ii) a platform for sharing of expert knowledge on characterisation, typology and dating of Roman Empire epigraphies/artefacts; (iii) dedicated data visualisations and analytics tools. Taking into consideration the design and development of such a computational infrastructure, the conceptual modelling and knowledge representation effort in EPNet has been planned to address three main problems: (i) structuring and accessing large collections of data through the Web, (ii) providing a formally defined, unambiguous, framework for analysing the data and exporting them in a way that can be further manipulated by computer-based simulation and complex network analysis tools, and (iii) making each collection of data integrable with other complementary data sources. *Here is where interdisciplinarity and epistemological transparency require advanced semantic-aware data management methodologies and technologies, scholars need access to the data in a distributed worldwide-scaled environment, and semantic-based data access and integration becomes a 'sine qua non' requisite.*

Given the above requirements, a semantic-based data management can account for discrete data in addition to qualitative interpretations. In particular, it enables scholars to retrieve information in a domain-centred and scholar-friendly way, thus supporting the identification of patterns and trends in this information and discover relationships between disparate pieces of it. Semantic technologies support EPNet in facing the main challenge of providing users with: (i) a running technology for accessing data in a way that is conceptually sound with their own domain knowledge; (ii) a semantically-transparent platform, ready to acquire and be complemented with new data from different sources; (iii) a theoretically grounded mechanism to homogenise information stored in different formats and according to different conceptualisations.

The rest of the paper focuses on the results of the formal knowledge representation in EPNet: the construction of the EPNet Conceptual Reference Model (CRM) and the EPNet ontology, which unambiguously represent all the relevant epigraphic information and domain expert knowledge about Roman Empire latin inscriptions in the way it is understood by scholars in this research area (see, Sec. 2). The EPNet Ontology eases the integration with other datasets and supports the user with the possibility of accessing data through a domain-centred conceptual layer and terminology that is much simpler and easy to manage that the full CRM. A few hints are provided on the EPNet *semantic data integration, data consistency* checking, and *conceptual data access* systems (see, Sec. 3). A preliminary, testing-oriented, interface is hyperlinked is also introduced. Section 4 concludes the paper.

2   Knowledge Representation and Data Management in EPNet

The first result of the knowledge acquisition and modelling effort in EPNEt has been the so-called EPNet CRM. Our job started from the analysis and systematic review of the "Corpus de epigrafa anfrica latina" database (http://ceipac.gh.ub.es), deployed and maintained by the members of the CEIPAC research team at the University of Barcelona (Spain), and already running at the beginning of the project. The preliminary investigation recollects all the available information associated to the old CEIPAC dataset, highlights drawbacks and limits in the existing Corpus thus offering useful suggestions for the future development of a new EPNet conceptual model and relational database. In terms of *integrity constraints*, for example, the old Corpus database was characterised by the explicit presence of only the primary keys, and some unique constraints. Foreign key constraints were present from a semantic point of view, but they were not explicitly defined in the database. Nonetheless, two major modelling tasks also emerged from the analysis of the Corpus: (i) extending the coverage of the exposed data w.r.t. the domain of interest (completeness), and (ii) complementing the characterisation of the main objects (e.g. the geographical entities) present in the initial dataset (accuracy). Both of them have been achieved through the specification of a conceptual reference model and the derived ontology, on the basis of which a new semantic-based data management system has been then implemented.

The EPNet Conceptual Reference Model (CRM): The specification of the EPNet CRM for the representation of epigraphic information and domain expert knowledge about Roman Empire latin inscriptions is meant to unambiguously represents the way the data are understood by scholars, how they are connected together and what is their coverage w.r.t. the literature of reference and the current research practices in the history of the Roman Empire. The CRM has been formally specified in the conceptual modelling language called 'Object Role

Modelling' (ORM2 [12]), and by means of NORMA, a data modelling tool for ORM2[12]. It is worth to mention here that NORMA is equipped with a sound and complete reasoning module for ORM2 modelling support tool providing automated reasoning services facilitating the conceptual modelling activity. By relying on existing OWL2 reasoners (e.g. HermiT, FaCT++), ORMiE provides: (i) *implicit constraints deduction*, and (ii) *automatic translation into OWL2*. Previous results on the formal semantics of ORM2 and its sound and complete encoding into the description logic ALCQI provide the theoretical grounding for the design and implementation of ORMiE [13]. Having specified the EPNet CRM in ORM2 gave us the advantage of relying on innovative database technologies for *semi-automatically* deploying the new relation EPNet database in fifth normal form. Moreover, the EPNet CRM has been defined according to the state-of-the-art formal ontological models and standards for representing cultural heritage objects and the relationships between them. In particular, in order to increase the interoperability of the model, and of the whole EPNet dataset, with other similar initiatives and data sources, the main section of the model results in a specialisation/extension of the well known CIDOC CRM[13].

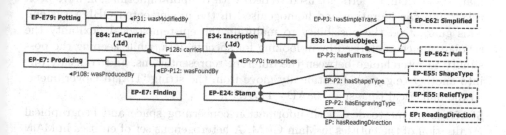

Fig. 1. A fragment of the orm2-based EPNet CRM where `Inscriptions` are related with the activities `Producing`, `Potting`, and `Finding`. `Stamps` are inscriptions characterised, among other, by their `Relief`, `Shape` type, and `ReadingDirection`. The model also shows that inscriptions are directly connected with 'simplified' and 'full' transcriptions, bringing information about their translation into contemporary languages, and the conservation status of them, respectively.

For the sake of the model maintenance, and according to the specific nature of the involved information, the EPNet CRM has been structurally organised into distinct interrelated sub-module. Moreover, according to the different aim of each sub-section, we relied on the import of existing standards for digital

---

[12] NORMA is an open source plug-in to Microsoft Visual Studio .NET freely downloadable from http://www.ormfoundation.org.
[13] The CIDOC Conceptual Reference Model provides a formal ontological model for describing the structure of cultural heritage objects and the relationships between them.

recording and publishing information on the Semantic Web, such as FaBiO[14] for the bibliographic references documenting the entities in the model, the Europeana Data Model (http://pro.europeana.eu/edm-documentation), and the EAGLE Matadata Model and domain-centred vocabularies (http://www.eagle-network.eu/about/documents-deliverables/). The following are the five main modules of the EPNet CRM:

MAIN deals with the representation of the main domain entities (e.g. inscriptions, amphoric types, associated epigraphic information), their properties (e.g. finding place, letter dimensions, archaeometric characterisation), and mutual relationships (see Fig. 1). Notice that an *ad-hoc* notational convention has been adopted in order distinguish: (i) concepts/relations that have been simply imported from CIDOC (the labelling format like 'E?' reports the CIDOC name of the concept/relation at hand), from (ii) concepts/relations that are meant to be specialisations of CIDOC elements (the 'EP-E?' indicates that the element is an EPNet specialisation of the CIDOC 'E?').

TIME offers a conceptual arrangement, driven by the experts, of the different modalities used to denote interval periods, dates, punctual instants of time, w.r.t. the given research domain. As usual in the historical research, the different formats the domain experts are used to deal with temporal information have been taken into consideration and homogenised in the implemented OBDA system. This guaranteed the possibility to maintain the epistemological flexibility the scholars are characterised by in looking for specific data, while keeping the possibility to interchange between the different representations, and translate one into another (e.g. to bidirectionally move from the string 'Trajan Government' to the numerical time-span '98AD - 117AD').

SPACE is meant to deal with information concerning space and geographical localisation of the entities in Main CRM. A heterogenous set of entities in MAIN brings a characterisation in terms of spatial localisation, from finding activities involved in the discovery of an artefact, till the relative position of an inscription with respect other stylistic and structural elements of the surface of an amphora. The SPACE module of the CRM has been, for this reason, subdivided into two distinct representation sub-modules: (i) a 'carrier-centred' one, used to represent the spatial relationships between the structural and the epigraphic components of an amphora (e.g., relative position of an inscription with respect to the amphora hands) and, (ii) a geographic-oriented one, which provides the elements for the representation of the location of a carrier finding, its production and potting, the 'function' of this location (e.g., civil settlement, legionary camp, fort) and the lat. and long. coordinates identifying it on a map. The geographic module, complemented by information coming from different sources (e.g., the Pleiades and Pelagios[15] dataset), offers the possibility to geo-localise

---

[14] The FRBR-aligned Bibliographic Ontology 'FaBiO' is an ontology for recording and publishing descriptions of entities that are published or potentially publishable, and that contain or are referred to by bibliographic references (http://vocab.ox.ac.uk/fabio).

[15] http://pelagios-project.blogspot.com

the domain entities, as well as to make a distinction, and a semantically sound mapping, between historical (e.g., roman provinces) and contemporary places. The role of the integration with the Pleiades dataset is further commented in Sec. 3.

DOCUMENTAL is a module devoted to the representation of the bibliographic information documenting the entities of interest (e.g. conference and workshop papers, books, web portals and digital encyclopaedia). As regards to this module, the above mentioned FaBiO ontology has been implemented, and customised according to the project requirements.

UPPER TYPING is the module dedicated to collect all the taxonomical structures characterising the concepts and relations in the Main CRM. Having all the taxonomies stored in a single place makes their management and successive extension a lot easier. Scholars with no technical background on formal modelling languages, even if supported by the automatic verbalisation facilities offered by NORMA, might experience difficulties in putting their hands in between axioms and constraints representation, labelling conventions, etc. in order to consistently modify them. Therefore, modularising the shared representation of their expert knowledge is finally a way to get them deeply involved in the process of effectively specifying it.

The EPNet CRM model, made of the above introduced five modules, beside being formally correct and consistent, is comprehensive enough to host all the information and knowledge elicited from the domain experts, and represents a definitive improvement in quality and granularity w.r.t. to the previously adopted informal data structure descriptions we faced at the beginning of EPNet project.

The EPNet Ontology. In order to support the user with the possibility of accessing data through a conceptual layer informed by the domain expert knowledge and terminology, a relevant part of the EPNet CRM introduced in the previous section has been transformed into an OWL-based ontology. The resulting ontology, written in a formal language whose expressivity stays within the OWL 2 QL profile[16], *modifies* and *extends* (by means of suitable concept/relation hierarchies) the vocabulary of the deployed database schema by re-introducing part of the domain specific terminology extracted with the support of the domain experts. Moreover, the ontology captures the domain knowledge by, at the same time, taking into consideration the available data and the user requirements in terms of data accessibility and usage. Differently from other cultural heritage projects dealing with semantic technologies, where the conceptualisation of the domain are expected to expose data structures that are suitable for a generic audience (from tourists visiting a museum or searching on the Web their favourite piece of art, till public administrations willing to open up their cultural resources and historic properties), the EPNet ontology has been specified in collaboration with experts of the history of the Roman economy with the

---

[16] http://www.w3.org/TR/owl2-profiles/

main aim of systematically collecting information to question standard narratives [14]. The characteristic trait of the EPNet CRM and ontology is that being 'functional to research'.

The ontology contains logic-based axioms which provide formal definitions for the concepts and the relations the experts make use of in conceptually classifying the 'living' entities of their research domain. As an example[17], the axioms below say that the concepts :Stamp and :TitulusPictus are both subconcepts of the concept 'inscription' (see Fig.1), while an :Amphora is a specialisation of the :InfCarrier concept. The :carriedBy relation links an :Inscription with its information carrier and, in a similar way, the domain and range of the :producedAt relation are specified to be the :InfCarrier concept and the :TimeSpan in which the existence of the carrier is historically attested, respectively. Last axiom in the example is for characterising the :hasName as a datatype property, i.e. a property having a range in a specific datatype (:String in this case).

```
      :Stamp rdfs:subClassOf  :Inscription .
:TitulusPictus rdfs:subClassOf  :Inscription .
     :Amphora rdfs:subClassOf  :InfCarrier .
    :carriedBy rdfs:domain :Inscription .
    :carriedBy rdfs:range  :InfCarrier .
    :producedAt rdfs:domain :InfCarrier .
    :producedAt rdfs:range  :TimeSpan .
      :hasName rdf:type owl:DatatypeProperty .
```

In addition, in order to expose the user to a domain knowledge-centred terminology, special axioms have been added to the ontology moving away from the technical jargon of the deployed database system. For instance, the following axiom introduces the engravedOn relation as the generalisation of the carriedBy relation:

```
:carriedBy rdfs:subObjectPropertyOf  :engravedOn .
```

Notice that the expressivity of the OWL 2 QL allows for the specification of disjointness constraints between concepts and functionality of the relations (see, Sec. 3).

3   The Benefits of Investing in Knowledge Representation

Having invested a considerable amount effort in knowledge acquisition and formal knowledge representation has provided the EPNet project with a host of opportunities that go far beyond the design and deployment of a new relational epigraphic database and the documentation of its data structure. Here below, we list some of the potential applications we are currently developing.

---

[17] A picture of the ontology can be found at http://136.243.8.213/obdasystem/, where a preliminary interface has been implemented mainly for testing purposes.

Semantic-Based Data Integration. There are several OBDA systems in both, the academia and the industry [8,15,16]. We work with -ontop-, a mature open-source system, which is currently being used in a number of projects. -ontop- allows for virtual data integration. In this approach, the ontology is connected to the data sources through a declarative specification given in terms of mappings that relate symbols in the ontology (classes and properties) to (SQL) views over data. The data remain in the sources and are accessed at query time. -ontop- does not modify the underlying databases. The classes and properties in the ontology, cluster different fragments of the databases into an homogenised well defined set of triples. On the basis of the -ontop- implementation, EPNet has been already integrated with the following two datasets:

- *Pleiades*[18], an open-access digital gazetteer for ancient history. It provides stable URIs for tens of thousands of geographic entities. The integration with Pleiades supports EPNet in tracing trade routes and economic connections on the Roman Empire territory in a more precise way and over a satisfactory picture of the past anthropic environment.
- The *Epigraphic Database Heidelberg*[19] (EDH), an on-line repository of Latin and bilingual inscriptions of the Roman Empire. EDH focuses on the recollection of epigraphs engraved mainly on monuments. This kind of epigraphic data is rich in information about distinguished personalities of the Roman society and their social connections. In this sense, EDH is a perfect complement for the EPNet dataset, as epigraphs on amphorae are brief and do not give much details on the role played by the persons being mentioned. Being able to correlate the names found on amphorae with the 'social networks' described in the monumental epigraphy should put us in a better position to understand the trade routes and the agents involved.

Data Consistency Checking. A logic based ontology language, such as OWL, allows ontologies to be specified as logical theories. This makes possible to constrain the relationships between concepts, properties, and data, on the one hand, and to automatically deal with the data consistency issue by reducing it to standard reasoning problem and then using already available tableau-based algorithms. Being able to apply data consistency checks over the project data is of particular interest in such a context, if you consider that the data are usually collected by non-experts and manually entered into a database system without the support of any specific data entry interface. In -ontop- the verification of the following types of constraints is automatically checked: (i) *Disjointness*: the intersection between classes should be empty. For instance, :MilitarCamp and :CivilSettlement must not have elements in common (disjoint properties can also be stated). (ii) *Functional Properties*: Individuals can be related to at most one element through a functional property. For instance, the property :hasShape is functional since every amphora can have at most a single shape.

---

[18] http://pleiades.stoa.org
[19] http://edh-www.adw.uni-heidelberg.de/

Conceptual Data Access. A preliminary user interface is available online[20] for testing the OBDA functionalities in EPNet. It provides users with a text area where to write SPARQL queries using the vocabulary of the ontology discussed in Sec. 2 (for the user's convenience, a summary of the ontology is provided by the interface). Following SPARQL syntax, users need to begin their queries with a prefix declaration[21]. After executing the query, the interface shows the SQL query that was send to the underlying RDBMS, and the result of the query in tabular form.

Technical details of the -ontop-based implementation of the services above, together with datasets-to-model mappings examples, can be found in [17].

4   Concluding Remarks

The paper introduces the results of the knowledge representation effort of the EPNet project. The resulting ontology enabled the implementation of -ontop-, an OBDA technology which helped us to deal in an efficient and sound way with data access, integration, and consistency issues. In order to develop an EPNet CRM and ontology, we specialised existing standards in the domain of epigraphy. In this direction, we also relied on the EAGLE Matadata Model, not only because it is already based on solid modelling standards such as CIDOC CRM and EpiDoc, but also pave the way to the possibility of a future integration in the Eagle federation of epigraphy databases.

Acknowledgments. *We thank the reviewers for their valuable comments. The work is partially funded by the ERC Advanced Grant n. ERC-2013-ADG 340828 and the EU under the large-scale integrating project (IP) Optique (Scalable End-user Access to Big Data), grant agreement n. FP7-318338.*

References

1. Hitzler, P., Krötzsch, M., Rudolph, S.: Foundations of Semantic Web Technologies. Chapman & Hall/CRC (2009)
2. Shadbolt, N., Berners-Lee, T., Hall, W.: The semantic web revisited. IEEE Intelligent Systems **21**(3), 96–101 (2006)
3. Domingue, J., Fensel, D., Hendler, J.: Handbook of semantic web technologies. Handbook of Semantic Web Technologies, vol. 1–2. Springer (2011)
4. Raghavan, P.: It's time to scale the science in the social sciences. Big Data & Society (2014)

---

[20] http://136.243.8.213/obdasystem/
[21] PREFIX:⟨http://136.243.8.213/obdasystem#⟩, and PREFIXrdf:⟨http://www.w3.org/1999/02/22-rdf-syntax-ns#⟩ in our case.

5. Schreiber, G., Amin, A.K., van Assem, M., de Boer, V., Hardman, L., Hildebrand, M., Hollink, L., Huang, Z., van Kersen, J., de Niet, M., Omelayenko, B., van Ossenbruggen, J., Siebes, R., Taekema, J., Wielemaker, J., Wielinga, B.J.: Multimedian e-culture demonstrator. In: Cruz, I., Decker, S., Allemang, D., Preist, C., Schwabe, D., Mika, P., Uschold, M., Aroyo, L.M. (eds.) ISWC 2006. LNCS, vol. 4273, pp. 951–958. Springer, Heidelberg (2006)
6. Meroo-Peuela, A., Ashkpour, A., van Erp, M., Mandemakers, K., Breure, L., Scharnhorst, A., Schlobach, S., van Harmelen, F.: Semantic technologies for historical research: A survey. Semantic Web Journal 588–1795 (2014)
7. Rodríguez-Muro, M., Kontchakov, R., Zakharyaschev, M.: Ontology-based data access: *Ontop* of databases. In: Alani, H., Kagal, L., Fokoue, A., Groth, P., Biemann, C., Parreira, J.X., Aroyo, L., Noy, N., Welty, C., Janowicz, K. (eds.) ISWC 2013, Part I. LNCS, vol. 8218, pp. 558–573. Springer, Heidelberg (2013)
8. Rodriguez-Muro, M., Rezk, M.: Efficient SPARQL-to-SQL with R2RML mappings. Journal of Web Semantics (2015)
9. van Harmelen, F., Lifschitz, V., Porter, B.: Handbook of Knowledge Representation. Elsevier Science, San Diego (2007)
10. Garnsey, P., Whittaker, C.: Trade and famine in classical antiquity. Supplementary volume - Cambridge Philological Society. Cambr. University Press (1983)
11. Cascio, E., Rathbone, D.: Production and Public Powers in Classical Antiquity. Cambridge Philological Society (2000)
12. Halpin, T., Morgan, T.: Information modeling and relational databases. MK (2010)
13. Franconi, E., Mosca, A., Solomakhin, D.: ORM2: formalisation and encoding in OWL2. In: Herrero, P., Panetto, H., Meersman, R., Dillon, T. (eds.) OTM-WS 2012. LNCS, vol. 7567, pp. 368–378. Springer, Heidelberg (2012)
14. Guldi, J., Armitage, D.: The History Manifesto. Cambridge University Press (2014)
15. Bishop, B., Kiryakov, A., Ognyanoff, D., Peikov, I., Tashev, Z., Velkov, R.: OWLIM: A family of scalable semantic repositories. Semantic Web **2**(1), 33–42 (2011)
16. Civili, C., Console, M., De Giacomo, G., Lembo, D., Lenzerini, M., Lepore, L., Mancini, R., Poggi, A., Rosati, R., Ruzzi, M., Santarelli, V., Savo, D.F.: MASTRO STUDIO: managing ontology-based data access applications. PVLDB **6**(12), 1314–1317 (2013)
17. Mosca, A., Remesal, J., Rezk, M., Rull, G.: A 'Historical Case' of ontology-based data access. In: Digital Heritage 2015 (2015, to appear)

A Pattern-Based Framework for Best Practice Implementation of CRM/FRBRoo

Trond Aalberg[1]([⊠]), Audun Vennesland[1], and Maliheh Farrokhnia[2]

[1] Norwegian University of Science and Technology, Trondheim, Norway
{Trond.Aalberg,Audun.Vennesland}@idi.ntnu.no
[2] Oslo and Akershus University College of Applied Sciences, Oslo, Norway
Maliheh.Farrokhnia@hioa.no

Abstract. The CIDOC Conceptual Reference Model and extensions to this model such as the FRBRoo, are important for semantic interoperability in the area of cultural heritage documentation. However, the real life use of such reference models is challenging due to their complexity as well as their extensive and detailed nature. In this paper we present a framework for sharing best practice knowledge related to the use of these models, which is inspired by the use of design patterns in software engineering. The main contribution is a framework for sharing and promoting knowledge and solutions for best practice that supplements existing documentation and is adapted to the needs of developers.

1 Introduction

Sharing and dissemination of semantic data is a major trend for memory institutions worldwide, but efficient reuse, integration and exploitation of this data requires a high level of semantic interoperability. The CIDOC Conceptual Reference Model [2] and its extensions, such as FRBRoo [7] which is based on the FRBR bibliographic model [10], are important as basis for semantic interoperability in this domain. They provide the authoritative guide for developing interoperable information models and they are the global models that local models can be mapped and explained through. As reference models they provide the formal and shared understanding of how to represent the facts recorded in this domain – expressed through a system of thoroughly documented classes and properties. However, the interpretation and use of these reference models when developing new application schemas or transforming data can be a challenge due to the extensive nature and complexity of the models. In particular, the FRBRoo model which both elaborates upon the entities introduced in FRBR as well as redefines them in the context of the CRM model and its object-oriented formalism, is perceived as difficult and has not been much utilised yet. This problem may be caused by a number of reasons such as the complexity of the FRBRoo model which makes implementors hesitate using the model or it can be that the domain itself is inherently resistant to new models because of the large amount of existing data and legacy systems. The original FRBR model is the

© Springer International Publishing Switzerland 2015
T. Morzy et al. (Eds): ADBIS 2015, CCIS 539, pp. 438–447, 2015.
DOI: 10.1007/978-3-319-23201-0_44

basis for the modernised cataloguing rules RDA and is now also available as an RDF schema[1] with properties that are close to the existing catalogue records. The FRBRoo model, on the other hand, does not currently have an application schema at the same level. Although the Erlangen implementation[2] of CRM and FRBRoo are direct translations of the models into RDF/OWL, it lacks the user defined types and direct support for properties that actually are coded in bibliographic records.

To facilitate more widespread use of CRM and FRBRoo in particular, there is a need for guidance of a kind that currently is missing in the documentation and forums where the models are discussed. Although existing documentation is highly elaborated, it essentially is a sequential documentation of distinct classes and properties. Implementors, on the other hand, often have questions where the answers are given by best practice knowledge rather than in the formal documentation, such as "how can I model this case", "when should I use this subtype rather than the super type" and "how can this model fragment be simplified and implemented". Additionally, implementors frequently need to find solutions to issues that are related but not in the scope of the formal documentation such as best practice mapping to existing RDF vocabularies or XML schemas or best practice simplifications that can be made when implementing complex model fragments. The need for such guidance is also evident when inspecting the mailing list of the CIDOC CRM Special Interest Group[3] which is a discussion forum engaging cultural heritage stakeholders having different roles. Here e.g. representatives from the cultural heritage sector formulate requirements and issues and the CRM experts respond by providing explanations addressing them. However, as per the nature of mailing list these explanations are limited to textual descriptions of the solution. Our belief is that providing an environment that enables a more persistent knowledge base as well as the possibility of expressing best practice solutions in a uniform manner and by using richer notation (e.g. graphical models) as contextualisation could improve the accessibility for a wider audience and hence foster more collaboration and engagement.

In this paper we present the use of patterns to facilitate and communicate best practice modelling, implementation and coding of CRM- and FRBRoo-based data. This approach is inspired by the well known concept of design patterns from Software Engineering. The patterns of interest to users of CRM and FRBRoo are model fragments for common use cases accompanied with a documentation explaining the pattern, the context for this pattern as well as other relevant information related to the the pattern. Additionally, our patterns include implementation guidelines which can be best practice implementations in RDF or XML Schemas, or best practice mappings to existing schemas. A pattern and all related information can be documented using a common template which makes it easy to create a catalogue of patterns using for example wiki-software.

---

[1] http://www.rdaregistry.info
[2] http://erlangen-crm.org
[3] http://www.cidoc-crm.org/mailing_list.html

2   Design Patterns

Software engineering design patterns have since the 1990's [5,9] been a lingua franca for software developers and help conceptualise more complex structures through a set of uniform pattern names and clarifications. Developers use design patterns to communicate ideas, implement software supported by the "bigger" picture provided by design patterns, and document their developments through a widely understood vocabulary [8]. Software engineering design patterns are typically classified into three types of patterns:

- Creational Patterns: These are patterns that enable a system to be independent of how objects are created, composed and represented. The aim is to give the system more flexibility in deciding which objects need to be created for a given case.
- Structural Patterns: These are patterns that focus on how classes and objects are composed (composition).
- Behavioural Patterns: These patterns are concerned with algorithms, the assignment of responsibilities between objects and how classes and objects communicate.

Our patterns are model fragments based on an already defined system of classes and properties, hence they can be defined as structural patterns according to the classification above. At the same time it is also naturally that the patterns are more usage and application oriented, so we should describe or illustrate the instantiation of the patterns in a proper manner.

In ontology engineering, some more recent research articles on design patterns are written by for example Blomqvist et al [1] and Gangemi and Presutti [6]. The result of this research includes a set of Ontology Design Patterns (ODPs) that re-uses many of the fundamental ideas used in software engineering design patterns. These patterns are shared in a design pattern catalogue[4] developed as a part of the NeOn project (2006-2010)[5]. As with the traditional SE design patterns, these ODPs are also classified into different pattern types:

- Structural ODPs: Compositions of logical constructs that solve a problem of expressivity
- Correspondence ODPs: Provide designers with solutions to the problem of transforming a conceptual model and for creating semantic associations between ontologies
- Content ODPs: Each content pattern covers a specific set of competency questions (requirements), which represent the problem they provide a solution for. Hence, Content ODPs represent best practice solutions for a set of competency questions.
- Reasoning ODPs: Reasoning OPs are oriented towards obtaining certain reasoning results, based on the behaviour implemented in a reasoning engine

---

[4] http://ontologydesignpatterns.org/wiki/Main_Page
[5] http://www.neon-project.org/nw/Welcome_to_the_NeOn_Project

- Presentation ODPs: Describe aspects related to usability and readability of ontologies from a user's perspective
- Lexio-Syntactic ODPs: Describe linguistic structures or schemas that permit to generalise and extract some conclusions about the meaning they express

The templates used for defining patterns in ontology engineering suits our purposes well, so in combination with the principles of traditional structural patterns we intend to build upon relevant elements of these templates in our development, especially reusing template elements from the Content ODPs for describing best practice patterns and the Correspondence ODPs for describing mappings with other relevant formats.

Such design patterns are however not stand-alone entities operating in isolation. On the contrary, design patterns are often interrelated and composed of many of the same elements. Hence, some systematic method for establishing and presenting the interrelationships among patterns is required. Falbo et al. [4] suggests organising design patterns using a pattern language as a network of interrelated domain-related patterns. This language aims to provide a more holistic view that gives guidance on what problems can arise in the domain, specify the order to address these problems and suggest the patterns that can help solve them. One of the possible notations for representing this process is UML activity diagrams where patterns are represented as activities and their ordering is defined by the control flow associations between the activities.

3   Examples

The pattern-based approach can be illustrated with examples from a project where the aim was to propose a mapping from FRBRoo to the Europeana Data Model [3]. This project analysed various models of bibliographic structures typically found implicit or explicit in library records, which then were mapped to structures based on the RDF-vocabulary for EDM. During this work we encountered many issues that needed clarification and informed decisions. One issue was the use of the different Work-classes defined in the model where different partners had different opinions on the level of subtyping that was needed in different contexts. Another issue was related to how to model the path between a work or an expression and their creators. We also encountered misunderstandings related to what classes and properties were needed to model simple patterns such as work with parts. When representing these models using the Erlangen RDF/OWL schema we also encountered issues such as how to code and represent values, how to implement local subtyping schemes that are out of the scope for the reference models to define (CRM E55 Types) as well as best practice simplifications that can be made.

Description of such issues as well as the best practice solution can fortunately be described in a concise and educational way considering a combination of design and implementation aspects. We have identified the following aspects as a starting point:

Model Fragments that represent best practice modelling of common use cases. Typically this can be presented by a generic model fragment limited to the main classes and properties that are needed to describe the use case. Model fragments need to be supplemented with textual description as well as specific examples. The presentation of models and examples should be based on well known formalism such as UML class and instance diagrams.

Formal Coding of the model fragment using an authoritative vocabulary that directly implements CRM and FRBRoo such as the Erlangen RDF/OWL implementation is often needed in addition to the graphical model itself. The main purpose of such a coding is to show the actual implementation of a model fragment as well as to exemplify a formal implementation that local implementations needs to be compliant with.

Compliant Implementations showing how the formal model and code can be simplified and implemented using existing vocabularies or in ad hoc local implementations. Such implementations can be described graphically using UML or coding examples in RDF. Rather than prescribing only one specific solution for implementing a model fragment, it is more interesting to guide the implementation of a model fragment. Relevant implementation examples for FRBRoo could be mapping to the EDM-format or the RDA-vocabulary.

Mapping Descriptions is a natural addition if local implementations are included in the pattern. Such mappings have to include a description of how structures are to be transformed between the global model and the local model. Such a mapping can be graphically presented by aligning models and showing the correspondence, or it can be more formally described using rules.

A sample pattern for representing a translation and the work it is a translation of is shown in figure 1. This pattern is explained using a class diagram (in the dotted box) and a concrete example expressed as an object diagram. This

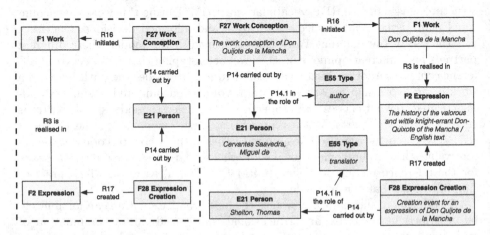

Fig. 1. Model of a translation using *R3 is realisation of* between work and expression

pattern illustrates the common understanding of a translation that is reflected in bibliographic records and the original FRBR model. This may represent best practice when there is no need to include a work entity for the translation itself in the database. The example in figure 2 shows a more elaborated modelling of the same translation using individual work for the translation as well as for the original text. This may represent best practice if there is a need for detailed recording of the derivation history. The example in figure 3 shows a pattern for titles of different subtypes and best practice coding of this in RDF using common vocabularies.

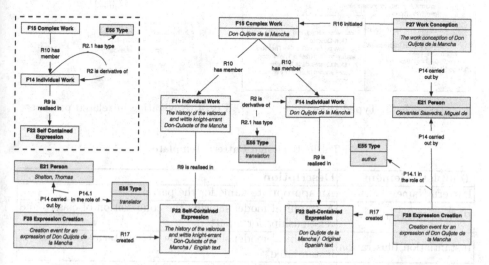

Fig. 2. Model of a translation using *R2 is derivative of* between works

4   Documenting Patterns

Patterns are typically organised in catalogues where each pattern is consistently described both graphically and with text. We have chosen to implement our pattern catalogue as a semantic wiki using the MediaWiki extension Semantic Mediawiki (SMW)[6] as our development platform. SMW enables semantic annotation of content that helps organise, retrieve and navigate in relevant patterns. For the graphical aspect we are using a PlantUML[7] extension to SMW that transforms textual descriptions into UML diagrams. SMW does not have any built-in UML modeling environment and to our knowledge there are no extensions offering such features, so PlantUML is a good substitute.

For our framework a suitable template should consist of a name for the pattern, use case scenarios and a description of the intent of the pattern, a graphical

---

[6] https://semantic-mediawiki.org/
[7] http://plantuml.sourceforge.net

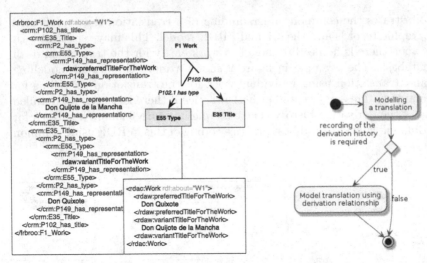

Fig. 3. Typed titles   **Fig. 4.** Navigating in related patterns

Table 1. Design Pattern Template

Template element	Description
Pattern Name	An appropriate name for the pattern
Pattern Illustration	A graphical model showing the composition of classes and relationships for a pattern
Instantiation Illustration	A graphical model showing how the pattern should be instantiated
Author	The author of the pattern
Intent	A description of the intent of the pattern
Domains	Domains relevant for this pattern
Competency Questions	Questions formulating the purpose the pattern aims to fulfill
Solution Description	A description of the solution the pattern provides (i.e. answering the competency questions)
Consequences	Any consequences from applying the pattern
Scenarios	(Real-life) Scenarios describing how the pattern is applied use case scenarios
References	Any references associated with the pattern
Examples	Concrete examples (e.g. code snippets) exemplifying the application of the pattern
Relevant Components	Components that are used as building blocks when constructing the pattern
Related Patterns	Other patterns that are related to this pattern
Modelling Issues	Any modelling issues the user needs to consider when applying the pattern

model fragment and authoritative implementation and mapping examples. Additional descriptions that may be needed are limitations and side effects as well as generalisations, shortcuts and simplifications that can be made to the pattern. From these requirements we have developed the design pattern template described in table 1. A prototype implementation of this template in SMW is shown in figure 5.

A pattern catalogue should also support an overarching structure for navigating in the patterns. Patterns can be of different levels and one pattern may be part of a more extensive pattern, or patterns may complement each other and represent alternatives that have to be chosen depending on the context. When defining design patterns SMW enables us to declare explicit relations between patterns and the inherent elements being used to compose the patterns. This way the user can easily browse related patterns and navigate in the catalogue using either a top-down approach (starting directly from the defined design patterns

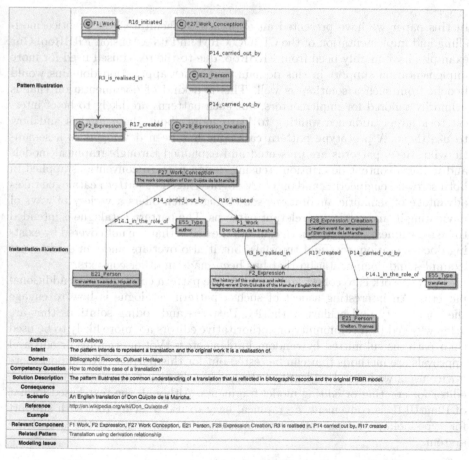

Fig. 5. Example from the Ontology Pattern Semantic Wiki

and then see elements associated with the pattern) or bottom-up (starting from the element-level and the discovering which patterns are relevant for that particular element). Furthermore, using semantics for describing more explicitly the characteristics and the relationships among patterns and elements enables us to specify how to navigate the patterns, similarly to the ideas from Falbo et al. [4]. Figure 4 illustrates how such a pattern navigation among related patterns could be implemented.

Users of a pattern catalogue will mainly be implementors such as system developers or other practitioners working with data structures. The kind of pattern catalogue we have in mind is intended to be a toolkit used when developing information models or coding schemes that have to be compliant with CRM/FRBRoo, but it should also be valuable for transformation projects and other projects related to creating FRBRoo/CRM compliant linked open data.

5   Conclusions and Further Work

In this paper we have presented an approach to facilitate best practice modelling and implementation of the CIDOC CRM and its extension FRBRoo. Our examples have mainly been from FRBRoo, due to the recognised need for more implementation support in this domain, but other application domains would benefit from such a solution as well. This is a kind of documentation that is primarily tailored for implementors, but such patterns are likely to be of interest to a larger audience wanting to learn more about these models and how to use them. A prototype pattern catalogue has been developed as a semantic wiki. Here, patterns are presented and explained through graphical models and uniform context descriptions reusing design pattern conventions applied in both software engineering and ontology engineering. The pattern catalogue takes advantage of semantic annotations something which offers a variety of ways of navigating in and retrieving relevant patterns. The pattern catalogue is intended to be a documentation system that meets the needs that are not covered by existing documentation, fora and tutorials, but it also overlaps and can easily reuse examples and documentation that has been made in other contexts.

Further work is to collect and populate the pattern catalogue with additional patterns. An interesting aspect of such a pattern catalogue is how to engage the community and achieve authority. Patterns and coding solutions that are acknowledged by a community or authoritative editors are more likely to be used than patterns submitted by random implementors. Voting systems or editorial approvals are methods that can be tested out for this purpose. To support more novice users we have also identified the need to develop a more specific purpose editor for creating or editing model fragments. Although the text-based format of PlantUML is convenient, a graphical easy to use drawing tool with support for type-lookup and automatic validation could contribute to the use of the system.

References

1. Blomqvist, E., Gangemi, A., Presutti, V.: Experiments on pattern-based ontology design. In: Proceedings of the Fifth International Conference on Knowledge Capture, K-CAP 2009, pp. 41–48. ACM, New York (2009)
2. CIDOC CRM Special Interest Group: Definition of the CIDOC Conceptual Reference Model (2011). http://www.cidoc-crm.org/docs/cidoc_crm_version_5.0.4.pdf
3. Doerr, M., Gradmann, S., LeBoeuf, P., Aalberg, T., Bailly, R., Olensky, M.: Final Report on EDM - FRBRoo Application Profile Task Force (2013). http://pro.europeana.eu/files/Europeana_Professional/EuropeanaTech/EuropeanaTech_taskforces/EDM_FRBRoo/TaskfoApplication%20Profile%20EDM-FRBRoo.pdf
4. de Almeida Falbo, R., Barcellos, M.P., Nardi, J.C., Guizzardi, G.: Organizing ontology design patterns as ontology pattern languages. In: Cimiano, P., Corcho, O., Presutti, V., Hollink, L., Rudolph, S. (eds.) ESWC 2013. LNCS, vol. 7882, pp. 61–75. Springer, Heidelberg (2013)
5. Gamma, E., Helm, R., Johnson, R., Vlissides, J.: Design Patterns: Elements of Reusable Object-oriented Software. Addison-Wesley Longman Publishing Co., Inc., Boston (1995)
6. Gangemi, A., Presutti, V.: Ontology design patterns. In: Staab, S., Studer, R. (eds.) Handbook on Ontologies. International Handbooks on Information Systems, pp. 221–243. Springer, Heidelberg (2009)
7. International Working Group on FRBR and CIDOC CRM Harmonisation: FRBR object-oriented definition and mapping from FRBRer, FRAD and FRSAD (version 2.1) (2015). http://www.cidoc-crm.org/docs/frbr_oo//frbr_docs/FRBRoo_V2.1_2015February.pdf
8. Riehle, D.: Lessons learned from using design patterns in industry projects. In: Noble, J., Johnson, R., Avgeriou, P., Harrison, N.B., Zdun, U. (eds.) Transactions on Pattern Languages of Programming II. LNCS, vol. 6510, pp. 1–15. Springer, Heidelberg (2011)
9. Schmidt, D.C.: Using Design Patterns to Develop Reusable Object-oriented Communication Software. Commun. ACM 38(10), 65–74 (1995)
10. The International Federation of Library Associations and Institutions: Functional Requirements for Bibliographic Records. UBCIM Publications - New Series, vol. 19 (1998)

Application of CIDOC-CRM for the Russian Heritage Cloud Platform

Eugene Cherny[1,2]($\boxtimes$), Peter Haase[1,3], Dmitry Mouromtsev[1],
Alexey Andreev[1], and Dmitry Pavlov[1,4]

[1] ITMO University, St. Petersburg, Russia
eugene.cherny@niuitmo.ru, ph@metaphacts.com, mouromtsev@mail.ifmo.ru,
aandreyev13@gmail.com, dmitry.pavlov@vismart.biz
[2] Åbo Akademi University, Turku, Finland
[3] Metaphacts GmbH, Walldorf, Germany
[4] Vismart Ltd., St. Petersburg, Russia

Abstract. This paper describes the usage of CIDOC-CRM ontology
for the online representation of cultural heritage data on, based on class
templates; and also describes the motivation for choosing the CIDOC-
CRM ontology as the basis for the Russian Heritage Cloud project, a
recent collaboration started between ITMO University and a number of
museums in Russia.

Keywords: Cultural heritage · Digital humanities · Semantic web ·
CIDOC-CRM

1 Introduction

The transfer of cultural heritage between generations is the key aspect of cultural
identity preservation. It plays major role in maintaining function of the culture
in the era of globalization, when traditional mechanisms of preservation and
transferring of cultural heritage are subject to change, challenging by the rapid
development of information technologies. Digital revolution forced content and
application developers to use new communication languages and explore different
mediums for transferring information to the end-users. In the modern world
information technologies undoubtedly are the most important medium for the
culture translation. These are the main motivation for our project we name the
Russian Heritage Cloud. One of the main long-term goal is to motivate museums
in Russia to publish their data in the semantic form.

The most effective and illustrative way to test the technology is to create a
system, therefore we decided to develop a system targeted to the museum as an
institution, whose primary goal is preservation of the cultural heritage and its
transfer to the people. We began to cooperate with two museums in St. Petersburg: The Russian Museum[1] and Peter the Great Museum of Anthropology and

---

[1] http://www.rusmuseum.ru/eng/home/

© Springer International Publishing Switzerland 2015
T. Morzy et al. (Eds): ADBIS 2015, CCIS 539, pp. 448–457, 2015.
DOI: 10.1007/978-3-319-23201-0_45

Ethnography (Kunstkamera) of the Russian Academy of Sciences[2]. The former has the biggest collection of Russian art in Russia, the latterbiggest collection of ethnographic artifacts. The primary goal of our research was stated as "To test and showcase the applicability and benefits of usage of semantic data to tackle the challenges of cultural heritage transfer in the digital era". The system is meant to deliver benefits to two different target groups: the museum experts working in the museum and regular museum attendants. These two groups greatly differ in their needs, but the system should cover the interests of both of them.

This paper describes the usage of Erlangen CRM/OWL[3] implementation of CIDOC-CRM ontology for the semantic representation of rmgallery.ru website, which contains selected artworks chosen for the official mobile application. The work has been done as part of the pilot project with The Russian Museum.

2   Choosing the Ontology

There are number of instruments that that are used to manage information in cultural institutions, including Dublin Core (DC) Metadata Elements [6] and DC Terms [5], Simple Knowledge Organization System (SKOS) [1], Functional Requirements for Bibliographic Record (FRBR) [2], Europeana Data Model (EDM) [4], CIDOC-CRM [3] and others. We will not describe them in detail as it had been done in a number of resources, for example [4] or [10], instead we briefly justify our choice of CIDOC-CRM.

First of all, most of the listed tools do not aim to represent full richness of cultural heritage data and focused in highly specific area. For example DC Core, DC Terms and SKOS, being highly usable for mapping museums' thesauri to the standardized machine readable format, could not add additional semantic meaning to the original data besides existing in the original data. FRBR is a model for describing bibliographic entities and it does not have semantics for describing abstract and conceptual objects, also it is not well suited for representing ethnographic artifacts, as they tend to have a big diversity of connections between concepts, which could not be described in bibliographic terms. EDM extensively reuse existing ontologies, including some of the listed above, but it has some issues concerning its effective usage [10].

CIDOC-CRM is a very abstract model developed from the ground up as a tool for representing explicit and implicit cultural heritage knowledge and providing a "semantic glue" to mediate between different cultural heritage information providers[4]. Being very abstract and flexible there could be difficulties in interoperation between users of this ontology, but advantages of using it outweighs this difficulties. For the general advantages of CIDOC-CRM and general ideas behind it we suggest the reader to refer to[8], for here we want to list a practical reasons for choosing CIDOC-CRM as the main ontology for our project:

---

[2] http://www.kunstkamera.ru/en/
[3] http://erlangen-crm.org/
[4] CIDOC-CRM official website: http://www.cidoc-crm.org/

1. CIDOC-CRM is standardized (ISO 21127:2006)—this provides great benefits for the future of the project as we are enthusiastic enough to take into account the possibility of integrating the most of Russian heritage providers into the single "heritage cloud". Using a well-defined standard in this regard is a big advantage.

2. CIDOC-CRM provides a lot of abstract concepts that could be used to describe different kinds of cultural heritage artifacts, for example, ethnographic knowledge is much harder to describe than fine art.

3. A lot of museums around the world have already implemented CIDOC-CRM for their data[5] which gives the possibility to integrate our work into the global cultural semantic web.

3   Usage of CIDOC-CRM

In this project we have the agreement with the The Russian Museum management to work with data from one of their sites: rmgallery.ru. The data provided by the museum was rather simple, consisting only of basic information about artwork (its title, size, materials it's made of, genre, creation date and textual annotation) and author (only appellation and textual annotation)—all of this information was represented in the form of HTML pages. That data does not contain rich set of information, basically it represents two main entities: authors and artworks. Artwork representation has the following information: title, type (drawing, painting, etc.), genre (still life, portrait, etc.), size, materials it consists of, textual annotation. Author information is much poorer, it consists of just person appellation and textual annotation.

The original data undergoes a transformation process the main goal of which is to structure initial information into an RDF data graph conforming with the CIDOC-CRM ontology, more exactly Erlangen CRM OWL implementation[6]. This process has been implemented with Ruby scripts[7], which were created to extract structured information from HTML pages and create RDF triples. Fig. 1 shows an example of the initial data representation in RDF and interlinking.

CIDOC-CRM is an event-centric model. The central part of semantic representation is the event of production of some object crm:E12_Production. It connects all other entities that are relevant to it: A creator is connected with the crm:P14_carried_out_by property, an artwork is connected with the crm:P108_has_ produced_property, creation time is connected with the crm:P4_has_time_span property. The artwork is represented as an instance of class crm:E22_Man-Made_Object (Fig. 2). While not shown in the diagrams, the person's biography and artwork description are associated with crm:E21_Person and crm:E22_Man-Made_Object respectively via the crm:P4_has_note property.

---

[5] Applications of CIDOC-CRM as listed on the official website: http://cidoc-crm.org/ uses_applications.html

[6] http://erlangen-crm.org/current-version

[7] https://github.com/ailabitmo/CultureCloud-Crawling

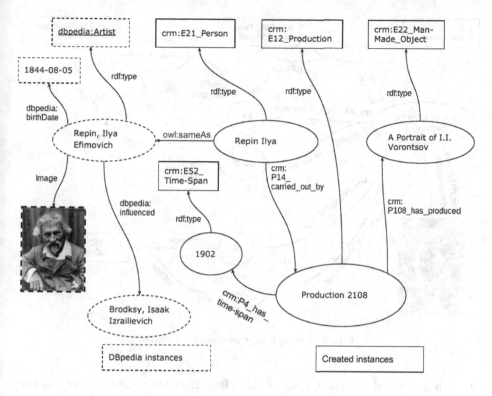

Fig. 1. Example of initial data representation and interlinking

All textual information in the dataset (names, titles, descriptions, etc.) were placed in two languages annotated with the corresponding language tag. As information about authors in the original data is rather poor, we decided to interlink authors with DBpedia where it was possible. This allowed us to add such essential information as author image or birth date. The interlinking process has been implemented in two steps: at the first step author names had been passed to the Wikipedia Search API, then remaining uninterlinked authors were processed using fuzzy string matching technique comparing authors' names from museum's data with individuals of class *dbpedia:Artist* available DBpedia. More detailed description of interlinking process and its evaluation can be found in [7].

3.1 Reusing Thesauri of The British Museum

The main principle of the knowledge management in our project is to avoid creating custom classes or entities in the ontology. The British Museum has published high-quality thesauri that could be used with any museum, thus we decided to reuse them.

The thesauri is based on the SKOS ontology combined with CIDOC-CRM, i.e. every concept belongs to some SKOS concept scheme and in the same time

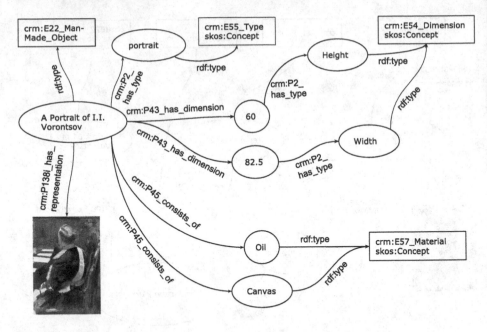

Fig. 2. Artwork representation example

it is an individual of some CIDOC-CRM class. Every thesaurus object has the skos:Concept type and one of CIDOC-CRM more specific type. For example, the material "oi" has types of skos:Concept and crm:E57_Material and is part of "BM MATERIAL" concept scheme. The "allegory/personication" subject has types skos:Concept and crm:E55_Type and is part of "BM SUBJECT" concept scheme. We used the latter for describing the genre of the artworks: Fig. 3 shows an example of this.

For some genres there were no appropriate entities in British Museum dataset (illustration, caricature, theatrical scenery), for these we created additional instances following the exact same scheme.

4   Template-Based Visualization

Our system is built on the metaphacts Knowledge Graph Platform[8] backed by Systap's Blazegraph[9] triple store. We have chosen this Platform as some members of our team are involved in its development, which makes it natural choice for our long-term project.

The metaphacts Knowledge Graph Platform allows to create templates for RDFS classes which could be used to represent any individual of these classes.

---

[8] http://metaphacts.com/
[9] http://blazegraph.com/

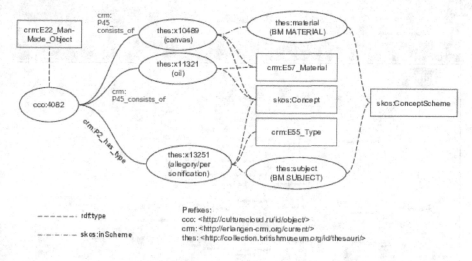

Fig. 3. Usage example of The British Museum thesauri

The template consists of a set of widget definitions configured with SPARQL-queries, which used to visualize the data. For example, on the main page[10] there are a PivotViewer widget, which present all artworks in the dataset in the relatively compact and structured form allowing to set up different filters (for example to show only artworks of the 20-th century), on the author page we put a timeline widget that displays the events author have relation to, etc. Template-based approach together with semantic-enabled widgets provide a meaningful and efficient way to visualize semantic data. Besides, a class template, once created, potentially could be reused on any other system which uses CIDOC-CRM.

In current implementation we have created the templates for two classes: ecrm:E21_Person and ecrm:E22_Man-Made_Object. The ecrm:E21_Person template contains the following information:

- Name.
- Author's bio.
- Additional information gathered by interlinking with DBpedia: dates of birth and death, art movement a person belongs to, influenced/influenced by information.
- Works by author represented in the PivotView and on the timeline widget.
- Ontology structure visualization and navigation based on OntoDia OWL diagramming tool[11] (see Fig. 4).

An example of the *ecrm:E21_Person* class visualization can be found on the Fig. 5.

---

[10] http://culturecloud.ru/
[11] http://ontodia.org/

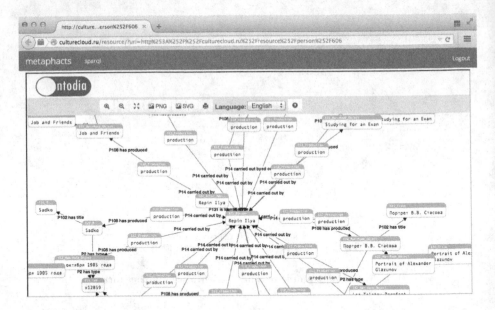

Fig. 4. Example of OntoDia's OWL diagram visualization

The ecrm:E22_Man-Made_Object template has less information due to our focus on the experiments with interlinking:

- Artwork title.
- Genre.
- Date of production.
- Size.
- Annotation for the artwork.

Both templates use CIDOC-CRM specific SPARQL queries, thus they can be used by any CIDOC-CRM-enabled application. Each piece of information is visualized with widgets, which could be, for example, Google Maps, PivotView with filters and others.

Here is the example of PivotView widget configuration that could be seen on the main page. In the CONSTRUCT clause we define the information, that could be used to filter the artworks by artist, date of creation and genre.

```
{{#widget: Pivot |
  query = 'CONSTRUCT {
    ?uri :creator ?artist .
    ?uri :date ?date .
    ?uri dbpedia:thumbnail ?img .
    ?uri :genre ?genre .
  } WHERE {
    ?creator ecrm:P131_is_identified_by ?appelation .
    ?appelation rdfs:label ?artist .
```

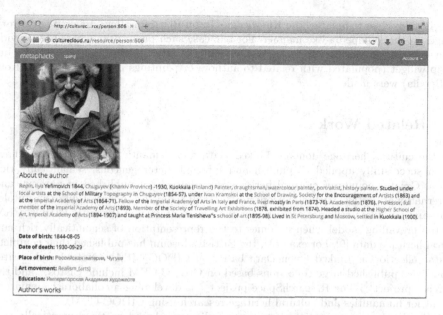

Fig. 5. Example of the ecrm:E21_Person class visualizatoin

```
?uri ecrm:P108i_was_produced_by ?production .
?production ecrm:P14_carried_out_by ?creator .
?production ecrm:P4_has_time-span ?span .
?span ecrm:P82a_begin_of_the_begin ?date .
?uri ecrm:P138i_has_representation ?img .
?uri ecrm:P2_has_type ?genre .
?uri rdfs:label ?label .
FILTER (LANG(?artist)='en') .
}'}}
```

Next example demonstrates usage of the map widget, that shows birth places of all authors, the information gathered after interlinking with DBpedia.

```
{{#widget: MapWidget |
  markers = '
    SELECT ?description ?link ?image ?lat ?lng
    WHERE {
      ?link a ecrm:E21_Person ;
        owl:sameAs ?dbp_person .
      ?dbp_person rdfs:label ?description;
        dbpedia:thumbnail ?image;
        dbpedia:birthPlace ?birthPlace .
      ?birthPlace <http://www.w3.org/2003/01/geo/wgs84_pos#long> ?lng .
      ?birthPlace <http://www.w3.org/2003/01/geo/wgs84_pos#lat> ?lat .
}'}}
```

Besides class-templates, the Knowledge Graph Platform allows to create arbitrary pages in the Wikipedia-like manner. For our case such pages were created to provide the site structure for user to navigate. Overviews of artists, paintings, genres and a big map widget (populated with related to authors or paintings places we gathered from DBpedia) were made.

5   Related Work

In the cultural heritage domain, Linked Data and Semantic Web technologies have been successfully applied to publish and interlink heterogeneous, semantically rich data. Great amounts of cultural heritage data have been published in national and international portals, such as Europeana[12]. As of today, a number of different ontologies and metadata schemes are used for the representation of the data. CIDOC-CRM is the prevailing model when it comes to the representation of semantically rich cultural heritage data [9]. For example, the British Museum has published their complete data collection as Linked Open Data based on CIDOC-CRM[13]. Notable other sites that have published large collections based on CIDOC-CRM include Claros[14] and the Arches project[15]. The ResearchSpace project[16] is developing a collaborative environment for humanities and cultural heritage research using CIDOC-CRM.

On the data consumptions side, new applications based on the semantically rich data have been developed that enable new forms of user experience. These range from supporting semantic search in portals to mobile applications. E.g., the SMARTMUSEUM [11] system utilizes an ontology-based representation of content descriptions as a basis for context-aware, on-site access to cultural heritage in a mobile scenario. Applying context reasoning and recommendation algorithms provide users with recommendations for sites, such as museums or buildings of architectural interest, and objects on those sites, such as sculptures or other works of art, and provides explanatory descriptions and multimedia content associated with individual objects. In comparison to the related solutions our project stands out as being an external service to heritage owners, which provides interlinking and search / representation facilities to end-users and third-party applications.

6   Conclusions and Future Work

In this paper we described the practical usage of CIDOC-CRM ontology for the Russian Heritage Cloud project and the system architecture for publishing semantic cultural heritage data. Template-based visualizations together with SPARQL-enabled widgets provide a convenient way for agile development of semantic cultural heritage applications.

Implementation of this project opens up new questions, most of them related to scaling the platform to bigger datasets. Together with performance issues arise new

---

[12] http://www.europeana.eu
[13] http://collection.britishmuseum.org
[14] http://www.clarosnet.org/XDB/ASP/clarosHome/
[15] http://www.getty.edu/conservation/our_projects/field_projects/arches/
[16] http://www.researchspace.org

ones, related to heterogeneous knowledge management and representation. Our partners already asked us for practical use-cases of the platform and semantic data representation in general, e.g. to make a tool for automatic thematic exposition creation, including on the physical media (CD/DVD). Solutions of those yet to be faced challenges will rely on the flexibility of CIDOC-CRM ontology.

Acknowledgments. This work was partially financially supported by Government of Russian Federation, Grant 074-U01.

References

1. Bechhofer, S., Miles, A.: Skos simple knowledge organization system reference. W3C recommendation, W3C (2009)
2. Condron, L., Tittemore, C.P.: Functional requirements for bibliographic records (2004)
3. Crofts, N., Doerr, M., Gill, T., Stead, S., Stiff, M.: Definition of the cidoc conceptual reference model. version 5.0.4. ICOM/CIDOC Documentation Standards Group. CIDOC CRM Special Interest Group (2011)
4. Doerr, M.: Ontologies for cultural heritage. In: Handbook on Ontologies, pp. 463–486. Springer (2009)
5. Initiative, D.C.M., et al.: Dcmi metadata terms, dcmi recommendation (2012)
6. Initiative, D.C.M., et al.: Dublin core metadata element set, version 1.1 (2012)
7. Mouromtsev, D., Haase, P., Cherny, E., Pavlov, D., Andreev, A., Spiridonova, A.: Towards the russian linked culture cloud: data enrichment and publishing. In: Gandon, F., Sabou, M., Sack, H., d'Amato, C., Cudré-Mauroux, P., Zimmermann, A. (eds.) ESWC 2015. LNCS, vol. 9088, pp. 637–651. Springer, Heidelberg (2015)
8. Oldman, D., Doerr, M., de Jong, G., Norton, B., Wikman, T.: Realizing lessons of the last 20 years: A manifesto for data provisioning & aggregation services for the digital humanities (a position paper). D-lib magazine **20**(7), 6 (2014)
9. Oldman, D., Doerr, M., de Jong, G., Norton, B., Wikman, T.: Realizing lessons of the last 20 years: A manifesto for data provisioning & aggregation services for the digital humanities (A position paper). D-Lib Magazine **20**(7/8) (2014). http://dx. doi.org/10.1045/july2014-oldman
10. Peroni, S., Tomasi, F., Vitali, F.: Reflecting on the europeana data model. In: Agosti, M., Esposito, F., Ferilli, S., Ferro, N. (eds.) IRCDL 2012. CCIS, vol. 354, pp. 228–240. Springer, Heidelberg (2013)
11. Ruotsalo, T., Haav, K., Stoyanov, A., Roche, S., Fani, E., Deliai, R., Mäkelä, E., Kauppinen, T., Hyvönen, E.: SMARTMUSEUM: A mobile recommender system for the web of data. J. Web Sem **20**, 50–67 (2013). http://dx.doi.org/10.1016/j.websem.2013.03.001

Designing for Inconsistency – The Dependency-Based PERICLES Approach

Jean-Yves Vion-Dury[1], Nikolaos Lagos[1], Efstratios Kontopoulos[2(✉)], Marina Riga[2],
Panagiotis Mitzias[2], Georgios Meditskos[2], Simon Waddington[3], Pip Laurenson[4],
and Ioannis Kompatsiaris[2]

[1] Xerox Research Centre Europe (XRCE), 38240, Meylan, France
{Jean-Yves.Vion-Dury,Nikolaos.Lagos}@xrce.xerox.com
[2] Information Technologies Institute, CERTH, 57001, Thessaloniki, Greece
{skontopo,mriga,pmitzias,gmeditsk,ikom}@iti.gr
[3] King's College London, London, UK
simon.waddington@kcl.ac.uk
[4] Tate, London, UK
Pip.Laurenson@tate.org.uk

Abstract. The rise of the Semantic Web has provided cultural heritage researchers and practitioners with several tools for ensuring semantic-rich representations and interoperability of cultural heritage collections. Although indeed offering a lot of advantages, these tools, which come mostly in the form of ontologies and related vocabularies, do not provide a conceptual model for capturing contextual and environmental dependencies contributing to long-term digital preservation. This paper presents one of the key outcomes of the PERICLES FP7 project, the Linked Resource Model, for modelling dependencies as a set of evolving linked resources. The proposed model is evaluated via a domain-specific representation involving digital video art.

Keywords: Digital preservation · LRM · Ontology · Digital video art · Dependency

1   Introduction

With the advent of the *Semantic Web*, Cultural Heritage (CH) researchers and practitioners have been gradually adopting the respective approaches and technologies in order to ensure the *semantic interoperability* between physical artefacts and intangible attributes. Work in this direction is mostly revolving around deploying relevant established ontologies and vocabularies. *CIDOC CRM* [3], acknowledged as an ISO standard (21127:2006), is arguably the most popular ontology for representing concepts and information in CH and museum documentation. Other similar resources include *Europeana*[1], a multilingual digital library for facilitating user access to an integrated content for European cultural and scientific heritage, and the *Getty vocabularies*[2] that

---

[1] www.europeana.eu/portal/
[2] www.getty.edu/research/tools/vocabularies/

© Springer International Publishing Switzerland 2015
T. Morzy et al. (Eds): ADBIS 2015, CCIS 539, pp. 458–467, 2015.
DOI: 10.1007/978-3-319-23201-0_46

provide structured terminology for works of art, architecture, material, culture, as well as artists, architects and geolocations.

Although the utility of the above vocabularies is indisputable in capturing descriptive metadata, they do not provide the conceptual model for capturing contextual and environmental dependencies that contribute significantly in *long-term digital preservation*, which refers to the process of adopting a series of managed activities necessary to ensure continued access to and re-use of digital materials for as long as needed [2]. This is the aim of *PERICLES*[3], a four-year integrated project focusing on representing and evaluating the risks for long-term digital preservation of digital resources. The project treats the environments in which digital objects are created, managed and used, holistically as *digital ecosystems*, rather than focusing on individual digital objects. The first aspect of this approach involves developing a domain-independent ontology, the *Linked Resource Model* (*LRM*), to model dependencies as a set of evolving linked resources. The second aspect is *preservation by design* that relies on capturing relevant context and environment of a digital object at source. The approach is *model-driven* and considers digital objects as generated and existing within an evolving continuum.

PERICLES addresses preservation challenges in two domains: (a) digital artworks (e.g. digital video artworks and software-based artworks); (b) experimental scientific and associated space operations data. To this end, the project will deliver two preservation prototypes, as well as a portfolio of models, services, tools and research that supports the development of practice related to the notion of preservation ecosystems and life-cycle management. Whilst the project's focus lies in the preservation of both cultural and scientific heritage, this paper is concerned solely with the former domain and presents early results in applying the PERICLES design principles to a specific challenge encountered by those working on the conservation of digital video artworks. In the rest of the paper, Section 2 describes related work for representing dependencies; Section 3 offers a description of the overall use case, followed by concrete descriptions of the LRM and its domain-specific representations in Sections 4 and 5, and the paper is concluded with final remarks and directions for future work.

2   Related Work: Existing Dependency Models

The *PREMIS Data Dictionary* [7] defines three types of relationships between objects: *structural*, *derivation* and *dependency*. From the PERICLES perspective, derivation and dependency relationships are the most relevant. A *derivation relationship* results from the replication or transformation of an object. A *dependency relationship* exists when one object requires another to support its function, delivery, or coherence. Examples include a font, style sheet, DTD or schema that are not part of the file itself. Objects can also be related to events through user-defined dictionaries of terms, and events can in turn be linked to agents that performed those events.

The *Open Provenance Model* (*OPM*) [9] introduces the concept of a provenance graph that aims to capture the causal dependencies between entities. The most

---

[3] www.pericles-project.eu/

relevant concept from our perspective is *process* that represents actions performed on or caused by artefacts, and resulting in new artefacts.

In a preservation context, [10] defines notions of *module*, *dependency* and *profile* to model use by a community of users. A *module* is defined as a software/hardware component or knowledge base that is to be preserved, and a *profile* is the set of modules that are assumed to be known. A *dependency relation* is then defined by the statement that module A depends on module B if A cannot function without B. For example, a `readme.txt` file depends on the availability of a text editor.

The authors of [11] also define the more specific notion of task-based dependency, expressed as Datalog rules and facts. For instance, `Compile(HelloWorld.java)` denotes the task of compiling 'HelloWorld.java'. Since the compilability of the latter depends on the availability of a compiler, this dependency can be expressed using a rule of the form: `Compile(X):- Compilable(X,Y)`, where the binary predicate `Compilable(X,Y)` denotes the appropriateness of Y for compiling X. This more formal approach enables various tasks to be performed, such as risk and gap analysis for specific tasks, possibly considering contextual information, such as user profiles.

In [6] the authors elaborate a more sophisticated approach (as compared to [11]). The original notion of *task* formerly associated with dependencies is now abstracted toward the notion of *intelligibility*, which allows for typing dependencies. The LRM approach goes one step further toward genericity, by allowing any kind of dependency specialization, the intelligibility being replaced by the notion of *intention*, which can be described informally or formally through additional properties. Moreover, LRM offers a much richer topology for dependency graphs through managing dependencies as instances instead of properties. By combining genericity and semantic refinement, we expect a tighter management of consistency criteria, whatever semantics might be potentially involved.

3   Case Study: Video Art and Preservation Challenges

Video as a medium is now a mainstream element of contemporary art practice. 'Video' is a broad term used to refer to a wide array of rapidly changing technologies, including the formats in which an artwork might be made, archived and displayed and the equipment used in its presentation. One of the key issues for artists' video relates to how changes that will impact the appearance of the video are tracked and managed over time. Video artworks typically require the following three main components to be displayed: (a) the media (i.e. the video file), (b) the equipment needed for display in the gallery, and, (c) the installation instructions. In recent years video has moved from tape to file-based delivery and storage.

The lack of consensus, tools and services around the preservation of high-value video within the broader preservation community has created a significant problem for previously well-managed collections. There is usually no retention schedule for works; rather it is assumed that they will be preserved in perpetuity. The planning horizon for heritage institutions is often expressed as 100 or 500 years and sometimes as forever.

As already mentioned in the introduction, within the PERICLES project the primary aim is to focus on modelling the risks related to the long-term digital preservation of digital objects in general, and, amongst them, the media elements of video art objects. One of the scenarios in which we are exploring the benefits of such modelling is ensuring the consistent playback of video files and it is this use case that is presented later in the paper (Section 5). In order to identify and understand the technical variables involved in the consistent playback of digital video files, Tate commissioned a report by Dave Rice [8]. Thus, we are not aiming to model the domain at the artwork level, but rather the specifics of dependencies between digital things within a system which forms part of the artwork. It is therefore a partial model related to the dependencies of some of the components of the artwork.

In this case, modelling enables the conservator to better understand the digital dependencies within the system and also helps identify areas where automation might be achievable. Modelling also facilitates communication with computer scientists and software developers who might provide tools to support activities related to long term digital preservation and the corresponding dependencies. The process of establishing any such model is the identification of the key entities, the relevant types of dependency and change and, finally, the study of scenarios regarding how this information might be used in the assessment of material that comes into the repository.

4   LRM: An Ontology Focusing on Change Management

The *Linked Resource Model* (*LRM*) is an upper level ontology designed to provide a principled way for modelling evolving ecosystems, focusing on aspects related to the changes taking place. This means that, in addition to existing preservation models that aim to ensure that records remain accessible and usable over time (e.g. see [1]), the LRM also aims to model how changes to the ecosystem, and their impact, can be captured. It is important to note here that we assume that a policy governs at all times the dynamic aspects related to changes (e.g. conditions required for a change to happen and/or impact of changes). As a consequence, the properties of the LRM are dependent on the policy being applied; therefore, most of the defined concepts are related to what the policy expects. At its core the LRM defines the ecosystem by means of participating entities and dependencies among them. A set of other properties and specialised entity types are also provided but they are all conditioned on what is allowed/required by the policy. The notion of policy is not further defined here, as it is out of the scope of this work. The main concepts of the static LRM are illustrated in Fig. 1 (the prefix pk refers to the LRM namespace) and discussed further below.

Resource. Represents any physical, digital, conceptual, or other kind of entity and in general comprises all things in the universe of discourse of the LRM Model[4]. A resource can be *Abstract* (c.f. AbstractResource in Fig. 1), representing the abstract part of a resource, for instance the idea or concept of an artwork, or *Concrete* (c.f. ConcreteResource in Fig. 1), representing the part of an entity that has a physical extension and can therefore be accessed at a specific location (a corresponding attribute called location is used to specify spatial information; for instance for a

---

[4] This definition is close to CIDOC CRM's Entity – we are exploring possible mappings.

`Digital-Resource`, which represents objects with a digital extension, this information can be the URL required to retrieve and download the corresponding bit stream). The above two concepts can be used together to describe a resource; for example, both the very idea of an artwork, as referred by papers talking about the artist's intention behind the created object, and the corresponding video stream that one can load and play in order to manifest and perceive the artwork. To achieve that, the abstract and concrete resources can be related through a specific `realizedAs` predicate, which in the above example could be used to express that the video file is a concrete realization of the abstract art piece. It should be noted that an abstract resource could be connected to one or several concrete resources; in that case, the concrete resources could be aggregated (a class `AggregatedResource` is defined in the LRM, although not shown in Fig. 1), and `realizedAs` could be used to connect the abstract resource to the aggregated one.

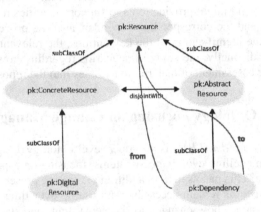

Fig. 1. Main concepts of the static LRM.

Dependency. The core concept of the static LRM is that of a *dependency*. An LRM `Dependency` describes the context under which change in one or more entities has an impact on other entities of the ecosystem. The description of a dependency minimally includes the intent or purpose related to the corresponding usage of the involved entities. From a functional perspective, we expect that dedicated policies/rules will further refine the context (e.g. conditions, time constraints, impact) under which change is to be interpreted for a given type of dependency. For instance, consider a document containing a set of diagrams that has been created using MS Visio 2000, and that a corresponding policy defines that MS Visio drawings should be periodically backed up as JPEG objects by the workgroup who created the set of diagrams in the first place[5]. According to the policy, the workgroup who created the set of JPEG objects should be able to access but not edit the corresponding objects. The classes and properties related to the `Dependency` class can be used to describe each such conversion in terms of its temporal information and the entities it involves along with their roles in the relationship (i.e. person making the conversion and object being converted).

---

[5] This example is adapted from a use case described in [1], pp. 52-53.

In addition, the LRM Dependency is strictly connected to the intent underlying a specific change. In the case described here the intent may be described as *"The workgroup who created the set of diagrams wants to be able to access (but not edit) the diagrams created using MS Visio 2000. Therefore, the workgroup has decided to convert these diagrams to JPEG format"* and it implies the following.

- There is an explicit dependency between the MS Visio and JPEG objects. More specifically, the JPEG objects are depending on the MS Visio ones. This means that if an MS Visio object 'MS1' is converted to a JPEG object, 'JPEG1', and 'MS1' is edited, then 'JPEG1' should either be updated accordingly or another JPEG object 'JPEG2' should be generated and 'JPEG1' optionally deleted (the use case is not explicit enough here to decide which of the two actions should be performed). This dependency would be particularly useful in a scenario where MS Visio keeps on being used for some time in parallel to the JPEG entities, which are in turn used for back up purposes.
- The dependency between 'MS1' and 'JPEG1' is unidirectional. Actually, JPEG objects are not allowed to be edited and, if they are, no change to the corresponding MS Visio objects should apply.
- The dependency applies to the specific workgroup, which means that if a person from another workgroup modifies one of the MS Visio objects, no specific conversion action has to be taken (the action should be defined by the corresponding policy).

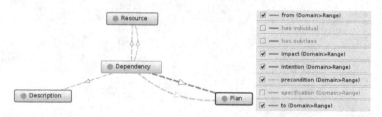

Fig. 2. A view of the Dependency concept in LRM.

To enable recording the intent of a dependency, we can relate in the LRM the Dependency entity with an entity that describes the intent via a property that we name "intention", as illustrated in Fig. 2. In Fig. 2, properties from and to indicate the directionality of the dependency.

The LRM model provides also concepts that allow recording when a change is triggered and what is the impact of this change on other entities. Let us take once more the above example: we need to be able to express the fact that transformation to JPEG objects is possible only if the corresponding MS Visio objects exist and if the human that triggers the conversion has the required permissions to do that (i.e. belongs to the specific workgroup). The impact of the conversion could be to generate a new JPEG object or update an existing one. The action to be taken (i.e. generate or update) in that case, would be decided based on the policy governing the specific operation. Assuming that only the most recent JPEG object must be archived, then the old one must be deleted and replaced by the new one (conversely deciding to keep the old JPEG object may imply having to archive the old version of the corresponding old MS Visio object as well).

Plan. The condition(s) and impact(s) of a change operation are connected to the `Dependency` concept in LRM via `precondition` and `impact` properties as illustrated in Fig. 2. These connect a `Dependency` to a `Plan`, which represents a set of actions or steps to be executed by someone/something (either human or software). The Plan can be used, thus, as a means of giving operational semantics to dependencies. Plans can describe how preconditions and impacts are checked and implemented (this could be for example defined via a formal rule-based language, such as SWRL). The temporally coordinated execution of plans can be modelled via activities. A corresponding `Activity` class is defined in LRM, which has a temporal extension (i.e. has a start and/or end time, or a duration). Finally, a resource that performs an activity, i.e. is the "bearer" of change in the ecosystem, either human or man-made (e.g. software), is represented by a class called `Agent`[6].

5   Domain-Specific Extension: The DVA Ontology

In this work we propose a domain-specific representation for modelling dependencies between different digital entities that impact our ability of preserving digital video art.For the representation of digital entities we re-use several constructs from CIDOC-CRM and CRMdig [4], ensuring, thus, *semantic interoperability* with other ontologies

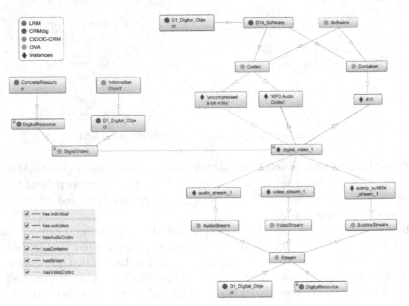

Fig. 3. Representation of a digital video resource, along with its descriptive details.

---

[6] Classes `Activity` and `Agent` relate to provenance information. We explored potential mappings between LRM and PROV (www.w3.org/TR/prov-o), a widely used resource for representing provenance information, but some PROV constraints (i.e. Activities cannot be Entities) are structurally incompatible with the LRM.

already aligned with these models. On the other hand, the mechanisms for representing dependencies are based on the aforementioned LRM. As already mentioned, we mostly focus on representing and evaluating the risks for long-term digital preservation. The adopted representation approach is presented under the scope of a specific challenge, i.e. sustaining consistent video playback.

As seen in Fig. 3, the digital video, i.e. one of the concrete resources of a digital video artwork, contains: (a) stream(s) for video, audio (optional) and subtitles (optional), (b) a codec, and (c) a container (or wrapper). Relations between these entities and a digital video are represented via properties `has-stream`, `has-container` and `has-codec`, respectively, with further categorization of the latter to `has-video-codec` and `has-audio-codec`.

Additionally, key LRM notions, such as `AggregatedResource` and `Dependency`, have been integrated and fully-adopted in the DVA ontology. As mentioned in Section 4, an abstract resource may be realised as one or more concrete resources; in the latter case, the concrete resources are aggregated into one `AggregatedResource` instance, while the abstract with the aggregated resources are connected via property `realizedAs`. Fig. 4 illustrates such an example, as represented in the DVA ontology.

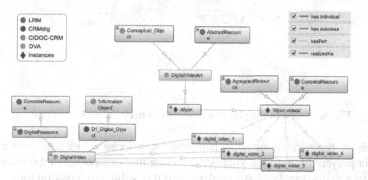

Fig. 4. Example of a digital video artwork with aggregated resource.

The concept of `Dependency` is adopted from LRM to represent relations between digital video artworks and complementary entities (i.e. media players, wrappers and relevant software). A specific challenge concerning the preservation of digital video artworks is to sustain the consistent playback of their video files. In this context, Fig. 5 displays the dependency of the playback activity to the digital video file.

Furthermore, since a digital video file is associated to a container, playing the container correctly depends on the usage of an appropriate media player. `PlayerDependency` involves the compatibility of media players with certain video containers. Specifically, a video container (e.g. 'AVI') depends on the media players supporting its playback. This classification offers the possibility to spot proper media players for a certain playback activity.

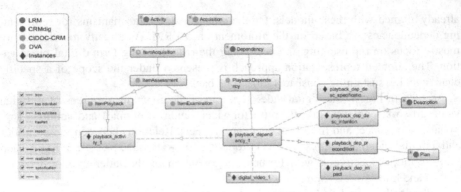

Fig. 5. A view of the playback dependency concept in the DVA ontology.

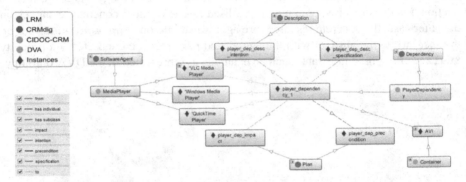

Fig. 6. A view of the player dependency concept in the DVA ontology.

The above representations allow handling various cases, like e.g. finding compatible media players for a digital video or detecting inconsistencies in the video's aspect ratio, colour matric etc. We have implemented this validation layer on top of the DVA ontology as *SPARQL rules (SPARQL Inferencing Notation - SPIN* [5]). In SPIN, SPARQL queries can be stored as RDF triples alongside the RDF domain model, enabling the linkage of RDF resources with the associated SPARQL queries, as well as sharing and re-using SPARQL queries. SPIN supports the definition of SPARQL inference rules that can be used to derive new RDF statements from existing ones through iterative rule application, serving as a ready-to-use framework. The presentation of the implemented SPIN constructs is outside the scope of this paper.

6   Conclusions and Future Work

This work presents early results in using a domain-independent ontology, the *Linked Resource Model* (*LRM*), to model and manage change in a cultural heritage setting. Viewing dependencies as complex constructs instead of simple links between resources, allows defining the semantics governing a change in terms of the intention underlying this change, the pre-conditions that should be satisfied to trigger it, and the corresponding resulting impact(s) on the ecosystem itself. We have illustrated via our case study that the LRM can be combined with CIDOC-CRM (and its CRMdig extension for modelling digital resources) and we are planning of exploring tighter integration in the near future. Of interest to us is also extending the LRM ontology in order to represent causality and temporal aspects.

Acknowledgements. This work was supported by the European Commission Seventh Framework Programme under Grant Agreement Number FP7-601138 PERICLES.

References

1. Australian Government Recordkeeping Metadata Standard Version 2.0 Implementation Guidelines (2010). www.naa.gov.au/records-management/publications/agrk-metadata-standard.aspx (accessed: Apr-15)
2. Digital Preservation Coalition: Digital Preservation Handbook. York, UK (2008). www.dpconline.org/advice/preservationhandbook (accessed: Apr-15)
3. Doerr, M.: The CIDOC CRM, an Ontological Approach to Schema Heterogeneity. Semantic Interoperability and Integration **4391** (2005)
4. Doerr, M., Theodoridou, M.: CRMdig: a generic digital provenance model for scientific observation. In: 3rd USENIX Workshop on the Theory and Practice of Provenance, TaPP 2011, Heraklion, Crete, Greece, June 20–21, 2011
5. Knublauch, H., Hendler, J. A., Idehen, K.: SPIN - overview and motivation. World Wide Web Consortium, W3C Member Submission, February 2011
6. Marketakis, Y., Tzitzikas, Y.: Dependency Management for Digital Preservation Using Semantic Web Technologies. Int. Journal on Digital Libraries **10**(4), 159–177 (2009)
7. PREMIS Data Dictionary for Preservation Metadata (Official Web Site), The Library of congress, USA. www.loc.gov/standards/premis/ (accessed: Apr-15)
8. Rice, D.: Sustaining Consistent Video Presentation. Tate Research Articles, March 2015. www.tate.org.uk/research/publications/sustaining-consistent-video-presentation (accessed: Apr-15)
9. The Open Provenance Model Core Specification (v1.1). http://eprints.soton.ac.uk/271449/1/opm.pdf (accessed: Apr-15)
10. Tzitzikas, Y.: Dependency management for the preservation of digital information. In: Wagner, R., Revell, N., Pernul, G. (eds.) DEXA 2007. LNCS, vol. 4653, pp. 582–592. Springer, Heidelberg (2007)
11. Tzitzikas, Y., Marketakis, Y., Antoniou, G.: Task-based dependency management for the preservation of digital objects using rules. In: Konstantopoulos, S., Perantonis, S., Karkaletsis, V., Spyropoulos, C.D., Vouros, G. (eds.) SETN 2010. LNCS, vol. 6040, pp. 265–274. Springer, Heidelberg (2010)

A Semantic Exploration Method Based on an Ontology of 17th Century Texts on Theatre: la *Haine du Théâtre*

Chiara Mainardi[1]([✉]), Zied Sellami[1,2], and Vincent Jolivet[1]

[1] Labex OBVIL - Université de la Sorbonne, 28 Rue Serpente, 75006 Paris, France
chiara85.mc@gmail.com, vincent.jolivet@paris-sorbonne.fr
[2] Laboratoire d'Informatique de Paris 6, 4 Place Jussieu, 75005 Paris, France
zied.sellami@lip6.fr

Abstract. This paper proposes a method to explore a collection of texts with an ontology depending on a particular point of view. In the first part, the paper points out the characteristics of the corpus, composed of 17th century French texts. In the second part, it explains the methodology to isolate the discriminant terms for the ontology creation. Furthermore, not only the projection of the ontology on the texts is pointed out, but also how to explore the corpus thanks to the defined perspective based on semantic fields.

1 Context

1.1 The Project *Haine du théâtre*

The Project *Haine du théâtre* (HdT)[1] is one of the many outstanding projects at the Labex OBVIL[2] in Paris. In this article, we present an account of the building of an ontology and a semantic search engine implementing it, an experience which is complementary to the ones developed at the ACASA team[3].

The Project HdT aims at analysing theatre debates in Europe by using scientific approaches and critical editions of polemical texts. The reflections of the scientific HdT team are mainly focused around the discovery of the circumstances and the arguments used in theatre controversies all across Europe, not limited to France, but also to England, Spain, Italy, and the Germanic area, from the last decades of the 16th century up to the beginning of the 19th century.

The corpus that we have used to start this investigation concerns the quarrels in France. The total collection of the HdT Project is, by now, constituted by 300 texts in the PDF format. Among those, in the last months we have done

---

[1] The directors of this Projet are François Lecercle and Clotilde Thouret. http://obvil.paris-sorbonne.fr/projets/la-haine-du-theatre.

[2] The Labex OBVIL (Laboratoire d'Excellence : Observatoire de la Vie Littéraire) is headed by Didier Alexandre. http://obvil.paris-sorbonne.fr

[3] The ACASA team (Cognitive Agents and Automated Symbolic Learning) is headed by Jean-Gabriel Ganascia. http://www-poleia.lip6.fr/ACASA/

© Springer International Publishing Switzerland 2015
T. Morzy et al. (Eds): ADBIS 2015, CCIS 539, pp. 468–476, 2015.
DOI: 10.1007/978-3-319-23201-0_47

the XML/TEI[4] critical edition of twenty texts and, consequently, we have used them for the semantic analysis[5]. Particularly, among the most important texts, the corpus[6] was composed of: the *Dissertation sur la condemnation des théâtres* (1666) by l'Abbé D'Aubignac, the *Traité de la Comédie et des spectacles* (1666) by the Prince of Conti, the *Traité de la Comédie* (1667 et 1675) by Pierre Nicole, the *Défense du traité de Mgr le Prince de Conti touchant la comédie et les spectacles ou la réfutation d'un livre intitulé Dissertation sur la condamnation des théâtres* (1671) by Joseph de Voisin and the *Traité des théâtres* by Philippe Vincent (1647).

The aim of the Project is to explore the grey areas of the controversy in order to outline a global overview of the central issues which led to these polemics, discovering where and how they began, their chronological discrepancies in the different countries, and the links between them and their contemporary resurgences.

In this direction, the first step was to figure out which authors wrote some texts in favour of the theatre and some others against. Therefore, a comparison among texts and passages by the same or different authors was needed.

1.2 Ontology

Ontologies (in the computer sense, not in the philosophical sense) are often used to represent a specification of domain knowledge that reflects a consensual agreement on the domain, or the knowledge required for a specific application. Ontologies are now essential in many applications (access to shared knowledge, semantic web, information retrieval, etc.). An ontology model is made of concepts structured by hierarchical and non-hierarchical semantic relations [1].

Building an ontology relies on elaborated processes to identify the right domain knowledge and to formalize it in a relevant way so that it could be usable by software applications and interpretable by humans [2].

Indeed, modeling assumes to make various decisions that organize, classify and describe the concepts of a specific domain according to the point of view that best suits the use of the ontology. Whatever the domain and the application, several possibilities exist to organize knowledge and it is difficult to estimate the "best" one [3].

In recent years, the use of text (as knowledge source) and language processing (as a technique for rapidly detecting linguistic clues of knowledge in text) have emerged as a solution to make this task easier [4]. Practically, text-based approaches rely on the ontology engineer to decide on how to organize knowledge in the ontology, but they provide him with data for conceptual structuring. For

---

[4] http://www.tei-c.org/index.xml

[5] First of all we transcribe the 17th century texts using an OCR software. Then, we correct and modernize the language manually, with a word processor. XML/TEI is generated from the ODT file thanks to a dedicated software (See http://obvil-dev. paris-sorbonne.fr/developpements/odt2tei/).

[6] http://obvil.paris-sorbonne.fr/corpus/haine-theatre/

example, some candidate phrases can become concept labels, classes of phrases can contribute to define concepts, lexical relations can serve as clues for semantic relations, etc.

Fundamental steps for the success of ontology building from linguistic resource include (*i*) the choice of a relevant corpus, (*ii*) the selection of appropriate NLP software tools, adapted to the corpus content and to the sought knowledge, (*iii*) the identification of relevant elements (terms, named entities, lexical relations...) among the large amount of linguistic data, and finally (*iv*) an appropriate formalization of these elements [5].

2 Our Semantic Approach to Explore the Corpus *Haine du théâtre*

2.1 Ontology of *Haine du théâtre*

Understanding what the different authors of the corpus think about theatre in every passage of their text is not obvious, but it is possible to discern if we consider the semantic fields related to the authorities quoted by the writer. With the word "authority" we mean the Latin "auctoritas": the ancient authors, such as Plato, Aristotle, Cicero, which were evoked by a certain writer to highlight his ideas and opinions about a certain matter. Their writings are quoted by the writer to underline his theory or point of view and are very often provided of the author's personal comment.

After these presuppositions, we started to study the semantic fields that revolved around the quoted authorities. As a result, a certain amount of semantic fields about the theatre debate came out. For example, the semantic field about a derogatory discourse on theatre, which we have called "the condemnation field".

Our subsequent idea was to register these semantic fields and their distinctive terms (linguistic signs) in an ontology. The choice of using an ontology was made since it would have allowed us to organize the knowledge of the studied domain of critical editions of polemical texts about theatre as a structured point of view on the subject.

We have then manually built[7] an ontology with 44 structured classes (concepts). Each class is described by several conceptual attributes (a definition, labels, linguistic signs and exact linguistic signs) and 2 functional attributes (semantic field and grammatical gender) necessary for the semantic analysis tool. A particular concept called "Autorité" (Authority) was instantiated (populated) with a list of the authorities cited in the corpus (see figure 1).

The creation of our ontology rests upon the semantic fields outlining the opinions about theatre of the different authors of our corpus. To figure out these semantic fields we did a lexicometric examination of the collocations around the authorities of the corpus. Particularly, we partitioned our collection of texts (into 17 parts, one for each text) and calculated, for each part, the number of proper

---

[7] We have used the free, open-source ontology editor called Protégé: http://protege. stanford.edu/

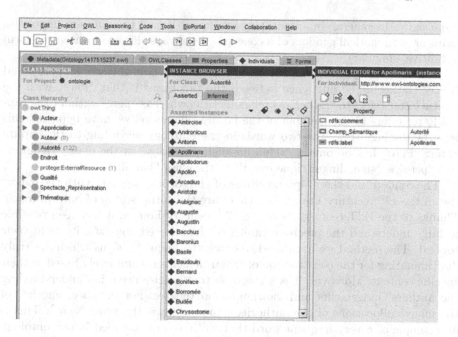

Fig. 1. Screenshot of the ontology *"Haine du théâtre"* developed on the Protégé ontology Editor.

nouns[8]. Then, we classified the most frequent proper nouns (and eliminated the non-pertinent ones, i.e. the ones which were not authorities) with more than 20 appearances in the whole corpus. After that, we studied the collocations[9] of each occurrence (i.e. each authority) with a threshold of 5 units (on the left and on the right from the pivotal term). We examined the meaning of every collocation referred to its context and we eliminated all the confusing and equivocal words.

For example, some ambiguous terms have been included when they were at the core of the debate about theatre. For instance, an important collocation of the authority Chrysostom[10] is the word "jeu" (game). This French word has several meanings[11], but in our corpus only the one with the sense of "acting" is used, and this is the reason why we decided to put the word in the "Spectacle" (performance) class. In addition, we have chosen to include in the ontology only the word "jeu" because this term (masculine, singular) would detect anyway the plural and its composed forms, like "Jeux Majuma" which is quite frequently

---

[8] For the textometric analysis (e.g. for proper nouns detection), we used TXM (http://sourceforge.net/projects/txm/).

[9] In TXM the collocations are called "co-occurrents".

[10] John Chrysostom (c. 349 - 407) was an important Church Father and Archbishop of Constantinople.

[11] It refers to "play" when its meaning is that of "divertissement" or of functioning; to "acting" then it refers to theatre and performances; to "game" when there are rules which define it, and so on.

employed[12]. In general, in an ontology we do not list all the different linguistics forms or grammatical genders of a concept. We always choose a consensual form (infinitive, singular).

Another example of Chrysotome's collocations are the word "père" and "Père" (father). The difference between the two terms is that "Père" with the capital letter refers to "father" as religious title, whereas "père" without the capital letter refers to the member of the family. Therefore, we have bypassed this issue by distinguishing the two words in the ontology according to the capital letter: "Père" has become a "signe linguistique exact" in the class "Religion" and "père" a "signe linguistique exact" in the class "Famille".

The context and the deep knowledge of the corpus, together with the expertise in the 17th century French language, are the cornerstones of our approach. Thanks to the HdT team skills in the 17th century French, it has been possible to fully understand the precise meaning of the selected linguistic items in their context. The method we have used to catch the linguistic signs which are truly discriminating for the construction of semantic fields is completely based in their specific context. Moreover, it is a response to a precise need, i.e. understanding the authors' judgements and viewpoints on theatre. For instance, another of the many collocations of the authority Chrysostom is the word "acte". This is an example of a very frequent word that we have not included in our ontology word list because it is not pertinent for the purpose of our ontology. Moreover, this word has different meanings, such as gesture, deed and act (as a section of play), which are all used in our corpus. When compared to the corpus context, the results would consequently have been distorted. Ambiguous and polysemous words like "act" have been rejected when they were not relevant for the comprehension of the controversy for or against the theatre.

Finally, we have created a list of pertinent terms for the ontology: we have divided all the co-occurrences into semantic fields. We have considered as a "field" only a group of more than three co-occurrents. Even if finding the name of the semantic classes has been strictly subjective, we can certify that the list of words and the division into fields are objective: they come from the textometric study of the corpus. Moreover, the detection of semantic spheres has been done according to the precise context in which every term was included. For example, the semantic field of "condamnation" (condemnation) contains words which are not usually employed nowadays to express a reprehension of something, in theatre in particular. Some examples are the terms "vanité" (vanity, triviality), "fureur" (rage, violence, passion), "impure" (impure, polluted, lascivious), "maudit" (wretched, damned) and "pernicieux" (pernicious) and their other equivalent grammatical forms.

2.2 Projection of the Ontology on the Corpus

This ontology corresponds to a structured lexicon of discriminant terms of the semantic fields related to theatre. To detect the significant linguistic signs of

---

[12] This term refers to acting, games and celebrations that people from Palestine used to fete. They were afterwards adopted by Greeks and Romans.

the corpus, we made use of the team's expertise on the domain and then placed these terms into the appropriate ontology concept. The idea we came up in the ontology exploration was to look at the relative importance of the various semantic fields in each text of the collection and then be able to understand *a priori* the content of a chosen text.

In order to reach this aim, we label automatically the corpus to pick out all the forms (lemmas) of the ontology discriminant terms. An automatic annotation tool based on the ontology and developed in java language enriches automatically the XLM/TEI sources by setting down some tags <term> which are useful for tracking and identify all the terms included in the ontology list. In the code, the attribute @key gives the identification key of the linguistic sign which is linked to the semantic field inquired by the attribute @type. The annotation result is available in the OBVIL website (see figure 2).

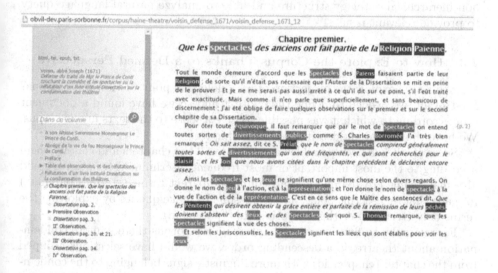

Fig. 2. Screenshot of a chapter annotation available on http://obvil-dev. paris-sorbonne.fr/corpus/haine-theatre/voisin_defense_1671/voisin_defense_1671_12

We notice that following this method we find in our collection of texts a high density of the linguistic signs recorded in our ontology. In addition, the combination of these terms and the semantic fields connected to them come into being a semantic descriptor of the corpus content. For example, the text by Voisin shows a bit more than 1550 discriminant terms for each chapter (17100 words in total).

Concretely, in Chapter 10^{13} of Voisin's text there is a overrepresentation of the semantic field of condemnation if compared to the defence field. Especially,

---

[13] http://obvil-dev.paris-sorbonne.fr/corpus/haine-theatre/voisin_defense_1671/ voisin_defense_1671_18

we find 9 occurrences which belong to the defence (morale positive) whereas there are 132 terms which belong to the condemnation (morale negative). After having read the chapter, whose title is "Que l'extrême impudence des Jeux Scéniques et des Histrions fut condamnée"[14] we can confirm that these descriptors reflect quite precisely the text content. This kind of semantic exploration method gives interesting and useful outcomes: without preconception and previous knowledge of the corpus, it was possible for us to determine whether a certain text or chapter was, for example, laudatory or derogatory about theatre.

As a consequence, we developed the idea to improve this method and, in the future, by implementing a semantic research engine. We were able to use existing search engines (like lucene search[15]) to query the corpus with the instances of concepts extracted from text. However, these search engines do not exploit the structure of the ontology. Also, it considers a user query as a set of independent words. Our goal is to develop a search engine able to exploit the hierarchical and non-hierarchical ontology structure and able to analyze natural language query to provide relevant results.

2.3 How to Explore the Corpus Thanks to a Defined Perspective

How could we exploit the projection of an axiological point of view on a corpus? First of all, we needed to define the context and so we have found a document unit according to which it was possible to give a sense to the texts in our corpus. We have thus given an identifier for each chapter.

Secondly, it has been possible to gather up all the chapters and classify them according to the most important or most frequent discriminant terms. We could also make some crossed queries and count the parent terms ("termes parents") for each chapter. At this point, it was possible to make queries by exploiting the terms only (term_id) or the semantic fields (parent_id).

For example, to find in which chapters of our corpus there is more condemnation about theatre, in a descending order, we would have selections grouped from the chapter (chapter_id) with more linguistic signs belonging to the condemnation semantic field (parent_id='Morale_Négative') to the chapter with less of them. The score of the discriminant terms of the semantic field about the "morale négative" (condemnation) is the index which allows to understand how deeply a chapter falls within the reprehension of theatre. The result of this query is that the first 10 chapters which revealed to be the most against theatre were all by Voisin and Conti (6 chapter by Voisin and three by Conti), with the exception of one chapter by Vincent. Particularly, according to gross scores, chapter 12[16] by Voisin, has 1563 linguistic signs related to the condemnation of theatre (on a

---

[14] "How the extreme impudence of scenic games and histrions was condemned", this title gives already the idea of the chapter's content and the prevalence of the condemnation semantic field.

[15] https://lucene.apache.org/core/

[16] "Que la représentation des Comédies et des Tragédies ne doit point être condamnée tant qu'elle sera modeste et honnête".

total of 74000 words), so the 2%. The chapter called "IVe siècle" (4th century) by Conti is one of the most reprehensive as it has 151 linguistic signs (on a total of 4300 words), so the 3,5%. Chapter 12^{17} by Vincent shows 140 linguistic signs (on a total of 7800 words), so the 1,7%. Our semantic search engine is already able to exploit the hierarchy of the ontology and the semantic text annotations. First results are promising; since it is very delicate to manipulate effective gross numbers, we are now improving the tool by implementing new statistic weighting techniques (based on chapter dimension and terms weighting).

This result permits us to understand that according to our corpus the authors who mostly condemned the theatre are Voisin and Conti. The fact that our results show a link between these writers confirm the efficiency of our method. In fact, Voisin wrote his treatise in order to defend Conti's one, as, among the other reasons, can be easily shown by the title: *Défense du traité de Mgr le Prince de Conti touchant la comédie et les spectacles [...].*[18]

3   Conclusions

In this paper we showed a method based on using an ontology to analyse automatically many texts together with a dual interest: documentary and linguistic. On the one hand, by improving a state of knowledge about the French 17th century language. On the other hand, refining the vocabulary linked to the theatre controversy.

The practice we are presenting and currently proceeding with is an iterative method that constantly analyses the results and their reliability. The process of creating the ontology is a very long and demanding one. Once the ontology is created, it needs to be adjusted and expanded when new texts added to the corpus contain new terms. In fact, at the core of the ontology there is the reliability of the discriminant terms (linguistic signs), but their importance as linguistic signs can evolve according to the collection of texts. The fact that we are working on a non-closed corpus constrains us to reflect on our expertise and its degree of reliability. We need to constantly verify the lexicon of our ontology list as being completely detached from preconceptions: the approval or disapproval of the method is faced to actual scenarios taken from the context of each text.

To explore a corpus with an ontology depending on a particular point of view, i.e. on the precise point of view of condemnation or approval of theatre, has revealed by now a successful conception. This approach can be reusable in other domains, as long as the precise perspective of the ontology is pointed out. Possibly, this method and this code are exploitable in other contexts, such as the corpus du Projet Apollinaire, Projet Molière ou Projet Historiographie théâtrale du Labex Obvil[19].

---

[17] "Que les raisons dont on essaie d'appuyer les Théâtres sont tout à fait futile".

[18] Transl. "Defence of the treatise by M. the Prince of Conti concerning the comedy and spectacles [...]".

[19] http://obvil.paris-sorbonne.fr/bibliotheque

References

1. Guarino, N., Oberle, D., Staab, S.: What is an ontology? In: Staab, S., Studer, R. (eds.) Handbook on Ontologies, pp. 1–17. International Handbooks on Information Systems. Springer, Berlin Heidelberg (2009)
2. Maedche, A.: Ontology learning for the Semantic Web, vol. 665. Kluwer Academic Publisher (2002)
3. Cimiano, P., Buitelaar, P., Völker, J.: Ontology construction. In: Indurkhya, N., Damerau, F.J. (eds.) Handbook of Natural Language Processing, 2nd edn, pp. 577–604. CRC Press, Taylor and Francis Group (2010)
4. Buitelaar, P., Cimiano, P., Magnini, B.: Ontology Learning from Text: Methods, Evaluation and Applications. Frontiers in Artificial Intelligence and Applications Series. IOS Press, Amsterdam (2005)
5. Aussenac-Gilles, N., Despres, S., Szulman, S.: The TERMINAE Method and Platform for Ontology Engineering from texts. In: Buitelaar, P., Cimiano, P. (eds.) Bridging the Gap Between Text and Knowledge - Selected Contributions to Ontology Learning and Population from Text, pp. 199–223. IOS Press (2008)

Combining Semantic and Collaborative Recommendations to Generate Personalized Museum Tours

Idir Benouaret[(✉)] and Dominique Lenne

Sorbonne Universités, Université de Technologie de Compiègne Heudiasyc – UMR CNRS 7253, Compiègne, France
{idir.benouaret,dominique.lenne}@hds.utc.fr

Abstract. Our work takes place in the field of support systems to museum visits and access to cultural heritage. Visitors of museums are often overwhelmed by the information available in the space they are exploring. Therefore, finding relevant artworks to see in a limited amount of time is a difficult task. Our goal is to design a recommender system for mobile devices that adapts to the users preferences and is sensitive to their contexts (location, time, expertise...). This system aims to improve the visitors' experience and help them build their tours on-site according to their preferences and constraints. In this paper we describe our recommendation framework, which consists in a hybrid recommendation system. It combines a semantic approach for the representation of museum knowledge using ontologies and thesauruses with a semantically-enhanced collaborative filtering method. A contextual post-filtering enables the generation of a highly personalized tour based on the physical environment, the location of the visitors and the time they want to spend in the museum. This work is applied to the Compiègne Imperial Palace museum in Picardy.

Keywords: Semantic web · Recommender systems · Context · Culturage heritage

1 Introduction

Research on mobile systems of assistance to the visit of museum sites was particularly active during the last years. Indeed, there is often very large collections put at the disposal of the general public. However, visitors cannot see all the artworks that are presented at the museum they are visiting and they do not always known what they like or where the artworks they would like are situated. The visitor of a museum is thus confronted to several problems. He generally has a limited amount of time to spend in the museum and does not know necessarily what he should see or what he is going to like. The tour he makes is thus generally not very thoughtful [1]. As a consequence, the visitor can waste his/her time in looking at artworks that do not interest him. Conversely, he can

© Springer International Publishing Switzerland 2015
T. Morzy et al. (Eds): ADBIS 2015, CCIS 539, pp. 477–487, 2015.
DOI: 10.1007/978-3-319-23201-0_48

miss artworks that would have been of interest for him. In [2], it has been shown that museum visitors may suffer from physical and/or psychological fatigue due to walking effort and information overload. It is also said that 28.9 % of visitors quit museum tours halfway. Among those, 21.0 % are due to physical tiredness and 20.1 % are due to the boredom when facing uninteresting artworks [2].

Information overload is a phenomenon that has invaded every field in our lives, from work activities to leisure time ones. One way to solve the problem is through the use of recommender systems. These systems have proved to be very satisfying in helping users access to desired resources [3]. During the last decades, recommender systems have found their way in the context of museums to help visitors finding relevant artworks to see. To propose a recommender system in the field of cultural heritage, we are confronted to some problems. Some of them are generic and concern all recommender systems such as Cold-start, Sparsity and Over-specialization [4], others are specific to the context of museums. In the case of museum visits, the physical environment brings an additional problem. It is important to suggest a visiting path to the user, the order of discovering the artworks is important so that there is an efficient way to walk from an artwork to another and avoid goings and comings in the physical space. In general, the user has a limited amount of time to spend in the museum. Suggesting the most interesting visiting path that better covers the user interests in the limited available time is a crucial task. These problems shows the importance of contextual information such as visitor's location, physical environment and time of visit.

The majority of existing works in recommender systems focus on recommending the most relevant items to the user and do not take into account any contextual information, such as time, location, weather and the company of other peoples. Traditionally approaches give recommendations based only on two type of entities, users and items, and do not put them into a context when providing recommendations.

The rest of this paper is organized as follows. Section 2 presents a brief literature review. In Section 3, we present our framework. In Sections 4 and 5, we present out semantic model of artworks and user context. Sections 6 and 7 present our recommender system, namely the semantic and collaborative approaches. Section 8 presents the generation of visiting paths. Section 9 concludes the paper and announces future research directions.

2   Related Work

In this section, we present a short literature review on recommender systems and context-aware recommender systems. We also present some existing work on museum artworks recommender systems and their limits.

2.1   Recommender Systems

As discussed in [4] [5], the most mature and widely used techniques for recommendation are content-based recommender systems and collaborative filtering.

Content-based recommender systems analyze item features and descriptions to identify items that are similar to those the user liked. They perform well when there are sufficient features for items. They also suffer from the cold-start problem, and additionally, they have the problem of over-specialization [6], which means the user is limited to get recommendations that are very similar to the items he already knows. Collaborative filtering recommender systems estimate the similarity between users in order to suggest unseen items to a user [7]. They work best when many users have similar taste. However, they tend to offer poor results when there are not enough user ratings (cold-start problem) and the number of items to rate far exceeds what a user can rate (sparsity problem). More often both recommender systems types are combined into so-called hybrid recommender systems that helps to reduce the limitation of each method used alone [5].

2.2 Context-Aware Recommender Systems

Unlike traditional recommender systems, context-aware recommender systems estimate the degree of utility of an item for a user as a function of not only items and users, but also context. CARSs are classified into three different classes according to how contextual information is incorporated in the recommendation process [8]: contextual pre-filtering, contextual modeling and contextual post-filtering.

In CARSs with contextual pre-filtering, context is used for selecting the relevant set of user data. Then, recommendation can be generated using any traditional recommendation algorithm on the selected data. In contextual post-filtering approaches, contextual information is ignored when generating recommendations and incorporated to adjust the resulting predictions to the user context. The adjustments can be made by filtering out items of a recommended list that are irrelevant, or adjusting a rating prediction according to the context. In contextual modeling approaches, contextual information is used as an explicit predictor of users ratings. These approaches use multidimensional algorithms, while contextual pre-filtering and post-filtering approaches can use traditional algorithms.

2.3 Recommender Systems for Cultural Heritage

Much work has been done on systems that support museum visits and assist the visitor with content adaptation and/or personalization. MyMuseum [9] is a web-based prototype virtual museum system, in which the user accesses a web page interface to input initial visit targets, which are considered as the user's preferences and then used to generate suggestions of art topics. The Hippie project [10] was one of the first systems using suggestions in the context of museum visits. Hippie uses the ICONCLASS taxonomy, an exhaustive classification of the different themes of western art. The user is characterized by scores of interest to the different themes of the taxonomy. When the user is inside the museum, the system detects its position using a radio localization and suggests artworks around him according to his/her interests. However, the artwork model in this project is poor, which limits the suggestions made to the user. Moreover, the context of the user only includes the location of the visitor.

The Chip project [11] is one of the most accomplished content-based recommender system for museums. The system recommends artworks and art topics that correspond to the user profile using semantic relations. The profile is based on the ratings the user gave to various artworks or their characteristics (artist, theme, style, etc). From the positive ratings, the system can suggest to the user artworks and related topics similar to those that have interested him. Furthermore, the number of recommendations is too high (all the recommendations are not relevant for the user). Therefore, the problem of information overload is not completely solved. The context as well as the opinion of other users is not taken into account.

Our work differs from existing ones mainly in two aspects. First, in presented works, only the visitor's background and the visitor's interests are mostly used. Previous public opinions on the artworks are often neglected. Public history ratings can be helpful in refining recommendation decision by methods of collaborative filtering. Second, previous works focuses on the recommendation of artworks singly. However, it could be very useful to recommend to a user a whole tour including interesting artworks to visit in an efficient order and in a limited amount of time.

3   Our Framework

The goal of our framework is to enhance the experience of museum visitors by suggesting the artworks that are the most interesting and tailored to them. To achieve this goal, we propose the following architecture (see figure 1). Our hybrid recommender systems uses three different methods: demographic, semantic and collaborative. Each one is well adapted to a specific visit step: first, when the user enters the system for the first time, as he has not rated any artworks, the demographic is well suited to this situation (cold-start).

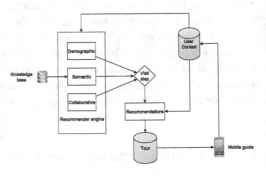

Fig. 1. Recommendation framework

Then, when the user has given few ratings on artworks, the semantic method is activated. Finally, when we better "know" the user the collaborative filtering approach is also integrated. These three methods will provide the visitor with sets of recommendations of artworks. However, the user can have a limited time to spend in the museum and it is important to walk through artworks in an efficient order to avoid round trips and physical tiredness. To this end, a contextual post filtering enables the generation of an optimized visiting path, based on the location of the visitor, the physical environment and the time the visitor wants to spend in the museum.

4   Semantic Model of Artworks

The use of a simple database to describe artworks is not sufficient. We need a rich description of artworks that allows us to take into account different aspects on which artworks will be compared. Thus, the artwork model (see figure 2) we propose uses the CIDOC-CRM (Center for Intercultural Documentation-Conceptual Reference Model) ontology[1]. To which we have to integrate the ICONCLASS taxonomy[2] which is a classification of themes of occidental art, the thesauruses ULAN[3] (Union List of Artist Names), AAT[4] (Art & Architecture Thesaurus) and TGN[5] (Thesaurus of Geographic Names). To achieve this integration, these vocabularies are expressed in the SKOS formalism. The CIDOC-CRM ontology provides a specific concept for these kind of situations: the CIDOC:E55.Type concept, This concept is specifically designed to be an interface between controlled vocabularies and other concepts of CIDOC-CRM.

Thus, we can calculate accurate semantic similarities between artworks. Thus recommending to the user artworks semantically related to those he liked in the past. The computation of the semantic similarity between artworks will be described more in detail in the section 6.

5   Context Model

We propose a semantic context model which is inspired from the definition proposed by Zimmermann et al. [12], specifying five categories which belongs to contextual information: individuality, activity, relation, temporality and location. We define five classes: a main class *User* and four other classes *Activity*, *Time*, *Location* and *Relation* (see figure 3). The model of context plays two important roles in our framework:

1. It is used for the recommendation process: namely the *User*, *Activity*, *Relation* classes. The *User* class has the following data-properties: *user_id, age, sex, language, level of expertise in art*; this information is entered by the user when starting using the system and are used in demographic recommendations. The *Activity* class captures the fact that a user gave a rating to an artwork. The *Relation* class refers to users who have made similar activities, these two classes are updated every time the user gives a new rating on an artwork, the recommendation process is described more in detail in the sections 6 and 7.

---

[1] http://www.cidoc-crm.org/
[2] http://www.iconclass.org/
[3] http://www.getty.edu/research/tools/vocabularies/ulan/
[4] http://www.getty.edu/research/tools/vocabularies/aat/
[5] http://www.getty.edu/research/tools/vocabularies/tgn/

2. It is used for generating tours with a contextual-post filtering method: namely the *Time* and *Location* classes, which capture the context of the visit in terms of the location of the user, the whole time he wants to spend in the museum, the time he spends seeing each artwork and the time remaining. The method for generating tours is described in section 8.

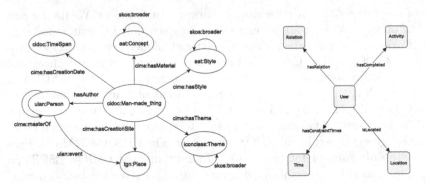

Fig. 2. Semantic artwork model   **Fig. 3.** Modeling user context

6   Semantic Based Recommendations

The aim of this approach is to recommend to the user artworks that are semantically related to those he liked before. For example, if a user liked the artwork *Mona Lisa*, other artworks made by the same artist, *Leonardo da Vinci*, might be recommended to him. We use information in the user context, more precisely the positive ratings given by the user, and the semantic model of artworks for the recommendation process. It is then necessary to define a measure to calculate the semantic similarity between two artworks. The general approach of similarity by properties proposed by [13] was followed. The principle of this approach is to determine the similarity between two instances by comparing the properties that they share. In our case, to calculate the similarity of two artworks we calculate the similarity of their styles, themes, periods, etc:

$$SIM(o_i, o_j) = \sum_k W_k * SIM_k(o_i, o_j) \tag{1}$$

where W_k is a weight representing the importance factor of the property k, with $\sum_k W_k = 1$. Indeed, semantic relations don't always have the same importance. We note also that the weights associated to the properties can vary between users. A user may prefer recommendations on the theme whereas another one could prefer recommendations on the artist.

$SIM_k(o_i, o_j)$ represents the similarity between object o_i and o_j with respect to the property k. For example if k represents the "*hasStyle*" property then $SIM_k(o_i, o_j)$ is the similarity between the two styles of artworks o_i and o_j. The difficulty in the construction of these similarity measure is the variety of types these properties can have. Properties may link instances hierarchically organized (e.g. *aat:Style*), instances not hierarchically organized (e.g. *ulan:Artist*) or datatype values (e.g. cidoc:BirthDate).

Let's note $P_k(o) = \{i_1, i_2, ..., i_n\}$ the set of instances that are linked to the artwork o with respect to the property k. We detail now the computation of the similarity according to a property k in the three cases cited above.

- **Instances hierarchically organized:** if the range of the property consists of a concept which instances are organized hierarchically, the value of the similarity is given by the proximity computation proposed by Wu & Palmer [14], this measure is in fact well adapted to determine the relatedness between two elements in a hierarchy:

$$SIM_{wu}(i,j) = \frac{2 \times depth(lcs(i,j))}{depth(i) + depth(j)} \tag{2}$$

where $lcs(i,j)$ represents the *least common subsummer* of i and j, $depth(i)$, respectively $depth(j)$, the distance in arcs between the root of the hierarchy and i, respectively and j. To calculate the average similarity between the sets $P_k(o_i)$ and $P_k(o_j)$, for each instance in $P_k(o_i)$, we determine the instance which is most similar in $P_k(o_j)$ according to the similarity of Wu & Palmer. Then, for each instance in $P_k(o_j)$ we determine the instance which is most similar in $P_k(o_i)$.

$$SIM_k(o_i, o_j) = \frac{\sum_{i \in P_k(o_i)} Max_{j \in P_k(o_j)} SIM_{wu}(i,j)}{2 \times |P_k(o_i)|}$$
$$+ \frac{\sum_{j \in P_k(o_j)} Max_{i \in P_k(o_i)} SIM_{wu}(j,i)}{2 \times |P_k(o_j)|} \tag{3}$$

- **Instances not hierarchically organized:** if the range of the property consists of a concept which instances are not organized hierarchically, the value of similarity is given by the computation of Jaccard measure. Indeed, this measure is well adapted to determine the relatedness between two sets.

$$SIM_k(o_i, o_j) = SIM_{jaccard}(P_k(o_i), P_k(o_i))$$
$$= \frac{|P_k(o_i) \cap P_k(o_j)|}{|P_k(o_i) \cup P_k(o_j)|} \tag{4}$$

- **Datatype values:** in our model, the only possible case of comparison between two literals is the case where these literals correspond to the type *date*. These dates correspond to the date of the creation of an artwork, the birth or death date of an artist, etc. We defined for this case the function:

$$Sim_{date}(d_1, d_2) = \begin{cases} \frac{100-d}{100}, & \text{if } 100 - d > 0 \\ 0 & otherwhise \end{cases}$$

We can now estimate whether a new artwork o not already rated might be interesting for the user u or not. To this end, we compute the similarity between o

and the artworks that match his preferences (artworks for which he gave positive ratings, 4 or 5 stars).

$$preferences(u) = \{o_i / u \ likes \ o_i\} \tag{5}$$

$$Pred(u, o) = \frac{\sum_{o_i \in preferences(u)} SIM(o_i, o)}{|preferences(u)|} \tag{6}$$

$|preferences(u)|$ is the number of artworks included in the set of user preferences. The formula takes into account the similarity between artworks in the set of user preferences and the one which is a candidate to the recommendation process. If the value of $pred(u, o)$ is upper than a threshold, then the artwork o is recommended to the user. With these predictions, a ranked list of recommendations is generated.

7   Collaborative Based Recommendations

This method recommends to the user artworks that similar users have liked. A basic approach to compute the similarity between users is to count the proportion of common artworks in their history ratings. The PCC (Pearson's Correlation Coefficient) is widely used in the literature.

$$Sim_1(a, b) = \frac{\sum_i (r_{ai} - \overline{r_a})(r_{bi} - \overline{r_b})}{\sqrt{\sum_i (r_{ai} - \overline{r_a})^2 (r_{bi} - \overline{r_b})^2}} \tag{7}$$

This measure is the one that is used in the methods of traditional collaborative filtering. However, its use introduces some problems. The similarity between two users can be significant only if both users have many artworks in common in their rating history. It is also known that this method suffers from the sparsity problem. Thus, we propose to integrate the semantic information existing between artworks when computing the similarity between users, to enhance the accuracy of collaborative filtering and avoid the problems cited above. The idea is that even if two users have few artworks in common in their user preferences, they may have liked semantically related artworks. We propose then this new similarity measure:

$$Sim_2(a, b) = max(sim(a * b), sim(b * a)) \tag{8}$$

with

$$sim(a * b) = \frac{\sum_{k_1} \max_{k_2}(SIM(o_{k1}, o'_{k2}))}{k_1} \tag{9}$$

and

$$sim(b * a) = \frac{\sum_{k_2} \max_{k_1}(SIM(o'_{k2}, o_{k1}))}{k_2} \tag{10}$$

where the user a rated artworks o_1 to o_{k1}, and the user b rated artworks o'_1 to o'_{k2}, and SIM represents the semantic similarity between artworks.

Finally, the similarity between two users is defined as the weighted average of Sim_1 and Sim_2.

$$Sim(a, b) = \sum_{k=1}^{2} \alpha_k Sim_k(a, b) \tag{11}$$

We can now estimate if a new artwork o not already rated might be interesting for the user u. The prediction is made using the K nearest neighbors.

$$Pred(u, o) = \frac{\sum_{i=1}^{K} r(u_i, o) * Sim(u_i, u)}{\sum_{i=1}^{K} Sim(u_i, u)} \tag{12}$$

where $r(u_i, o)$ is the rating given by the user u_i to the artwork o and $Sim(u_i, u)$ is the similarity between the neighbor user u_i and u. If the value of $Pred(u, o)$ is upper than a threshold, then the artwork o is recommended to the user.

8   Generation of Tours

We suggest a tour to the user based on the results of the recommender system, using a contextual post-filtering and information about the museum environment. We define the museum tour as a directed graph, where the nodes represent artworks and the edges the time needed to reach an artwork from another one. With the hypothesis that a user visits only one time the same artwork, the problem is then formalized as an orienteering problem (OP) [15]. Formally, a set of N vertexes correspond to the artworks to be visited. A score S_i correspond to the pertinence of the artwork i. Note also that the tour must start from vertex 1 and ends at vertex N. t_{ij} represents the time needed to walk from artwork i to artwork j, t_i represents the time the visitor takes to look at the artwork i. Not all artworks can be visited since the available time of the visitor is limited, we note this $Tmax$. The goal is to determine a path, limited by $Tmax$ so that the visitor visits the most relevant artworks in order to maximize the total collected score of artworks.

Making use of the notation given above, the problem can be formulated as an integer problem. The following variable of decision is used: $x_{ij} = 1$ if a visit to artwork i is followed by a visit to artwork j, 0 otherwise.

$$Max \sum_{i=2}^{N-1} \sum_{j=2}^{N} S_i x_{ij} \tag{13}$$

$$\sum_{j=2}^{N} x_{1j} = \sum_{i=1}^{N-1} x_{iN} = 1 \tag{14}$$

$$\sum_{i=1}^{N-1} x_{ik} = \sum_{j=2}^{N} x_{kj} \leq 1, \forall k = 2, ..., N-1 \tag{15}$$

$$\sum_{i=1}^{N-1} \sum_{j=2}^{N} t_{ij} x_{ij} + t_j \leq Tmax \tag{16}$$

The objective function (13) is to maximize the total collected score (pertinence of artworks). Constraint (14) guarantees that the path starts from vertex 1 and ends in vertex N. Constraint (15) ensures the connectivity of the path and guarantees that every artwork is visited only one time. Constraint (16) ensures the respect of the limited time of the visit.

9 Conclusion and Future Work

We presented in this article the context of the visit of museum and the possibilities of enrichment of this one by means of techniques of recommendation. We proposed an hybrid approach based at the same time on the semantic modeling of the museum domain and on a semantically-enhanced collaborative filtering by trying to face the traditional limits of the systems cited in the state of the art. We plan to realize an experiment of our system in the museum of the imperial palace of Compiègne. The main objective is to collect on a large scale real data of museum visits. These data will allow us to realize first tests on the prototype and to get some feedback on the satisfaction of visitors, as well as the evaluation of museum experts on the quality of the recommendations and proposed tours.

Acknowledgments. This work is supported by the Regional Council of Picardie, under the CIME project.

References

1. Kuflik, T., Stock, O., Zancanaro, M., Gorfinkel, A., Jbara, S., Kats, S., Sheidin, J., Kashtan, N.: A visitor's guide in an active museum: Presentations, communications, and reflection. J. Comput. Cult. Herit. **11**(1–11), 25 (2011)
2. Jeong, J.H., Lee, K.H.: The physical environment in museums and its effects on visitors satisfaction. Building and Environment, 963–969 (2006)
3. Resnick, P., Varian, H.R. Recommender systems. Communications of the ACM, 56–58 (1997)
4. Adomavicius, G., Tuzhilin, A.: Toward the next generation of recommender systems: A survey of the state-of-the-art and possible extensions. IEEE Trans. on Knowl. and Data Eng., 734–749 (2005)
5. Burke, R.D.: Hybrid recommender systems: Survey and experiments. User Model. User-Adapt. Interact., 331–370 (2002)
6. Balabanović, M., Shoham, Y.: Fab: Content-based, collaborative recommendation. Commun. ACM, 66–72 (1997)
7. Resnick, P., Iacovou, N., Suchak, M., Bergstrom, P., Riedl, J.: Grouplens: an open architecture for collaborative filtering of netnews. ACM, pp. 175–186 (1994)
8. Adomavicius, G., Tuzhilin, A.: Context-aware recommender systems. In: Recommender Systems Handbook, pp. 217–253. Springer (2011)
9. Bright, A., Kay, J., Ler, D., Ngo, K., Niu, W., Nguid, A.: Adaptively recommending museum tours, pp. 29–32 (2005)
10. Oppermann, R., Specht, M.: A nomadic information system for adaptive exhibition guidance. Archives and Museum Informatics, 127–138 (1999)
11. Wang, Y., Stash, N., Aroyo, L., Hollink, L., Schreiber, G.: Using semantic relations for content-based recommender systems in cultural heritage, 16–28 (2009)
12. Zimmermann, A., Lorenz, A., Oppermann, R.: An Operational Definition of Context. In: Kokinov, B., Richardson, D.C., Roth-Berghofer, T.R., Vieu, L. (eds.) CONTEXT 2007. LNCS (LNAI), vol. 4635, pp. 558–571. Springer, Heidelberg (2007)

13. Pirró, G., Euzenat, J.: A Feature and Information Theoretic Framework for Semantic Similarity and Relatedness. In: Patel-Schneider, P.F., Pan, Y., Hitzler, P., Mika, P., Zhang, L., Pan, J.Z., Horrocks, I., Glimm, B. (eds.) ISWC 2010, Part I. LNCS, vol. 6496, pp. 615–630. Springer, Heidelberg (2010)
14. Wu, Z., Palmer, M.: Verbs semantics and lexical selection. In: 32nd Annual Meeting on Association for Computational Linguistics, pp. 133–138 (1994)
15. Vansteenwegen, P., Souffriau, W., Van Oudheusden, D.: The orienteering problem: A survey. European Journal of Operational Research, 1–10 (2011)

A Novel Vision for Navigation and Enrichment in Cultural Heritage Collections

Joffrey Decourselle[1], Audun Vennesland[2], Trond Aalberg[2],
Fabien Duchateau[1(✉)], and Nicolas Lumineau[1]

[1] LIRIS, UMR5205, Université Claude Bernard Lyon 1, Lyon, France
{joffrey.decourselle,fabien.duchateau,nicolas.lumineau}@liris.cnrs.fr
[2] Norwegian University of Science and Technology, 7491 Trondheim, Norway
{audun.vennesland,trond.aalberg}@idi.ntnu.no

Abstract. In the cultural heritage domain, there is a huge interest in utilizing semantic web technology and build services enabling users to query, explore and access the vast body of cultural heritage information that has been created over decades by memory institutions. For successful conversion of existing data into semantic web data, however, there is often a need to enhance and enrich the legacy data to validate and align it with other resources and reveal its full potential. In this visionary paper, we describe a framework for semantic enrichment that relies on the creation of thematic knowledge bases, i.e., about a given topic. These knowledge bases aggregate information by exploiting structured resources (e.g., Linked Open Data cloud) and by extracting new relationships from streams (e.g., Twitter) and textual documents (e.g., web pages). Our focused application in this paper is how this approach can be utilized when transforming library records into semantic web data based on the FRBR model in the process that commonly is called FRBRization.

Keywords: Cultural heritage · Data integration · Semantic web · Linked open data · Entity linking · Ontology and entity matching

1 Introduction

The last decade has seen a significant effort towards the use of semantic web data and related technologies. Linked Open Data (LOD) can be seen as "the Semantic Web done right" according to Tim Berners-Lee, with hundreds of interconnected knowledge bases (KBs) containing structured and semantic data [3]. However, the creation of reusable Linked Data from legacy data such as library records requires more than the transformation into new formats. The data often has to be transformed into acknowledged models (or type vocabularies) and needs to be correctly aligned with other resources before it appears as linked data.

This work has been partially supported by the French Agency ANRT (http://www.anrt.asso.fr), the company PROGILONE (http://www.progilone.fr), a PHC Aurora funding (#34047VH) and a CNRS PICS funding (#PICS06945).

© Springer International Publishing Switzerland 2015
T. Morzy et al. (Eds): ADBIS 2015, CCIS 539, pp. 488–497, 2015.
DOI: 10.1007/978-3-319-23201-0_49

In the cultural heritage domain, the model in the Functional Requirements for Bibliographic Records (FRBR) [14] aims at representing data from cultural institutions with clear semantics [20], and it also offers benefits for improving search and visualization and new possibilities for semantic enrichment of cultural entities [5,8]. To be widely adopted in cultural institutions, the FRBR model must be accompanied with a transformation process for converting legacy MARC data. The potential of FRBR lies in the relationships between entities, which unfortunately are rarely available in existing catalogues. Thus, it is necessary to enrich FRBRized collections with additional information from external data sources. Some relevant sources are already available on the Semantic Web, but a vast body of knowledge is still only available as text in documents (e.g., web pages). To facilitate the enrichment task needed in the FRBRization and other enrichment processes, the LOD and unstructured documents can be exploited. For instance, consider the Norwegian writer Henrik Ibsen. General information about this author are stored in knowledge bases such as DBpedia, VIAF or Freebase, uncommon facts are spread in fans web pages, and news about exhibitions related to his works might be available on streaming media such as Twitter. To provide a complete view of Ibsen's artistic life, it is necessary to aggregate this complementary, inconsistent and/or redundant knowledge from multiple heterogeneous data sources.

In this paper, we propose a generic framework for enriching FRBR collections. Our vision is to create thematic knowledge bases (TKBs) which gather relevant, reliable, and fresh information about a cultural topic (e.g., an artist, a work). The main objective of these TKBs is to help end-users and librarians discovering new knowledge. To build these TKBs, the idea is to exploit both types of data sources: the LOD, which is simpler to browse due to semantics but limited in terms of content, and the Web, with large amount of information but rather difficult to extract and with variable quality. Our framework aims at organizing the different processes involved in the building of a TKB, which are related to the following research areas: entity linking, information extraction, ontology/schema matching and entity matching. In addition, we explain how these processes should be adapted in the context of the cultural heritage domain, and we demonstrate the benefits of the TKB by presenting a use case.

In the rest of this paper, we first describe related work in Section 2. Then, Section 3 provides details about our framework for building thematic knowledge bases. Next, we illustrate the use of our framework with an enrichment scenario about *Natalie Dessay* (Section 4). We conclude by outlining future work.

2 Related Work

This work is at the crossroads of four research domains, namely entity linking, information extraction, and ontology and entity matching. We briefly present each of them in this section, and we also describe related projects.

Entity linking is the task of finding the corresponding entity (in a knowledge base) for a given mention (i.e., words used for labelling the entity). For instance, when the term *Tolkien* is found in a document, the objective is to decide whether this refers to *dbpedia:J._R._R._Tolkien* or to *dbpedia:Christopher_Tolkien*. Due to the emergence of knowledge bases, that enable a long-term disambiguation, the named entity recognition community moved to entity linking [7,11]. In our context, this task could be adapted to take into account the FRBR entity (from which the mention is extracted) and the application domain, which constrains the search of the corresponding entity in the knowledge base to the subset of entities related to cultural heritage [27].

Information extraction deals with the extraction of facts (i.e., relationships between two entities) from textual documents. Many issues arise since this task is at the crossroads of various research domains such as natural language processing, named entity recognition and data integration [22,26]. Typical problems consist of entity linking (i.e., detecting and disambiguating entities based on their textual mentions and surrounding sentences), and discovering the type of relationship holding between two entities (e.g., by using generic patterns to represent sentences). Many approaches are able to extract facts (usually triples) from documents by considering the quality aspect (i.e., extraction of true facts) and the performance aspect (i.e., processing a large set of documents) [6,23,28]. A significant difference deals with the type of extraction: open information extraction means that new relationships can be created while the "closed" paradigm is limited to set of predefined relationships [9]. In our context, an open solution seems more interesting from the user point of view but more difficult to implement. Existing relationships between FRBR entities may be exploited to learn patterns rather than building them from textual documents.

Schema and ontology matching aims at solving the heterogeneity issues of data sources at the schema/ontology level by discovering semantic links (i.e., correspondences). These research fields have been largely studied in the literature [1,10]. Traditionally, all possible pairs of elements are compared using similarity measures (e.g., Levenhstein distance), and the selection of the correspondences is performed using a decision maker (e.g., a threshold). As illustrated by the most recent challenges of Ontology Alignment Evaluation Initiative[1], traditional matching approaches cannot improve quality results any more. Thus, the new trend is to rely on complementary information, either from instances [16,18,25] or from user interactions. Similarly, we benefit from user feedback and these validations, as well as reuse of existing correspondences, need to be smartly integrated in the ontology matching process.

Entity matching, also known as record linkage, deals with the discovery of corresponding elements at the instance level, for example entities or records [15, 17]. The comparison of pairs (of elements) can be performed in a similar fashion as in the schema/ontology matching. However, the large amount of instances mainly requires a pre-processing step named blocking. Elements that share some common values (for a subset of their properties) are placed in the same block, and the comparison of pairs is applied inside every block in order to improve performance [4]. In our context, the selection of the best blocking key may be

---

[1] http://oaei.ontologymatching.org/

chosen according to statistics applied to the FRBRized collection. We may also benefit from the interconnections of LOD knowledge bases, which share common properties and may already link corresponding entities.

Related Projects. The previously described research domains mainly focus on a single issue, and propose generic solutions to solve them. In the context of semantic enrichment for cultural heritage, we need to tackle the same issues but each solution can be adapted to benefit from the FRBR model. Europeana[2] is the closest project to our work and it shares some common goals in terms of enrichment [13]. However, it aims at creating a centralized authoritative source while we believe that cultural institutions should be responsible for managing their resources. In Europeana, a first proposition for enrichment is based on machine-learning algorithms to extract relevant added values [2]. Later, an automatic enrichment is proposed, but limited to four properties, for instance places (with links to Geonames) or agents (with links to DBPedia). Since the enrichment is performed at large scale, the frequency of errors can only be estimated: it reaches 1.5% of the dataset, which still represents more than 15,000 errors [24]. In our proposition, we combine user interactions and reuse of validated knowledge to favour a high quality enrichment. Besides, our work is one step beyond by proposing the integration of textual and streaming contents.

3 Framework for Building Thematic Knowledge Bases

In this section, we introduce our framework for building thematic knowledge bases (TKBs). Note that the model of the TKB is out of scope of this paper, but we expect it as open as possible according to user requirements. Indeed, the representation of basic properties for the cultural heritage entities is covered in the FRBR specifications. For additional information, it is either possible to use existing ontologies (e.g., Linked Open Vocabularies[3]) or to create a specific one.

As illustrated in Figure 1, our framework includes four main processes (square boxes) and uses as input a mention (e.g., the title of a FRBR Work, the name of an Agent). Each process can be seen as a black box, and we describe each of them in the rest of this section.

3.1 Entity Linking

In the cultural heritage domain, many artistic works can be found on the LOD. The exploitation of this resource at the first place is therefore relevant. The entity linking process uses a mention as input and it detects the LOD entities related to this mention (one entity per LOD knowledge base). Contrary to existing entity linking approaches [7,11,27], this process has two specificities in our framework. First, entity linking approaches traditionally use a mention with surrounding terms (e.g., sentence) while our input mention is part of a FRBR entity (mainly

---

[2] http://www.europeana.eu/
[3] http://lov.okfn.org/dataset/lov/

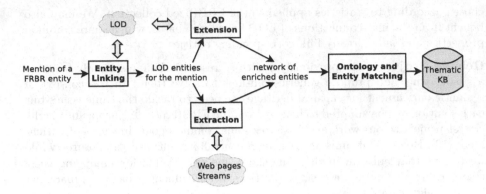

Fig. 1. Framework for building thematic knowledge bases

a Work or an Agent). Secondly, we are not limited to search on a single LOD knowledge base. This means that we can exploit the interconnection between LOD knowledge bases (e.g., *owl:sameAs* predicates) to improve the accuracy of the entity linking [12]. For instance, linking the mention *"Henrik Ibsen"* to both LOD entities *dbpedia:Henrik_Ibsen* and *freebase:/m/03pm9* reinforces the confidence that the linking is correct since the two LOD entities are connected through a property *owl:sameAs*. The discovered LOD entities are then processed in parallel by the LOD extension and the fact extraction.

3.2   LOD Extension

The LOD extension process aims at discovering new entities related to the LOD entities which represent the mention. As LOD knowledge bases are reputed for their good quality, the goal of this process is to integrate reliable information in the TKB. The main challenge lies in the level of extension. LOD knowledge bases, especially general ones, may contain facts that are either too broad or not useful for semantic enrichment or user navigation. Similarly, some properties can include a long list of values (e.g., *owl:sameAs*, *rdf:type*) and should be filtered to avoid overloading the TKB. A possible solution for the LOD extension could be an iterative process in which a limited number of facts is added and evaluated at each iteration until the result is satisfying, for instance in terms of consistency. Such process should also take into account the FRBR context. The relationships between FRBR entities and the FRBR attributes can be associated to LOD properties. Back to our running example, the Ibsen DBpedia entity enables us to extend to *dbpedia:The_Wild_Duck*, one of his play, or to *dbpedia:August_Strindberg*, another novelist who influenced Ibsen. With LOD extension, we gather new facts about our initial mention. Note that some facts may be redundant (e.g., provided by two KBs) and the cleaning is performed in the ontology/entity matching process.

3.3   Fact Extraction

In addition to the LOD extension, our framework exploits unstructured documents with variable quality, such as web pages. The main motivation is to gather uncommon facts (i.e., that are not present in the LOD). In our context, the initial mention and the LOD entities obtained from the entity linking step are used to identify interesting documents and detect sentences about the entity. To enable enrichment in the TKB, new relationships and properties can be added to the FRBR model, thus promoting the use of an open information extraction tool (see Section 2). Another notable difference deals with human intervention: cataloguers who validate a discovered fact for enriching a TKB also provides feedback for future fact extraction. This means that the sentence (or pattern) can be stored and marked as reliable, the predicate of the new fact is validated, etc. Finally, microblogging data sources such as Twitter contain abbreviations, notations (e.g., hashtags) which require specific solutions. The FRBR collection is helpful for learning patterns: if an Agent entity does not have a given property that other Agent entities own, we can learn the patterns or sentences in which this fact could be detected. In our running example, we could extract from this blog[4] the fact *<dbpedia:Henrik_Ibsen, relationship:acquaintanceOf, dbpedia:James_Joyce>*, where two DBpedia entities representing famous novelists are linked through the *acquaintanceOf* property (from the *relationship* vocabulary[5]). At the end of this task, the initial LOD entities have been enriched with new facts, and new entities may have been created. This additional knowledge can include redundancies, that the next task is in charge of cleaning.

3.4   Ontology and Entity Matching

In this last process, we perform ontology and entity matching to clean extracted information. In the same LOD entity, different properties can represent the same concept, which in turn may be redundant with a FRBR attribute (e.g., the *name* of a FRBR agent is equivalent to the LOD properties *dbpprop:name* and *foaf:name*). As our framework is more suited to an open extraction approach, it produces facts whose predicate needs to be mapped to an existing one to avoid redundancies. Solving these issues requires both ontology matching and entity matching, two research areas traditionally considered separately [1,10]. In our framework, the idea is to combine ontology and entity matching in order to clean the "network" of enriched entities resulting from the previous steps. Correspondences at the ontology level are needed to perform entity matching. Conversely, we believe that entity matching results can reinforce the discovery of new correspondences at the ontology level. A basic ontology matching approach can be used as a bootstrap process to detect correspondences between concepts and properties, then entity matching is performed to discover entity correspondences. An iterative combination of both matching processes enables

---

[4] http://blog.bookstellyouwhy.com/archive/2015/03 (article from March 18th)
[5] http://vocab.org/relationship/

the refinement of existing correspondences and the discovery of more complex ones. When possible, extracted relationships have to be mapped to FRBR relationships or attributes by relying on ontology matching too. In case of conflicting correspondences, information provenance may be useful since LOD extension is considered as more reliable than fact extraction. Note that the correspondences at the ontology level needs to be stored for reuse. Data fusion (a.k.a. augmentation), which consists in merging redundant or complementary information, is optional in our context. The cultural heritage expert may decide how to select or merge information from the TKB. In the Ibsen example, we could obtain during LOD extension the entities *dbpedia:The_Wild_Duck* and *viaf:312333678*, which both represent the novel *Wild Duck*. By applying an entity matching process, a correspondence between the two entities is discovered and only one of them is added to the TKB (and a property *sameAs* is used to link to the discarded entity). At the end of the matching process, the TKB is built and ready for use, as illustrated in the next section with a real-world scenario.

4 Building a TKB about *Natalie Dessay*

A TKB is useful both for experts who need to enrich their original collections and for library users who can explore it for finding new resources related to their initial query. Since an implementation of our framework is currently in progress, this section explains how it can be applied in a semantic enrichment scenario.

Let us describe an example about the well-known French singer *Natalie Dessay*. A librarian has generated a FRBR entity for the Agent *Natalie Dessay*, but there is almost no information in the original records about the singer. Thus she decides to build a TKB to enrich the FRBR entity. As shown in Figure 2, general information about *Natalie Dessay* are stored in LOD knowledge bases such as DBpedia[6] or Freebase[7], uncommon information could be found in textual websites[8], and news about her activities (e.g., concerts, TV appearance) might be available on streaming media such as Twitter[9].

Instead of manually querying and browsing these multiple data sources, the librarian simply runs an implementation of our framework with the input mention *Natalie Dessay* and the corresponding FRBR Agent entity. During the first step, the mention is linked to its corresponding entities in DBpedia and Freebase (respectively *dbpedia:Natalie_Dessay* and *fb:m.0cfmsz*). The search engines of both general knowledge bases ranks the correct entity at the top. Since both LOD entities are already linked through a *owl:sameAs* predicate, it increases the confidence in this discovery. The LOD extension consists in gathering facts from the LOD entities, e.g., the triples <*dbpedia:Natalie_Dessay, dbpprop:birthDate, '1965-04-19'*> and <*fb:m.0cfmsz, fb:people.person.date_of_birth, '1965-04-19'*>. To avoid overloading the TKB, mainly with properties which have numerous

---

[6] *dbpedia:Natalie_Dessay*, http://dbpedia.org/page/Natalie_Dessay

[7] *fb:/m/0cfmsz*, https://www.freebase.com/m/0cfmsz

[8] http://blogclarabel.canalblog.com/archives/2014/12/09/31081571.html

[9] https://twitter.com/n2cfan

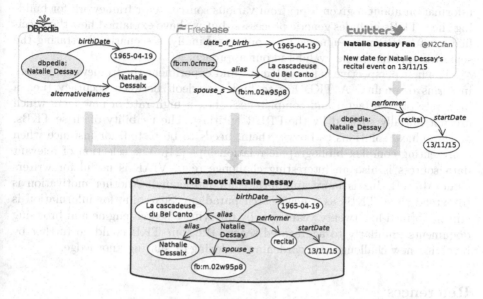

Fig. 2. From data sources to the thematic knowledge base about *Natalie Dessay*

values, we decide to apply a filter based on the provenance and to only store properties and values from the *DBpedia* and the *Freebase* ontologies. The LOD extension also provides alternatives mentions using alias properties such as *dbp-prop:alternativeNames* or *fb:common.topic.alias*, thus enabling us to collect the extra mention *Nathalie Dessaix*. These alternatives mentions are exploited during the third process, fact extraction. In the documents containing the mentions, we discover a tweet about an upcoming recital of Nathalie Dessay. As shown in Figure 2, information extraction tools are needed to transform natural language into triples. Note that new predicates such as *performer* can either be available in the FRBR model or based on an external ontology (e.g., *schema.org*). In the ontology and entity matching step, cleaning is performed to remove redundancies or to reduce heterogeneity in the TKB. Mappings are performed at the ontology level (e.g., between the properties *dbpprop:birthDate* and *fb:people.person.date_of_birth*), or at the entity level (e.g., to check that *Laurent Naouri*, the husband of Natalie Dessay, is not represented by two entities in the TKB). Mappings can be validated by the user and reused later, specifically those at the ontology level. When the TKB about *Natalie Dessay* is constructed, the librarian can select the relevant facts to enrich the FRBR initial entity.

5 Conclusion

This paper introduces a new framework for enriching cultural heritage collections. It is based on the notion of thematic knowledge bases, which aggregates information about a given topic from various sources. Our framework for building these TKBs includes generic processes, but we have explained how the specificities of FRBR may enhance each process. Finally, a scenario illustrating the semantic enrichment has validated our approach.

The first perspective is to implement and experiment the framework, mainly in terms of quality. A TKB has to find a tradeoff between accuracy (i.e., a high rate of true facts) and completeness (i.e., a high rate of new facts which are not initially covered by the FRBR entities). The usability of these TKBs, and graphical solutions to browse them, needs to be tested, for instance when representing complex bibliographic relationships [21]. The selection of relevant data sources is also an interesting challenge (e.g., VIAF is useful for writers while MusicBrainz is more appropriate for musicians). Another motivation is to extend these TKBs as a new search paradigm. Searching for information is still performed like twenty years ago, by querying a search engine and browsing documents. Similarly to aggregated search [19], our TKB could go further by involving new challenges for gardening, indexing or sharing knowledge.

References

1. Bellahsene, Z., Bonifati, A., Rahm, E. (eds.): Schema Matching and Mapping. Data-Centric Systems and Applications. Springer (2011)
2. Berardi, G., Esuli, A., Gordea, S., Marcheggiani, D., Sebastiani, F.: Metadata Enrichment Services for the Europeana Digital Library. In: Zaphiris, P., Buchanan, G., Rasmussen, E., Loizides, F. (eds.) TPDL 2012. LNCS, vol. 7489, pp. 508–511. Springer, Heidelberg (2012)
3. Bizer, C., Heath, T., Berners-Lee, T.: Linked Data - The Story So Far. International Journal on Semantic Web and Information Systems 5(3), 1–22 (2009)
4. Böhm, C., de Melo, G., Naumann, F., Weikum, G.: Linda: Distributed web-of-data-scale entity matching. In: International Conference on Information and Knowledge Management, CIKM 2012, pp. 2104–2108. ACM (2012)
5. Buchanan, G.: FRBR: Enriching and integrating digital libraries. In: JCDL (June 2006)
6. Carlson, A., Betteridge, J., Kisiel, B., Settles, B., Hruschka Jr., E.R.H., Mitchell, T.M.: Toward an architecture for never-ending language learning. In: Proceedings of Artificial Intelligence (AAAI 2010) (2010)
7. Dai, H.J., Wu, C.Y., Tsai, R., Hsu, W.L.: From entity recognition to entity linking: a survey of advanced entity linking techniques. In: The 26th Annual Conference of the Japanese Society for Artificial Intelligence, pp. 1–10 (2012)
8. Dickey, T.J.: FRBRization of a Library Catalog: Better Collocation of Records, Leading to Enhanced Search, Retrieval, and Display. Information Technology & Libraries 27, 23–32 (2008)
9. Etzioni, O., Banko, M., Soderland, S., Weld, D.S.: Open information extraction from the web. Communication of ACM 51, 68–74 (2008)
10. Euzenat, J., Shvaiko, P.: Ontology matching, 2nd edn. Springer (2013)
11. Hachey, B., Radford, W., Nothman, J., Honnibal, M., Curran, J.R.: Evaluating entity linking with wikipedia. Artificial Intelligence 194, 130–150 (2013)

12. Halpin, H., Hayes, P.J., McCusker, J.P., McGuinness, D.L., Thompson, H.S.: When owl:sameAs Isn't the Same: An Analysis of Identity in Linked Data. In: Patel-Schneider, P.F., Pan, Y., Hitzler, P., Mika, P., Zhang, L., Pan, J.Z., Horrocks, I., Glimm, B. (eds.) ISWC 2010, Part I. LNCS, vol. 6496, pp. 305–320. Springer, Heidelberg (2010)
13. Haslhofer, B., Momeni, E., Gay, M., Simon, R.: Augmenting europeana content with linked data resources. In: Proceedings of the 6th International Conference on Semantic Systems, I-SEMANTICS 2010, pp. 40:1–40:3. ACM (2010)
14. IFLA: Functional requirements for bibliographic records: final report. Tech. rep., IFLA (February 2009)
15. Ioannou, E., Rassadko, N., Velegrakis, Y.: On Generating Benchmark Data for Entity Matching. Journal on Data Semantics 2(1), 37–56 (2013)
16. Isaac, A., van der Meij, L., Schlobach, S., Wang, S.: An Empirical Study of Instance-Based Ontology Matching. In: Aberer, K., et al. (eds.) ASWC 2007 and ISWC 2007. LNCS, vol. 4825, pp. 253–266. Springer, Heidelberg (2007)
17. Köpcke, H., Thor, A., Rahm, E.: Evaluation of entity resolution approaches on real-world match problems. PVLDB 3(1), 484–493 (2010)
18. Lacoste-Julien, S., Palla, K., Davies, A., Kasneci, G., Graepel, T., Ghahramani, Z.: Sigma: Simple greedy matching for aligning large knowledge bases. In: International Conference on Knowledge Discovery and Data Mining, KDD 2013, pp. 572–580. ACM (2013)
19. Lalmas, M.: Aggregated search. In: Melucci, M., Baeza-Yates, R. (eds.) Advanced Topics in Information Retrieval, The Information Retrieval Series, vol. 33, pp. 109–123. Springer (2011)
20. Le Bœuf, P.: FRBR: Hype or cure-all? Introduction. Cataloging & Classification Quarterly 39(3-), 1–13 (2005)
21. Merčun, T., Žumer, M., Aalberg, T.: Presenting and Exploring the Complexity of Bibliographic Relationships. In: Chen, H.-H., Chowdhury, G. (eds.) ICADL 2012. LNCS, vol. 7634, pp. 63–66. Springer, Heidelberg (2012)
22. Nakashole, N., Weikum, G., Suchanek, F.M.: Discovering and exploring relations on the web. PVLDB 5(12), 1982–1985 (2012)
23. Parameswaran, A., Garcia-Molina, H., Rajaraman, A.: Towards the web of concepts: extracting concepts from large datasets. Proceedings of VLDB Endowment 3, 566–577 (2010)
24. Stiller, J., Petras, V., Gäde, M., Isaac, A.: Automatic Enrichments with Controlled Vocabularies in Europeana: Challenges and Consequences. In: Ioannides, M., Magnenat-Thalmann, N., Fink, E., Žarnić, R., Yen, A.-Y., Quak, E. (eds.) EuroMed 2014. LNCS, vol. 8740, pp. 238–247. Springer, Heidelberg (2014)
25. Suchanek, F.M., Abiteboul, S., Senellart, P.: Paris: Probabilistic alignment of relations, instances, and schema. Proc. VLDB Endow. 5(3), 157–168 (2011)
26. Suchanek, F.M., Sozio, M., Weikum, G.: SOFIE: A Self-Organizing Framework for Information Extraction. In: International World Wide Web conference (WWW 2009). ACM (2009)
27. Takhirov, N., Duchateau, F., Aalberg, T.: Linking FRBR Entities to LOD through Semantic Matching. In: Gradmann, S., Borri, F., Meghini, C., Schuldt, H. (eds.) TPDL 2011. LNCS, vol. 6966, pp. 284–295. Springer, Heidelberg (2011)
28. Takhirov, N., Duchateau, F., Aalberg, T.: An Evidence-Based Verification Approach to Extract Entities and Relations for Knowledge Base Population. In: Cudré-Mauroux, P., Heflin, J., Sirin, E., Tudorache, T., Euzenat, J., Hauswirth, M., Parreira, J.X., Hendler, J., Schreiber, G., Bernstein, A., Blomqvist, E. (eds.) ISWC 2012, Part I. LNCS, vol. 7649, pp. 575–590. Springer, Heidelberg (2012)

Improving Retrieval of Historical Content with Entity Linking

Max De Wilde[✉]

Information Science Department, Université libre de Bruxelles (ULB),
Avenue F.D. Roosevelt 50 – CP 123, 1050 Brussels, Belgium
madewild@ulb.ac.be

Abstract. The relevance of Named-Entity Recognition and Entity Linking for cultural heritage institutions is evaluated through a case-study involving the semantic enrichment of historical periodicals. A language-independent approach is proposed in order to improve the search experience of end-users with the mapping of entities to the Linked Open Data (LOD) cloud. Preliminary results show that a precision rate of almost 90% can be achieved with very little fine-tuning, while an increase in recall remains necessary.

1 Introduction

Faced with budget cuts, Libraries, Archives and Museums (LAMs) are compelled to adopt a creative mindset toward content management. Funding bodies expect short-term results and encourage cultural heritage institutions to gain more value out of their existing data by linking them to external knowledge bases. The LOD Around The Clock (LATC)[1] initiative of the European Commission, for instance, explicitly helps institutions to publish quality Linked Data on the Web.

In this context, Named-Entity Recognition (NER) and Entity Linking (EL) have attracted attention since they allow small institutions to enrich their collections with semantic information at a relatively low cost. For LAMs, the perspective of freely reusing existing knowledge to map their collections to the Web represents a great opportunity.

This paper evaluates the relevance of NER and EL to improve information retrieval from documents in a multilingual archive of OCRized Belgian periodicals. We focus on historical locations since they constitute the main interest of our users, but the methodology described can be extended to any type of content. Several configurations are compared in order to determine their added value for archive retrieval, with a special emphasis on the handling of non-English data.

The remainder of this paper is structured as follows: Section 2 discusses related work, sections 3 and 4 present our case-study and workflow respectively, Section 5 provides preliminary results and Section 6 concludes the paper.

---

[1] http://cordis.europa.eu/project/rcn/95552_en.html

© Springer International Publishing Switzerland 2015
T. Morzy et al. (Eds): ADBIS 2015, CCIS 539, pp. 498–504, 2015.
DOI: 10.1007/978-3-319-23201-0_50

2   Related Work

Many cultural institutions have experimented with NER over the last decade. The Powerhouse Museum has implemented OpenCalais within its collection management database,[2] although no evaluation of the entities has been performed. The authors of [1] also explore NER in order to create a faceted browsing interface for users of large museum collections, while [2] offers an interesting evaluation of the extraction of actors, locations and events from unstructured text in the collection management database of the Rijksmuseum in Amsterdam.

In the specific domain of archives, [3] compares the results of several NER services on a corpus of mid-20th-century typewritten documents. A set of test data, consisting of raw and corrected OCR output, is manually annotated with people, locations, and organizations. This approach allows a comparison of the F-score of the different NER services against the manually annotated data. Their methodology was generalized for LAMs by [4] in the context of the *Free Your Metadata* project.[3] The BBC also set up a system to connect its vast archive with current material through Semantic Web technologies [5].

More recently, [6] compared the performance of various entity classifiers on the DEREKO corpus of contemporary German [7] which they say exhibits a "strong dispersion [with regard to] genre, register and time". However, the authors later concede that newspaper texts are largely prevailing and that "relatively few texts reach back to the mid-20th century", which casts doubt over the actual strong temporal dispersion of this corpus. Moreover, although the study of NER in German is particularly challenging due to its use of capital letters for all common nouns, their evaluation remains monolingual and does not offer any insights as to how the classifiers would perform on a linguistically diverse corpus.

On a larger scale, the Newspaper Aggregation and Indexing Plan for Europeana Newspapers, which produced metadata for 18 million pages of news and OCRized full-text for around 10 million pages, also included a NER and EL component performed by the National Library of the Netherlands.[4] A new website[5] was launched in December, 2014, allowing users to cross-search and reuse over 25 million digital items and over 165 million bibliographic records. Entities are disambiguated with a local ontology and the whole collection can also be queried through an API.

Finally, [8] and [9] considered how to adapt the EL task to cultural heritage content, but both focus exclusively on English data and do not take advantage of the multilingual structure of the Semantic Web. We therefore propose to go a step further toward the full-scale semantic enrichment of an archive with a case-study involving historical newspapers, building upon state-of-the-art techniques detailed in the thorough survey provided by [10].

---

[2] http://www.freshandnew.org/2008/03/opac20-opencalais-meets-our-museum-collection-auto-tagging-and-semantic-parsing-of-collection-data/

[3] http://freeyourmetadata.org/

[4] http://blog.kbresearch.nl/2014/03/03/ner-newspapers/ reported on a preliminary experiment on Dutch, French and German, accounting for about half the corpus.

[5] http://www.theeuropeanlibrary.org/

3   Dataset

The *Historische Kranten*[6] project involved the digitization, OCR processing and online publication of over a million articles compiled from 41 Belgian newspapers published between 1818 and 1972. Articles are written in Dutch, French and English and focus mainly on the city of Ypres and its neighbourhood.

Currently, only full-text indexing has been performed on the periodicals, which means that searches for particular mentions in the corpus suffer from both noise and silence. For instance, a query on the string "Rabelais" returns correct results about Franois Rabelais:

Example 1. Science sans conscience, disait Rabelais, n'est que ruine de l'me.[7]

But one also gets hits that are not relevant in this context (noise):

Example 2. Alexandre-Gustave Eiffel is in zijn woning, rue Rabelais, te Parijs overleden.[8]

Moreover, interesting results are lost due to OCR errors (silence):

Example 3. 1553: Overlijden van *Rabela-ls*, Fransch schrijver en opvoedkundige.[9]

In order to get a clearer picture of the interests of users, we tracked individual queries on http://www.historischekranten.be/ with Google Analytics over a 4-year period, yielding 124,510 results. About 4,200 unique keywords were used at least three times, of which the ten most popular are shown in Table 1.

Table 1. Top ten keywords

# Term	Hits	Type	DBpedia URI
1. Zillebeke	398	Place	http://dbpedia.org/resource/Zillebeke
2. Westouter	252	Place	http://nl.dbpedia.org/resource/Westouter
3. Passchendaele	250	Place	http://dbpedia.org/resource/Passendale
4. oorlog	178	Concept	http://dbpedia.org/resource/War
5. Reninghelst	163	Place	http://dbpedia.org/resource/Reningelst
6. Ieper	148	Place	http://dbpedia.org/resource/Ypres
7. Hollebeke	108	Place	http://dbpedia.org/resource/Hollebeke
8. moord	102	Concept	http://dbpedia.org/resource/Death
9. Watou	93	Place	http://dbpedia.org/resource/Watou
10. Bikschote	83	Place	http://nl.dbpedia.org/resource/Bikschote

We can see that locations are especially favored by the users, which prompted us to focus preliminary work on this type of entities. Most are related to the First World War since Ypres was the scene of four major battles during that conflict.

---

[6] http://www.historischekranten.be/
[7] Science without conscience, said Rabelais, is but the ruin of the soul.
[8] Alexandre-Gustave Eiffel died at his home in Rabelais Street, Paris.
[9] 1553: Death of Rabelais, French writer and educator.

4   Methodology

The intuition behind Entity Linking is that knowledge bases can be leveraged to perform a full disambiguation of entities through URIs. For example, Rabelais can be disambiguated with http://dbpedia.org/resource/Franois_Rabelais which includes the alternative label "Alcofribas Nasier" (one of his anagrammatic pseudonyms) but excludes information about American composer Akira Rabelais (which has his own unique URI: http://dbpedia.org/resource/Akira_Rabelais) or the main-belt asteroid named after the French humanist (http://dbpedia.org/resource/5666_Rabelais).

In order to overcome the issues of noise and silence highlighted in Section 3, we designed a system to extract place mentions from a subset of documents and link them to DBpedia [11]. We first generated a random sample of 50 documents distributed over the three languages (1430 chars per doc on average) and annotated mentions of places with DBpedia URIs, which gave us a gold-standard reference of 200 locations (4 per doc on average).

In parallel, we extracted all French and Dutch labels from a DBpedia dump from August 2014,[10] giving us 942,505 and 674,849 labels respectively. These were combined in various ways (see Section 5), transformed into a Python dictionary[11] and mapped to their corresponding URIs, providing a fast lookup mechanism kept in memory. A schematisation of this dictionary is shown below:

```
{
  "Bruxelles"@fr => <http://dbpedia.org/resource/Brussels>,
  "Oostende"@nl => <http://dbpedia.org/resource/Ostend>,
  "Derde Slag om Ieper"@nl => <http://dbpedia.org/resource/Battle_
    of_Passchendaele>
}
```

The next step was to tokenize the documents from the sample with the NLTK[12] WordPunctTokenizer and to perform a simple greedy lookup of entities up to three tokens in length against our dictionary. For the entities present in the dictionary, a query forwarded to the DBpedia SPARQL endpoint[13] retrieved the various associated types and checked if `dbpedia-owl:Place`[14] was one of them.[15] If this was the case, the longest match was chosen and annotated with offset and length in addition to the URI. The first five results are shown in Table 2.

---

[10] http://wiki.dbpedia.org/Downloads

[11] In Python, `dict` is a data structure allowing to map key to values: https://docs.python.org/2/library/stdtypes.html#mapping-types-dict.

[12] http://www.nltk.org/

[13] http://dbpedia.org/sparql

[14] http://dbpedia.org/ontology/Place

[15] Our methodology can therefore easily be adapted to other entity types, such as Persons and Organizations for instance, with a single line change in the SPARQL query.

Table 2. Annotation scheme

File	Offset	Length	URI
gsc.txt	187	11	http://dbpedia.org/resource/Bouvancourt
gsc.txt	199	6	http://dbpedia.org/resource/Fismes
gsc.txt	561	4	http://dbpedia.org/resource/Pvy
gsc.txt	631	9	http://dbpedia.org/resource/Yorkshire
gsc.txt	1076	6	http://dbpedia.org/resource/Trigny

5   Results

Following the evaluation method described in [12], we used the *neleval* tool[16] to compute the number of true positives, false positives and false negatives generated by our system against the gold-standard reference, along with the classic measures of precision, recall and F-score. Two variants of the evaluation are provided thanks to the material available at http://sites.google.com/site/entitylinking1: simple entity match (ENT), i.e. without alignment, and strong annotation match (SAM) which is stricter on entity boundaries [13].

Tables 3 and 4 present the results for ENT and SAM respectively. The best figures, indicated in **bold**, match the state-of-the-art performance achieved by popular open source EL systems such as AIDA[17], Babelfy[18] and Wikipedia Miner[19] according to the benchmarking study produced in [12].

Table 3. Entity match (ENT)

Lang.	True pos.	False pos.	False neg.	Precision	Recall	F1-score
nl	19	15	70	.559	.213	.309
fr	45	11	44	**.804**	.506	.621
fr+nl	37	17	52	.685	.416	.517
nl+fr	46	13	43	.780	**.517**	**.622**

Table 4. Strong annotation match (SAM)

Lang.	True pos.	False pos.	False neg.	Precision	Recall	F1-score
nl	33	21	169	.611	.163	.258
fr	99	12	103	**.892**	.490	**.633**
fr+nl	78	23	124	.772	.386	.515
nl+fr	100	15	102	.870	**.495**	.631

---

[16] https://github.com/wikilinks/neleval

[17] http://www.mpi-inf.mpg.de/yago-naga/aida/

[18] http://babelfy.org

[19] http://wikipedia-miner.cms.waikato.ac.nz/

The four versions of the system that were compared correspond to variations in the creation of the label dictionary:

1. nl: Dutch labels only
2. fr: French labels only
3. fr+nl: French labels complemented by Dutch labels
4. nl+fr: Dutch labels complemented by French labels

The difference between (3) and (4) is due to the inner workings of Python dictionaries: if two labels are identical from one language to another, the second to be loaded erases the first. For instance, the French label `"Lige"@fr` predictably corresponds to the URI http://dbpedia.org/resource/Lige, but the Dutch label `"Lige"@nl` redirects to the homonymy page http://dbpedia.org/resource/Lige_ (disambiguation), decreasing the recall if combined in that order. Precision is consistently ahead of recall, with a peak between 80% and 90% for the former and about 50% for the latter, yielding F-scores just over 60%.

6   Conclusion and Future Work

In this paper, we showed how Entity Linking could easily be adopted by LAMs in order to semantically enrich their collections. We are looking forward to integrate our findings into the *Historische Kranten* project's Web interface in order to improve the search experience of the users. We plan to interact with them to get feedback about the relevance of entities extracted and of automatic related search suggestions based on semantic relatedness, which are currently quite random and of poor quality.

English DBpedia labels were deliberately left out from this study in order to focus on Dutch and French, but a fully language-independent system would obviously need to incorporate them, along with labels from other languages, relying on a fallback mechanism when an entity does not exist in a specific language. To address the issue of low recall, with the combination of several knowledge bases instead of DBpedia only. For place names, the aggregation of GeoNames[20] and GeoVocab[21] looks promising.

The precise impact of Optical Character Recognition quality on NER output also remains to be evaluated. In similar work on Holocaust testimonies, [3] found that "manual correction of OCR output does not significantly improve the performance of named-entity extraction". In other words, even poorly digitized material with lots of OCR mistakes could be successfully enriched to meet the needs of users. The confirmation of these findings would mean a lot to institutions that lack the funding to perform first-rate OCR on their collections or the manpower to curate them manually.

---

[20] http://www.geonames.org/
[21] http://geovocab.org/

References

1. Lin, Y., Ahn, J.W., Brusilovsky, P., He, D., Real, W.: ImageSieve: Exploratory Search of Museum Archives with Named Entity-Based Faceted Browsing. Proceedings of the American Society for Information Science and Technology **47**, 1–10 (2010)
2. Segers, R., van Erp, M., van der Meij, L., Aroyo, L., Schreiber, G., Wielinga, B., van Ossenbruggen, J., Oomen, J., Jacobs, G.: Hacking History: Automatic Historical Event Extraction for Enriching Cultural Heritage Multimedia Collections. In: Proceedings of the 6th International Conference on Knowledge Capture (K-CAP), Banff, Alberta, Canada (2011)
3. Rodriquez, K.J., Bryant, M., Blanke, T., Luszczynska, M.: Comparison of Named Entity Recognition Tools for Raw OCR Text. In: Proceedings of KONVENS 2012, Vienna, pp. 410–414 (2012)
4. van Hooland, S., De Wilde, M., Verborgh, R., Steiner, T., Van de Walle, R.: Exploring Entity Recognition and Disambiguation for Cultural Heritage Collections. Digital Scholarship in the Humanities **30**, 262–279 (2015)
5. Raimond, Y., Smethurst, M., McParland, A., Lowis, C.: Using the Past to Explain the Present: Interlinking Current Affairs with Archives via the Semantic Web. In: Alani, H., et al. (eds.) ISWC 2013, Part II. LNCS, vol. 8219, pp. 146–161. Springer, Heidelberg (2013)
6. Bingel, J., Haider, T.: Named Entity Tagging a Very Large Unbalanced Corpus: Training and Evaluating NE Classifiers. In: Proceedings of the 9th International Conference on Language Resources and Evaluation (LREC), Reykjavik, Iceland (2014)
7. Kupietz, M., Belica, C., Keibel, H., Witt, A.: The German Reference Corpus DeReKo: A Primordial Sample for Linguistic Research. In: Proceedings of the 7th International Conference on Language Resources and Evaluation (LREC), Valletta, Malta (2010)
8. Agirre, E., Barrena, A., De Lacalle, O.L., Soroa, A., Fernando, S., Stevenson, M.: Matching Cultural Heritage Items to Wikipedia. In: Proceedings of the 8th International Conference on Language Resources and Evaluation (LREC), pp. 1729–1735 (2012)
9. Fernando, S., Stevenson, M.: Adapting wikification to cultural heritage. In: Proceedings of the 6th Workshop on Language Technology for Cultural Heritage, Social Sciences, and Humanities, pp. 101–106. ACL (2012)
10. Shen, W., Wang, J., Han, J.: Entity Linking with a Knowledge Base: Issues, Techniques, and Solutions. IEEE Transactions on Knowledge and Data Engineering **27**, 443–460 (2015)
11. Bizer, C., Lehmann, J., Kobilarov, G., Auer, S., Becker, C., Cyganiak, R., Hellmann, S.: DBpedia - A Crystallization Point for the Web of Data. Web Semantics: Science, Services and Agents on the World Wide Web **7**, 154–165 (2009)
12. Ruiz, P., Poibeau, T.: Combining Open Source Annotators for Entity Linking through Weighted Voting. In: Proceedings of the 4th Joint Conference on Lexical and Computational Semantics (*SEM), Denver, CO, USA (2015)
13. Cornolti, M., Ferragina, P., Ciaramita, M.: A Framework for Benchmarking Entity-Annotation Systems. In: Proceedings of the 22nd International Conference on the World Wide Web, pp. 249–260 (2013)

Disambiguation of Named Entities in Cultural Heritage Texts Using Linked Data Sets

Carmen Brando[1]([✉]), Francesca Frontini[1,2], and Jean-Gabriel Ganascia[1]

[1] Labex OBVIL. LiP6, UPMC. CNRS, 4 place Jussieu, 75005 Paris, France
{Carmen.Brando,Francesca.Frontini,Jean-Gabriel.Ganascia}@lip6.fr
[2] Istituto di Linguistica Computazionale CNR, Pisa, Italy
Francesca.Frontini@ilc.cnr.it

Abstract. This paper proposes a graph-based algorithm baptized REDEN for the disambiguation of authors' names in French literary criticism texts and scientific essays from the 19th century. It leverages knowledge from different Linked Data sources in order to select candidates for each author mention, then performs fusion of DBpedia and BnF individuals into a single graph, and finally decides the best referent using the notion of graph centrality. Some experiments are conducted in order to identify the best size of disambiguation context and to assess the influence on centrality of specific relations represented as edges. This work will help scholars to trace the impact of authors' ideas across different works and time periods.

Keywords: Named-entity disambiguation · Centrality · Linked data · Data fusion · Digital humanities

1 Introduction

Semantic enrichment of cultural heritage texts is a current practice in Digital Humanities (DH). The TEI annotation standard[1] enables the tagging of persons, geographical places and organizations, that can be later used for the automatic creation of search indexes. TEI also permits the linking of portions of text to external resources. While many of the steps in the production of quality digital editions require manual check, natural language processing tools may speed up the process to a great extent. Named Entity Recognition and Classification (NERC) algorithms automatically detect entities in texts and assign them to a given class (e.g. persons, places). Named Entity Linking (NEL) [9] disambiguates the detected mentions by pointing them to a unique reference, for instance, a Universal Resource Identifier (URI) in Linked Data (LD) sets.

NEL may be difficult because an entity is usually mentioned in the text in ambiguous forms. For instance, the mention "Goncourt" can refer to any of the two Goncourt brothers, Edmond or Jules. At the same time Jules de Goncourt can be referred to in the text as "Goncourt", "J. Goncourt", "J. de Goncourt", etc. In order to automatically retrieve all passages in a text where Jules de

---

[1] http://www.tei-c.org/index.xml

© Springer International Publishing Switzerland 2015
T. Morzy et al. (Eds): ADBIS 2015, CCIS 539, pp. 505–514, 2015.
DOI: 10.1007/978-3-319-23201-0_51

Goncourt is mentioned, it is necessary not only to annotate all these mentions as a named entities (NE) of the class person, but to provide them with a unique key that distinguishes them from those of other people, in this case those of Edmond. The bibliographic identifier "Goncourt, Jules de (1830-1870)", as well as the link <http://www.idref.fr/027835995> are examples of such a key.

Linked Data [3] are a standardized way of publishing knowledge in the Semantic Web and many of the available data sets are of great interest for the DH [12]. LD principles such as interlinking and vocabulary reuse facilitate manipulation of data from heterogeneous sources. The formalised knowledge published in the form of LD can provide the background information required to disambiguate NEs in a given context by means of reasoning. In addition, such external knowledge remains available in the annotation by means of the referencing, and can be accessed at later stage to perform advanced queries.

The paper presents an experiment on the linking of authors in corpora of French literary criticism and essays from the 19th century[2]. We shall first present previous approaches to NE disambiguation, then the proposed graph based disambiguation algorithm, baptized REDEN, which includes strategies to consistently handle multiple LD sets. We describe the experiments carried out along with their results and finally give some conclusions and suggestions for further improvement of the algorithm.

2   Related Work

Graph-based approaches to NEL are unsupervised algorithms relying on existing knowledge bases (e.g. the Wikipedia article network, Freebase, DBpedia, etc.). Reasoning can be performed through graph analysis operations. It is thereby possible to at least partially reproduce the actual decision process with which humans disambiguate mentions. In particular, these approaches build a graph out of the candidates available for each possible referent in a given context then use the relative position of each referent within the graph to choose the correct referent for each mention. The graph is built for a context (such as a paragraph) containing possibly more than one mention, so that the disambiguation of one mention is helped by the other ones.

More specifically, the notion of graph centrality is used for the disambiguation, following similar approaches in Word Sense Disambiguation [11]. In NEL, mentions take the place of words, and Wikipedia articles that of WordNet synsets. Centrality measures may be performed on the Wikipedia structure in order to use the rich set of relations to disambiguate mentions. More specifically, in [7] English texts were disambiguated using a graph that relied only on English Wikipedia, and was constituted of the links and of the categories found in Wikipedia articles. For instance, the edges of the graph represent whether ArticleA links to ArticleB or whether ArticleA has CategoryC. Here too "local" centrality is then used to assign the correct link to the ambiguous mention.

---

[2] http://obvil.paris-sorbonne.fr/corpus/critique/

Another WSD based tool for NEL is proposed by the DBpedia Graph Project[3]. However, it is highly dependent on DBpedia structure and only links to this broad-coverage data set and not to other domain-specific ones.

Centrality is an abstract concept, and it can be calculated by using different algorithms[4]. In [11], the experiment was carried out using the following algorithms: *Indegree, Betweenness, Closeness, PageRank*, as well as with a combination of all these metrics using a voting system. Results showed the advantage of using centrality with respect to other similarity measures.

In opposition to the approaches mentioned above, REDEN, described in the next section, uses multiple LD sets as knowledge base, appropriate fusion strategies are thus crucial.

3 Our Disambiguation Approach

The acronym REDEN, in French, stands for Disambiguation and Referencing of Named Entities. First, we illustrate how REDEN works then we describe briefly the algorithm and the most challenging steps.

3.1 Illustrative Example

Let us consider the following paragraph excerpt of a French text of literary criticism entitled "Une thèse sur le symbolisme" written by Albert Thibaudet (1874-1936) and published in 1936:

*[...] lorsqu'il s'est mis réfléchir sur le sens de son œuvre. C'est à cette place que l'on situerait par exemple, chez **Corneille** ou **Victor Hugo**, les discours sur le poème dramatique, ou **William Shakespeare**. La troisième partie du livre est consacrée aux maîtres du symbolisme, qui sont, d'après **M. Barre**, **Verlaine**, **Mallarmé** et **Moréas***

In bold, we see seven mentions automatically that were recognized by a NER algorithm, that need now be linked to an identifier. For each mention, the NEL algorithm selects candidate URIs from multiple linked data sets (e.g. Eugéne Delacroix described by DBpedia, Henri Delacroix by BnF). URIs are obtained from an automatically built index of authors surface forms. Thanks to these URIs, it is possible to retrieve the corresponding RDF graphs from the multiple LD sets (DBpedia, BnF), then combine them into a single graph by fusing homologous individuals thanks to *SameAs* links.

An excerpt of the resulting candidates of the nine NE mentions from the example are listed below[5]. Conveniently, the mentions William Shakespeare and

---

[4] For a discussion of the notion of centrality see also [10]
[5] In this example we identify candidates by distinguishable personal information instead of URI for readability sake. For the same reason only the candidates containing more than six edges in the resulting fused graph are listed.

Victor Hugo are unambiguous, in other words have a single candidate, i.e. Victor Hugo (1802-1885) and William Shakespeare (1564-1616), respectively.

Candidates (**Corneille**) = Pierre Corneille (1606-1684), Thomas Corneille (1625-1709), ...
Candidates (**M. Barre**) = Raymond Barre (1924-2007), ...
Candidates (**Verlaine**) = Paul Verlaine (1844-1896), Georges Verlaine (1871-1926), ...
Candidates (**Mallarmé**) = André Mallarmé (1877-1956), Stéphane Mallarmé (1842-1898), ...
Candidates (**Moréas**) = Jean Moréas (1856-1910), ...

We prune the resulting graph so that it contain only those RDF predicates (i.e. edges) involving at least two candidates of different mentions, because we want only the predicates that play an important role in the disambiguation process. Calculating the centrality (e.g. DegreeCentrality) for every candidate would give us the best candidates for the four mentions. Figure 1 shows an excerpt of the resulting graph where the chosen and correct mention candidates are marked in bold.

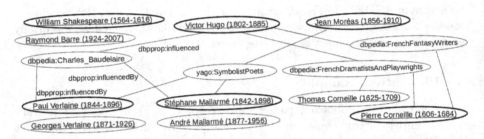

Fig. 1. Excerpt of the chosen URIs (in bold) for the nine candidates (underlined); all edges represent *rdf:type* links except those marked differently.

Notice that all mention candidates share a common *rdf:type* edge to the vertex of the general category *dbpedia:Person*[6]. We can observe that the vertices *yago:Symbolist Poets*, *dbpedia:Charles_Baudelaire*, *dbpedia:French Dramatists And Playwrights* are the vertices that influenced the most the centrality measure. Remarkably, the brothers Pierre and Thomas Corneille could both be convincing candidates for the "Corneille" mention, however, it is Pierre Corneille who shares an extra link to vertex *dbpedia:French Fantasy Writers* and *Victor Hugo*. Also notice that the algorithm chose Raymond Barre for the mention "M. Barre" which is incorrect. The real M. Barre does not exist in any of the LD sets, experts do not know his identity.

3.2   Brief Description of the Algorithm

With REDEN[7], we propose a graph-based, centrality-based approach that is particularly adapted for Digital Humanities; in fact it can exploit RDF sources

---

[6] Some of the chosen candidates share *rdf:type* edge to more specific categories such as *yago:Poet110444194* or *dbpedia:Writer*. These are not shown for readability sake.

[7] Code source and useful resources can be found here: https://github.com/cvbrandoe/REDEN

directly (thus allowing users to choose domain adapted knowledge bases) and takes TEI-annotated texts as input. The disambiguation algorithm processes a file where NE mentions are already annotated (e.g. using the tag <persName>). Possible referents from all mentions in a given context and for a given class (e.g. person) are retrieved from an index built on the fly from selected LD sources. The fusion of homologous individuals across different sources is performed thanks to *owl:sameAs* or *skos:exactMatch* predicates; a graph is created where RDF objects and subjects are vertices and RDF predicates represent edges. Irrelevant edges are removed from the graph before calculating centrality: only edges which involve at least two vertices representing candidate URIs are preserved. Thanks to the selected centrality measure, the best connected candidates for each mention are chosen as referents and an enriched version of the input TEI file is produced. In [5,6], we described the steps of the algorithm in detail. In the following subsections, we discuss the most challenging steps of the algorithm.

3.3   Candidate Identification and Retrieval Using Linked Data

In the context of 19th century French essays, while some of the most famous authors have rich entries in broad-coverage ontologies, other less known ones are only present in domain-specific knowledge bases. Thus the ideal linking algorithm for cultural heritage, and in general for the Digital Humanities, needs to combine heterogeneous sources.

More specifically, for our purposes DBpedia and BnF seem to be the most relevant sources. DBpedia describe authors reusing widely-accepted vocabularies (e.g. foaf, skos); authors are linked to each other by semantic relations such as *InfluencedBy*, and, indirectly, by being linked to the same concept, such as *Romanticism*. BnF entries list all authors of books ever published in France; their entries contain information on date of birth and death, gender, works authored. For instance the BnF entry for Voltaire[8] gives several alternative names such as Franois-Marie Arouet (Voltaire's real name), Wolter, Good Naturd Wellwisher.

The combination of these two sources seems to provide a sufficient coverage for a corpus of 19th century French essays, thus the algorithm retrieves candidate identifiers (URIs) of NE mentions from these LD sets using a domain-adapted crawler [5] that performs SPARQL queries. Such queries use widely-accepted properties such as names (*foaf:name, foaf:familyName, skos:altLabel*) for filtering and retrieving exact matches from mentions. Other complementary properties such as foaf:gender help to match mentions containing honorific titles (e.g. M. Vigny, Madame De Stael, etc.). To reduce waiting time of query response retrieval, an index per class (e.g. Person) is built and updated regularly. It lists automatically generated forms and their associated URIs such as : surname only (Rousseau), initials + surname (J.J. Rousseau, JJ Rousseau, ...), title + surname (M. Rousseau, M Rousseau), etc. This procedure ensures the retrieval of at least one candidate URI for most mentions. At the same time, the mass of information present in the BnF repository will generate several homonyms and make most mentions ambiguous; thus good disambiguation becomes crucial.

---

[8] http://data.bnf.fr/11928669/voltaire/

3.4   Graph Fusion from Linked Data Sets

As seen previously, our algorithm requires the construction of the relevant graph. This graph must represent domain knowledge while avoiding at best redundant and conflicting information and must possess relevant knowledge to the disambiguation process. In the presence of multiple LD sources, it is thus important to dispose of appropriate mechanisms to handle proper data fusion and make data usable by an application [1,8], for our needs, an NLP application. Crucially, an important step is the merging of RDF predicates in the presence of multiple individuals having different URIs pointing to the same real World entity. For doing so, there are two necessary conditions that must comply.

Condition#1. The first one is the existence of a relation that explicitly links two equivalent representations of a real World entity. Fortunately, the compliance of LD principles such as the existence of equivalence relations between homologous individuals enables the automatic interlinking of multiple LD data sets. For what concerns our datasets, mappings between different resources is rich, although it may follow different strategies. So for instance broad coverage resources such as DBpedia and Yago make use of *SameAs* relationships; but other relations such as *skos:exactMatch* may be also used with the same semantics. Biblioteconomic datasets like BNF or other libraries also strongly rely on domain specific standards such as idref, viaf, or ISNI for interlinking.

Condition#2. The second one is the choice of a reference resource URI of mention candidates which can be the one used by one of the involved sources. We choose to use BnF as reference source, and thus BnF URIs to designate fused mention candidates, because it proved to be the most complete source for our purposes[6]. More importantly, BnF links its entry to the DBpedia one when existing, thus making it very easy to connect the two resources in one knowledge graph. Besides, BnF entries also list the author's Idref, which is the official identification system used by French universities and higher education establishments to identify, track and manage the documents in their possession. BnF URIs are thus obtained during the phase of Candidate identification and retrieval described in subsection 3.3.

In this manner, our fusion process can be described more detailed as following. Given a mention M (e.g. Hugo) which corresponds to a real World entity E, and has the candidates C_1, C_2, ... (e.g. Victor Hugo, François Victor Hugo). Each candidate possesses a set of URIs from several LD sources (e.g. Victor Hugo from BnF, Victor Hugo from DBpedia, and so on). In particular, let us name the LD sources such as LD_{ref}, LD_1, D_2, ... where LD_{ref} is chosen to be the reference source (Condition #2). It is straightforward to obtain the corresponding RDF graphs and convert them in equivalent undirected and unweighted graphs [2] named $G_ref=\{V_{ref}, EG_{ref}\}$, $G_1=\{V_1, EG_1\}$, $G_2=\{V_2, EG_2\}$, ... Vertices V represent URIs of mention candidates of the source as well as instantiated ontology concepts (e.g. *dbpedia:Writer*) and data-typed literals (*bio:birth*). Edges EG designate binary and labeled relations (e.g. *rdf:type*).

The aforementioned entity E is represented as E_{ref} in graph G_{ref}, as E_1 in G_1, E_2 in G_2, ... It exists a *sameAs* relation (Condition#2) defined among E_ref and its equivalent E_1, E_2, ... In this manner, the iterative fusion of G_{ref}, G_1, G_2, ... give us a graph G_f where E_{ref} identifies the entity E which is product of the fusion. E_{ref} inherits the set of edges along with the corresponding vertices in which their homologous in the other sources are objects of the relation (in RDF terminology). Moreover, it is also possible to a posteriori assign weights to these edges based on user preferences, in particular, the higher the weight is attributed to an edge, the more priority is given to this edge during centrality calculation.

4   Experiments and Results

This section describes the experiments settings used to test our proposal as well as preliminary results which are encouraging. In this experiment, in order to evaluate the performance of the algorithm, the disambiguation assumes that authors mentions are correctly identified and the right class (in this case *Person*) has been assigned.

4.1   General Experiments Settings

The test corpora consists of a French text of literary criticism entitled "Une thèse sur le symbolisme" (A thesis about Symbolism) published by Albert Thibaudet in 1936, and a scientific essay entitled "L'évolution créatrice" (Creative Evolution) written by Henri Bergson and published in 1907. Both texts are quite rich in NE mentions particularly concerning authors, but are different in style and in the density of references. Mentions in text concerning authors were manually annotated by experts; URIs assigned to mentions are those from Idref[9] The resulting test corpora contains 1027 (Thibaudet) and 277 (Bergson) manually annotated mentions of person entities. We measure the correctness rate of the proposed NEL approach in terms of the attribution of the right URI to a mention with respect to the URI manually assigned by humans. We chose *DegreeCentrality*[4] as centrality measure because it has empirically proved in previous work [6] to be the most satisfying one in our domain. In previous experiments [5], we also compared the correctness rates obtained by REDEN and a widespread NEL tool using. REDEN performed similarly as state of the art graph-based NEL tools do in journalistic texts. Here, we perform two new experiments.

4.2   Context Analysis

The first experiment aims at determining what is the optimal mention context for a corpus, i.e. paragraph, chapter or whole text. Larger contexts will normally contain more mentions, thus produce larger graphs with more candidates,

---

[9] http://www.idref.fr

more links between them. Intuitively this additional information could help disambiguation, but also introduce noise. Notice that within the same context all identical mentions will receive the same referent. Results in table 1 show that a larger graph tends to produce better results.

Table 1. Correctness rate of the NEL algorithm with DegreeCentrality using different contexts. "Density" columns list the (average) number of mentions for each context.

	Par. Precision	Density	Ch. Precision	Density	Text Precision	Density
Thibaudet	0.74	3.60	**0.85**	121.8	0.82	1027
Bergson	0.67	0.76	0.65	76.2	**0.70**	277

Only for Thibaudet the whole text as graph underperforms the results obtained using the chapter. This is due to the higher density in mentions, which makes the chapter a better context. Clearly, this may vary for texts with different density of NE mentions per paragraph and with different levels of topic cohesion. Thus our algorithm allows for the user to customize the context of disambiguation.

4.3   Relation Analysis

In the second experiment, we investigate to what extent relations (edges in the graph) can influence graph centrality calculation and thus improve correctness rate. As said previously, our approach enables setting up weights to the edges according to user preferences. Increasing the weight of a relation can be translated into assigning importance to that relation during centrality calculation. Intuitively some relations such as *dbpedia-owl:influence* (and its variants *dbpedia-owl:influences* and *dbpedia-owl:influenceBy*) as well as *rdf:type*, may have an important impact.

Assigning relation weights requires knowledge of the corpus as well as the state of LD data sets, here BnF and DBpedia. In order to facilitate this task to a user, supervised learning of relation weights from a training set could be used, with the purpose of adjusting values to the domain of the corpus and its context size. Our experiment here aims also at determining whether this is necessary.

We consider for this experiment the two data sets described above and choose the ideal disambiguation context found beforehand for each one, i.e. the paragraph for Thibaudet and the whole text for Bergson. For sake of completeness, we selected the 50 more frequent relations in the set of graphs for each corpus. We significantly increased the weight of each of the 50 relations, one at a time. We consider that a relation has influenced the centrality calculation when the algorithm makes fewer mistakes thus correctness rate is improved. Figure 2 present these results.

As you can see, adding weight to frequently used relations only rarely results in an improvement. In general, an adequate balance between mention density

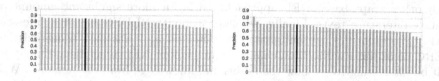

Fig. 2. Correctness rate obtained by adding a weight to the 50 most common relations, one at a time. Results are ranked by decreasing correctness rate. The bar in red represents the best results obtained without weights. Thibaudet to the left and Bergson to the right.

for a given context and a number of relations involved is enough for the algorithm to choose the best candidate for each mention. The mere presence of some relations such as as *dbpedia-owl:influenceBy* and *rdf:type* is enough to obtain optimal results, no weights are needed. Only few exceptions can be found to this observation. For instance, a significant improvement of the results for Thibaudet (correctness rate raises to **0.87**) is obtained by enhancing *foaf:gender*, presumably because most authors mentioned are male, (the same relation produces no effects for Bergson). On the other hand, the relation *foaf:depiction*[10] contributes to ameliorate correctness rate for Bergson (up to **0.81**). In general, both relations are systematically present for all candidates and are inherited from the BnF dataset. In the first case, the *foaf:gender* makes reasonable meaning considering the times when these texts were written. For a contemporary text, it may introduce errors. In the second case, the relation *foaf:depiction* can be considered as a criterion for indirectly identifying the most renowned candidates, namely the ones whose entries are extensively described (i.e. a picture was provided). This is particularly true for biblioteconomic datasets such as BnF. In the presence of an otherwise poor context, this information helps to disambiguate a mention.

5   Conclusions and Future Work

We presented an algorithm to perform NE disambiguation by referencing author mentions to broad-coverage and domain-specific Linked Data sets, DBpedia and BnF, respectively. We set up a procedure which crawls for RDF data of people represented in these LD sets. This procedure can be generalised to other classes of NE (e.g. places) only by modifying the corresponding SPARQL query. Furthermore, our fusion procedure enables for the constitution of a non-redundant graph which is well-suited for centrality calculation. We performed experiments on French literary criticism texts and scientific essays from the 19th century with promising results.

Ongoing development will result in a linking tool that can efficiently disambiguate persons and other classes (notably places and organization) in a large historical corpus of French essays. Further experiments will compare REDEN

---

[10] A common use of depiction is to indicate the contents of a digital image.

with other graph-based NEL approaches using a more significant amount of French Literature texts. Finally, we shall evaluate REDEN in a real life scenario, where entities are automatically detected and classified using existing NERC algorithms without manual checking before NEL is applied. Finally, it will be interesting to see how the additional knowledge from the semantic Web can be used to make advanced queries on the corpus, finding concepts and ideas (such as *Evolution* in Bergson) that are linked to authors mentioned in the text.

Acknowledgments. This work was supported by French state funds managed by the ANR within the Investissements d'Avenir programme under reference ANR-11-IDEX-0004-02 and by an IFER Fernand Braudel Scholarship awarded by FMSH.

References

1. Auer, S., Bryl, V., Tramp, S.: Linked Open Data - Creating Knowledge Out of Interlinked Data. In: Auer, S., Bryl, V., Tramp, S. (eds.) Linked Open Data. LNCS, vol. 8661, pp. 1–17. Springer, Heudelberg (2014)
2. Baget, J.F., Chein, M., Croitoru, M., Fortin, J., Genest, D., Gutierrez, A., Leclère, M., Mugnier, M.L., Salvat, E.: RDF to Conceptual Graphs Translations. In: 3rd Conceptual Structures Tool Interoperability Workshop: 17h International Conference on Conceptual Structures. LNAI, vol. 5662, p. 17. Springer, Moscow, Russia (2009)
3. Bizer, C., Heath, T., Berners-Lee, T.: Linked data - the story so far. Int. J. Semantic Web Inf. Syst. 5(3), 1–22 (2009)
4. Freeman, L.C.: A set of measures of centrality based on betweenness. Sociometry, 35–41 (1977)
5. Frontini, F., Brando, C., Ganascia, J.G.: Domain-adapted named-entity linker using linked data. In: Proceedings of the 20th International Conference on Applications of Natural Language to Information Systems in Conjunction with the 1st Workshop on Natural Language Applications: Completing the Puzzle (accepted, 2015)
6. Frontini, F., Brando, C., Ganascia, J.G.: Semantic web based named entity linking for digital humanities and heritage texts. In: Proceedings of the First International Workshop Semantic Web for Scientific Heritage at the 12th ESWC 2015 Conference, pp. 77–88 (2015). http://ceur-ws.org/Vol-1364/
7. Hachey, B., Radford, W., Curran, J.R.: Graph-Based Named Entity Linking with Wikipedia. In: Bouguettaya, A., Hauswirth, M., Liu, L. (eds.) WISE 2011. LNCS, vol. 6997, pp. 213–226. Springer, Heidelberg (2011)
8. Laudy, C., Ganascia, J.G.: Information fusion using conceptual graphs: a tv programs case study. In: ICCS, pp. 158–165 (2008)
9. Rao, D., McNamee, P., Dredze, M.: Entity linking: Finding extracted entities in a knowledge base. In: Multi-source, Multilingual Information Extraction and Summarization, pp. 93–115. Springer (2013)
10. Rochat, Y.: Character Networks and Centrality. Ph.D. thesis, University of Lausanne (2014)
11. Sinha, R.S., Mihalcea, R.: Unsupervised graph-basedword sense disambiguation using measures of word semantic similarity. ICSC 7, 363–369 (2007)
12. Van Hooland, S., De Wilde, M., Verborgh, R., Steiner, T., Van de Walle, R.: Exploring entity recognition and disambiguation for cultural heritage collections. Literary and Linguistic Computing (2013)

First International Workshop on Information Systems for AlaRm Diffusion (WISARD 2015)

Abduction for Analysing Data Exchange Policies

Laurence Cholvy[✉]

ONERA Toulouse, Toulouse, France
laurence.cholvy@onera.fr

Abstract. This paper addresses the question of checking the quality of data exchange policies which exist in organizations in order to regulate data exchanges between members. More particularly, we address the question of generating the situations compliant with the policy but in which a given property is unsatisfied. We show that it comes to a problem of abduction and we propose to use an algorithm based on the SOL-resolution. Our contributions are illustrated on a case study (This research was supported by ONERA.).

1 Introduction

Organizations are structured groups of atomic or complex entities which have to cooperate in order to achieve a specific goal. Moreover, in order to cooperate efficiently, these entities have to exchange information. For instance, the different entities must have a common understanding of the current situation but each of them only has a partial view of it. Thus exchanging information is necessary.

Organizations in which data exchanges are not constrained are rare. Most of the time, data exchanges are ruled by a policy for security reasons (confidentiality for instance) or for efficiency (communication of relevant information for instance). For instance, in military coalitions, command and control units of different countries need to share information in order to get a common representation of the crisis situation and then take relevant decisions to achieve their mission. They also have to cope with amounts of pieces of partial information, with short information processing time limits. Moreover, such information sharing takes place in a high risk environment [4]: countries involved in a coalition are not necessary allies; trust relation between them may change over time; trust relations may be not symmetric between countries; people may change their role in the organization of the coalition, and so change their "need to know". In such conditions, there is quite a big threat of violating information security properties, such as confidentiality (no unauthorized divulging of secrete information) or availability (information must be available according to users' rights). This may have disastrous consequences for each country's national security. So, in order for users to trust an information exchange system, it is necessary to control and regulate information exchanges within the coalition and this can be done by expressing a *data exchange policy* [5].

A data exchange policy is a means to specify a data exchange system through the organization, aiming at increasing the system reliability and, as a consequance, the trust of its members. But this can be achieved only if the data

© Springer International Publishing Switzerland 2015
T. Morzy et al. (Eds): ADBIS 2015, CCIS 539, pp. 517–526, 2015.
DOI: 10.1007/978-3-319-23201-0_52

exchange policy is of high quality. In particular, it must be *consistent* [6], it must be as *complete* as possible [7] and it must be *applicable* and *minimal* [8]. Consistency ensures that in any case the policy will not lead to a conflict; completeness ensures that there are as few legal uncertainties as possible; applicability and minimality ensure that none of the rules of the policy is unapplicable or redundant.

Checking consistency, completeness, applicability and non redundance on a given policy is a first step towards guaranteeing its quality. But it is not enough and some other properties must generally also be checked. They are more "context-dependent". For instance, it must be checked that a policy in a given organization guarantees that a secretary may read any mail sent to its manager. Or it must be checked that it guarantees that nobody can access mails sent to managers.

Checking these properties is the main topic of our research. More precisely, instead of checking if a given property is satisfied, we aim to know the cases when this property is not satisfied. For instance, instead of checking if, when the policy is applied, a secretary may read any mail sent to its manager (which is a yes-no question), we want to know the situations compliant with the policy in which a secretary cannot read mails sent to its manager. Knowing these problematic situations is of course very informative and helpful for updating the policy and increase its quality.

In this paper, this general problem is restricted as follows: (1) the policies are sets of rules which express the conditions under which it is obligatory (or forbidden) to send some data to somebody else; (2) we aim to find the cases when some people may or may not access a given data.

This paper is organized as follows. Section 2 describes a case study which will be used all along the paper in order to illustrate our contributions. Section 3 describes the formal language used to model all the needed concepts. Section 4 shows that we face a problem of abduction, or equivalently a problem of consequence finding. Then it describes an algorithm for generating the cases when a given context-dependent property is not satisfied. Section 5 applies this algorithm to the case study. Finally, section 6 concludes this paper.

2   A Case Study

The case study we consider is inspired from the SSA (European Space Surveillance) system [1]. More precisely, the organization which is considered here is made of several entities, called *authorities*, which can be *military* authorities or not and which can be *allied* or not. They possess their own satellites and sensors to get information about space environment. Pieces of information they may exchange are here called *measures*. They are for instance images got from satellites.

The policy we consider is made of the following sentences:

- *(1) A military authority must send the measures it owns to any allied military authority.*

- (2) A military authority must not send the measures it owns to another military authority if they are not allied. But it must send it a blurred measure.
- (3) A military authority must not send the measures it owns to a non-military authority. But it must send it a blurred measure.

The three different analysis we want to make are the following:

- (A1) We want to know the cases when a military authority accesses a measure which is owned by a non allied authority.
- (A2) We want to know the cases when a non military authority accesses a measure which is owned by a military authority.
- (A3) We want to know the cases when a non military authority accesses a non blurred measure owned by a military authority.

3 Formal Modelling

3.1 A Formal Model

The formal language we define for expressing data exchange policy is a first order logic language, defined as usual by some variable symbols $(x, y, z...)$, constant symbols, function symbols, and predicate symbols. Among the predicate symbols, there are the following:

- $osend$ is a ternary predicate symbol so that $osend(x, y, z)$ means that x must send information z to y.
- $fsend$ is a ternary predicate symbol so that $fsend(x, y, z)$ means that x must not send information z to y.
- $send$ is a ternary predicate symbol so that $send(x, y, z)$ means that x sends information z to y.
- $access$ is a binary predicate symbol so that $access(x, z)$ means that x may access information z (or x can read z, or x may be informed about z).

Definition 1. *A data exchange rule is a formula of the form:* $\forall x_1..x_n\ \psi(x_1, ...x_n) \rightarrow osend(x, y, z)$ *or* $\forall x_1..x_n\ \psi(x_1, ...x_n) \rightarrow fsend(x, y, z)$ *so that:*
$x \in \{x_1, ..., x_n\}$ *and* $y \in \{x_1, ..., x_n\}$ *and* $z \in \{x_1, ..., x_n\}$
$\psi(x_1, ...x_n)$ *is a formula in which* $x_1...x_n$ *are free and which is written without* $osend$ *nor* $fsend$ *predicates.*

Definition 2. *A data exchange policy is a finite set of data exchange rules.*

As for the analysis to be made, we restrict them to find the cases when some people may (or not) access a given data. This leads to the following definition:

Definition 3. *An analysis is a pair of the form* $<?, \varphi(a_1, ...a_n) \rightarrow access(a_1, a_2) >$ *or* $<?, \varphi(a_1, ...a_n) \rightarrow \neg access(a_1, a_2) >$ *in which* $a_1... a_n$ *are new constant symbols and* φ *a closed[1] first order formula written without* $osend$ *nor* $fsend$.

---

[1] There is no free variable

$<?, \varphi(a_1, ...a_n) \rightarrow access(a_1, a_2) >$ must be read "what are the cases when if $\varphi(a_1, ...a_n)$ is true, then $access(a_1, a_2)$ is true. i.e., what are the cases when, if $a_1...a_n$ satisfy φ then a_1 accesses a_2 ?. $<?, \varphi(a_1, ...a_n) \rightarrow \neg access(a_1, a_2) >$ must be read "what are the cases when is $\varphi(a_1, ...a_n)$ is true, then $\neg access(a_1, a_2)$ is true. i.e., what are the cases when, if $a_1...a_n$ satisfy φ then a_1 does not access a_2 ?

Finally, we consider some domain constraints, which express information considered as true in the context.

Definition 4. *A domain constraint IC is a consistent formula which is written without osend nor fsend.*

3.2   Application to the Case Study

The three rules of the case study policy are modelled by the following five sentences:

(1) $\forall x \forall y \forall z \ \ mil(x) \wedge mil(y) \wedge owns(x, z) \wedge allied(x, y) \rightarrow osend(x, y, z)$
(2.a) $\forall x \forall y \forall z \ \ mil(x) \wedge mil(y) \wedge owns(x, z) \wedge \neg allied(x, y) \rightarrow fsend(x, y, z)$
(2.b) $\forall x \forall y \forall z \ \ mil(x) \wedge mil(y) \wedge owns(x, z) \wedge \neg allied(x, y) \wedge blurred(z, t) \rightarrow$
       $osend(x, y, t)$
(3.a) $\forall x \forall y \forall z \ \ mil(x) \wedge \neg mil(y) \wedge owns(x, z) \rightarrow fsend(x, y, z)$
(3.b) $\forall x \forall y \forall z \ \ mil(x) \wedge \neg mil(y) \wedge owns(x, z) \wedge blurred(z, t) \rightarrow osend(x, y, t)$

In this context we could add for instance: $IC = \forall x \forall y \ \ allied(x, y) \rightarrow allied(y, x)$, which would express the symmetry of the relation of alliance. Or $IC = \forall x \ \neg blurred(x, x)$ expressing that, once blurred, the measure is different.

Analysis (A1) is modelled by:
$<?, (mil(a) \wedge \neg allied(b, a) \wedge owns(b, m)) \rightarrow access(a, m) >$

Analysis (A2) is modelled by:
$<?, (\neg mil(a) \wedge mil(b) \wedge owns(b, m)) \rightarrow access(a, m) >$

Analysis (A3) is modelled by:
$<?, (\neg mil(a) \wedge mil(b) \wedge owns(b, m) \wedge (\forall x \neg blurred(x, m))) \rightarrow access(a, m) >$

3.3   Working Hypothesis

For generating the situations compliant with the policy in which a given property is not satisfied, we have to make some working assumptions about the organization. The first working hypothesis is that the members of the organization respect the policy (i.e., they send information if it is required, they do not send information if it is forbidden). The second working hypothesis is that agents may access some information if and only if it has been sent by some other agent inside the organization.

This comes to consider W, the set of the four following formulas:

$$\forall x \forall y \forall z \quad osend(x, y, t) \rightarrow send(x, y, z)$$
$$\forall x \forall y \forall z \quad fsend(x, y, z) \rightarrow \neg send(x, y, z)$$
$$\forall x \forall y \forall z \quad send(x, y, z) \rightarrow access(y, z)$$
$$\forall x \forall y \forall z \quad access(y, z) \rightarrow \exists x \; send(x, y, z)$$

We insist on the fact that these assumptions will not necessarily be satisfied in real organizations with this policy. For instance, it may happen that in the organisation, one agent does not fulfill its obligations and/or its prohibitions (then violating the first and/or the second assumption); or it may happen that, because the communication chanel is broken, a piece of information sent to an agent, is never received so cannot be accessed (violation of the third assumption); or, an agent reveices a piece of information from somewhere outside the organization (violation of the third assumption). However, we have to consider these working hypothesis for making the analysis because they express the links between the required behavior (obligation or prohibition to send a piece of information) and the analysed behavior (possibility of accessing or not).

4   An Algorithm for Analysing Policy Under the Working Hypothesis

4.1   Analysis vs Abduction

We aim to know the cases when some people accesses (or not) some data. This section shows that this comes to a problem of abduction or, equivalently, to a problem of consequence finding.

Let P be a policy and IC an domain constraint as modelled in section 3.1, and let W be the set of working hypothesis as defined in section 3.3.

Consider the analysis: $<?, \varphi(a_1, ...a_n) \rightarrow access(a_1, a_2) >$ [2]. It means that we want to know the possible cases when $\varphi(a_1, ...a_n) \rightarrow access(a_1, a_2)$. This can be answered by finding formulas F so that:

1. $P \cup W \cup \{IC\} \cup \{F\} \models \varphi(a_1, ...a_n) \rightarrow access(a_1, a_2)$
2. $W \cup \{IC\} \cup \{F\}$ is consistent
3. F is not written with $osend$ and $fsend$ predicates.

Condition 1 expresses that we want to find cases in which $\varphi(a_1, ...a_n) \rightarrow access(a_1, a_2)$. Conditions (2)and (3) expresses that these must be possible real cases. Indeed, any F such that $W \cup \{IC\} \cup \{F\}$ is inconsistent satisfies condition 1 but does not represent a possible case. And any F which is written with $osend$ or $fsend$ does not characterize a real case.

---

[2] The approach is similar for the analysis $< \varphi(a_1, ...a_n) \rightarrow \neg access(a_1, a_2) >$. Just consider $\neg access(a_1, a_2)$ instead of $access(a_1, a_2)$ all along section 4

Condition 1 characterizes *a problem of abduction.*

But, because condition 1 is equivalent to condition 1.*bis* below, we can equivalently say that we face *a problem of consequence finding.*

1.*bis.* $P \cup W \cup \{IC\} \cup \{\varphi(a_1, ...a_n), \neg access(a_1, a_2)\} \models \neg F$

The problem of consequence finding consists in generating the formulas which are logical consequences of a given set of formulas. But, because this set is infinite (due to the transitivity of the logical consequence relation), this problem cannot be automatically solved. It is then reformulated by: generate all the clauses which are logical consequences of a given set of clauses and which are minimal for subsumption.

Because the consequences $\neg F$ we want to generate *must belong to a given language* (condition (3)) and *must satisfy a given condition* (condition (2)), we decide to apply SOL-resolution [3]. Indeed, this rule has specially been designed for generating clauses which are consequences of a given set of clauses (*not necessarily Horn clauses*), which belong to a given language and which satisfy a given condition.

4.2   SOL-Resolution

Let us here quickly recall SOL-resolution defined by Inoue in [3].

A production field PF is defined by a language L_{PF} and a condition $Cond_{PF}$.
A clause belongs to PF iff it belongs to language L_{PF} and it satisfies $Cond_{PF}$.
A production field PF is stable if any clause which subsumes a clause belonging to *PF* belongs to *PF* too.
A structured clause is a pair $< C, D >$ where C and D are clauses.

Let Σ a set of clauses, C a clause and *PF* a production field. *A SOL deduction* of a clause S from Σ with C as top clause and production field *PF* is a sequence of structured clauses $D_0, D_1...D_n$, such that:[3]

1. $D_0 = < \Box, C >$
2. $D_n = < S, \Box >$
3. For each $D_i = < Pi, Qi >$, $P_i \cup Q_i$ is not a tautology.
4. For each $D_i = < Pi, Qi >$, $P_i \cup Q_i$ is not subsumed by a clause $P_j \cup Q_j$, where $D_j = < P_j, Q_j >$ is a structured clause such that j<i.
   This rule is not applied if D_i is generated form D_{i-1}, by application of the rule 5(a)i.
5. $D_{i+1} = < P_{i+1}, Q_{i+1} >$ is generated from D_i according to the following steps :
   (a) Let l be the left-most literal of Q_i. P_{i+1} and R_{i+1} are obtained by applying one of the following rules :

---

[3] In the following, $\Box$ denotes the empty clause.

i. (**Skip**) If $P_i \cup \{l\}$ belongs to PF, then $P_{i+1} = P_i \cup \{l\}$ et R_{i+1} is the new ordered clause obtained by removing l from Q_i.

ii. (**Resolve**) If there is a clause B_i in Σ such that $\neg$ k $\in B_i$ and l and k are unifiable with most general unifier θ, then $P_{i+1} = P_i\theta$ and R_{i+1} is an ordered clause obtaining by concatening $B_i\theta$ and $Q_i\theta$, framing $l\theta$ and removing $\neg$ k θ.

iii. (**Reduce**) If either,

A. P_i or Q_i contains an unframed literal k different from l (**factoring**),

   or

B. Q_i contains a framed literal $\neg$ k (**ancestry**)

such that l and k are unifiable with most general unifier θ, then $P_{i+1} = P_i\theta$ and R_{i+1} is obtained from $Q_i\theta$ by deleting $l\theta$.

(b) Q_{i+1} is obtained from R_{i+1} by deleting every framed literal not preceded by an unframed literal in the remainder (**truncation**).

Soundness and completeness of SOL-resolution. Let Σ be a set of clauses and C a clause. Let PF be a stable production field.

(1) Let $< S, \square >$ be a structured clause generated by a SOL deduction from Σ, with $< \square, C >$ as top clause, and with production field PF. Then, S is a logical consequence of $\Sigma \cup \{C\}$ and it belongs to PF.

(2) Let T be a clause which is a logical consequence of the set $\Sigma \cup \{C\}$, but not a consequence of Σ only, and which belongs to PF. Then, there is a clause S which subsumes T such that $< S, \square >$ is generated by a SOL deduction, from Σ with $< \square, C >$ as top clause, with production field PF.

4.3   An Algorithm which Applies SOL-Resolution

Let us here consider the following production field:

$$PF_{IC} = \{L \setminus \{osend, fsend\} \ , \ \{c : \neg c \wedge W \wedge IC \text{ is consistent}\}\}$$

We can prove that PF_{IC} is stable. So we can apply SOL-resolution to our problem.

More specifically, we have implemented an algorithm which builds the tree of the SOL-deductions from the set of clauses of $P \cup W \cup \{IC\} \cup \{\varphi(a_1, ...a_n)\}$ with $< \square, \neg access(a_1, a_2) >$ as top clause. The tree is bread-first developped. At each level, the Skip rule is first applied, then the Resolve rule, and finally, the Reduce rule. The logical consequences of this level are generated and presented to the user. The user is then asked to resume the generation or to stop. Finally, if $l_1(x_{1_1}, ...x_{1_m}) \vee ... \vee l_n(x_{n_1}, ..., x_{n_m})$ (the x_{i_j} are free variables) is a generated clause, then the corresponding answer is: $\exists x_{1_1} .... \exists x_{n_m} \neg l_1(x_{1_1}, ...x_{1_m}) \wedge ... \wedge \neg l_n(x_{n_1}, ..., x_{n_m})$[4].

Notice that we have chosen to generate answers level by level and ask the user to resume the proof or to stop, because in somes cases, the tree is infinite. For instance if clauses of $P \cup W \cup \{IC\} \cup \{\varphi(a_1, ...a_n)\}$ are recursive, then there may

---

[4] Remember that, according to condition 1.*bis*, we have to take the negation of clauses generated by the algorithm to get the final answers.

be an infinite set of answers. At a given level, when it considers that generated anwers are sufficienly informative, the user can then stop their generation.

Finally, it must be noticed that the satisfiability test in the (Skip) rule which aims to test if $\neg c \wedge W \wedge IC$ *is consistent* is not yet implemented. This is why, some answers are generated even if they should be deleted by this test, and the user must itself remove them for the moment. (see for instance, the first answer in section 5.1). In the same way, the replacement of an answer by another one equivalent but simplier is left to the user (see for instance, the second answer in section 5.1).

5   Application to the Case Study

In this section, we apply the previous algorithm to the case study.

5.1   Analysis (A1)

We want to know the cases when a military authority accesses a measure which is owned by a non allied authority.

Here $\varphi(a_1...a_n)$ is: $mil(a) \wedge owns(b, m) \wedge \neg allied(a, b) \wedge \neg allied(b, a)$

The top clause is $< \Box, \neg access(a, m) >$

The answers which are generated at the first level by the algorithm are:

- $\exists x \exists y\ mil(x) \wedge owns(x, y) \wedge blurred(y, m) \wedge \neg mil(a)$. Literally, this expresses that a may access m if m is a blurred measure of a measure owned by a military authority and if a is not a military authority. This answer should be deleted because it is inconsistent with our assumptions. Indeed, a is a military authority. As mentionned previously, testing inconsistent answers is not yet included in the algorithm; this is why such an answer appears.
- $\exists x\ owns(x, m) \wedge mil(x) \wedge allied(x, a) \wedge mil(a)$. Notice that the last conjunct of this answer is a true condition because a is a military authority. Thus this answer is equivalent to $\exists x\ owns(x, m) \wedge mil(x) \wedge allied(x, a)$ which expresses that a may access m if m is a measure owned by a military authority allied with a. This case happens when the measure m is not only owned by b but is also owned by another military authority, say c, which is allied to a.
- $\exists x \exists y\ mil(x) \wedge \neg allied(x, a) \wedge owns(x, y) \wedge blurred(y, m)$ This third answer expresses that a may access m if m is a blurred measure of a measure owned by a military authority which is not allied to a.

5.2   Analysis (A2)

We want to know the cases when a non military authority may access a measure which is owned by a military authority.

Here $\varphi(a_1...a_n)$ is: $mil(b) \wedge owns(b, m) \wedge \neg mil(a)$

The top clause is still $< \square, \neg access(a, m) >$

The consistent answers which are generated at the first two levels by the algorithm are:

- $\exists x \; owns(b, x) \wedge blurred(x, m)$
  This expresses that a may access m if m is a blurred measure of a measure owned by b.
- $\exists x \exists y \; mil(x) \wedge owns(x, y) \wedge blurred(y, m)$. This expresses that a may access m if m is a blurred measure of a measure owned by a military authority.

5.3   Analysis (A3)

(A3) is a slightly different formulation of (A2). Here, we are not interested in blurred measures. I.e, *we want to know the cases when a non military authority accesses a non-blurred measure which is owned by a military authority.*

Here $\varphi(a_1...a_n)$ is: $mil(b) \wedge owns(b, m) \wedge \neg mil(a) \wedge (\forall x \neg blurred(x, m))$

The top clause is still $< \square, \neg access(a, m) >$

The previous two answers will be eliminated because they will be inconsistent with our new assumption. That means that according to the policy, a non military authority never accesses a non-blurred measure which is owned by a military authority.

6   Conclusions

In this paper, we have presented an algorithm for generating the situations compliant with a data exchange policy but in which a given property is unsatisfied.

The main interest of this algorithm is that, instead of generating problematic situations, it generates first order formulas which give an intensional definition of the problematic situations. This is much more informative. However, this is paid by an high complexity.

Furthermore, this algorithm is not entirely automatic for the moment. In particular, the test which could eliminate some unsatisfiiable answers is not implemented. This could be done by using a SAT solver [2]. But this will augment the complexity of the method.

As for future work, we plan to generalize this work and study policies which are not restricted to obligations and prohibitions, but which also express permissions. This will lead us to change the working assumptions.

Acknowledgments. I Thank Claire Saurel for discussions we had about this paper.

References

1. Donath, T., Schildknecht, T., Martinot, V., Del Monte, L.: Possible European systems for Space Situational Awareness. Acta Astronautica **66**(9–10), 1378–1387 (2010)
2. Schaefer, T.: The complexity of satisfiability problems. In: Proceedings of the 10th Annual ACM Symposium on Theory of Computing, San Diego, California. pp. 216–226 (1978)
3. Inoue, K.: Linear resolution for consequence finding. Artificial Intelligence **56**, 301–353 (1992)
4. Phillips, C.E., Ting, T.C., Demurjian, S.A.: Information sharing and security in dynamic coalitions. In: SACMAT, pp. 87–96 (2002)
5. Cholvy, L., Garion, C., Saurel, C.: Information Sharing Policies For Coalition Systems. In: NATO RTO-IST Symposium on Dynamic Communications Management, Budapest (October 2006)
6. Cholvy, L., Cuppens, F.: Analysing consistency of security policies. Proceedings of IEEE Symposium on Security and Privacy, Oakland (May 1997)
7. Cholvy, L., Roussel, S.: Reasoning with an Incomplete Information Exchange Policy. In: Mellouli, K. (ed.) ECSQARU 2007. LNCS (LNAI), vol. 4724, pp. 683–694. Springer, Heidelberg (2007)
8. Cholvy, L., Delmas, R., Polacsek. T.: Vers une aide la spécification d'une politique d'échange d'informations dans un SI. 30ième congrès Informatique des Organisations et des Systèmes d'Informations (INFORSID 2012), Montpellier (Mai 2012)

ADMAN: An Alarm-Based Mobile Diabetes Management System for Mobile Geriatric Teams

Dana Al Kukhun[✉], Bouchra Soukkarieh, and Florence Sèdes

IRIT, Université Paul Sabatier, 118 Route de Narbonne 31062, Toulouse, France
danakukhun@gmail.com, {soukkari,sedes}@irit.fr

Abstract. In this article, we introduce ADMAN an alarm-based diabetes management system for the disposal of Mobile Geriatric Teams MGT. The system aims at providing a form of remote monitoring in order to control the diabetes rate for elder patients. The system is multidimensional in a way that it resides at the patient mobile machine from a side, the doctor's mobile machine from another side and can be connected to any other entity related to the MGT that is handling his case (e.g. dietitian).

1   Introduction

One of the main objectives behind the initiation of Mobile Geriatric Teams is to provide specialized services for elder patients at hospitalization phases and to minimize the number of hospitalization days especially at the recovery phase. Since Mobile Geriatric Teams MGT can follow the patient wherever he is, a special follow up technique usually takes place in order to return patients to their homes or elder centers and monitor their stability with the help of their families or the centre's staff. The Mobile Geriatric Teams then make follow up visits and trace the evolution of each case.

Diabetes [12] is a chronic disease that occurs either when the pancreas does not produce enough insulin or when the body cannot effectively use the insulin it produces. Hypoglycemia, or raised blood sugar, is a common effect of uncontrolled diabetes and leads to serious damage to many of the body's systems over time, especially the nerves and blood vessels. This is why it is important to keep good glucose levels.

Not controlling diabetes can also complicate medical cases and influence patient's recovery. Usually diabetic patients need extensive care and daily follow up in order to stabilize the case and ensure recovery. Many works discussed the economic burden of hospitalization costs when other solutions could exist. For instance, the works of [2] introduced a study dedicated to avoid the admission of diabetes patients in UK hospitals, other works like [8] discussed the influence of Mobile Geriatric teams on the length of hospital stay among elder patients.

In France, the MGT were the core of many studies, such as [3] that aimed at evaluating the performance of the MGT Service at the Midi Pyrénées region and their efficiency in taking care of elder patients in charge. Other research works like [1] have proposed adaptive accessibility while accessing to information sources during the mobility of the MGT members.

© Springer International Publishing Switzerland 2015
T. Morzy et al. (Eds): ADBIS 2015, CCIS 539, pp. 527–535, 2015.
DOI: 10.1007/978-3-319-23201-0_53

In this paper, we aim to help in solving the remote monitoring problem in another direction where we propose a web-based mobile application that helps doctors in following up with their diabetes patients during the recovery period and that alerts them in emergent cases. The application is multi-dimensional in a way that serves the patient from a side, the medical team from another and other sides whenever needed (dietitian, etc.). The application can be downloaded at the patient side where he, a family member or a personal assistant could enter his daily diabetes rates and at the MGT's side where they could monitor the patient's progress or retreat. Other sides such as family members, care specialists and the dietitian can also have access to the system through their mobiles.

The remainder of this paper is structured as follows: in Section 2, we present some examples of web applications and multi-tier systems for diabetes management. In section 3, we describe the architecture of our ADMAN system detailing its database and its functionality. In section 5, we compare the proposed system with related works. Finally, section 5 concludes the paper with brief remarks.

2   State of the Art

In this section we will preview the different research works that presented systems to solve the problem of diabetes monitoring, starting with server based mobile applications and continuing with multi-tier systems that connect the patient with the doctor.

2.1   Mobile Applications for Diabetes Management

With the evolution of mobile health, nowadays, we can find more than 1,000 iOS and Android applications listed on the Apple App Store and Google Play. These applications are specifically designed for diabetes patients or for healthcare professionals treating diabetes.

Through an analytical study, we have followed examples of the evolution of mobile diabetes management software, the first type of mobile applications dedicated for diabetes is concerned in tracking the diabetes measurements through time in order to improve the patient's consciousness and self-management of diabetes glucose readings.

Glucose Buddy [5] is an example of these applications where it allows users to manually enter their glucose measurements, carbohydrate consumption, insulin dosages, and other activities. The system allows entering contextual information related to the measurement time (e.g. before breakfast, after breakfast, etc.).

iGBStar [13] is another example of diabetes management applications that is connected with a hardware solution, where you can connect your mobile to the blood glucose meter and the application makes it easier to collect, view, share blood glucose readings and store their values.

Glooko [4] is another mobile, cloud based diabetes management system that serves patients by seamlessly, easily and quickly downloading blood glucose readings into their mobile devices and accounts. It Saves time and eliminates errors from manual entry.

We have noticed that the presented examples were mainly interested in monitoring diabetes at the patient side for enhancing the patient's self control. These systems have taken into consideration the patient and his diabetes level's evolution but they haven't classified them or connected them in a seamless way with the doctor or any other specialist (e.g. dietitian). Also, these systems didn't consider cases and scenarios where a family member could be interested to be connected and alarmed in the cases of Hypoglycemia or severe insulin increase.

Next, we will present some of the systems dedicated for connecting the patient results with the doctor in order to ensure diabetes rate monitoring.

2.2   Multi-tier Systems for Diabetes Management

In this section, we overview another type of systems for diabetes management that were present in the literature. Their objective is to share the glucose blood readings with the doctor and healthcare professionals in order to enable him to monitor the patient case remotely.

The works of [11] have proposed a Patient monitoring framework that is dedicated for the assistance of diabetes patients. They present a module-based application, divided in 2 main parts, the specialist and the patient. The platform provides statistics created to offer the doctor the progress of each patient and suggestions, in charge of giving the doctor some pieces of advice based on the statistics of the patient.

The application at the patient's side enables communications, diet enhancements and suggestions. The suggestion module lays on the diet module to present suggestions about beneficial food.

The works of [10] have presented a pervasive healthcare platform for diabetes management and that employs sensors for monitoring the patient's heart rate and physical activity. The main goal of the system is to empower patients to manage their life style with respect to diet and exercise hence lowering their risk factors for metabolic syndrome, and to enable physicians to be more effective in helping their patients reach their goals.

The prementioned works have connected the patient results to the doctor using conventional internet applications. Knowing that nowadays, people mostly interact using their mobile devices and that these applications did not adapt with the mobility of the Mobile Geriatric Teams, we have proposed a mobile application that aims at monitoring the patient's diabetes rates remotely by the doctor or specialist.

3   Contribution

The increasing number of heterogeneous applications and devices used for communication in the internet claims numerous efforts to improve the interoperability [9]. Interoperability is a key feature enabling the interaction between distributed components. In the domain of healthcare, interoperability is the ability of different information technology systems and software applications to communicate, exchange data, and use the information that has been exchanged [6]. Interoperability means the ability of health information systems to work together within and across organizational bounda-ries in order to advance the health status of, and the effective delivery of healthcare for, individuals and communities [7].

The interoperability ensured by the Web server is the principal reason that moti-vated us to use it in our system.

In this part, we will start by detailing the architecture of our system ADMAN then, we will describe the components of the ADMAN xml database and finally, we will explain the functionalities of the ADMAN system.

3.1   Architecture of Our Application ADMAN

In this section, we will present our system architecture (see fig. 1) that aims to ensure that each measured insulin result is connected between the MGT and the patient.

Our architecture is composed of three main entities:

- Patient: This entity is an elder patient who needs to be taken care of either by a family member or an assistant. The surrounding environment of the patient is important; it helps the family member or nurse to deliver his in-sulin measurements by using the diabetes measurement machine and his mobile device that contains our ADMAN application.
- Server: this entity is the core of our system. It is responsible for all the computational activities needed to be done in order to connect between the patient and the MGT in a real time basis. The entity can analyze the measured results and sort them within predefined rates in order to judge the condition of the patient (hypoglycemia, normal, severe increase).
- MGT: this entity contains the Mobile Geriatric Team that is composed of a secretary, nurse, nursing assistant and physician (generalist or special-ist). In our case, the system is mainly interested in being downloaded to the doctor's machine and could also be downloaded to external parties (e.g the dietitian's machine and any other family member that is interest-ed to monitor the patient's case).

The functionality within our architecture goes as follows:

(1) As the patient wakes up, he should measure his diabetes rate (1a), this measurement can also be taken by a family member or an assistant (1b).
(2) The rate value is then explicitly inserted to our system by the patient (2a) or by the care specialist or any family member (2b). In an advanced sce-nario, this rate will be implicitly inserted from the machine (2c).
(3) The measured rate will be stored to an xml database (c.f. fig. 3). The file contains the reading information for this particular patient, for more de-tails, see section 3.2.
(4) The results will be transferred to the web server in order to be shared with the doctor's machine (3).
(5) This web server will take the measured rate and compare it with different intervals (c.f. fig 4 and section 3.3 for the algorithm) in order to return a textual value that describes the patient's status. The returned value is shared with both sides (4a, 4b) (the doctor and the patient who will view the analysis of the measured result). For more details, see paragraph 3.3

(6) The system will enable both sides to view other analytical reports of the calculated values over time (weekly report, monthly report and a 3 months report).

(7) In the case of Hypoglycemia or severe insulin rate, the doctor and patient will receive push notifications to alarm them for the danger of the case; this alarm will be placed in the message area (4a, 4b).

(8) After receiving the alarm, the Mobile Geriatric Team would make the appropriate action and contact the patient through a notification (5a, 5b) this aims at dealing with the patient's case.

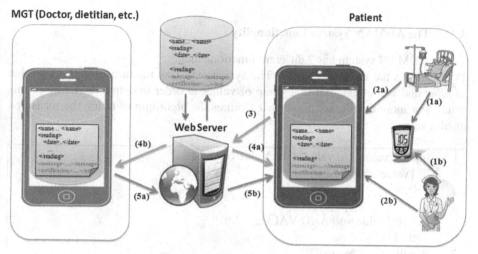

Fig. 1. ADMAN Architecture

The implementation of our system will depend on the REST service, the http protocol as an information channel, logging on to the system will be secured by https.

3.2   The ADMAN Database

As shown in fig. 2, the database that our ADMAN system employs is an xml database [14] which is interoperable and light in size. This file contains information about the patient such as his name, reading (date, time and value), message and notification.

```
<patient id = "">
        <name> ... </name>
        <reading>
                <date> ... </date>
                <time> ...</time>
                <value> ... </value>
        </reading>
        <message alarm = ""> ... </message>
        <notification> ... </notification>
</patient >
```

Fig. 2. XML File of the ADMAN database

- The first element (name) is the account name of the patient.
- The second element represents the glucose measurement (reading) that the user enters explicitly; the system enters the date and time implicitly.
- The third element is the (message) which is filled by the web server side.
- The forth element is the (notification), which is inserted to the system by the Geriatric Team specialist.

The database existing at the server will be related to the patient's side and will also be residing at the Mobile Geriatric Team's side, the MGT will eventually manage different accounts for different patients.

3.3   The ADMAN System Functionality

Our ADMAN system has 2 different functionalities:

1) It analyzes the measured results. The system will take the measured insulin value as input and will compare it to a group of values in order to output the rating of this result. For more details, see fig. 3 that features the algorithm of rating the measured insulin value.

```
1    Input:   value    // value is the Value of diabetes
2    IF {value <60}
3    THEN
4    Result ←  "Hypoglycemia";
5    ELSIF {value >60 AND VALUE < 99]}
6    THEN
7    Result  ←  "Normal";
8    ELSIF {value >100 AND VALUE < 125]}
9    THEN
10   Result  ←  "Disturbed value";
11   ELSIF {value >126 AND VALUE < 140}
12   THEN
13   Result  ←  "Mild increase";
14   ELSIF {value >141 AND VALUE < 200}
15   THEN
16   Result  ←  "Moderate Increase";
17   ELSIF {value >200 }
18   THEN
19   Result  ← " Severe Increase";
20   Endif
21     Return Result ;
22     IF {Result  = "Hypoglycemia"    or   Result =" Severe Increase"   }
23     THEN
24     Alarm = on;
25     Endif
```

Fig. 3. Algorithm for analyzing the patient's diabetes measurements.

This step will also include passing a message containing the analyzed measured result to both sides the MGT and the patient. Finally, it will also alarm the MGT and the patient in extreme cases.

2) In case of extreme cases, it will pass a message (notification) from the doctor to the patient to help him to react.

4   Use Case

Taking the example of the patient "John Smith" who is aged 75 years and who is recovering from hospitalization. A nurse will be passing by him on a daily basis in order to check up his insulin rate measurements and to do some other checkups.

Our application is downloaded on different machines related to different people who are interested in following up the patient's case (Doctor, dietitian, family member).

On the first day, the nurse measures the patient's diabetes rate. Supposing that the inserted measurement of his insulin value is equal to *90*, then the ADMAN system will rate it as a *normal value* and will include this result within the xml file that is sent to both sides.

On the second day, the insulin value turned out to be equal to 57, then the ADMAN system will rate it as a *Hypoglycemia* and will send this result and an alarm to the different entities related to the patient.

Here, the doctor will order the nurse to give the patient some sugar and tell her to re-measure the diabetes rate after 15 minutes. Also, he might think of rescheduling the insulin dose taken by the patient. The dietitian can then take an action and send a notification including some directions to the patient. The family member would call in order to check up on his case.

Any action taken can be included within the notification tag of our database. Fig 4 shows an example of a filled xml file coming from the doctor's side and the dietitian side and that is being received by the patient and other users.

```
<patient id = "777">
      <name> John Smith </name>
      <reading>
            <date> 2015-04-14 </date>
            <time> 08:00:00</time>
            <value> 57 </value>
      </reading>
      <message alarm = "on"> "Hypoglycemia" </message>
      <notification id = "1"> "Please take some sugar and re-measure the diabetes
                              rate after 15 minutes, Please take a lower dose of
                              insulin at night" </notification>
      <notification id =   "2"> "Please make sure that you eat some carbohydrates at
                              dinner " </notification>
</patient >
```

Fig. 4. An XML file completed by our ADMAN system.

This way, the notes taken from this experience would be shared with the family member who will make sure that they will be applied.

5   ADMAN vs. Related Works

In this section, we will show a comparison in Table 1 between the capabilities of our ADMAN system and the capabilities of other systems that were presented in previous works as solutions for diabetes management.

Table 1. Characteristics of our ADMAN system Vs. other systems.

System / Criteria	Glucose Buddy [5]	iGBStar [13]	Patient monitoring framework [11]	Pervasive healthcare platform [10]	ADMAN
Mobile App.	X	X			X
Reporting results to the doctor			X	X	X
Reporting results to family					X
Reporting to a dietitian				X	X
Notifications for severe cases					X

6   Conclusion

Many systems were proposed to encounter the challenge of diabetes monitoring but none of these systems considered to channel the doctor with the patient in the form of a simple mobile application. In this paper, we have presented the ADMAN system that helps the mobile Geriatric Team members to remotely monitor the daily insulin measurements for diabetic patients. Our objective is to minimize the hospitalization period for elder people while providing them with distant medical surveillance. Our system will also help patients to optimize their self-control and monitor themselves.

One of the limits to the success of our system is the accuracy of the measured result taken from the machine (normally the MGT depends on values checked up at the hospital).

In future work, our system will take into consideration that different doctors can treat the same patient. Another perspective is taking into consideration the user profile and his medical doses in order to automatically measure the insulin doses to be taken after each measurement.

References

1. Al Kukhun D., Sèdes F.: La mise en œuvre d'un modèle de contrôle d'accès adapté aux systèmes pervasifs. Application aux équipes mobiles gériatriques. Dans : Document numérique, Hermès **12**(3), 59–78 (2009)
2. Allan, B., Walton, C., Kelly, T., Walden, E., Sampson, M.: Admissions avoidance and diabetes: guidance for clinical commissioning groups and clinical teams. Diabetes UK (2013)
3. Arthus, I., Montalan, M.A., Vincent, B.: Quels outils pour piloter la performance d'une Equipe Mobile de Gériatrie. Journal d'Economie Médicale **27**(1–2), 43–59 (2009)
4. Glooko: Unified Platform for Diabetes Management. http://glooko.com/ (last consulted in April 11, 2015)
5. Glucosebuddy. http://www.glucosebuddy.com/ (last consulted in March 20, 2015)
6. HIMSS Dictionary of Healthcare Information Technology Terms, Acronyms and Organizations, 2nd edn., 190 (2010)
7. HIMSS Dictionary of Healthcare Information Technology Terms, Acronyms and Organizations, 3rd edn., 75 (2013)
8. Launay, C., de Decker, L., Hureaux-Huynh, R., Annweiler, C., Beauchet, O.: Mobile Geriatric Team and Length of Hospital Stay Among Older Inpatients: A Case-Control Pilot Study. J. Am. Geriatr. Soc. **60**(8), 1593–1594 (2012)
9. Moritz, G., Zeeb, E., Golatowsk, F., Timmermann, D., Stole, R.: Web Services to Improve Interoperability of Home Healthcare Devices. IEEE-Pervasive Computing Technologies for Healthcare, London, 1–4 (2009)
10. Nachman, L., Baxi, A., Bhattacharya, S., Darera, V., Deshpande, P., Kodalapura, N., Mageshkumar, V., Rath, S., Shahabdeen, J., Acharya, R.: Jog Falls: A Pervasive Healthcare Platform for Diabetes Management. In: Floréen, P., Krüger, A., Spasojevic, M. (eds.) Pervasive 2010. LNCS, vol. 6030, pp. 94–111. Springer, Heidelberg (2010)
11. Villarreal, V., Laguna, J., López, S., Fontecha, J., Fuentes, C., Hervás, R., de Ipiña, D.L., Bravo, J.: A Proposal for Mobile Diabetes Self-control: Towards a Patient Monitoring Framework. In: Omatu, S., Rocha, M.P., Bravo, J., Fernández, F., Corchado, E., Bustillo, A., Corchado, J.M. (eds.) IWANN 2009, Part II. LNCS, vol. 5518, pp. 870–877. Springer, Heidelberg (2009)
12. World Health Organization. http://www.who.int/mediacentre/factsheets/fs312/en/index.html (last consulted in March 2015)
13. www.bgstar.com/ (last consulted in March 20, 2015)
14. XML, W3C, XML Base W3C Recommendation 28 January 2009, the last version: http://www.w3.org/TR/xmlbase (last consulted in March 2015)

An Architectural Roadmap Towards Building an Alarm Diffusion System

Sumit Kalra[✉], T.V. Prabhakar, and Saurabh Srivastava

Computer Science Department, IIT Kanpur, Kanpur 208016, India
sumitk@cse.iitk.ac.in

Abstract. Alarm Diffusion Systems(ADS) have complex, exacting and critical requirements. In this work we aim to provide a software architecture perspective towards ADS. We look at both functional and quality requirements for an ADS and also attempt to identify certain quality attributes specific to an ADS and attempted to provide a set of architectural tactics to realise them. We also propose a Reference Architecture for designing such systems. We have provided ample examples to support our inferences and take a deeper look at a case study of the Traffic Collision Avoidance System(TCAS) in aircrafts.

1 Introduction

Alarms are fairly common in most of the systems built today. They could be as simple as a *Check Engine Warning LED* on Car Dashboards, or as subtle as indicating textitAir speed of an aircraft is dangerously low. The typical building blocks of any Alarm System may include sensing/measuring relevant parameters, applying rules/inferences for detecting an alarm, diffusing an alarm among recipients and accumulating feedback, learning/adapting the system from collected statistics, etc. In particular, Every Alarm Diffusion System (ADS) has to diffuse the alarms among a set of *Recipients*. These recipients could be human or machine. The scenarios for an ADS could be varying, depending upon the type of recipients, the communication channels, applied levels of confidentiality, severity of alarms etc.

In most cases, the diffusion part is handled via specialized software. These software are built to take care of above mentioned requirements and constraints. An essential aspect of designing any software is articulating requirements, both explicit, and implied, and mapping them into software components which form the overall system. A System Architect performs this crucial step in the process, called *System Architecture*. In this work, we will take a deeper look at this step of the Development Process. We will discuss the *Quality Attributes* [8] applicable to an ADS, and a set of Architectural Tactics, to provide hints at achieving them. We will list out plausible scenarios for specific requirements, with the help of examples.

The rest of the paper is organized as follows. In Section 2 we discuss the requirements associated with an ADS. In particular, we discuss the various scenarios, which the system may have to handle. We also define some Quality

© Springer International Publishing Switzerland 2015
T. Morzy et al. (Eds): ADBIS 2015, CCIS 539, pp. 536–546, 2015.
DOI: 10.1007/978-3-319-23201-0_54

Attributes that can be associated with an ADS along with common architectural tactics [2] that may be used to realise them. In Section 3 we attempt to provide a reference architecture for an ADS. We aim to do so by highlighting overall aspects of the system using *Component-and-Connector View* [3] [13]. We then present a small case study on a collision avoidance system for aircrafts, called TCAS in Section 4. The aim is to analyze TCAS from the perspective of alarm diffusion. Finally, we present our conclusion in Section 5.

2   Requirement Analysis for an ADS

The design of every system is initiated by enlisting the goals it is expected to achieve [12]. This includes detailing the functional aspects of the system like data structures, algorithms, operations, policies, communications etc. The other set of requirements associated with a system involve measuring quality parameters, and adhering to certain quality expectations. Although the requirements associated with any ADS, functional or quality related, could be fairly specific from one application to other, we are interested in bringing out aspects which cross-cut a wide range of implementations.

We make certain assumptions while doing so. First, we assume that the steps prior to alarm diffusion are relatively straightforward. It implies any hardware like transducers or software logic required to detect the alarm conditions can be implemented without much ado. This assumption relieves us of the considerations related to a wide array of sensors and embedded software modules, leaving us with more space for concentrating on diffusion issues of the system. Second, we assume that the ADS is primarily a software intensive system [7]. This means that the system expects minimal manual intervention in the diffusion process, mostly restricted to defining diffusion policies, and the system is capable enough to implement the same seamlessly. This assumption limits the processes that are subject to be *managed* and not *engineered*[14].

With these assumptions, we now formalise the requirements of an ADS. We divide our requirements into Functional and Quality related issues as discussed before.

2.1   Functional Considerations

The Functional Aspects of a system deal with the implementation of protocols and operations that the system is expected to fulfill. For an ADS, this would involve specifying information such as the communication mediums, alarm priorities, recipient hierarchies, as well as other requirements detailing the preliminary steps. These requirements may vary significantly from one implementation to another. We will iterate over some possible requirements - and their plausible variations.

Diffusion Protocol: An ADS can be seen as a Rule-based system [5], which is aware of how an alarm should be diffused. It is fairly possible that the ADS has to choose different diffusion strategies, for different set of alarms. We briefly go through a set of possible *Diffusion Classes*, in which an alarm may lie:

- **All-or-None**: The diffusion protocol for this class of alarms is that the alarm should either reach all the recipients, or none. Consider a scenario of an army, which wants to attack an enemy post from all sides. All its stationed units shall attack in coordination with each other. It could be catastrophic if some units get this message, and launch the offensive, while others didn't. We term this diffusion policy as *Atomic Diffusion*, since the diffusion should either succeed to all recipients, or none at all - very much like a transaction in a database.

- **Exactly m out of n**: The diffusion protocol for this class of alarms is that the alarm should reach exactly m recipients, out of the total n possible recipients. Special cases of this will be when $m = 1$ or $m = n$. Consider the sample scenario of a city where multiple road accidents are reported. Exactly one ambulance must reach to each location. With limited number of ambulances, sending more than one ambulance to a location, results in waste of resources and unavailability for some more accident locations. The message for reaching to one accident location thus, must reach exactly one ambulance, no less, no more.

- **At most m out of n**: Under this diffusion protocol, the alarm should reach to a number between 0 and m recipients out n. Consider a social networking website, which limits the reach of *free posts* to a small percentage of user's friend network (to encourage users to enroll for *paid posts*). In such cases, the critical aspect of the system is to put an upper bound on the number of recipients of the post. The system doesn't care about the lower bound.

We do not claim the above list is exhaustive, but most of the diffusion policies in a real life ADS would fall in one of these categories.

2.2 Quality Aspects

We build systems keeping certain quality considerations in mind. These quality constraints may often not be stated explicitly. For example, a system built to calculate taxes, will have detailed description of tax slabs, rebates, allowed deductions etc. as a part of the requirement specifications. The speed of computation is often determined by human factors.

Quality attributes defines "the fitness and suitability of a product"[1]. The architectural community has identified other quality attributes of a system such as availability, interoperability, modifiability, security, testability, usability and others[8]. All these quality attributes are achieved through Architectural Tactics[2]. Architectural Tactics are broad structural and behavioural decisions, which when applied to a system, can realise certain quality attributes. Often a trade-off is required to be made between one or more quality attributes, i.e. not all of them may be realised in a system simultaneously. We now identify some quality attributes specific to an ADS. For each of the quality attributes, we have specified the requirements using stimulus, artifact, response and response measure[3] as follows:

- **Trackability:** In some cases, diffusing an alarm may not be a simplistic process. Take for example, the diffusion of a Post by a Facebook Page, to its followers. Just an attempt to put the post in their timelines is not enough. Facebook keeps a track of the reach of these posts too. In certain scenarios which requires "Follow-up" alarms, it is important to track which recipients, out of the intended recipients, received a particular alarm. Trackability is a measure of the extent, to which an ADS keeps itself informed about the reach of the diffused alarms. Follow-up alarms make sense only for those recipients who received the previous one. The analogy of a "follow-up" alarm is Facebook sending you a notification, in case somebody comments on a post you liked.

  The tactics for achieving trackability may include usage of the *Observer Pattern*[6]. Another tactic is to achieve trackability is to make use of the *Publish-Subscribe pattern*[6] where the recipients themselves subscribe for follow-up alarms or continuously pull alarm information from the ADS to check if intended alarm has occurred or not. In table 1, we expressed the quality requirements of a system for trackability.

Table 1. Trackability General Scenario

Scenario Portion	Possible Values
Stimulus	Alarm, Multistage alarm, follow-up alarm, negation of alarm
Artifact	Communication channels, temporary storage, diffusion processes and policies
Response	Track the recipients for subsequent alarms: Log the alarm diffusion and diffuse subsequent alarms to only logged recipients (people or systems)
Response Measure	Trackability percentage (e.g., 99.999%) Proportion or rate of a certain class of alarms that the system tracks successfully
Tactics	ADS track the recipient *or* Recipients subscribe or pull the follow up alarms

- **Idempotency:** Certain aspects of a Distributed System are also applicable to an ADS such as re-transmission of certain packets or messages, due to communication failure or timeouts. An ADS may diffuse the same alarm, multiple times, and yet, the recipients shall be able to process the alarm only once. Idempotency is the ability of a system to handle repeated alarms appropriately. Idempotency requires an understanding between the recipient, and the ADS. The ADS may adopt a protocol for such cases, and the recipients shall abide by the same.

  Tactic to realise idempotency is to use a versioning system for the alarms. The alarms follow a non-decreasing sequencing system which assigns a unique

ID to every distinct alarm. The details related to idempotency are discussed
in table 2.

Table 2. Idempotency General Scenario

Scenario Portion	Possible Values
Stimulus	Repeated Alarms, Identical alarms
Artifact	Communication channels, temporary storage, diffusion processes
Response	Discard the identical alarm or repeated transmission of the alarm
Response Measure	Filtering accuracy (rate of successfully identifying duplicate alarms) Number of times duplicated alarms are diffused by the system and cost of duplicate diffusion
Tactics	Versioning of Alarms

- **Aptness:** A diffusion protocol may define more than one acceptable outcomes for diffusion. For example, in the *All* or *None* protocol, it is acceptable
to not diffuse the alarm to any of the recipients. However, it is desirable to
achieve the *All* state more often than the *None* state. In such a case, the
ability of a system to exhibit one accepted behaviour over the other may be
an indication of system's quality. The aptness is the measure of a system to
adopt the more desirable outcome of a diffusion protocol as compared to the
other states. Quality requirements for aptness are specified in table 3.

Table 3. Aptness General Scenario

Scenario Portion	Possible Values
Stimulus	Alarm
Artifact	Intended recipients
Response	Among multiple acceptable diffusion policies, some are preferred over others
Response Measure	The number of times the preferred diffusion policy is chosen
Tactics	Re-transmission of Alarms Synchronous Communication

3   Reference Architecture for Alarm Systems

After skimming through issues related to designing an ADS, we now propose a
Reference Architecture for ADS. A reference architecture does not attempt to
build a system per se, rather, but bring out the important building blocks which
assembled to build a specific architecture, for the given domain [3] [13].

A general procedure followed at an ADS is depicted in Figure 1. The process mainly has three steps. The detection of an alarm, using some alarm metadata (data from sensors etc.), followed by its classification [9] based on its features. The second step is the diffusion of the alarm to intended recipient groups. Finally, a set of post-diffusion steps, are taken to deal with cases such as follow-up alarms, or negation alarms. A proposed reference architecture for an Alarm System is show in Figure 3. An ADS receives an alarm from preliminary stages, which implements protocols for triggering alarms. We refer it as Alarm Trigger System (ATS). The ATS is responsible for sensing the environment, via hardware or software components, and detect events, which are alarms using trigger protocols. The ATS is not a part of ADS and hence, we will not discuss it further. Within the ADS, the Information Dissemination & Tracking System (IDTS) is the one which receives the alarm from the ATS. After IDTS receives the alarm, it uses Diffusion Policies to diffuse information from local storage such as intended recipient groups, privacy policies, trackability options to deploy etc. The alarm is then passed on the Diffusion Network built of one or more varying communication media and intermediary hardware. The recipients could be different devices, human beings, or social media platforms. A local storage may also be at the helm of IDTS, to keep operational information such as trackability data of some previous alarms. ADS may improve itself over the period of operation. For example, it may behoove the ADS to change its diffusion policies, or recipient groups for some alarms, based on the feedback from recipients or by direct sensing of the environment itself. The Adaptation Component performs heuristics over available data and may change the diffusion policies for future diffusion.

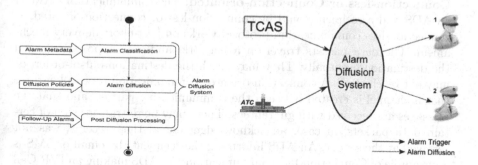

Fig. 1. Alarm Diffusion Process

Fig. 2. Envisioning an ADS for avoiding mid-air collisions

An important aspect of the ADS is the Diffusion Network. The diffusion network could either be a homogeneous, or a heterogeneous mix of various hardware and software entities. As part of reference architecture, we will discuss some aspects of the Diffusion Network related to architecture of the system.

Initiation: Diffusion process of an alarm can be initiated in two ways:

- **Push-based ADS**: A push-based ADS is one, in which the alarm is *pushed* by ADS to the recipients. ADS is responsible to inform the recipients an alarm, mostly within a limited time window. The recipients must be in *listener* mode, to receive messages from the ADS. The ADS either guarantee the receipt of the alarm, or adopt a best-effort delivery mechanism. An example of a push-based ADS can be a natural calamity warning system used by a Government Agency to warn citizens of an area via SMS or Email.
- **Pull-based ADS**: In pull-based ADS, alarm is *pulled* by the recipients, via sending *queries*. The ADS receives information about various alarms from the preliminary stages, and keeps a track of them locally without diffusing them. The interested recipients send periodic queries to ADS to know if any alarms have triggered for them. The ADS replies the recipients alarms intended to them triggered recently (definition of recent may be different for different implementations). ADS can choose to be benevolent in nature, sending negative replies to recipient if alarms haven't triggered recently, or simply ignore the queries, in case the response is negative. An example of a pull-based ADS can be a Web Server/Mobile App duo, which notifies user whenever a goal is scored in a game of Football (the Web Server is the ADS, different instances of the Mobile App are the recipients).

In addition, there can be **Hybrid ADS** which implements both push-based and pull-based services, depending on alarm severity or recipient capabilities.

Communication and Topology: Another aspect of an ADS is to manage communication links with recipients. We will briefly discuss related issues.

- **Connection-less or Connection-oriented**: The communication between an ADS and a recipient could be connection-less or connection-oriented. A connection-less communication protocol works on best-effort delivery mechanism. The data packets travel on a network, and are supposed to reach the destination eventually. They may reach the destination out-of-order, or may not reach at all. A connection-oriented communication protocol ensures that a channel is created between the communicating parties, and that the messages are received with guarantees. There may be re-transmission of certain data packets, in case, an acknowledgement of their receipt, was not received at the sender. An ADS informing the recipients by email or SMS is an example of Connection-less communication. An ADS making an IVR Call to prompt user to provide his Credit Card credentials to complete a purchase, could be considered an example of Connection-oriented communication.
- **Single or Multiple Hops**: Another aspect of ADS is *hop distance* of recipients. An ADS may be built on a hierarchical basis. For example, there could be a National-level ADS, which diffuse alarms to one or more State-level ADS, which in turn, may diffuse the alarm further to recipients or to District-level ADS. The advantage of this topology is the distribution of diffusion load across multiple systems. Different diffusion protocols can be used at different levels. An example of such a system could be national and regional remote sensing agencies. The national agency have access to current

satellite data, which can be processed to know about alarming conditions in one or more regions and diffuse it to the respective regional agencies.

- **Link Redundancy**: Communication links are prone to failures due to power outages, natural calamities, vandalism etc. The ADS may have to establish redundant links with some or all recipients in these scenarios. An example could be usage of two different kind of links - say Optical Fibres, as well as Satellite Communication, for alarm diffusion. The ADS may use Optical Fibres when available, as it'll be faster. In case of an eventuality, ADS uses Satellite Communication, which is slow but available with higher probability.

The proposed reference architecture can act like a reference to architect an ADS. Not all the elements presented here may be applicable in all scenarios. We have proposed the reference architecture for wide range of plausible functional and quality requirements adhering to an ADS.

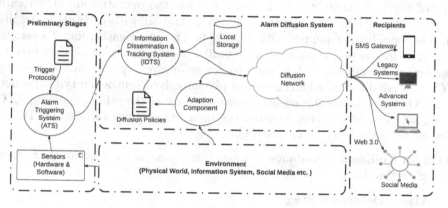

Fig. 3. Reference Architecture of Alarm System

4   Case Study: Issues Related to TCAS Operation

In this section, we will study Traffic Collision Avoidance System(TCAS), critically important system used for aviation safety. TCAS is an aircraft collision avoidance system, used in modern airliners to avoid the mid-air collisions. Although the diffusion of alarms in TCAS are not of much importance, the overall integration of TCAS within the aviation process is of importance to anyone interested in building an ADS.

Introduction to TCAS: The necessity of a TCAS was felt in 1950's, after a mid air collision in the Grand Canyon [10]. TCAS is a set of hardware and software components which can communicate and negotiate with other when planes are in close vicinity to avoid any collisions [11]. When two planes are close, the TCAS can alert both about the potential danger, and provide specific instructions to avoid the collision.

TCAS and Its Environment: Although TCAS is an efficient way to avoid mid-air collision, the situation becomes complicated because of overall aviation process. A typical flight involves a series of communication between three different entities: the flying crew, the on-board computer(s) and the Air Traffic Control (ATC). TCAS is part of on-board computers. All three entities can, at least in theory, initiate the process of avoiding a mid-air collision, independently of each other. For example, a pilot may visually see another aircraft in vicinity, and decide to change the altitude of the plane to avoid the collision in very short time. Or, The ATC can observe on their radar if two planes are on a collision path and directs to change their altitude to maintain minimal distance. If the two entities fail to do so, TCAS will generate warnings like "Descend" or "Climb" on the screens of the pilots, to avoid the collision. An internal protocol, negotiates the process between the two planes, and ensures that neither of the two planes get the same warning - climb or descend.

Failure of TCAS: We now look at a tragic incident took place in the German city of berlingen in July 2002 in order to view the operation of TCAS with respect to an ADS. Two airliners collided in mid-air with each other despite both of them being equipped with TCAS [4]. The investigation found a serious flaw that led to the accident - the overall system was ambiguous on how to handle(avoid) collision. While both planes were given specific instructions by TCAS to avoid the collision, one of the plane received contradictory instruction from ATC. The communication between the crew of the other plane and ATC could not succeed and they followed the instructions given to them by TCAS. The result of the same was catastrophic.

ADS for Collision Avoidance: We can visualise a Collision Avoidance System as a potential ADS.

- **Alarm Detection Stages:**
    1. Collect altitude information of nearby planes.
    2. Extrapolate their paths.
    3. Find out if any of them can potentially collide with the plane or not.
- **Alarm Trigger:**
    1. The pilot can visually see another plane in vicinity, and realise they are too close to each other.
    2. The ATC realise that two aircrafts are on a collision trail, with the help of their radars.
    3. The TCAS sensed that a nearby plane's trajectory will invite a collision.
- **Protocol:**
    1. The pilots should communicate with each other, and decide manoeuvres that will avoid the collision (practically not possible in most cases).
    2. The ATC should provide instructions to both the planes to change their altitudes, to avoid the collision. Typically, one plane should either climb or descend, and the other should do the opposite.
    3. Both the pilots shall listen to the TCAS warnings on their planes, and follow the instructions - either to climb or descend - to avoid the collision.

We now analyze probable flaw in this process. Ruling out manoeuvres by pilot communication, which of the other two protocols have a higher precedence?

In case a pilot receives instructions from both the TCAS and ATC, which instruction shall be processed by him? In case of the berlingen tragedy, one of the pilots followed instructions of TCAS, while the other followed the instructions of ATC.

Now we model an ADS for avoiding collisions between aircrafts. Let us assume that all the humans involved in the system, behave according to a predefined standard protocol. The system thus looks roughly similar to Figure 2.

The diffusion protocols discussed in 2.1, are also applicable in this case for ADS. An example of the **All-or-None** protocol can be applied to instructions (alarms) issued by the ATC. The instructions should either reach both planes, or none. It may be catastrophic, as it was in the above mentioned accident, for the message to reach one, and not the other plane.

In a nutshell, required instructions given to the pilots of both the planes as alarms and model the whole collision avoidance system as presented, we can study and take inferences from it, which can be applied to other ADS. One inference that we drew from our modelling is that an ambiguity in Diffusion Protocols can wreck havoc on an ADS. Another inference is that an additional communication channel between the ADS and the recipient can prove critical in certain scenarios, when the primary channel for communication is not available.

5   Conclusion and Future Work

Alarm diffusion systems are complex and sophisticated. An architecture perspective enables us to tackle their complexity in a systematic way. The delineation of functional and quality requirements is helpful in building such systems. A reference architecture suggests a base line implementation. A pluggable, extensible and adaptable ADS is a byproduct Quality attribute requirements warrant careful analysis - the inter-dependencies would be interesting. A detailed study of design patterns applicable to an ADS will be useful in enriching the pool of architecture knowledge for building alarm diffusion systems.

References

1. Firebrand Architect. Quality attributes. http://www.softwarearchitectures.com/qa.html
2. Bachmann, F., Bass, L., Klein, M.: Deriving architectural tactics: A step toward methodical architectural design. Technical report, DTIC Document (2003)
3. Bass, L.: Software architecture in practice. Pearson Education India (2007)
4. Brooker, P.: The überlingen accident: Macro-level safety lessons. Safety Science 46(10), 1483–1508 (2008)
5. Buchanan, B.G., Shortliffe, E.H., et al.: Rule-based expert systems, vol. 3. Addison-Wesley Reading, MA (1984)
6. Gamma, E., Helm, R., Johnson, R., Vlissides, J.: Design patterns: elements of reusable object-oriented software. Pearson Education (1994)
7. Hilliard, R.: Recommended practice for architectural description of software-intensive systems. IEEE 12, 16-20 (2000). http://standards.ieee.org

8. ISQS ISO. Iso/iec 25010". 2011. Systems and software engineering–Systems and software Quality Requirements and Evaluation (SQuaRE)–System and software quality models (2011)
9. Liu, B., Hsu, W., Ma, Y.: Integrating classification and association rule mining. In: Proceedings of the 4th (1998)
10. Murphy, G.K.: The grand canyon midair collision: A stimulus for change. The American journal of forensic medicine and pathology **11**(2), 102–105 (1990)
11. Rich, R.S., Anderson, M.W.: Traffic alert and collision avoidance coding system. US Patent 5, 636, 123 (June 3, 1997)
12. Sommerville, I., Sawyer, P.: Requirements engineering: a good practice guide. John Wiley & Sons Inc. (1997)
13. Taylor, R.N., Medvidovic, N., Dashofy, E.M.: Software architecture: foundations, theory, and practice. Wiley Publishing (2009)
14. Prabhakar T.V.: cs654 software architecture class notes. Dept. of CSE, IIT Kanpur. https://cse.iitk.ac.in/users/tvp/teaching.html

Information Exchange Policies at an Organisational Level : Formal Expression and Analysis

Claire Saurel[(⊠)]

ONERA Toulouse, Toulouse, France
claire.saurel@onera.fr

Abstract. This paper starts from a logical framework intended to define and analyse information exchange policies for critical information systems. A layer is introduced to express organisational information exchange policies at abstract level. Properties are defined within this organisational layer, in particular information permeability through organisations. More efficiency is expected for policies expression and analysis.

1 Introduction

In critical information systems, such as spatial awareness systems or crisis management systems, one challenge is to enable users to exchange sensitive information in order to work together, while preserving information confidentiality. An other one is to prevent irrelevant information diffusion, for sake of efficiency. In order to get users' trust, one can previously define an information exchange policy as a requirement to be enforced by the system : it can be seen as a regulation, ie a set of rules which specify authorized, permitted or forbidden information diffusion within the system. *It is forbidden for security officers to send information about a nuclear accident to journalists* is an example of such a rule.

In fact, information exchange policies are not intended for individual agents, but rather for agents as role holders in organisations in which they have to collaborate. So organisations have to specify information exchange policies specifically dealing with their specific information of interest, for the roles they have defined. Roles are not only to be seen as sets of individual agents ponctually employed in organisations : the definition of an organisation relies its structuring roles, often defined by a set of tasks, and associated ressources, or/and obligations, permissions or prohibitions associated to these tasks [6] [8] [1] [7]. Furthermore, since employees of an organisation usually change more often than its structure, the policy update effort ought to be cheaper if it is defined at an organisational level.

The formal framework PEPS [5] has been developped to help users [1] to specify information exchange policies with expected properties. A policy is defined as

---

[1] From here to the end of the paper, *user* will mean : person in charge of information exchange policy definition.

© Springer International Publishing Switzerland 2015
T. Morzy et al. (Eds): ADBIS 2015, CCIS 539, pp. 547–556, 2015.
DOI: 10.1007/978-3-319-23201-0_55

a set of rules stating under which conditions which *agents* have to, are allowed to, or are forbidden to send information to other *agents*. So we want here to provide PEPS with an organisational layer defined upon the existing one, in order to enable users to easily express policies and check properties only at this organisational level for sake of clarity, while taking advantage of PEPS's efficiency.

This paper is organized as follows. Section 2 describes the starting framework PEPS. Section 3 introduces a layer in PEPS to express *organisational* information exchange policies. Section 4 proposes sets of exchange rights transmission axioms between roles or organisations. Section 5 defines some general and specific properties in an organisational context. The last section concludes the paper.

2   PEPS : A Logical Framework to Express and Analyse Agent-Oriented Information Exchange Policies

PEPS [5] has two components : a many-sorted first-order logical language to express information exchange policies between agents, and a SAT-solver to check expected properties on their formal expression. Its main primitive concepts are information, agents, and deontic modalities associated to a *send* action. Given an information exchange policy, PEPS can check pre-defined or domain specific properties, since all of them have been expressed within the PEPS language.

The PEPS language is defined as a quadruple $\{Sort, Var, Fun, \sigma\}$, respectively denoting sorts identifiers, variables identifiers, functions identifiers and signatures. Predefined sorts include : $\mathcal{A}$ (agents), $\mathcal{I}$ (informations), and $\mathcal{T}$ (topics). Predicates and constants are special functions. Predefined predicates include $K(a, i)$ (the agent a knows information i) and $Topic(i, t)$ (information i deals with the topic t). Normative predicates have also been defined : $O_{Send}(a, b, i)$ (respectively : $F_{Send}(a, b, i)$, $P_{Send}(a, b, i)$) means that the agent a is obliged (respectively : forbidden, permitted) to send information i to the agent b. An axiom classically binds obligation and permission to send. As an example, the rule *"Every agent a who knows i dealing with the topic Nuclear is obliged to send i to t he agent Martin"* can be expressed with PEPS as :

$$\forall a, \forall i, Topic(i, Nuclear) \wedge K(a, i) \implies O_{Send}(a, Martin, i) \ .$$

3   PEPS-ORG : A Layer Above PEPS to Express and Analyse Organisational Information Exchange Policies

In order to express in the PEPS framework information exchange rules between roles within organisations in the PEPS framework, we define sorts to express primitive organisational concepts : $\mathcal{O}$ and $\mathcal{R}$ respectively denote organisations and roles . Following [1] we introduce the predicate $Enpower(org, a, r)$ wich means that an organisation *org* enpowers an agent a to the role r. Note that, as in [1], an agent can be empowered to several roles, or in several organisations.

For a given application, a set of PEPS formulas using this *Enpower* predicate will describe a part of the domain knowledge about an organisation.

We also need to express two kinds of concepts : applicative domain constraints about roles in organisations, and information exchange rules between roles.

3.1 Role Constraints

We define constraints about roles and organisations to specify how many individual agents can play a given role, or to precise if some roles can be enpowered by several agents. We introduce the two following predicates for PEPS-ORG :

- *Exclusive*(r, o) means[2] that in the organisation o the role r can be fulfilled by atmost one agent.
- *Incompatible*(r, o, r', o') means[3] that no agent may be employed both to r in o, and to r' in o'. Of course we state that $\neg Incompatible(r, o, r, o)$.

For instance, the constrainst *"nobody can fulfill the roles of SecurityOfficer in both the organisation ON and in an other one"* is expressed as:

$\forall o,\ \ Distinct^{org}(o, ON) \implies Incompatible(SecurityOfficer, ON, SecurityOfficer, o)$

For a given application, Σ^{org} denotes the set of role constraints formulas.

3.2 Organisational Information Exchange Policies

Since PEPS doesn't provide inheritance between sorts, it is not possible to use its normative predicates O_{Send}, P_{Send} and F_{Send}. But the language will gain clarity by using distinct normative predicates for organisational and agent levels.

We introduce normative predicates to exclusively deal with the organisational role level : $O^{org}_{Send}(r, o, r', o', i)$, $P^{org}_{Send}(r, o, r', o', i)$ and $F^{org}_{Send}(r, o, r', o', i)$ respectively express obligation , permission and prohibition for a role r in an organisation o (ie, to agents affected to the role r) to send an information i to a role r' in an organisation o' (ie, to agents affected to the role r'). We choose to specify the organisations r and r' belong to, because we don't have role identifiers unicity hypothesis : the rights [4] defined for a role *Secretary* may be different in two organisations; moreover, it looks legitimate to enable an organisation to specify to which organisation it wants an information to be or not to be sent.

Classically, the following axiom says that every role who is obliged to send an information to another one is also permitted to do it.

$(D^{org})\ \ \ \forall r1, \forall r2, \forall o1, \forall o2, \forall i, O^{org}_{Send}(r1, o1, r2, o2, i) \implies P^{org}_{Send}(r1, o1, r2, o2, i)$

As in [5], an organisational information exchange rule is defined as:

---

[2] *Exclusive*$(r, o) \equiv \forall a, b,\ \ \ Enpower(o, r, a) \land Enpower(o, r, b) \implies \neg Distinct(a, b)$.
[3] *Incompatible*$(r, o, r', o') \equiv \forall a, b,$
  $(Distinct^{org}(o, o') \lor Distinct^{role}(r, r')) \land (Enpower(o, r, a) \land Enpower(o', r', b) \implies$
  $Distinct(a, b))$ where $Distinct^{role}, Distinct^{org}$ and $Distinct$ are predefined predicates.
[4] ie, obligations, permission or prohibitions for roles to send information.

Definition 1 (organisational information exchange rule). *It is a closed* PEPS *formula of any of the following three kinds :*

$$Qx_1, \ldots, Qx_n, \ (\phi \implies O^{org}_{Send}(r_1, o_1, r_2, o_2, i))$$
$$Qx_1, \ldots, Qx_n, \ (\phi \implies P^{org}_{Send}(r_1, o_1, r_2, o_2, i))$$
$$Qx_1, \ldots, Qx_n, \ (\phi \implies F^{org}_{Send}(r_1, o_1, r_2, o_2, i))$$

where :

- *Q is a logical quantifier $\forall$ ou $\exists$;*
- *$x_1, \ldots, x_n$ are all variables of ϕ, r_1, o_1, r_2, o_2, and i ;*
- *ϕ is a formula without any quantifier and any* PEPS-ORG *normative predicate;*
- *r_1, r_2 are terms of the sort $\mathcal{R}$, o_1, o_2 are of the sort $\mathcal{O}$; i is of the sort $\mathcal{I}$.*

Given ON and OJ two organisations respectively employing security officers and journalists, an example of such a rule is :

$$\forall i : \mathcal{I}, \ Topic(i, Nuclear) \implies F^{org}_{Send}(SecurityOfficer, ON, Journalist, OJ, i).$$

We denote PEPS-ORG the restriction of PEPS to the above formula of sections 3.1 and 3.2, only dealing with roles, organisations and information. Note that in PEPS-ORG we cannot express rules which regulate information diffusion between roles and individual agents : if you need it, first wonder if the agents wouldn't denote the role they fulfill, and reconsider the organisation structure.

Definition 2 (organisational information exchange policy). *An organisational information exchange policy OEP is a set of organisational information exchange rules.*

Definition 3 (organisational information exchange policy specification). *An organisational information exchange policy specification is a couple*
$$OEPS = \langle \Sigma, OEP \rangle \ where \ \Sigma \ describes \ all \ constraints \ of \ the \ application$$
domain [5] *such that $\Sigma^{org} \subset \Sigma$* [6]*, and OEP is an organisational information exchange policy, Σ^{org} and OEP being together expressed within* PEPS-ORG.

3.3   Information Transmission Rights Between Roles and Agents

Given an organisational exchange policy expressed within PEPS-ORG, it remains useful to precise how individual agents' diffusion rights are defined from it. We will only consider applications in which the axioms stated below are relevant.

Axiom 1 (Transmission of obligation from roles to agents)
$$\forall a, b, \ \forall r1, r2, \ \forall o1, o2, \ \forall i,$$
$$K(a, i) \wedge Enpower(o1, a, r1) \wedge Enpower(o2, b, r2) \wedge O^{org}_{Send}(r1, o1, r2, o2, i)$$
$$\implies O_{Send}(a, b, i)$$

Similarly, we define axioms for transmission of permission and prohibition from roles to agents. Note that as actions are performed by agents and not by

---

[5] ie, information which are true in the applicative context.
[6] Where Σ^{org} denotes applicative role constraints (see section 3.1, and then 4.2).

roles [6], the K predicate (know) only appears in PEPS formulas, but not in PEPS-ORG.

4 Rights Transmission

4.1 Rights Inheritance Within Organisations in Information Security

Following [2], we will consider two kinds of right transmission : within a role-based hierarchy, and within an organisation-based hierarchy.

The RBAC (Role-Based Access Control) model [8] was introduced to control access rights in organisations, therefore restricted to permissions. An interesting idea is to consider that roles can be ordered within a hierarchy (for instance, an administrative authority relation) and to state that rights are inherited from less preferred roles to more preferred ones (for instance, with an administrative authority relation, let the role *director* be preferred to the role *employee*[7]). This trick spares the burdersome work of defining rights for each role, and makes the formal expression of a policy more concise.

Unfortunately this simplification sometimes gives unexpected results, depending of the nature of the hierarchical order relation [4] [3] [2]. It is irrelevant for both a less technically specialized role and administrative authority preferred role (for instance, a director who would also be a doctor) to inherit for all rights of its more technically specialized but not administrative authority preferred roles (for instance, an employee who would be a surgeon) : directors would inherit rights they might be unable to exploit. Moreover, in [8] preferred roles inherit permissions from all less preferred roles even if useless for their tasks : this violates the cautious separation of powers principle. It also looks strange that preferred roles inherit prohibitions from less preferred ones, in case of some authority order relation.

To overcome these difficulties, [2] suggests to define the direction of permissions and obligations transmission according to the real nature of the order relation between roles : from less specialized to more specialized roles for a technical specialization order, and from "junior" to preferred "senior" roles for an administrative authority relation. The prohibitions transmission direction is similar for a technical specialization order, but it is from senior roles towards junior ones for an administrative authority relation - in order to avoid junior ones to get more power than senior ones. Some works even suggest to more closely control the rights transmission direction [3].

A second transmission process is the organisation-based rights inheritances. [2] suggests that both permissions and prohibitions may be transmitted from an organisation towards its component sub-organisations, when the concerned roles are defined in the sub-organisations.

---

[7] In the case of an administrative authority hierarchical relation, *director* is said to be a *senior role*, compared to *employee* which is said to be a *junior role*

4.2   Exchange Rights Transmission Within Organisations in PEPSORG

Following [2] as a fair compromise between common sense and expressiveness efficiency, we have defined three predicates in PEPS-ORG. Let $r1, r2$ be two roles, and $o, o1, o2$ be organisations,

- $Specializes(o, r1, r2)$ means that $r1$ technically specializes $r2$ in o,
- $Manages(o, r1, r2)$ means that $r1$ has administrative authority on $r2$ in o.
- $Composed(o1, o2)$ means that $o2$ is a component organisation of $o1$.

For a given application, formulas built with $Specializes$ and $Manages$ respectively define technical specialization and administrative authority orders between roles. Formulas built with $Composed$ define organisational composition order between organisations. All of them are added to Σ^{org} as context role constraints.

In order to cope with application specificities and needs, we propose the following axioms. Let $r1, r2, r3, r'$ be roles, and $o, o', o1, o2, o3$ be organisations :

- For information transmission permission :

Axiom 2 $\forall r1, r2, r', \ \forall o, o',$
$Specializes(o, r1, r2) \wedge P_{Send}^{org}(r2, o, r', o', i) \implies P_{Send}^{org}(r1, o, r', o', i)$

Axiom 3 $\forall r1, r2, r', \ \forall o, o',$
$Manages(o, r1, r2) \wedge P_{Send}^{org}(r2, o, r', o', i) \implies P_{Send}^{org}(r1, o, r', o', i)$

and, supposing that roles defined in $o1$ are also defined in $o2$:

Axiom 4 $\forall o1, o2, o3,$
$\forall r, r3, Composed(o1, o2) \wedge P_{Send}^{org}(r, o1, r3, o3, i) \implies P_{Send}^{org}(r, o2, r3, o3, i)$

- For information transmission prohibition :

Axiom 5 $\forall r1, r2, r, \ \forall o, o',$
$Specializes(o, r1, r2) \wedge F_{Send}^{org}(r2, o, r', o', i) \implies F_{Send}^{org}(r1, o, r', o', i)$

Axiom 6 $\forall r1, r2, r, \ \forall o, o',$
$Manages(o, r1, r2) \wedge F_{Send}^{org}(r1, o, r', o', i) \wedge \neg Specializes(o, r1, r2)$
$\implies F_{Send}^{org}(r2, o, r', o', i)$

and, supposing that roles defined in $o1$ are also defined in $o2$:

Axiom 7 $\forall o1, o2, o3, \ \forall r, r3,$
$Composed(o1, o2) \wedge F_{Send}^{org}(r, o1, r3, o3, i) \implies F_{Send}^{org}(r, o2, r3, o3, i)$

For a given application, we propose to the user to choose a set (H^{org}) of axioms among 2...7 for role-based[8] and organisation-based hierarchies.

---

[8] Note that for information transmission prohibition, in case of both technical specialization and administrative authority orders defined between two roles, priority is given to the technical specialization relation.

5 Properties for Organisational Information Exchange Policies

When an organisational exchange policy *OEP* can be expressed within PEPS-ORG, it looks more efficient to analyse its expression within PEPS-ORG because the number of entities (here, roles and organisations) to be investigated by PEPS's solver is a priori smaller than the number of agents to be considered at PEPS level. A policy analysis here consists in checking with the PEPS solver some expected properties expressed within PEPS-ORG . Following [5] we have formally redefined some general properties (inconsistency, completeness, applicability and minimality); we have also defined one more specific property in order to prevent some kinds of unauthorized information leaks in organisations.

5.1 Inconsistency

Following [11], we consider that $OEPS = \langle \Sigma, OEP \rangle$ is inconsistent if there might be a situation such that it would be both permitted and forbidden for an agent to send an information to an other one, because its role(s). Either inconsistency is inherent to information exchange rules defined in *OEP* for the sender role, or it may occur when it is possible for an agent to fulfill two distinct sender roles.

Definition 4 (O_{role}-inconsistency).   An organisational exchange policy specification *OEPS* is O_{role}-inconsistent if $\exists o1, o2\ \exists r1, r2\ \exists i, \exists \phi$,
$$(\Sigma, OEP, (D^{org}), (H^{org}), \phi) \models P_{Send}^{org}(r1, o1, r2, o2, i) \land F_{Send}^{org}(r1, o1, r2, o2, i)$$
where ϕ is a PEPS formula without any normative predicate, which describes a kind of possible situation.

It is the policy specifier's task to state if such an exhibed kind of situation ϕ is realistic in the considered application. If it is, any agent fulfilling $r1$ in $o1$ would have to face up a dilemma if it knew i: unless some priorities device between conflicting rules has been specified in order to solve the conflict [10], *OEP* has to be modified in order to get consistency.

Definition 5 (O_{orga}-inconsistency). An organisational exchange policy specification *OEPS* is O_{orga}-inconsistent if $\exists o1, o2, o3\ \exists r1, r2, r3\ \exists i, \exists \phi$,
$$(\Sigma, OEP, (D^{org}), (H^{org}), \phi) \models$$
$$P_{Send}^{org}(r1, o1, r3, o3, i) \land F_{Send}^{org}(r2, o2, r3, o3, i) \land \neg Incompatible(r1, o1, r2, o2)$$
where ϕ is a PEPS formula without any normative predicate, which describes a kind of possible situation.

In other words, any agent fulfilling the roles $r1$ in $o1$ and $r2$ in $o2$ in a situation corresponding to ϕ would have to face up with a dilemma if it knew i. In order to restore consistency, one can either add a domain role constraint in Σ (more exactly, in Σ^{org}) to make both $r1$ in $o1$ and $r2$ in $o2$ incompatible, or modify *OEP*, or definitively prevent any occurence of ϕ if possible.

5.2 Completeness

An exchange policy specification is complete if in any cases, it specifies if agents are obliged, permitted or prohibited to send information to others as far as they know it : it is often expensive and useless to require this absolute property. For a given organisation, it is relevant to limit the scope of completeness to topics of interest [9]. So we will first define a very restricted version of completeness : for a given role of an organisation , OEPS is complete for a given topic if it is complete for information dealing with this topic.

Let Ω be a set of organisations for which OEPS has been specified, o be an organisation of Ω, and T be a given topic (constant of the sort $\mathcal{T}$).

Definition 6 ($O_{role} - T$-completeness).   OEPS is $O_{role} - T$-complete in Ω for the role r in the organisation o if

$\forall o' \in \Omega, \ \forall r', \ \forall i, \ (\Sigma, OEP, (D^{org}), (H^{org})) \models$
$(Topic(i, T) \implies O_{Send}^{org}(r, o, r', o', i) \vee P_{Send}^{org}(r, o, r', o', i) \vee F_{Send}^{org}(r, o, r', o', i))$

Suppose now that Ω shares a domain-specific ontology Θ of topics.

Definition 7 ($O_{role} - \Theta$-completeness).   OEPS is $O_{role} - \Theta$-complete in Ω for the role r in o, if for each topic of Θ, it is $O_{role} - T$-complete in Ω for r in o.

Definition 8 ($O_{orga} - \Theta$-completeness).   OEPS is $O_{orga} - \Theta$-complete in Ω for the organisation o, if it is $O_{role} - \Theta$-complete in Ω for each role r in o.

5.3 Information Permeability Through Organisations

We want to detect at the PEPS-ORG layer the possibility of unauthorized diffusion through organisations, because of the compatibility of two roles in two organisations.

Let Ω be a set of organisations for which an organisational information exchange policy specification OEPS has been defined.

Definition 9 (O-information permeability).
OEPS is O-information permeable for Ω if, though it is not O_{role}-inconsistent,

$\exists i, \ \exists o_1 \in \Omega, \exists o_n \in \Omega, \ \exists r_1, r_n,$
$(\Sigma, OEP, (D^{org}), (H^{org})) \models F_{Send}^{org}(r_1, o_1, r_n, o_n, i) \wedge TransP_{Send}^{org}(r_1, o_1, r_n, o_n, i)$
$TransP_{Send}^{org}(r_1, o_1, r_n, o_n, i)$ means that, under OEPS, i could be transmitted from r_1 in o_1 to r_n in o_n thanks to some relays between organisations. $TransP_{Send}^{org}$ is step by step defined as follows :

1. Authorized tranmission of information within an organisation: $TransInterne^{org}$
    $\forall o, \forall r, r', \forall i,$
    $TransInterne^{org}(o, r, r, i)$
    $P_{Send}^{org}(r, o, r', o, i) \implies TransInterne^{org}(o, r, r', i)$
    $P_{Send}^{org}(r, o, x, o, i) \wedge TransInterne^{org}(o, x, r', i) \implies TransInterne^{org}(o, r, r', i)$
    ie, that according to the policy, the transmission of i is step by step allowed from the role r to the role r' inside the organisation o.

2. Exhibition of an information link between two organisations : $Link^{org}$

$\forall o, o', \quad \forall r, r', \quad \forall i,$

$\neg Incompatible(r, o, r', o') \lor P^{org}_{Send}(r, o, r', o', i) \implies Link^{org}(r, o, r', o', i)$

ie, i could be sent from o to o', either because an agent who would not clearly separate its functions could play roles in both organisations, or because this transmission is permitted.

3. Authorized information diffusion within and between some organisations : $TransP^{org}_{Send}$

$\forall o, o_1, o_2, o_n, \quad \forall rs, rr, rs_1, rr_1, rs_2, rr_2, rs_n, rr_n, \quad \forall i,$

$TransInterne^{org}(o, rs, rr, i) \implies TransP^{org}_{Send}(rs, o, rr, o, i)$

$TransInterne^{org}(o_1, rs_1, rr_1, i) \land Link^{org}(rr_1, o_1, rs_2, o_2, i) \land TransP^{org}_{Send}(rs_2, o_2, rr_n, o_n, i)$

$\implies TransP^{org}_{Send}(rs_1, o_1, rr_n, o_n, i)$

ie, i could be send from a first sender role rs_1 in o_1 to a last recipient role rr_n in o_n, via some obligation or permission exchange rules.

In order to prevent such an information leak to occur, one can either make incompatible some roles between organisations of Ω, thus modifying the domain constraints in Σ^{org}, or change the organisational information exchange policy OEP.

So we can reason within PEPS about possibility of information leak without doing it on individual agents belonging to organisations, and we can then propose beforehand organisational solutions.

6   Conclusion and Future Works

Critical information system are shared among or within organisations. We have explained why it would be more efficient to define their information exchange policies at an organisational level.

Starting from the PEPS framework, we have developped PEPS-ORG, an independant layer above the native language, to express and analyse exchange policies specifications at this organisational level. Main points of our contribution are : first, applicative domain constraints about roles in organisations, and organisational exchange policies can be both distinctly expressed within this layer. Second, expression of policies can be more concise by using some sets of right transmission axioms between roles or organisations, according to the application. The PEPS-ORG langage could yet be improved : for instance, to express temporal constraints for role acting .

Within PEPS-ORG, we have defined general and classical properties about policies, but also a more specific property about information leak. We expect to take advantage of the PEPS solver to easily analyse organisational policies : this remains to be experimented. Our approach is careful : if a property is not verified by an organisational policy, we can suggest to modify the policy, or some constraints in the organisation, in order to get the expected property verified; but in fact, we cannot know if a real information will really cause a pessimistic situation with violated expected properties.

More than an enrichment for the PEPS language, we have proposed a methodology to formalize information exchange through organisation, which ought to make users wonder on the real nature of the organisation structure, then getting more relevant and precise definitions of organisations.

Acknowledgments. We thank Laurence Cholvy, Rémi Delmas and Thomas Polacsek for their comments and suggestions which helped us to improve this paper. This work was funded by the ONERA project *MAPEIS*.

References

1. Abou El Kalam, A., Benferhat, S., Miège, A., El Baida, R., Cuppens, F., Saurel, C., Balbiani, P., Deswarte, Y., Trouessin, G.: Organization based access control. In: IEEE 4th International Workshop on Policies for Distributed Systems and Networks (Policy 2003), Lake Come, Italy, June 4–6, 2003
2. Cuppens, F., Cuppens-Boulahia, N., Miège, A.: Héritage de privilèges dans le modèle Or-BAC : application dans un environnement réseau. In: SSTIC 2004 : Symposium sur la Sécurité des Technologies de l'information et des Communications (2004)
3. Feldmeier, C.J.: Limiting hierarchical inheritance of permissions in access control models. In: ISA 767 Secure Electronic Commerce (2006)
4. Crampton, J.: On permissions, inheritance and role hierarchies. In: 10th ACM Conference on Computer and Communication Security (2003)
5. Delmas, R., Polacsek, T.: Formal methods for exchange policy specification. In: Salinesi, C., Norrie, M.C., Pastor, Ó. (eds.) CAiSE 2013. LNCS, vol. 7908, pp. 288–303. Springer, Heidelberg (2013)
6. Pacheco, O., Carmo, J.: A Role Based Model for the normative specification of organized collective agency and agent interaction. Journal of Autonomous Agents and Multi-Agent Systems **6**, 145–184 (2003)
7. Glassey, O., Chappelet, J.L.: Comparaison de trois techniques de modélisation de processus : ADONIS, OSSAD et UML. Working paper of l'IDHEAP/14, UER Management public-Systèmes d'informations, Lausanne (2002)
8. Sandhu, R., Coyne, E.J., Feinstein, H.L., Youman, C.E.: Role-based access control models. IEEE Computer **29**(2), 38–47 (1996)
9. Cholvy, L., Garion, C., Saurel, C.: Information sharing policies for coalition systems. In: NATO RTO-IST Symposium on Dynamic Communications management, Budapest, October 2006
10. Cuppens, F., Cholvy, L., Saurel, C., Carrère, J.: Merging regulations: analysis of a practical example. In: Data and Knowledge Fusion, Special issue of International Journal of Intelligent Systems, vol. 16. J. Wiley and Sons Pub. (2001)
11. Cholvy, L., Cuppens, F.: Analyzing Consistency of Security Policies. IEEE Symposium on Security and Privacy, pp. 103–112 (1997)

Critical Information Diffusion Systems

Rémi Delmas$^{(\boxtimes)}$ and Thomas Polacsek

ONERA, 2 avenue Édouard Belin, 31055 Toulouse, France
{remi.delmas,thomas.polacsek}@onera.fr

Abstract. Today, individuals, companies, organizations and national agencies are increasingly interconnected, forming complex and decentralized information systems. In some of these systems, the very fact of exchanging information can constitute a safety critical concern in itself. Take for instance the prevention of natural disasters. We have a set of actors who share their observation data and information in order to better manage crises by warning the more competent authorities. The aim of this article is to find a definition to such kind of systems we name Critical Information Diffusion Systems. In addition, we see why Critical Information Diffusion Systems need information exchange policies.

Keywords: Information system · Information exchange policies · Alarm system · Critical · CIDS · Critical information diffusion system

1 Introduction

In this paper, we focus on a category of applications and systems in which agents exchange information and, sometimes, a particular kind of information: an alert (or we use, sometimes, the word alarm). An alert is a piece of information representing the fact that an imminent danger has been identified, possibly accompanied with context information detailing the nature of danger. As we will see, the fact that an alert is not sent to the right agent or group of agents, is sent with insufficient context elements, or is not sent at all, can have critical consequences.

A first example of systems where agents share information and where alarms are generated is collaborative *Space Situation Awareness* (SSA) systems [6]. In these systems, space observation capabilities belonging to different nations, agencies and satellite operators are mutualized in order to build a complex data and information gathering, analysis and alert diffusion system. The mission of an SSA system is to warn relevant agents when situations of potential collision between orbiting objects, or between orbiting objects and space junk, are detected. The investment to build and launch a satellite is in tens or hundreds of millions of dollars, which is a strong incentive to avoid space collisions. An SSA system must, in case of potential collision, send relevant alerts and associated information to the right agents (nations, satellite operators) so as to allow them to avoid the collision.

© Springer International Publishing Switzerland 2015
T. Morzy et al. (Eds): ADBIS 2015, CCIS 539, pp. 557–566, 2015.
DOI: 10.1007/978-3-319-23201-0_56

We can also consider information sharing in the context of *Global Earth Observation and Surveillance Systems* (GEOSS). One of the many possible applications of an Earth observation system is the surveillance, monitoring and management of geohazards. In this case, we have a collaboration between satellite data providers, seismic data providers, geohazard scientific communities and geohazard monitoring agencies, forming a global Earth-wide network. The main goal of such a system is to ensure that information about natural disasters will always reach the relevant authorities in the right location. The role of these relevant authorities is to identify risks from the collected data so that population protection measures can be taken in due time.

Disease and Health Surveillance Systems (DHSS) are also worth considering. Here, field data, such as disease case reports, is collected and sent to epidemiology experts who monitor it and try to predict possible epidemic or pandemic situations. When epidemic situations become likely to happen, experts alert the competent authorities, like intergovernmental organizations or international bodies, which assess the severity of the identified risk and take appropriate sanitary and medical countermeasures. In areas struck by poverty, infections spread rapidly and it can be extremely difficult to contain them. In an age where mass transportation systems allow people to reach any point of the globe within 24h, health issues have become global, which requires governments and health agencies to use a global approach based on information sharing, to be able to quickly issue public health recommendations in response to the risk [12].

Far from crisis prevention and management, we find interconnected systems in enterprises. Take for instance the *Extended Enterprise* (EE) context [4] [9], which denotes an association of firms sharing a common business objective, usually the manufacture or the marketing of a complex product, and which, through alliances such as consortiums, share resources and knowledge to design and build the product. In this context, agents are stakeholders, large firms and their subcontractors, etc. which must exchange sensitive information such as advanced specifications, designs models, financial data, etc. Such exchanges occur through interconnected *Information Systems* (IS), federated technical platforms or even, sometimes, through more trivial means like simple file exchanges between individuals. In addition, each firm can be involved in several subsystems based on agreements, partnerships, subcontracting, or by being subsidiaries of different industrial groups, etc. In such a context, modifications of certain data, such as a product specification, a design model, validation results, etc. must be communicated to the relevant partners to guarantee the success of the enterprise.

2   Information Systems with Diffusion Requirements

Our presentation of space awareness, geohazard, disease surveillance and extended enterprise systems immediately reveals a common and salient feature: some designated agents of the system, have particular information exchange needs: they must acquire all available information related to a given topic in order to be able to accomplish their mission properly, and acquiring only a partial knowledge of an ongoing situation can greatly impair the performance of the system as a whole. Furthermore, it is not a fixed agent or group of agents which must be kept informed, but rather a specific agent or group of agents which can be different depending on the information and context. For instance, in the SSA system, the relevant agent for a given alert is the operator in charge of changing a given satellite's trajectory to avoid a collision. Since many companies operate satellites, the relevant agent can be different for each alarm. In GEOSS systems, data acquired by a given nation must reach the competent authorities in the right geographic location. The extended enterprise can also be geographically spread, and responsibilities over the whole design and manufacturing process are distributed among partners. Therefore, we are dealing with systems where agents can exchange information and where a specific context-dependent agent or group of agents must absolutely be warned depending on the situation: any information that is acquired by another agent (a kind of information we call alarm for convenience) must be communicated to the relevant agent or group of agents. We argue that all these requirements are actually instances of a generic *need-to-know* requirement. The need-to-know concept captures the idea that "agents of the system must know certain information to be able to fulfil their respective missions". However, we propose to reformulate this requirement in an *information diffusion* requirement, the *need-to-share*, which captures the notion that "if information is needed by an agent and another agent possesses this information, then the latter must provide it to the former". This formulation stresses the liveness aspect of information sharing: The need-to-know of some agents results in a requirement for other agents, which are then in charge of timely information delivery.

The information systems we discussed in the examples also have in common their heterogeneous nature and decentralized architecture. Heterogeneous in the sense the agents of the system can be of different kinds: nations, agencies, non-governmental organizations, individuals, *etc.*, can be in charge of different missions requiring to acquire specific kinds of information and can operate under possibly different sets of information sharing rules; Decentralized in the sense that no single agent of the system can acquire a truly total knowledge of all information that exists in the system at any given point in time.

3 The Need for Confidentiality

In the previous section we identified a first salient characteristic of the information systems we consider: the need-to-share requirements, according to which designated agents must absolutely be provided with information related to some topic of interest to fulfil their mission. In a paradoxical way, totally free, unrestricted information sharing in such systems is almost never the case in practice. The different parties cooperating in system must respect their own confidentiality constraints, to prevent any risk of leakage of private, sensitive or even strategic information.

SSA systems are a very obvious instance of that problem. Space, once only occupied by a handful of pioneers, has now become a shared resource populated by satellites belonging to an increasing number of nations and companies with potentially conflicting objectives and interests. Consequently, such organizations can be reluctant to share information in an globalized SSA system which could reveal the exact nature of objects orbiting the Earth, of their manœuvre capabilities and trajectories, regarded as strategic information by their owners. Moreover, space observation means are most often owned by national agencies or by military forces and regarded as strategic assets. This makes their owners reluctant to share information which could directly or indirectly reveal their real capabilities and performance. Today most experts of the SSA problem agree on the idea that plenty of space observation data is already available, and that it is really the lack of data and information sharing agreements between nations which prevents a better level of space situational awareness and management of collision risks. It is very likely that SSA will remain in its current state unless means of preserving confidentiality while diffusing alarms are found, sufficiently convincing to lessen the reluctance of stakeholders to participate in such global cooperations.

Earth observation and geohazard monitoring is also a global issue, which involves many nations all over the globe. In addition to notifying the competent authority when they detect a risk, agents of a GEOSS system must not communicate this information to the general public, mainly to avoid mass panic movements which could be harmful to the population. The same issue arises in disease and health monitoring systems. With the proliferation of social media, improper information release to the public, or the lack of official public recommendations leaves the gates wide open to rumours and misinformation, which can result in hazardous consequences. These constraints can be viewed as a form of confidentiality the system must fulfil. Moreover, fringe health surveillance and disease surveillance systems involve most of the time intelligence agencies or/and military forces, which purpose is to detect outbreaks attributable to biologic terrorism [14] [16]. The information circulating in such systems acquires a strategic importance and confidentiality requirements also apply.

Confidentiality constraints are also stringent in the extended enterprise context. Almost all industrial cooperations are run under non-disclosure agreements, competitors can be forced to interact and share design data and products as subcontractors of a same stakeholder, different subsidiaries of a same company can even be competing against each other in some markets. In the smartphone design and manufacturing domain for instance, a company S can market its own product, while providing a competitor A with various essential parts essential for the manufacturing of a competing product. So agents operating in an extended enterprise context must typically obey different sets of information sharing restrictions, which can be contradictory, incomplete or redundant, and makes the information management a really complex issue.

4   Critical Information Diffusion Systems

We propose to regroup the systems discussed in the previous sections under the term *Critical Information Diffusion System* (CIDS): decentralized information exchange systems consisting of agents, possibly heterogeneous, where some exchanges are of critical nature to some agents (safety critical, strategically critical, economically critical, *etc.*) hence becoming mandatory to allow the fulfilment of their missions, and in which confidentiality requirements, potentially conflicting the critical diffusion requirements, must also be satisfied.

We use the term "critical" in its nuclear, aeronautics or rail domain acceptation. A requirement is critical if failing to satisfy it (here, keeping a designated agent aware of certain information) makes it impossible to avoid unwanted, hazardous or catastrophic consequences, such as satellite destruction by collision, the violation of strategic secrets in the case of SSA, or the rapid spread of an epidemic threatening large numbers of human lives in the case of disease monitoring systems. We can extend the notion of criticality to encompass risks of economic and financial disruptions, which can result in important social issues in the society such as the pauperization which can result from job loss. In that case, information diffusion in the extended enterprise can also be viewed as critical, in that they condition the efficiency of the enterprise and its economic success. Therefore, from now on we consider information systems where diffusion is regarded as critical *in some way*, that is, where failing to properly exchange information can have important unwanted and harmful consequences.

We think that failing to handle properly the conflicts arising from having to handle antagonist diffusion (need-to-share) and confidentiality requirements in a same system can constitute a genuine obstacle to the adoption and deployment of such systems by nations, authorities, enterprises. On a side note, in systems with no confidentiality requirements, systematic information broadcasting becomes a simple way to meet the need-to-share requirement. However such open and unrestricted systems are seldom encountered in practice.

5   Information Exchange Policies for CIDS

So, the challenge when designing CIDS is to reconcile two antagonist needs: firstly, ensuring that agents always receive the information they need to perform their designated mission; Secondly, ensuring that no sensitive information will be released in an uncontrolled manner. For that purpose, it is necessary to be able to express clearly and unambiguously these requirements. In the context of CIDS, we must express under which conditions an agent has the obligation, the permission, or the interdiction to communicate information to other agents in the system. Therefore, we can draw a parallel between specifying *whether an agent has the right to communicate or not* and expressing *system security policies*, more precisely *access control policies*.

5.1   Access Control Policies

The *DoD* Orange Book [1] gives the following definition of a security policy: *"set of rules that are used by the system to determine whether a given subject can be permitted to gain access to a specific object."*. This definition is very general, and if we consider IS, we can focus on the access control, basically the regulation of access of the agents to information sources such as databases, files, network links, etc. But access control framework are typically focused on specifying permissions, i.e. what an agent *can do*, whereas in CIDS we also have the *obligation to do*: the obligation to warm the relevant authorities, which is used to capture the liveness requirement of the information flow.

Besides access control, some works add normative aspects to security policy model [3] [13] [2]. But, even though they express the obligation, the permission and the prohibition for a subject to perform an action on a given object, they only consider actions of an agent on an object. Policies hence regulate actions corresponding to transitive divalent verbs, for instance: "agent A shall not read database D". In the context of CIDS, we are not concerned with regulating actions in general, but rather by specific actions corresponding to trivalent verbs such "exchange something with someone", "tell someone something", "give someone something", "send someone something". Since existing access control policy frameworks focus on divalent verbs, it is not possible to capture typical CIDS diffusion requirements with them.

5.2 PEPS: Information Exchange Policies Specification and Analysis

To manage information sharing in CIDS, we propose the use of formal information exchange policies. Based on core concepts similar to that of security policies, an information exchange policy specifies the set of rules that regulates how information must be exchanged in the CIDS. We also support an early use of formal methods in the CIDS design process to validate the policy specifications. A formal specification framework enables early and automatic verification tasks aiming at discovering logical inconsistencies or incompleteness issues (and other logical issues discussed later), in a collection of complex requirements. It is with these formalization and verification goals in mind we undertook the design and implementation of an information diffusion policy specification framework, consisting of a formal specification language named PEPS[1] , accompanied by a tool named PEPS-analyzer for automatic semantic analysis. Indeed, the later specification errors are discovered, the costlier they are to fix, and the greater are the risks of either violating critical requirements or allowing unwanted information diffusion. Both the language and the tool are fully described in [7] [8], we only provide an quick overview of their features in the next paragraphs.

The PEPS Language. Deontic logic is a formal framework dedicated to the expression of obligation and related concepts and to reasoning on them [5]. To enable this, deontic logic defines the notion of *obligation* through a modal operator O. The logical formula Op means "it ought to be that the proposition p is true." A key property of this type of logic is that a proposition may be required to hold by the norms, without actually holding in the real world. For example, in France, although it is forbidden to communicate election results before all polling

---

[1] PEPS is a recursive acronym for *Peps for Exchange Policy Specification*

places are closed, some media do. In addition to the notion of obligation, come the notions of *permission* and *prohibition*. In standard deontic logic, permission and prohibition are directly defined in terms of obligation: Asserting "p is permitted", is equivalent to asserting that "it is not obligatory that p is false", and asserting "p is forbidden" is equivalent to asserting that "it is obligatory that p is false". In addition, we have the natural property that "if p is obligatory then p is allowed" which corresponds to the D axiom in deontic logic.

However, the use of deontic logic, and modal logics in general, can be perceived as rather cumbersome to use to express requirements by system designers. Indeed, as McCarthy said in [15] "If modal logic helps to manipulate concepts, it is difficult to handle by ordinary people and remains a tool limited only to logicians". Furthering this argument, Edmonds explains in [10] that formal logical models are unusable by people given the complexity of *formal logical languages* for non-logicians. One might however object that only natural language can be understood by everybody, and even this is not true, given all issues related to natural language ambiguities.

Away from the controversy, we choose to model information exchange policies for CIDS in many-sorted first order logic [11], for two reasons: (1) MSFOL is formal language yet relatively easy to use for requirements expression for average users (compared to modal logics) (2) MSFOL benefits from a great support from automatic theorem provers, SMT-solvers in particular, a kind of fully automatic and highly effective theorem provers requiring very little user interaction. At the present time, modal logic provers are less efficient than traditional logic provers such as SAT solvers or SMT-solvers [18].

The PEPS Framework and PEPS-analyzer Tool. If the PEPS language allows us to specify requirements without ambiguity from the early stages of information diffusion systems design, the main added value of using PEPS lies in the associated PEPS-analyzer tool. This tool offers an interactive support to policy designers for the debugging, verification and validation of policies. Several generic formal properties are built into the analyzer, capturing the notions of inconsistency, incompleteness and redundancy. These properties are automatically instantiated and analyzed for the provided policy to yield verification conditions, which are then analyzed by an SMT-solver. When a property is falsified, the tool presents an automatically generated counter-example to the designer. A counter-example is a concrete scenario which assigns values to the specification variables and is a significant help to locate and understand the cause of the problem, fix it and ultimately obtain a policy specification that is logically sound.

A policy is *inconsistent* if cases exists in which both the obligation (or permission) and interdiction to send a given piece of information to a target agent apply to a same source agent. Its dual property, *incompleteness*, characterizes situations in which the policy does not tell an agent what to do with information it possesses with respect to another agent. Other generic properties such as the *applicability* of rules and the *minimality* of the specification help identifying over-specification and rule redundancy cases. Please refer to [7] for full details on these properties.

Last, *awareness* properties formalize the notion that a designated agent (or group of agents) must be informed by other agents on designated information topics, and the properties of *restriction* of certain information topics with respect to designated agents (or groups of agents) allow to encode confidentiality requirements. Verifying awareness and restriction properties on a specification allows to discover automatically cases, if any, in which the diffusion and restriction properties contradict each other. The PEPS language was recently extended with the notion of information filtering operator, which can be used to resolve contradictions between awareness and restriction requirements. Please refer to [8] for full details. Together, awareness and restriction properties correspond to the core defining features of CIDS identified in this paper.

6   Branching out to other Domains

We are conscious that it is easy to make mistake to consider that if all you have is a hammer, everything looks like a nail. However in this section we try to show, through examples derived from the social networks, that the specific issues identified for CIDS in the previous sections can be encountered in a broader category of information system.

Recently, the micro-blogging platform Twitter announced a new alert diffusion service[2], purposely designed to help the police, emergency services, government organization, non-profit organizations, *etc.* diffuse safety critical alerts to the population. We can imagine cases in which one or more individuals in possession of some disaster-related information will contact a governmental agency, which will in turn decide or not, according to some complex criterion, to use the Twitter alert service to warn some other agency or the population. Once an alert is released in the wild, its diffusion can continue without control, by individuals themselves. Examples of misuse of such mass-diffusion systems have already been witnessed in the past (fake child abduction alerts, fake imminent tsunami rumours, *etc.*), and one can legitimately ask the question of how such mass-reaching information systems will be managed, regulated and protected against misuse: drowning such systems, initially meant to enhance the management of certain situations, in uncontrolled noise can effectively render them either useless or worse, harmful. The characteristics of this problem fit the CIDS notion introduced in this paper: In that case we view the blogging platform as a part of a larger system which is itself a CIDS, used as a means of warning concerned another category of agents, the people, with critical information.

Still illustrating the various situations in which required diffusion acquires a critical nature, we can cite [17], in which the authors describe how accounts on a major social network can be compromised, by exploiting obsolete web-based email addresses used as authentication tokens by the social network in question. Once identified, obsolete web-mail addresses (which are available to new users

---

[2] https://about.twitter.com/products/alerts

again) are reactivated by the attacker, who thereby gains total control over the email account. Then, a password reset is initiated on the social network login page, the new password being sent to the compromised email address, resulting in the attacker gaining total control over the social network account. Infiltration techniques such as this one could be somehow mitigated if users could specify that the email expiration event must result in a warning message being sent to all services using this email as authentication token to inform them not to trust this email anymore. This also fits the CIDS concepts since failing to meet an information diffusion requirement entails unwanted consequences such as identity theft, which can become a critical issue.

7   Conclusion

In this paper, after identifying the category of Critical Information Diffusion Systems (CIDS), which for the large part underpin alarm systems, we gave a brief reminder about the formal information exchange policy specification framework PEPS.

So far, we focused on cores concepts of exchange policies and we would like to investigate the modelling of organisations involved in the policies in more details. We will determine if some ideas from OrBAC (Organisation Based Access Control) theory like roles can be transposed in our framework while preserving the core properties. Technically speaking, such extensions are not a problem because PEPS is extensible by nature, the real question will be to find modelling concepts adapted to the problematic posed by CIDS.

Another very important question is that of the implementation of such policies. If we provide ways to ensure that an exchange policy specification is correct, we must also guarantee that these properties are preserved by the implementation. Research must be conducted to determine how essential properties of policies are retained once the policy is implemented on information systems.

References

1. Department Of Defense Standard Department Of Defense Trusted Computer System Evaluation Criteria (1985)
2. Barhamgi, M., Benslimane, D., Oulmakhzoune, S., Cuppens-Boulahia, N., Cuppens, F., Mrissa, M., Taktak, H.: Secure and privacy-preserving execution model for data services. In: Salinesi, C., Norrie, M.C., Pastor, Ó. (eds.) CAiSE 2013. LNCS, vol. 7908, pp. 35–50. Springer, Heidelberg (2013)
3. Bieber, P., Cuppens, F.: Expression of confidentiality policies with deontic logic. In: Meyer, J.J.C., Wieringa, R.J. (eds.) Deontic Logic in Computer Science: Normative System Specification, pp. 103–123. John Wiley & Sons, Inc. (1993)
4. Browne, J., Zhang, J.: Extended and virtual enterprises similarities and differences. International Journal of Agile Management Systems 1(1), 30–36 (1999)
5. Castaneda, H.N.: Thinking and doing. D. Reidel, Dordrecht (1975)
6. Dal Bello, B.R.: Managing risk in space. Federation of American Scientists, Public Interest Report, Winter (2011)

7. Delmas, R., Polacsek, T.: Formal methods for exchange policy specification. In: Salinesi, C., Norrie, M.C., Pastor, Ó. (eds.) CAiSE 2013. LNCS, vol. 7908, pp. 288–303. Springer, Heidelberg (2013)
8. Delmas, R., Polacsek, T.: Need-to-share & non-diffusion requirements verification in exchange policies. In: Zdravkovic, J., Kirikova, M., Johannesson, P. (eds.) CAiSE 2015. LNCS, vol. 9097, pp. 151–165. Springer, Heidelberg (2015)
9. Dyer, J.H., Singh, H.: The Relational View: Cooperative Strategy and Sources of Interorganizational Competitive Advantage. The Academy of Management Review 23(4), 660–679 (1998)
10. Edmonds, B.: How formal logic can fail to be useful for modelling or designing MAS. In: Lindemann, G., Moldt, D., Paolucci, M. (eds.) RASTA 2002. LNCS (LNAI), vol. 2934, pp. 1–15. Springer, Heidelberg (2004)
11. Gallier, J.H.: Logic for Computer Science: Foundations of Automatic Theorem Proving, chap. 10, pp. 448–476. Wiley (1987)
12. Heymann, D.L., Rodier, G.R.: Hot spots in a wired world: {WHO} surveillance of emerging and re-emerging infectious diseases. The Lancet Infectious Diseases 1(5), 345–353 (2001)
13. Kalam, A.A.E., Baida, R.E., Balbiani, P., Benferhat, S., Cuppens, F., Deswarte, Y., Miege, A., Saurel, C., Trouessin, G.: Organization based access control. In: IEEE 4th International Workshop on Policies for Distributed Systems and Networks Proceedings, POLICY 2003. IEEE (2003)
14. Mandl, K.D., Overhage, J., Wagner, M.M., Lober, W.B., Sebastiani, P., Mostashari, F., Pavlin, J.A., Gesteland, P.H., Treadwell, T., Koski, E., Hutwagner, L., Buckeridge, D.L., Aller, R.D., Grannis, S.: Implementing syndromic surveillance: a practical guide informed by the early experience. Journal of the American Medical Informatics Association 11, 141–150 (2004)
15. McCarthy, J.: Modality, si! modal logic, no!. Studia Logica 59(1), 29–32 (1997)
16. Meynard, J., Chaudet, H., Texier, G., Ardillon, V., Ravachol, F., Deparis, X., Jefferson, H., Dussart, P., Morvan, J., Boutin, J.: Value of syndromic surveillance within the armed forces for early warning during a dengue fever outbreak in french guiana in 2006. BMC Med. Inf. & Decision Making 8, 29 (2008)
17. Parwani, T., Kholoussi, R., Karras, P.: How to hack into facebook without being a hacker. In: Carr, L., Laender, A.H.F., Lóscio, B.F., King, I., Fontoura, M., Vrandecic, D., Aroyo, L., de Oliveira, J.P.M., Lima, F., Wilde, E. (eds.) WWW (Companion Volume), pp. 751–754. International World Wide Web Conferences Steering Committee / ACM (2013)
18. Sebastiani, R., Vescovi, M.: Automated reasoning in modal and description logics via sat encoding: the case study of k(m)/alc-satisfiability. J. Artif. Intell. Res. (JAIR) 35, 343–389 (2009)

A Case Study on the Influence of the User Profile Enrichment on Buzz Propagation in Social Media: Experiments on *Delicious*

Manel Mezghani[1,2(✉)], Sirinya On-at[2], André Péninou[2], Marie-Françoise Canut[2], Corinne Amel Zayani[1], Ikram Amous[1], and Florence Sedes[2]

[1] Department of Computer Science, Sfax University, MIRACL Laboratory, Sfax, Tunisia
mezghani.manel@gmail.com,
{corinne.zayani,ikram.amous}@isecs.rnu.tn
[2] Toulouse Institute of Computer Science Research (IRIT), University of Toulouse,
CNRS, INPT, UPS, UT1, UT2J, 31062, Toulouse Cedex 9, France
{sirinya.on-at,andre.peninou,
marie-francoise.canut,florence.sedes}@irit.fr

Abstract. The user is the main contributor for creating information in social media. In these media, users are influenced by the information shared through the network. In a social context, there are so-called "buzz", which is a technique to make noise around an event. This technique engenders that several users will be interested in this event at a time *t*. A buzz is then popular information in a specific time. A buzz may be a fact (true information) or a rumour (fake, false information). We are interested in studying buzz propagation through time in the social network *Delicious*. Also, we study the influence of enriched user profilesthat we proposed [2] to propagate the buzz in the same social network. In this paper, we state a case study on some information of the social network *Delicious*. This latter contains social annotations (tags) provided by users. These tags contribute to influence the other users to follow this information or to use it. This study relies onthree main axes: 1) we focus on tags considered as buzz and analyse their propagation through time 2) we consider a user profile as the set of tags provided by him. We will use the result of our previous work on dynamic user profile enrichment in order to analyse the influence of this enrichment in the buzz propagation. 3) we analyse each enriched user profile in order to show if the enrichment approach anticipate the buzz propagation. So, we can see the interest of filtering the information in order to avoid potential rumours and then, to propose relevant results to the user (e.g. avoid "bad" recommendation).

1 Introduction

In social media users are influenced by their information shared through the network. In a social context, there is so-called buzz that is a technique to make noise around an event. A buzz is popular information in a specific time. This technique engenders that several users will be interested in this event at a time *t*. A buzz may be a fact (true information) or a rumour (fake,false information).

© Springer International Publishing Switzerland 2015
T. Morzy et al. (Eds): ADBIS 2015, CCIS 539, pp. 567–577, 2015.
DOI: 10.1007/978-3-319-23201-0_57

Based on the definition of [5], a rumour is defined as *"an unverified proposition for belief that bears topical relevance for persons actively involved in its dissemination"*. According to [4], a rumour is characterized by its rapidly spread. However, rumour detection is a crucial problem since it requires additional background knowledge to verify information/proposition.

In this paper, we propose to study a buzz that could be a potential rumour.We are interested in studying the propagation through time of the buzz in the social network *Delicious*(more precisely a dataset of *Delicious* [3]). Also, we are interested in studying the influence of dynamic enrichment of users profiles proposed in [2], to propagate the buzz through time.

The dynamic enrichment approach considers the temporal dynamics of the social network. In fact, the user profile enrichment is done according to each period of time. It is not an accumulation of previous enrichment in previous periods. This enrichment approach takes into consideration the popularity, the freshness of information (a tag) and the similarity of users annotating (tagging) the same resource in a specific period of time.

In this paper, we make a case study on some information of the social network *Delicious* that contains social annotations (tags) provided by users. These tags contribute to influence other users to follow this information or to use it.

This study relies on three main axes:

1. Wefocus on tags considered as buzz and analyse their propagation through time.

2. We consider a user profile as the set of tags provided by him. We will use the result of our previous work on temporal user profile enrichment in order to analyse the influence of this enrichment in the buzz propagation.

3. We analyse each enriched user profile in order to show if the enrichment approach anticipate the buzz propagation.

This paper is structured as follows. First, we give an overview of the dynamic enrichment approach. Second, we detail the dataset used, we study some cases of buzz propagation through time with and without the enrichment approach, and also we analyseif the enrichment approach anticipates the buzz propagation. Finally, we conclude and give some perspectives.

2   Overview of the Dynamic Enrichment Approach

In this section, we give an overview of our approach for enriching users profiles already detailed in [2].The dynamic evolution of the user profile is treated by enriching users' interests with tags deemed relevant for each period of time. In fact, in social environment, the user consults the resources stored in the network, communicates and interacts with other users to find the information he needs. Enrichment in this context is done by analysing the environment of the user to detect relevant interests (relevant tags).

---

[1]www.delicious.com

The relevance of an interest is usually calculated from the frequency of use of the tag at a given time. Frequency periodicallyvaries. This change has already been treated by [1], through the concept of "temperature". This notion is interesting since it models the popularity of a term over time.

The user profile is constructed in an implicit way, using the list of tags assigned by the user. The user profile is enriched with tags (considered as his interests) in each period of time in order to reflect the current interests of the user.

The first step consists in dividing the database in each period of time.The choice of thisperiodenables us to detect the evolution of the user interests between two successive periods. This latter, should be consistent with the quantity of data presented in the social network. By dividing the database, we obtain temporal information of the user activity in each period like his neighbours, his tags and the tagged resources.

The second step consists in calculating the temperature of each resource in a given period. In order to calculate this attribute, we propose a formula that takes into consideration several parameters: the freshness of a tag associated to the resource, the similarity of the users who tagged the resource and the number of tags associated with the resource (popularity). The temperature of the resource varies through time. It may increase or decrease. We consider that a resource is interesting if its temperature increases.

The third step consists in detecting the resources where temperature increases over time. After calculating the temperature of each resource, we consider only the resourceswhere temperature value is increasing between two periods of times (this reflect the interest of the user with this resource). However, in social networks that are characterized by the amount of the resources, we can have a lot of resources where temperature is increasing and then their treatment can be complex. So, in order to overcome such a problem, we should keep only the most relevant resources to the user. That's why we analyse the content of the resources and more precisely their metadata (we consider that the resources are semi-structured data). In our work, we use the metadata as the descriptors of the content of the resource, in order to filter the most relevant tagged resources. We attribute a weight for the tags associated with the resources. This weight is calculated according to the degree of correspondence of the tags with the metadata of the associated resource.

The fourth step consists in enriching the user profile with the tags associated with the resources. After calculating the weight of the tags associated with the most interesting resources, we enrich in this step the user profile with tags that reflect the best the user interests. In fact, the more the tag has a higher weight, the more it reflects the content of the resource and then, the more it reflects the user interests. So, we choose from the result of the previous step, tags that are more interesting to the user. A tag is stated as a potential interest if its weight is higher than a giventhreshold.

As a result of this approach, we have an enriched profile in each period of time.

3   Case Study on *Delicious* Dataset of Buzz Propagation Through Time

In this section, we first present the dataset used in our experiment. Second, we analyse the evolution of the top-10 buzz (popular tags) through time. Third, we analyse the influence of the enrichment approach on the top-10 buzz propagation. Finally, we analyse if the user profile enrichment anticipates buzz propagation.

3.1   *Delicious* Dataset

The *Delicious* dataset contains social networking, bookmarking, and tagging information. The temporal interval of activity of the dataset varies between November 2003 and October 2010.It provides information about the user's friend relationships and the tagging relation information <U, T, R>. The users U are described through their ID (e.g. userID=8). The resources R are described through their ID, title and URL (e.g. 1 IFLA - The official website of the International Federation of Library Associations and Institutions http://www.ifla.org/). The tags T are described through their ID and value (e.g. 1 collection development). This dataset is extracted from [3]. We present some statistics of the data present in this dataset: 1867 users, 69226 URLs and 53388 tags. Also, the tagging behaviour is provided according to the time information. This behaviour implies that we know a tag is used in a specific period of time. An example of temporal tagging behaviour is shown in table 1.

Table 1. An example of the temporal tagging behaviour

userID	bookmarkID	tagID	day	month	year	hour	minute	second
8	1	1	8	11	2010	23	29	22

3.2   Buzz Evolution Tracking

In this section, we present the evolution of the selected tags considered as buzz on *Delicious* social network between the year 2003 and 2010. In this work, we consider the top-10 of the most popular tags on the whole dataset as the studied buzz.

Table 2. The top-10 of the most popular tags on *Delicious* between the year 2003 and 2010

Tag	Design	Tools	Video	Education	Webdesign	Web	Inspiration	Art	Web20	Google
Popularity	4060	2929	2236	2041	1907	1733	1723	1691	1653	1648

The evolution of each tag is presented as a graph of its popularity (number of use) along the temporal axis. In this study, we use the month granularity to study the evolution of each buzz. The visualization graph is presented in figure 1.

We can see that most of the studied tags represent the buzz characteristic: their popularity increases slightly in the beginning and then explodes at a time point and declines after that. We observe that, the most popularity period of all studied tags is around September 2010 - October 2010.

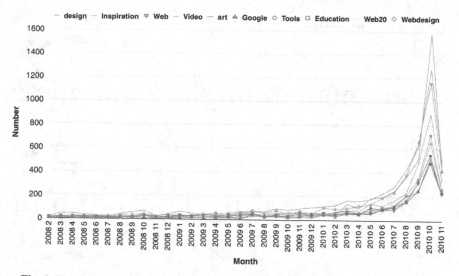

Fig. 1. The evolution of the top-10 popular tags on *Delicious* between 2003 and 2010

3.3   Analysis of the Influence of User Profile Enrichment on Buzz Propagation

To analyse the impact of user profile enrichment on buzz propagation, we are interested in studying the correlation between the buzz propagation in the dataset and

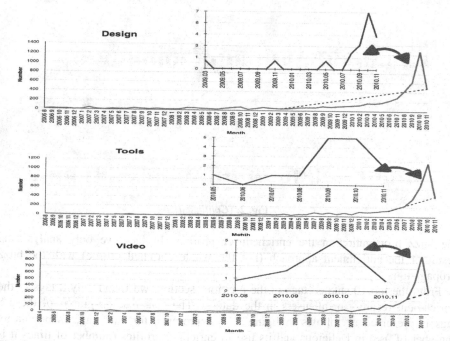

Fig. 2. (Blue)The buzz propagation in the dataset and (Red) the buzz propagation in the enriched user profiles

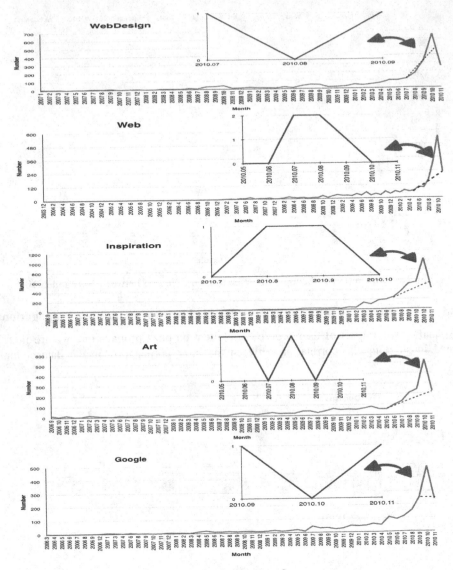

Fig. 2. (*Continued*)

the buzz propagation in the enricheduser profiles. In fact, we only analyse the result of the enrichment approach (not the whole enriched profile) with the buzz propagation.

From the top-10 studied tags in the previous section, we found only 8 tags in the enrichment results (for all users in the dataset).The visualization graphs of these 8 tags is presented as follows in figure 2. It represents, for each tag, its popularity (number of use) in Delicious and its use in enriching profiles (number of times it is used to enrich a profile)

The graphs above show that the studied tags are mostly retained in user profile after the enrichment process in the period in which they become popular. For example, the tag *Google* is retained in user profile between September 2010 - November 2010, the period in which the tag is more tagged by the whole users of this social network.

This analysis demonstrates that the popularity of the tags can be an important factor to the user profile enrichment process. If the tag is a buzz during a period, it has more chance to be extracted in the user profile enrichment process for this period.Thus, the user profile enrichment process can contribute to the buzz (as potential rumours) propagation in social networks.

3.4   Is User Profile Enrichment Approach Anticipating Buzz Propagation?

All along the previous analysis, we have analysed the buzz propagation in the whole network, independently of the user profile. In this section, we analyse each enriched user profile in order to show if the enrichment approach anticipate the buzz propagation. For each buzz found in the enrichment result, we detail the associated userID, the enrichment date, the number of occurrence of the tag (buzz) for the user before the enrichment date, the number of occurrence of the tag (buzz) for the user after the enrichment date and the date of the first use after enrichment by the user. These results are detailed in the tables 3, 4, 5, 6, 7, 8, 9 and 10.

Table 3. Analysis of the tag **design** according to each user

UserID	Enrichment Date	Occurrence before enrichment date	Occurrence after enrichment date	Date of the first use after enrichment
1094	8/10/2010	9	1	03/11/2010
1113	27/08/2010	11	51	29/08/2010
1113	29/08/2010	14	48	30/08/2010
16915	30/03/2009	1	6	16/04/2009
24802	10/11/2009	9	4	18/11/2009
8315	31/05/2010	1	2	24/06/2010
62070	20/09/2010	10	0	-
9960	28/09/2010	0	0	-
51543	30/09/2010	9	3	09/10/2010
2032	01/10/2010	6	0	-
8691	12/10/2010	32	38	13/10/2010
3233	21/10/2010	12	1	22/10/2010
1296	26/10/2010	0	0	-
11699	09/10/2010	0	3	01/11/2010
1701	09/10/2010	10	2	13/10/2010
15728	09/10/2010	29	17	11/10/2010
13222	03/11/2010	3	0	-
8452	04/11/2010	5	0	-
6067	04/11/2010	7	0	-

Table 4. Analysis of the tag **Tools** according to each user

UserID	Enrichment Date	Occurrence before enrichment date	Occurrence after enrichment date	Date of the first use after enrichment
8315	31/05/2010	1	20	01/06/2010
6120	31/07/2010	1	5	24/08/2010
46715	29/08/2010	1	6	12/10/2010
35745	16/09/2010	18	30	17/09/2010
11699	24/09/2010	2	34	26/09/2010
1328	29/09/2014	0	0	--
7396	19/10/2010	2	5	20/10/2010
2315	21/10/2010	7	2	27/10/2010
8554	22/10/2010	1	13	25/10/2010
70894	27/10/2010	12	6	01/11/2010
1505	29/10/2010	11	0	29/10/2010
13102	05/11/2010	8	2	06/11/2010
23135	06/11/2010	16	0	06/11/2010

Table 5. Analysis of the tag **Video** according to each user

UserID	Enrichment Date	Occurrence before enrichment date	Occurrence after enrichment date	Date of the first use after enrichment
74708	16/08/2010	7	9	23/08/2010
13084	14/09/2010	6	8	15/09/2010
4742	30/09/2010	2	4	06/10/2010
6796	12/10/2010	16	1	21/10/2010
8452	20/10/2010	4	0	--
11690	20/10/2010	7	0	--
8775	21/10/2010	2	1	21/10/2010
1701	21/10/2010	12	2	04/11/2010
12847	02/11/2010	2	0	--

Table 6. Analysis of the tag **Webdesign** according to each user

UserID	Enrichment Date	Occurrence before enrichment date	Occurrence after enrichment date	Date of the first use after enrichment
6120	23/07/2010	2	0	--
9660	28/09/2010	1	3	01/10/2010

Table 7. Analysis of the tag **Web** according to each user

UserID	Enrichment Date	Occurrence before enrichment date	Occurrence after enrichment date	Date of the first use after enrichment
13973	16/07/2010	2	12	02/08/2010
12506	05/07/2010	1	0	--
1113	27/08/2010	1	17	29/08/2010
1113	29/08/2010	3	15	31/08/2010
13084	14/09/2010	1	0	--

Table 8. Analysis of the tag **Inspiration** according to each user

UserID	Enrichment Date	Occurrence before enrichment date	Occurrence after enrichment date	Date of the first use after enrichment
1113	27/08/2010	7	22	29/08/2010
51543	30/09/2010	7	2	9/10/2010

Table 9. Analysis of the tag **Art** according to each user

UserID	Enrichment Date	Occurrence before enrichment date	Occurrence after enrichment date	Date of the first use after enrichment
31272	25/05/2010	8	5	09/06/2010
11962	30/06/2010	2	55	02/07/2010
10567	26/08/2010	1	1	09/11/2010
1701	27/10/2010	6	1	06/11/2010
8452	04/11/2010	4	0	--

Table 10. Analysis of the tag **Google** according to each user

UserID	Enrichment Date	Occurrence before enrichment date	Occurrence after enrichment date	Date of the first use after enrichment
1505	05/09/2010	2	1	13/09/2010
11853	05/11/2010	7	0	--

From this analysis, we notice that:

1. Regarding the occurrence before/after the enrichment date: it varies according to different cases. In fact, we notice that users who used the tag only before the enrichment are about 23.63 %, the users who used it only after the enrichment are about 1.81 %, the users who used it before and after the enrichment are about 69.09 % and the users who never used the tag and we have enriched their profile with this tag are about 5.45 %. So, we can conclude that the enrichment approach is somehow dependant with the previous activity of a user. However, the amount of these buzz found in the enrichment results is relatively low comparing to their popularity in the initial dataset.

2. Regarding the date of the first use of a tag after enrichment: this date aims to show the ability of the enrichment approach to anticipate the buzz. The bigger isthe interval between the enrichment date and the first use of the buzz, the buzz is antici-pated. According to these tables, we notice that the minimum value of anticipation is 0 day (we enrich the same day of a current activity) and is about 4,8 % of all cases. The maximum value of anticipation is 75 days (associated to the userID=10567 in table 9). The average value of anticipation is 9 days and the median value of anticipa-tion is 5 days.

4   Conclusion

In this paper, we have made a case study on some information of the social network *Delicious*. This latter, contains social annotations (tags) that are provided by the user. These tags contribute to influence the other users to follow this information or to use it.

This study relies on three main axes:

1. we have focused on tags considered as buzz and we have analysed their propaga-tion through time. In this analysis, we have noticed that the number of users in the network influences the propagation. The more active a user isin specific periods of time, the more the buzz is present in these periods.

2. we have considered a user profile as the set of tags provided by him. We have used the result of our previous work on temporal user profile enrichment, in order to analyse the influence of this enrichment in the propagation of the buzz. We have no-ticed that the enrichment process contributes to propagate the buzz in almost all the cases (8 tags about 10 were found in the enrichment result). Thus, the enrichment contributes to propagate the buzz in the network.

3. we have also analysed each enriched user profile in order to show if the enrich-ment approach anticipate the buzz propagation. So, we can see the interest of filtering the information in order to avoid potential rumours and then, to propose relevant re-sults to the user (e.g. avoid "bad" recommendation). We have found that the enrich-ment approach is somehow dependant with the previous activity of a user. Also, the amount of these buzz found in the enrichment results is relatively low comparing to their popularity in the initial dataset. The anticipation varies from 0 day to 75 days. And the average is 9 days.

As perspectives, in order to reduce the buzz propagation, that may be potential ru-mours, we should take into consideration a buzz filtering process before applying our enrichment approach. Also, we plan to enlarge this case study more than 10 tags. Thus, to study the evolution of the other buzz and also the influence of the enrichment approach on the buzz propagation.

References

1. Manzat, A., Grigoras, R., Sèdes, F.: Towards a user-aware enrichment of multimedia meta-data. In: Proceedings of the CEUR Workshop on Semantic Multimedia Database Technolo-gies (SMDT 2010), Saarbrcken, Germany, vol. 680, pp. 30–41 (2010)

2. Mezghani, M., Zayani, C.-A., Amous, I., Péninou, A., Sèdes, F.: Dynamic enrichment of social users interests. In: IEEE Eighth International Conference on Research Challenges in Information Science (RCIS), pp. 1–11. IEEE (2014)
3. Cantador, I., Brusilovsky, P., Kuflik, T.: 2nd Workshop on Information Heterogeneity and Fusion in Recommender Systems (HetRec 2011). Proceedings of the 5th ACM conference on Recommender systems, RecSys. ACM, New York (2011)
4. Hashimoto, T., Kuboyama, T., Shirota, Y.: Rumor analysis framework in social media. In: 2011 IEEE Region 10 Conference on TENCON 2011, pp. 133–137 (2011)
5. Rosnow, R.L., Kimmel, A.J.: Rumor. In: Kazdin, A.E. (ed.) Encyclopedia of Psychology, vol. 7, pp. 122–123. Oxford University Press, New York (2000)

Author Index

Aalberg, Trond   438, 488
Abbasi, Maryam   20
Abdellaoui, Sabrina   119
Afrati, Foto   165
Ahmed, Waqas   346
AlBalawi, Alia   279
AlBdaiwi, Bader   3
Algergawy, Alsayed   392
Al-Najran, Noufa   12
AlSaawi, Bariah   279
AlTassan, Ghada   279
Amous, Ikram   567
Andreev, Alexey   448
Aouine, Amina   127
Apiletti, Daniele   243
Attanasio, Antonio   229

Babalou, Samira   392
Bambia, Mariem   52
Bascuñana, Alejandro   235
Belo, Orlando   28
Ben Kraiem, Maha   68
Benouaret, Idir   477
Bidoit, Nicole   218
Boughanem, Mohand   52
Bouju, Alain   403
Brando, Carmen   505

Canaval, Sandra Gómez   186, 207
Canut, Marie-Françoise   567
Čech, Přemysl   135
Chernishev, George   97
Cherny, Eugene   448
Cherouana, Amina   127
Cholvy, Laurence   517
Colazzo, Dario   218

Dahanayake, Ajantha   12, 268
Davarpanah, S. Hashem   392
De Wilde, Max   498
Decourselle, Joffrey   488
Deguchi, Yutaka   36
Delmas, Rémi   557
Dessloch, Stefan   88

Diab, Hassan   76
Dromard, Juliette   197
Duchateau, Fabien   488

Eddine, Saidouni Djamel   414
Ethier, Jean-Francois   76

Faiz, Rim   52
Fakeerah, Zaynab   279
Farrokhnia, Maliheh   438
Feki, Jamel   68
Frontini, Francesca   505
Furtado, Pedro   20

Gaidukovs, Andrejs   382
Ganascia, Jean-Gabriel   505
Gargouri, Faiez   145
Garza, Paolo   243
Gomes, Claudia   28
Gorawska, Anna   358
Gorawski, Marcin   358
Gripay, Yann   36
Grošup, Tomáš   135
Guetmi, Nadir   371

Haase, Peter   448
Haberstroh, Alexander   291
Heuer, Andreas   259
Hichem, Houassi   414
Htoo, Htoo   60
Hu, Yong   88

Iftikhar, Nadeem   108
Imine, Abdessamad   371

Jaafar, Nouf   268
Jakunschin, Jevgenij   259
Jallet, Louis   229
Jolivet, Vincent   468

Kalra, Sumit   536
Khadraoui, Abdelaziz   127
Khnaisser, Christina   76

Khouri, Selma 119
Khrouf, Kaîs 68
Kirikova, Marite 382
Kompatsiaris, Ioannis 458
Kontopoulos, Efstratios 458
Kozmina, Natalija 334
Kruliš, Martin 305
Kukhun, Dana Al 527

Labidi, Taher 145
Lagos, Nikolaos 458
Lánský, Jan 135
Laurenson, Pip 458
Lavoie, Luc 76
Lenne, Dominique 477
Lis, Damian 358
Liu, Xiufeng 108
Lokoč, Jakub 135
Lorenzo, Manuel 235
Lotito, Antonio 229
Lumineau, Nicolas 488

Mahdaoui, Latifa 127
Mainardi, Chiara 468
Malki, Jamal 403
Mara, Alexandru 175
Marchand-Maillet, Stéphane 305
Marques, Ricardo 28
Martins, Pedro 20
Mechaoui, Moulay Driss 371
Meditskos, Georgios 458
Messaoud, Maarouk Toufik 414
Mezghani, Manel 567
Mitzias, Panagiotis 458
Momani, Zaid 165
Monjas, Miguel-Ángel 235
Mosca, Alessandro 427
Moško, Juraj 135
Mouromtsev, Dmitry 448
Mozo, Alberto 175, 186, 207
Mtibaa, Achraf 145

Nader, Fahima 119
Niedrite, Laila 334
Nordbjerg, Finn Ebertsen 108
Novikov, Boris 153
Nyunt, Naw Jacklin 60

Ohsawa, Yutaka 60
Oliveira, Bruno 28
On-At, Sirinya 567
Ordozgoiti, Bruno 186
Osella, Michele 229
Osipyan, Hasmik 305
Owezarski, Philippe 197

Pavlov, Dmitry 448
Penicina, Ludmila 382
Péninou, André 567
Petit, Jean-Marc 36
Pivert, Olivier 44
Polacsek, Thomas 557
Prabhakar, T.V. 536
Prade, Henri 44
Pulvirenti, Fabio 243
Raab-Düsterhöft, Antje 259

Rafik, Mahdaoui 414
Ravat, Franck 68
Remesal, José 427
Rezk, Martin 427
Riga, Marina 458
Roudière, Gilles 197
Ruà, Francesco 229
Rubio, Bruno Ordozgoiti 207
Rull, Guillem 427

Sánchez, Patricia 235
Santos, Vasco 28
Sartiani, Carlo 218
Saurel, Claire 547
Scuturici, Vasile-Marian 36
Sèdes, Florence 527, 567
Sein, Myint Myint 60
Sellami, Zied 468
Skopal, Tomáš 135
Solimando, Alessandro 218
Solodovnikova, Darja 334
Soukkarieh, Bouchra 527
Srivastava, Saurabh 536
Stasinopoulos, Nikos 165
Strohm, Peter 291
Suzuki, Einoshin 36

Teste, Olivier   68
Thalheim, Bernhard   3

Ulliana, Federico   218

Vennesland, Audun   438, 488
Vincent, Cécile   403
Vion-Dury, Jean-Yves   458

Waddington, Simon   458
Wannous, Rouaa   403
Wojciechowski, Artur   321

Yarygina, Anna   153
Yoshida, Naofumi   251

Zayani, Corinne Amel   567
Zhu, Bo   175
Zimányi, Esteban   346

Printed in the United States
By Bookmasters